Hong Kong

T0362195

The International Library of Social Change in Asia Pacific
Series Editor: John Clammer

Titles in the Series

Singapore
Garry Rodan

Cambodia
Sorpong Peou

People's Republic of China, Volumes I & II
Frank N. Pieke

Hong Kong
Benjamin K.P. Leung

Political Change in East Asia, Volumes I & II
Peter W. Preston

Hong Kong
Legacies and Prospects of Development

Edited by

Benjamin K.P. Leung

University of Hong Kong, Hong Kong

Routledge
Taylor & Francis Group

LONDON AND NEW YORK

First published 2003 by Ashgate Publishing

Reissued 2018 by Routledge
2 Park Square, Milton Park, Abingdon, Oxon OX14 4RN
711 Third Avenue, New York, NY 10017, USA

Routledge is an imprint of the Taylor & Francis Group, an informa business

Copyright © Benjamin K.P. Leung 2003.
For copyright of individual articles please refer to the Acknowledgements.

All rights reserved. No part of this book may be reprinted or reproduced or utilised in any form or by any electronic, mechanical, or other means, now known or hereafter invented, including photocopying and recording, or in any information storage or retrieval system, without permission in writing from the publishers.

Notice:
Product or corporate names may be trademarks or registered trademarks, and are used only for identification and explanation without intent to infringe.

Publisher's Note
The publisher has gone to great lengths to ensure the quality of this reprint but points out that some imperfections in the original copies may be apparent.

Disclaimer
The publisher has made every effort to trace copyright holders and welcomes correspondence from those they have been unable to contact.

A Library of Congress record exists under LC control number: 2001048701

ISBN 13: 978-1-138-72651-2 (hbk)
ISBN 13: 978-1-138-72649-9 (pbk)
ISBN 13: 978-1-315-19134-8 (ebk)

Contents

PART IV   ECONOMY AND SOCIETY

PART V   SOCIAL ISSUES AND SOCIAL POLICY

Acknowledgements

The editor and publishers wish to thank the following for permission to use copyright material.

American Anthropological Society for the essay: Siumi Maria Tam (1997), 'Eating Metropolitaneity: Hong Kong Identity in *yumcha*', *Australian Journal of Anthropology*, **8**, pp. 291–306.

Asian Journal of Public Administration for the essay: Paul Wilding (1997), 'Social Policy and Social Development in Hong Kong', *Asian Journal of Public Administration*, **19**, pp. 244–75.

Blackwell Publishing Limited for the essay: Ed Snape and Andy W. Chan (1997), 'Whither Hong Kong's Unions: Autonomous Trade Unionism or Classic Dualism?', *British Journal of Industrial Relations*, **35**, pp. 39–63.

Cambridge University Press for the essay: Ming K. Chan (1997), 'The Legacy of the British Administration of Hong Kong: A View from Hong Kong', *China Quarterly*, **151**, pp. 567–82. Copyright © 1997 School of Oriental and African Studies.

Stephen W.K. Chiu and David A. Levin (1999), 'The Organization of Industrial Relations in Hong Kong: Economic, Political and Sociological Perspectives', *Organization Studies*, **20**, pp. 293–321. Copyright © 1999 Stephen W.K. Chiu and David A. Levin.

Government and Opposition for the essay: Lau Siu-kai (1999), 'The Rise and Decline of Political Support for the Hong Kong Special Administrative Region Government', *Government and Opposition*, **34**, pp. 352–71.

Haworth Press, Inc. for the essay: Nelson W.S. Chow (1996), 'The Chinese Society and Family Policy for Hong Kong', *Marriage and Family Review*, **22**, pp. 55–72. Copyright © 1996 The Haworth Press.

Institute of Southeast Asian Studies for the essay: Lo Shiu Hing (1998), 'Political Parties, Élite–Mass Gap and Political Instability in Hong Kong', *Contemporary Southeast Asia*, **20**, pp. 67–87. Copyright © 1998 Institute of Southeast Asian Studies. Reproduced here with the kind permission of the publisher, Institute of Southeast Asian Studies, Singapore www.iseas.edu.sg/pub.html.

Institute of International Relations for the essay: Alvin Y. So (1997), 'Hong Kong's Embattled Democracy: Perspectives from East Asian NIEs', *Issues & Studies*, **33**, pp. 63–80.

Journal of Comparative Family Studies for the essay: Hoiman Chan and Rance P.L. Lee (1995), 'Hong Kong Families: At the Crossroads of Modernism and Traditionalism', *Journal of Comparative Family Studies*, **26**, pp. 83–99.

M.E. Sharpe, Inc. for the essay: Choi Po-king (1998), 'The Politics of Identity: The Women's Movement in Hong Kong', *Chinese Sociology and Anthropology*, **30**, pp. 65–74. Copyright © 1998 M.E. Sharpe, Inc.

Sage Publications Ltd for the essay: Eric Kit-wai Ma (1998), 'Reinventing Hong Kong: Memory, Identity and Television', *International Journal of Cultural Studies*, **1**, pp. 329–49. Copyright © 1998 Sage Publications Ltd. By permission of Sage Publications Ltd.

Taylor & Francis Ltd for the essays: Paul Morris (1997), 'School Knowledge, the State and the Market: An Analysis of the Hong Kong Secondary School Curriculum', *Journal of Curriculum Studies*, **29**, pp. 329–49. Copyright © 1997 Taylor & Francis; Lau Siu-kai (1997), 'The Fraying of the Socio-Economic Fabric of Hong Kong', *Pacific Review*, **10**, pp. 426–41. Copyright © 1997 Routledge; Siu-lun Wong and Janet W. Salaff (1998), 'Network Capital: Emigration from Hong Kong', *British Journal of Sociology*, **49**, pp. 358–74. Copyright © 1998 London School of Economics; Gordon Mathews (1997), 'Hèunggóngyàhn: On the Past, Present, and Future of Hong Kong Identity', *Bulletin of Concerned Asian Scholars*, **29**, pp. 3–13. http@//www.tandf.co.uk/journals

University of California Press for the essays: Anthony B.L. Cheung (1997), 'Rebureaucratization of Politics in Hong Kong: Prospects after 1997', *Asian Survey*, **37**, pp. 720–37. Copyright © 1997 The Regents of the University of Chicago; Stephen W.K. Chiu and Ching Kwan Lee (1997), 'After the Hong Kong Miracle: Women Workers under Industrial Restructuring', *Asian Survey*, **37**, pp. 752–70. Copyright © 1997 The Regents of the University of California.

Every effort has been made to trace all the copyright holders, but if any have been inadvertently overlooked the publishers will be pleased to make the necessary arrangement at the first opportunity.

Series Preface

Asia Pacific has emerged in the last two decades as one of the most significant regions of the world economically, politically and in terms of its social and cultural diversity, creativity and global impact. For these reasons, as well as for the intrinsic interest of its languages and civilizations, the Asia Pacific region has attracted enormous interest from managers, strategic studies specialists, and scholars of comparative sociology, religion, political institutions and change, cultural studies and post-colonial studies, and from tourists, development specialists and those concerned with the management of social change and ethnic diversity. In particular it is a region experiencing and initiating vast social and cultural changes, many of which will prove to be of great significance for the world in general.

The International Library of Social Change in Asia Pacific represents an important publishing initiative designed to bring together in easily accessible and carefully selected form the most significant published journal essays on contemporary Asia and of processes of social change throughout the Asia Pacific region. The series makes available to researchers, teachers, students, journalists and both area specialists and the reader needing up to date comparative material, an extensive range of essays which are indispensable for obtaining an overview of the fundamental social institutions and social processes in this vital region of the world.

The literature on Asia Pacific has proliferated with the expanding interest in the region and the number of journals providing coverage of the area has correspondingly expanded. This makes it increasingly difficult for students, scholars and practitioners to keep abreast with the materials which are essential to an accurate understanding of this complex and pivotal zone. *The International Library* addresses this by making available for research and teaching the key essays from specialist journals worldwide, including those from within the region which are often less accessible internationally.

Each volume is edited by a recognized authority who has selected a group of the best journal articles in their field of special competence and provided an informative introduction giving an overview of the country or field and the relevance of the articles chosen. The original pagination is retained for ease of reference and citation. The essays are selected on the basis both of their empirical coverage of countries and trends, and for their reflection of theoretical developments in the social sciences so that each volume represents in-depth analysis, breadth of coverage and the nature of current thinking and contemporary methods of social and cultural analysis. It is hoped that the result will be an accessible, up to date and authoritative introduction and will make available to scholars, diplomats, students and business people the latest scholarship on a dynamic and rapidly changing sector of the global society and economy.

JOHN CLAMMER
Series Editor
Sophia University, Tokyo

Introduction

Hong Kong is a society of contrasts and paradoxes. The wide variety of the city's contrasting and yet fluid and intermingling social and cultural images – east and west, local and colonial, modern and traditional, extravagant and frugal – has earned it the epithet 'a cultural kaleidoscope' (Siu, 1997). Contrasts and paradoxes abound, too, in the territory's political and economic landscapes. For illustration, we can consider some of the more obvious examples. Politically, before the handover, Hong Kong as a colony and as a rapidly modernizing society should have been highly susceptible to disturbances and disorder. Yet the colony had a remarkable record of political stability, a phenomenon which many have attributed to the political passivity of the local Chinese community (Lau, 1982; Miners, 1975; Shively, 1972; Wong and Lui, 1993). This view was certainly valid and justly prevalent until the early summer of 1989, when over a million of the population took part in the demonstrations and marches in support of the protesting students in Tiannanmen Square. Collective actions of such magnitude and lofty direction stand in sharp contrast to the astute observation that 'while, on the other side of the border, a civil war of world importance might rage, people in Hong Kong were able to pursue their own small personal wars, undeterred by greater events' (Coates, 1975, p. 4). Economically, many a Marxist would consider Hong Kong to be a highly exploitative capitalist regime, and yet the society has had one of the lowest rates of industrial conflict in the modern world (England, 1989; Leung, 1991). And while the government declared its economic policy to be that of 'positive *non-interventionism*', economist Youngson judged that 'Hong Kong and laissez-faire have only an occasional acquaintance' (Youngson, 1982).

I started with 'contrasts and paradoxes' intending to use these to highlight some distinguishing and intellectually intriguing aspects of Hong Kong society. These 'contrasts and paradoxes' will also serve as a guiding thread to some of the major issues covered by the essays in this volume. For instance, Hong Kong as a 'cultural kaleidoscope' would beg the question: what is the *Hong Kong* identity? (see Chapters 5–8 this volume). Paradoxes in the territory's politics would lead us to inquire into the colonial government's strategies for buttressing its legitimacy and promoting stability on the one hand (Chapters 2 and 4), and the conditions conducive to the population's political passivity and political activism on the other (Chapters 1–4, 9, 12, 21, 22). Paradoxes in the economy would draw our attention to such issues as the relationship between the state and the capitalist class in the territory's economic development (Chapters 2, 4, 17, 18) as well as trade unionism and industrial relations in Hong Kong (Chapters 15, 16). There are, of course, inquiries and considerations in this collection which go beyond the reach and direction of our guiding thread, but following this thread would nevertheless place us in the vicinity of their thematic and analytic concerns. Part I of this volume, to which we now turn, provides a sketch of the historical and sociopolitical contexts within which the pertinent issues have arisen and have been approached and tackled.

Profiles of Change and Development

Lui and Wong's opening essay on the 'Hong Kong Experience' (Chapter 1) is a discursive account of Hong Kong's major social and economic changes, as well as their impact on the population, since the early 1970s. Their thesis is premised on their view that changes in the economic and social make-up of Hong Kong around the early 1970s gave birth to experiences and beliefs which were *authentically Hong Kong*, and which have subsequently shaped the mentality and lifestyle of the Hong Kong people. This 'Hong Kong experience' is the authors' synonym for the Hong Kong way of life, and their study is an attempt to tease out its central features as an educated and judicious guide to the Chinese government's metaphorical promise of 'The horses will race as usual and the people will dance as usual' in post-1997 Hong Kong. But to better understand what is *authentically* Hong Kong, we need to contrast it with what is *not*. This takes us beyond, or rather before, the subject of Lui and Wong's inquiry. It takes us to Hong Kong in the 1950s and 1960s – a time when the ethos of the population was shaped largely by the 'refugee experience'.

The 'refugee experience' began shortly after the end of the Second World War when, to escape the prospect of communist rule, large numbers of refugees from China flooded into Hong Kong. This influx of refugees continued through the 1950s and 1960s to the extent that, in these two decades, the structure and functioning of the society, as well as the lives and times of the Hong Kong people, had to be understood in terms of the refugee experience and mentality. Thus the 'don't rock the boat' refugee mentality offered an important clue to the explanation of the colony's political stability (Shively, 1972; Miners, 1975). Refugee capital and entrepreneurship on the one hand, and cheap and industrious refugee labour on the other, were crucial facilitating factors in accounts of Hong Kong's successful industrialization (Wong, 1988; Chau, 1989; England, 1989). The refugees' survival strategies in using family and kin resources for socioeconomic advancement to compensate for the government's lack of support framed studies of the form and ethos of the Hong Kong Chinese family (Lau, 1981; Salaff, 1995). The ensuing familial ethos, termed utilitarian familism, was the principal depoliticizing factor in a seminal analysis of the viability of Hong Kong's sociopolitical system characterized by a lack of integration between a secluded alien government bureaucracy and its colonized subjects (Lau, 1982). The 'refugee experience' also informed explanations of Hong Kong's anomalous popular culture at the time, characterized by the prevalence of the language, themes and artistic people of the mainland rather than the indigenous community (Leung, 1996, pp. 63–65). In short, Hong Kong in the 1950s and 1960s was very much a society of refugees and, as such, its way of life to a large extent reflected the values, concerns, and predicament of a yet unsettled refugee population. The 'refugee experience' is in this sense not an *authentic* 'Hong Kong experience'.

What, then, is the authentic 'Hong Kong experience'? What changes in Hong Kong led Lui and Wong (Chapter 1) to the view that 'The "Hong Kong experience" . . . did not come to fruition until the 1970s' (p. 15)?

We may note, first of all, that the 'Hong Kong experience' in Lui and Wong's discourse is meant to capture the experiences and mentality of the Hong Kong people who regard Hong Kong as their home rather than a 'lifeboat' in the sea of political turmoil. Several factors (some of which are covered only implicitly in Lui and Wong's essay) contributed to the emergence and consolidation of a sense of attachment to Hong Kong.

Demographically, the post-war generation of local-borns were coming of age, and for them Hong Kong had literally been their home. Socio-politically, Governor MacLehose's community-building programme, predicated on the policy of 'caring for the people' and geared to the building of a new better Hong Kong on the 'four pillars' of housing, education, social welfare, and medical and healthcare, had been highly effective in forging a sense of belonging to the territory. Economically, Hong Kong in the 1970s was a society of increasing prosperity and affluence, and the attendant expansion in opportunities for socioeconomic advancement further strengthened the population's commitment to the territory. This was also a time when the Hong Kong identity gradually took shape. In short, the Hong Kong experience came to fruition amidst a social climate of attachment, optimism and rising expectations.

The focus of Lui and Wong's inquiry is the kind of ideology and moral outlook that emerged among the Hong Kong Chinese in the course of their experience of increasing affluence and expanding opportunities. That *collective* experience, the authors argue on the basis of the pertinent survey findings, has culminated in the Hong Kong Dream – the belief that Hong Kong is a land of abundant opportunities and that the 'room at the top' is accessible to those who are able and industrious. Yet, the authors add, with reference to findings from their social mobility studies, the *individual* experiences of a substantial portion of the population have fallen far short of that collective expectation, for class inequality in mobility opportunities still bars the majority of the lower socioeconomic groups from moving up the social hierarchy. The ensuing chasm, or paradox, between social ideology and personal experience has produced strain, the attempt to resolve which, the authors argue, has impacted further on the mentality and moral outlook of the Hong Kong Chinese. In this regard, Lui and Wong observe that, while the Hong Kong people are aware of class inequality and injustice, they have nevertheless preferred to resort to individualist strategies rather than collective actions to resolve the strain. The outcome is intensified efforts at self-improvement, and an instrumental and practical moral orientation which sidesteps the issues of class inequality and injustice while ironically seeking equality of opportunity and freedom in pursuit of the Hong Kong Dream.

But one can seek to understand the 'Hong Kong experience' from perspectives other than Lui and Wong's perspective of increasing opportunities and class inequality. We can examine this experience in the context of Hong Kong's political circumstances and social policy. With this objective in mind, we now turn to Paul Wilding's study of social policy and social development in Hong Kong (Chapter 2).

Among the host of factors which, in Wilding's view, circumscribed and shaped the territory's social policy and social development, the most crucial were the political ramifications of Hong Kong as a colony. Underlying Wilding's analysis is thus the thesis that the colonial government's policy-making was geared to buttressing its legitimacy and promoting political stability on one hand, and sustaining the goodwill of the Chinese government on the other; for the threat to its survival inhered both in the colonized population as well as their motherland across the border. To achieve this dual objective, Wilding argues, the colonial government pursued the following strategies in policy-making. It placed the promotion of economic prosperity as a top priority in its policy agenda, as economic prosperity was taken by the local population as a sign of good government. As far as possible it avoided introducing radical policy and political changes, for fear of overstepping into the sensitivities of the Chinese government. It targeted its provision of goods and services to the objectives of crisis resolution and legitimacy enhancement. Wilding captures the Hong Kong way of social policy with an apt characterization: 'The Hong Kong government's primary concerns have always been economic and political rather than social' (p. 45).

It is within the parameters of the twin concerns of economic prosperity and political stability that Wilding conducts his analysis of the development of social policy in Hong Kong. The aim to promote economic prosperity, which quite inevitably gave the business class the lion's share of influence on social policy contributed to the government's *laissez-faire* policy orientation before the 1970s, and its 'positive non-interventionist' approach thereafter. But, as Wilding points out, both *laissez-faire* and non-interventionism are misnomers in the Hong Kong context, for the government in its attempts to promote political stability, especially after the traumatic disturbances of the mid-1960s, did play an active and significant role in the provision of public goods and services such as housing, education and medical care. Indeed, as Wilding rightly observes, given Hong Kong's fragile export-oriented economy which depended heavily on overseas investment, political stability, together with measures such as low tax rates and low labour costs, were essential in attracting the inflow of capital and enhancing the economy's competitiveness in international markets.

These policy orientations of the government, Wilding contends, were consonant with the values and expectations of the Hong Kong people who, as Lui and Wong have also noted (Chapter 1), were primarily concerned with relying on individual efforts to get ahead in the free market and who demanded little more than a stable environment in which to do so. But the *paradox* inherent in the government's positive non-interventionist policy bred contradictory beliefs among the population; they believed in self-reliance and supported *laissez-faire*, but they also expected the government to look after their basic needs. And, as the government continued to expand its welfare role in the past couple of decades, the Hong Kong people's values and expectations underwent a concomitant change – demands arose for more education, better housing and, in connection with 1997, a more representative government. In at least this respect then, social policy development has shaped the 'Hong Kong experience'. Wilding's main concern, however, is to understand how the government's social policy coped with the 'Hong Kong experience', and in the light of that understanding, he judges it to be 'a very particular model . . . of good governance' (p. 48).

Hong Kong's social and economic climate from the early 1970s, evidenced by the first two essays in this collection, can be described as one of optimism and rising expectations. The same, however, cannot be said without qualifications about its political climate. Indeed the next essay by Kuan (Chapter 3) maintains that a past predicament of the Hong Kong people – 'escape from politics' – has been eclipsed only in recent years through the emergence of a political society. In the context of our discussion, Kuan's study of the territory's political development provides another perspective on the 'Hong Kong experience'.

Kuan's main thesis is that the emergence and consolidation of a political society in the local community is paramount to Hong Kong's political development which, the author explains, is a process of enhancement of political capability and democratization. It is evident from the author's definition of political society as 'an ensemble of autonomous groups mediating between the polity and society' (p. 63) that his study adopts the pluralist perspective in reviewing the prospects and possibilities of the territory's political development. Kuan's contention is that economic and social-political changes since the early 1980s have made capability enhancement and democratization urgent issues on Hong Kong's political agenda. On the one hand, the population's rising expectations, the society's increasing complexity, as well as its rapidly expanding economic ties with the mainland, have severely taxed the government's capability. In such circumstances, Kuan argues, the government's old style of 'politics by consultation', which in practice meant seeking and representing the views of a narrow circle of elites, would prove to be gravely inadequate for coping with the wide variety and large volume of demands from the population. On the other, the 1984 Joint Declaration and the 1990 Basic Law awakened the Hong Kong people to the importance of having a form of government that would safeguard their freedom and autonomy after the handover. Political necessity thus directed the population to the path of democratization. It is in view of such circumstances and developments that Kuan sees democratization as offering the solution to the territory's vexing political problems, for democratization not only helps to guard against undue interference in the form of 'political appointments' and 'authoritarian mandates' from the future sovereign power, but also enhances political capability through providing institutionalized and broadly-based channels for the articulation and resolution of the divergent and conflicting interests of the community. But to realize these benefits of democratization, Kuan adds, a political society is indispensable.

In Hong Kong, the political society has a relatively short history. Its roots, according to Kuan, can be traced to the community's budding political activism in the 1970s, when demographic and social-political changes inculcated rising expectations and a wave of pressure group politics. The impact of these pressure groups on politics was, however, only peripheral as they had no institutionalized access to, or representation in, the territory's major policy-making bodies. As such, pressure groups, in Kuan's view, constituted no more than the foundation of a political society. But the Joint Declaration and the Basic Law with the ensuing programme of gradual democratization eventually gave pressure groups the impulse and opportunity to develop into political parties. At the same time, pro-China political groups and alliances grew in strength under the patronage of the Chinese government while, in counterbalance, pro-democracy forces gained political ascendance through mass support. The outcome of political competition and mobilization in the course of Hong Kong's democratization, Kuan notes, is the making of a political society which signals an end to the population's past predicament of 'escape from politics', and which provides the bulwark, in the form of institutionalized political groupings and a politically attentive and concerned public, for safeguarding the integrity and interests of the community.

Kuan's study, as we have seen, brings us up to the eve of Hong Kong's return to China. It is appropriate at this juncture, as a conclusion to Part I on change and development, to consider the 'assets and liabilities' which Hong Kong has inherited from its colonial past. For this review, we now turn to Ming K. Chan's essay on the legacy of the British Administration of Hong Kong (Chapter 4).

Chan's discussion centres on the four aspects which the territory's last Governor Patten claimed to be the exemplars of British achievement in Hong Kong. Foremost among these is the *rule of law* which, despite its importance to the Hong Kong people as a guardian of order and justice after the handover, has nevertheless been found by Chan to be a deficient legacy. Chan considers the most unsatisfactory element in this connection to have been the use of English as the legal language in a context where the overwhelming majority of the population are Cantonese speakers. The colonial government's late and slow attempts at localization in respect of legal language, personnel and the common law, together with unhealthy precedents set by anti-Chinese and anti-grassroots court rulings, as well as a host of blemishes in the form of corrupt practices and unprofessional conduct, have, in Chan's view, left the Hong Kong

SAR legal system much to remedy and rectify. Nor does the *civil service*, despite its highly efficient and relatively uncorrupt reputation, deserve unreserved acclaim. Among its flaws, Chan pinpoints the elitist and expatriate-dominated facet of its structure and practice, resulting in a civil service wanting in responsiveness and accountability. In addition, the colonial government's halfhearted attempts at localization have handicapped local personnel in acquiring the administrative experience and expertise so essential for self-government after 1997.

Chan then scrutinizes the government's proclaimed policy of promoting *economic freedom with limited government*, which has long been held to be the key to Hong Kong's economic dynamism and success. He debunks it as a myth which served, on the one hand, to legitimize the government's limited provision of public goods and services and, on the other, to camouflage its biased intervention in favour of British and big business interests. The legacy of this so-called policy of 'positive non-interventionism', Chan argues, has been the territory's mounting social inequality, and a *modus operandi* for the pro-China business elites to simulate the success of their pro-British predecessors. But Chan's harshest critique of the colonial legacy pertains to *democracy*, which the British government had denied the Hong Kong people until the eve of the handover. This eleventh-hour effort at democratization was, in Chan's view, motivated by political considerations in which the political rights of the Hong Kong people were not a priority. What Hong Kong inherited from colonial rule, Chan contends, is only a half-baked democratic system, in which a small circle of elites – another legacy of colonial politics – can continue to have the lion's share of power under the new political master.

In the light of the above overview of the circumstances and factors shaping Hong Kong's social development and its way of life, we can now proceed to an examination of the culture and identity of the Hong Kong people.

Culture and Identity

A discussion on culture and identity is appropriately begun with anthropologist Helen Siu's portrayal of Hong Kong as a 'cultural kaleidoscope' (Siu, 1997, p. 145), for it not only aptly captures the quintessence of the territory's cultural landscape but also begs the question of its identity. In a similar vein, Abbas characterizes Hong Kong as a port – 'a passageway through which goods and currencies and cultures flow' (Abbas, 1996, p. 13) – where everything is in transit, in a state of flux. Hong Kong acquires these features, both writers point out, through its position as a world metropolis and a globalized capitalist economy. The global influences impacting on the territory's social and cultural life and shaping the identity of its people are indeed captured with imagination and ingenuity in Tam's opening essay (Chapter 5) in Part II.

The objective of Tam's study is to throw light on the Hong Kong culture and identity via the population's most popular foodway *yumcha* – 'the practice to eat various hot and cold foods, in a restaurant, when Chinese tea is served' (p. 48, fn. 2). In its traditional 'Guangzhou' form, *yumcha* is a leisurely activity consisting of the eating of a couple of delicately prepared food items called *dimsum*, which complements the tea-drinking. Its development in Hong Kong into a practice whereby a huge quantity of *dimsum* of an international make-up are offered and consumed in haste in eating places extending well beyond the conventional teahouse venue reflects, in Tam's view, the cosmopolitan diversity, hybridity, openmindedness and fast pace that characterize the city's social life. The Hong Kong-style *yumcha*, Tam adds, is also symbolic of the identity of the Hong Kong people who have paradoxically sought to retain their Chineseness (through enthusiastic adherence to this Chinese food custom) in a colonial setting, while at the same time asserting their distinction (through their globalized style of *yumcha*) from the mainland Chinese. For, as Tam so shrewdly observes, in the minds of the Hong Kong people, '[t]o be local is to be both Chinese and Western ... *as well as* to be non-Chinese and non-Western .... It is a sense of being helplessly international' (pp. 109–10).

But inherent in the Hong Kong identity is also a sense of pride and achievement which the territory's social and economic advancements over the past few decades have nurtured. This sense of pride, Tam contends, has not only led the Hong Kong people to see themselves as superior to, and hence different from, their compatriots across the border, it has also prompted the territory's emigrants to strive to retain their identity in foreign countries. The latter phenomenon is evidenced in the prevalence of Hong Kong-style *yumcha* among the territory's overseas immigrant community, for whom the practice is like returning home for a family reunion. In short, in Tam's account, *yumcha* has become the Hong Kong people's identity tag, in which are encoded the most distinguishing features of their culture and lifestyle.

Gordon Mathews in Chapter 6 takes a different approach to the subject of Hong Kong identity, the development and character of which he studies on the basis of interviews and mass media and scholarly reports. Focusing on the identity issue of whether Hong Kong people have considered themselves to be *a part of* China or *apart from* China, Mathews notes that, historically, competing views have emerged depending on the political stance and allegiance of the information source. Thus there is the claim that Hong Kong before the British takeover was but a barren island and that the Hong Kong Chinese under colonial rule had hardly any other concerns, let alone national identity, than just making a living. Against this is the view that, long before it became a British colony, Hong Kong was already a busy trading post and part of China, and that the Hong Kong Chinese had always considered the mainland to be their motherland. But what appears evident, Mathews contends, was the emergence of a Hong Kong identity in the 1970s, when the territory's affluence and cosmopolitanism contributed to a Hong Kong way of life distinctly different from that on the mainland. That identity then went through twists and turns in conjunction with the momentous developments in China and Hong Kong.

China's economic reforms and open-door policies since the late 1970s, Mathews observes, have substantially strengthened the Hong Kong people's economic and social ties with the mainland, inculcating in the process an increasing *Chinese* identity among the Hong Kong people. That identity was reinforced by the Sino-British Joint Declaration of 1984 which announced Hong Kong's return to the motherland in 1997. Then came the Tiannanmen Square Incident which instilled among the Hong Kong people widespread fears that the Joint Declaration's promise of 'one country, two systems' was no guarantee of the preservation of their cherished way of life after the handover. In other words, the Hong Kong people understood that they would soon have a Chinese national identity, but they were worried about the political implications of that identity. In Mathews' view, it was in this context that a new identity gradually emerged among the population – namely, a Chinese identity in the *cultural* sense, which signals an endeavour to preserve and uphold the best in the Chinese cultural heritage. As such, Mathews concludes optimistically, the Chineseness embodied in this identity, with its base in Hong Kong, may serve as the model for a reconstruction of the Chinese way of life throughout Greater China.

The allusion to the Chinese cultural heritage above raises the question of whether, and if so how, cultural traditions can survive in the highly modernized society of Hong Kong. The next essay by Chan and Lee (Chapter 7) provides an answer, arguing that traditional *and* modern values and practices have merged in a paradoxical, yet harmonious, coexistence in the Hong Kong Chinese way of life. This merging of opposites, in the authors' view, is achieved first and foremost within the domain of the family which combines the 'best of both worlds' to pave the way for a coherent social life. One core facet of traditionalism in Chan and Lee's analysis is religious beliefs and practices, or what the authors call 'the sacred presence', exemplified by the custom of ancestor-worship, the observance of religious festivals and rituals, the belief in feng shui (or spatial harmony), and a wide variety of astrological practices. On the secular level, the authors identify Confucian ethical imperatives, such as respect for the elderly, loyalty to the family and kin group and order and propriety in daily behaviour, to be other core ingredients of traditionalism. Modernism, on the other hand, is in Chan and Lee's view quintessentially represented in the pragmatic, instrumental outlook of the Hong Kong people. The merging of the traditional value of familism and the modern pragmatic outlook, the authors point out, is already evident in the Hong Kong people's familial ethos which Lau (1981) has appropriately termed 'utilitarian familism'. But the structure and processes of the Hong Kong family furnish further illustrations of this merging.

In respect of structure, Chan and Lee note that the predominant form of family in Hong Kong is the nuclear family which fits well into the society's highly modernized and urbanized setting. But underneath this modern façade of a relatively independent and isolated nuclear family, one sees in reality the prevalence of traditional familial practices such as mutual assistance and frequent interactions with the wider kin-group. The Hong Kong Chinese family, in Chan and Lee's view, is indeed a 'modified nuclear family', for not only in structure but also in processes, its modern features are modified by the persistence of traditional values and practices. Thus the modern trend towards equality among family members is tempered by the traditional ethical imperative of piety towards parents and seniors. The modern pragmatic instrumental outlook of using the family and kin-group for material interests and utilitarian purposes coexists with the traditional emphasis on loving care and emotional support among family members. And contrary to the view that the modern family has by and large lost its religious function, the sacred presence, in the form of ancestor-worship, attention to feng shui for family fortunes, belief in 'predestined affinity' as a source of family bonding and, most of all, reunions on festival occasions, remains an integral part of family life in Hong Kong. It is such an assimilation of the modern and the traditional, the West and the East, within the family, the authors contend, that lays the foundation for the Hong Kong experience of hybridity and globalization in the larger social world.

The final essay in Part II returns to the issue of Hong Kong identity. Entitled 'Reinventing Hong Kong', this essay by Ma (Chapter 8) illuminates the forces and considerations circumscribing and impacting on the re-creation of Hong Kong's collective memory and identity in a popular television programme on the eve of the handover. Although, in essence, a study of the political and organizational constraints on a mass media production, the essay also tells us much about the way in which the Hong Kong identity is being steered and recast in consonance with the territory's return to the motherland.

The television programme in question, entitled *Hong Kong Legend*, for which Ma served as a presenter, consisted of 39 hour-long episodes on Hong Kong's social life and development and was shown at prime time on the territory's most popular television channel. According to Ma, three dominant motifs can be discerned in the most telling first 13 episodes of the series. The first motif, re-creating a *politically correct past*, is achieved through a selective 'remembering' of Hong Kong people's remarkable ability to turn crises into blessings, as well as a selective 'forgetting' of the social frustrations and discontents which fermented such crises. In addition, the episodes highlight the affinity, but downplay the antipathy, between the Hong Kong people and the mainlanders. This positive image of Hong Kong's past is strengthened through the second motif of casting the territory as an archetype of *unfailing capitalism*, as a land of abundant opportunities where the able and industrious will prosper. The problem of social inequality and the deprivations of the underclass are glossed over, and the overall effect is a model of economic success which China may well follow in its reforms. If the first two motifs are meant to inculcate an identity of the Hong Kong people as resilient and resourceful and as having close bonds with the mainland Chinese, the third motif seeks to impart to the territory a local flavour through ironically accentuating its *cosmopolitan* make-up. As such, while Chineseness is emphasized, the Hong Kong identity retains its uniqueness, a befitting complement to the 'one country, two systems' prescription.

Ma then continues with an analysis of the factors and circumstances culminating in the programme's lopsided portrayal of Hong Kong's character and achievement. The television station's interest in the mainland market, the programme's sponsorship by an enterprise with strong ties to China and the organizational and procedural prerogatives of the television company are, in Ma's view, constraining influences on the way in which the programme was conceived, produced and presented. In this regard, the study furnishes a perspective from which one can ruminate on the forces circumscribing Hong Kong's development upon its reunion with China.

Recent Trends in Political Development

There are issues perhaps more urgent and engaging than that of identity in the course of Hong Kong's return to the motherland and, among these, political development arguably ranks at the top. Alvin So's opening essay (Chapter 9) in Part III explores, through a comparative analysis of the political and social-economic conditions in Hong Kong, Taiwan and South Korea, why the Hong Kong experience of democratization has been less fulfilling than that in the other two Asian countries. Criticizing the power dependency explanation on the grounds that it views, too simplistically, Beijing's and London's imposition of a prescribed agenda on a dependent polity with a politically alienated population as the cause of the territory's slow progress in democratization, So offers an alternative explanation in terms of the interplay of societal forces within the community. He notes the political ascendance of middle-class professionals whose pro-welfare and pro-democracy orientation has won them the support of the grassroots population, but, at the same time, he draws attention to a countervailing politically conservative alliance between Beijing and big business which is intent on tempering the pace and scale of democratic development in Hong Kong. The outcome is an embattled democracy in which the contestants disagree and debate on the desirability of different models of democratization, and this lack of unanimity has contributed to the territory's slow democratic progress.

So then contrasts the Hong Kong case with that of Taiwan and South Korea. Characterizing Taiwan as a party-state, he sees its democratization from the mid-1980s as the outcome of the dominant KMT's endeavour to strengthen its legitimacy in the context of the country's declining international status, and in the face of mounting challenge from an opposition party. The KMT believed, So points out, that it could rely on its established power and resources to win in democratic elections. In short, Taiwan's democratization has succeeded because it has been a top-down state-sponsored undertaking supported by the population. But South Korea's experience, the author observes, has been that of democratization from the bottom. In this case, an authoritarian state providing support and protection for a small number of business groups deemed as highly exploitative by the population culminated in a mass movement that led to the downfall of the government in 1987, followed by the establishment and consolidation of a democratically elected regime in the 1990s.

Hong Kong's democratization has faltered, So concludes, because, on the one hand, the state with its ally, big business, has had no genuine commitment to the cause – this makes it different from Taiwan; on the other hand, a non-authoritarian state, coupled with a relatively satisfied population, have obviated the outbreak of a mass pro-democracy opposition movement – this makes it different from South Korea. Hong Kong's democratization, the author concludes, has been a process of 'muddling through'. Some of its political consequences are reviewed in the next essay (Chapter 10) by Anthony Cheung.

Cheung's main thesis is that, despite the growth of party politics and the emergence of an elected legislature during the last years of colonial rule, Hong Kong's senior civil servants are very likely to retain their towering role in local politics after the handover. Government by senior civil servants, most of British racial stock, was a product of the colonial era, when an efficient administrative staff with the support of coopted local leaders were able to sustain the legitimacy of colonial rule. The opportunity existed, Cheung notes, for the elected legislature to take political precedence over the civil bureaucracy in the course of Hong Kong's democratization, especially in view of the fact that the last Governor Patten was steering the civil bureaucracy towards greater accountability to the legislature and the public. Yet, in Cheung's view, this opportunity was obliterated on the one hand by Britain's reluctance to depart from what seemed to have been a reliable model of colonial rule, and on the other by China's intent to replicate this 'executive-led system' after the handover. In addition, the author points out, Patten's public-sector reforms, in devolving greater managerial power to professional civil servants and strengthening the policy-making role of senior administrators, had the effect of enhancing the political prominence of the top echelons of the civil bureaucracy.

But the question remains as to why the Chinese government would choose to rely on a crop of senior civil servants who had been groomed under colonial patronage and who had supported Patten's political reforms which China denounced. Cheung's view is that to counteract the 'Patten heritage', the position of the chief executive was given to the shipping magnate Tung Chee-hwa, apparently with the prior consent of the civil service, but, other than that, the Chinese government had no better alternative than to provide a 'through train' for all top civil servants, for it was essential that the civil service felt secure under the new sovereign power. The Chinese government understood well that the civil bureaucracy was not only highly regarded by the local population for its efficient administration, but was also viewed by the business sector, the pro-China and the pro-democracy factions, as well as the public as a 'politically neutral' candidate acceptable to all. Most of all, in the view of the

Chinese government, an 'executive-led' political system would effectively curb the influence of the pro-democracy politicians.

Hong Kong's top civil servants, Cheung contends, will be in a strong position even in contention for power with the chief executive who, lacking a solid political basis in the form of popular mandate or party support, will inevitably have to depend on the civil service to establish his authority. But a potent challenge to the civil bureaucracy's power, as Cheung sees it, will come from the elected legislature and the public on the one hand, and demands from Beijing on the other. Yet ironically, the civil bureaucracy will require the support of both – the former to guard against undue interference from the central government and the latter to confront challenges from the local community.

In Chapter 11 Lau Siu-kai examines the way in which the post-handover Hong Kong Special Administrative Region (HKSAR) government has attempted to grapple with the problems of political interplay between the civil service, the chief executive and the legislature and, in the light of his survey findings from a territory-wide sample of Hong Kong Chinese, offers explanations for the population's declining support for the new regime.

Lau points out that, at the time of the handover, the HKSAR government faced three inherent difficulties which it had to surmount in order to win the population's support. First, the community still harboured a nostalgia for colonial governance, and a substantial portion of them were not enthusiastic about the regime change. Second, the majority of the population had reservations about the HKSAR government as they believed that it would place the interests of the Chinese government above those of the Hong Kong people. Finally, most of the population preferred a more democratic government than the one which the Basic Law had prescribed for the Special Administrative Region. To overcome these difficulties, Lau observes, one of Chief Executive Tung Chee-hwa's strategies was to pledge to the community that he would regard as his top priority the solution of social and economic problems – such as housing, education, employment, and the elderly – which the colonial government failed to tackle satisfactorily. Tung also referred to his close connections with the Chinese government as being contributory to better understanding between Hong Kong and Beijing. On top of all these, he made an appeal to the community to resort to traditional Chinese values such as harmony, benevolence, modesty and obligations to the community as the basis for building a better Hong Kong. In short, as Lau rightly remarks, Tung's overall strategy was to shift the population's attention from politics to livelihood which was Hong Kong people's main concern, and from colonial ethos to traditional Chinese virtues which the local population fervently endorsed. His initial performance as chief executive accordingly won him high popular support.

The turning point in Tung's popularity and the HKSAR government's popularity came with the arrival of the Asian financial crisis which dealt a crippling blow to the Hong Kong economy. Concomitantly, several crises occurred which cast doubt on the government's administrative capability. Tung's government, Lau argues, had banked on good performance and a vibrant economy as the bulwark of its legitimacy and authority, but the turn of events was driving it towards bankruptcy. Lau further observes that Tung, with his firm belief in benevolent leadership, was averse to mass politics, and his inattention to networking with the territory's political groupings and his lack of communication with the public had the effect of widening the gap between the government and the people. Tung failed to convince the people that his was a benevolent government. Support for the government plunged as the population turned their attention to politics – to demands for a more efficient, responsive and democratic government. Ironically, Tung's government arrived at a place that it had tried hard to avoid.

The population's resort to political means to redress their grievances in the circumstances of a widening gap between the government and the people, according to Lo's arguments in the next essay (Chapter 12), puts the territory at the risk of political instability. Lo's thesis is based on the contention that political parties, which perform the role of stabilizing the political system through providing an institutionalized channel for the communication and resolution of diverse interests, have been denied their proper political function in the HKSAR government. In the first place, Lo points out, no political party – however broad the basis of its popular support – can form a government under Hong Kong's executive-led polity which confers the policy-making power to the chief executive and the civil bureaucracy. The situation is aggravated by the rivalry among the territory's political parties – with pro-democracy parties competing with the pro-Beijing and pro-business parties as well as competing among themselves for electoral success – which has resulted in their inability to amalgamate into powerful political parties capable of effectively confronting the executive branch of the government. In addition, the territory's electoral system has been designed to preclude the most popularly supported, pro-democracy parties from exercising a dominant influence in policy-making. Instead, Lo goes on to argue, the pro-Beijing and pro-business 'patron-client' type of parties, which have weak popular support, have been given a prominent role in the running of the government. They have been enabled, through indirect elections by way of 'functional constituencies' and the 'Election Committee', to capture the majority of seats in the legislature. Moreover, they have gained privileged access to the territory's highest policy-making body, the Executive Council, whose members are all appointed by the chief executive. Further, when the government consults the interest groups in civil society, it far more often seeks the views of business groups than those of the grassroots. The outcome, Lo contends, is a government working in partnership with the pro-business and pro-Beijing factions – a government alienated from the majority of the population.

Lo observes that the territory's 1966–67 riots were evidence that the elite–mass gap would render the polity highly susceptible to widespread popular unrest. The colonial government had learnt a painful lesson from these disturbances and duly implemented a series of democratic reforms which subsequently restored its legitimacy. But the HKSAR government, Lo argues, has reversed these reforms at a time when the population's demands for democracy are rising and when the income gap between the rich and the poor is increasing. Economic distress under the Asian financial crisis, in Lo's view, may converge with social and political discontents to drive the population towards using protests and even rioting to redress their grievances. A remedy in the short run would lie in the government's ability to improve the people's livelihood, but the long-term solution, Lo remarks in conclusion, is to restore to political parties their due role and power in Hong Kong's political system.

From the discussion so far in this section, we can see that Hong Kong after the handover has faced a host of new problems and challenges, some of which, such as the Asian financial crisis, were chance occurrences while others had their roots in the system prescribed for the Hong Kong Special Administrative Region. It is befitting, as a conclusion to Part III, to tease out systematically the factors and conditions circumscribing the government and political life in post-handover Hong Kong in order to assess more realistically the prospects and possibilities of the territory's future development. We turn accordingly to Steve Tsang's essay (Chapter 13) for illumination.

Tsang's main thesis is that despite the Chinese government's promise of preserving Hong Kong's existing system and way of life after the handover, there will be some inevitable changes because the territory's significance to its new sovereign government is vastly different from its previous significance to its old sovereign government and also because the two sovereign governments are of a very different political make-up. Britain, according to Tsang, had only a peripheral interest in Hong Kong and therefore generally allowed the Hong Kong government and its people to pursue their preferred way of life. In contrast, China sees Hong Kong as the spearhead of its economic modernization and as the 'one country, two systems' model for Taiwan's reunion with the motherland. The Chinese leadership, Tsang argues, therefore has a keen interest in making sure that developments in Hong Kong do not stray from its expectations. This, coupled with the Chinese leadership's 'interventionist' and 'control' ethos, means that Hong Kong will not have the high degree of autonomy prescribed in the Basic Law. In Tsang's view, the political imperative of the Chinese leadership takes precedence over other considerations in the running of the HKSAR government. In this regard, he points out, the imperative of using Hong Kong as a 'one country, two systems' model is evidenced by Beijing's appointment of the chief executive to an unusually high bureaucratic rank, in an apparent attempt to deter interference from the lower-ranking heads and party cadres of provincial governments. Beijing's imperative of monitoring and, if necessary, intervening in Hong Kong affairs, in other words, is tempered by its imperative of demonstrating to Taiwan the viability of the 'one country, two systems' principle. This balance, Tsang believes, will be upset to Hong Kong's disadvantage if it departs too far from Beijing's expectations.

Beijing's expectations are: first, a restrained pace of democratic development in Hong Kong such that it will not be a threat to China's sovereign authority over the territory or to the communist leadership on the mainland; second, an efficient and capable Hong Kong civil service and police force that will maintain order and stability; and, finally, a vibrant Hong Kong economy that will continue to serve as the locomotive for China's economic development. Of these expectations, Tsang views the maintenance of order and stability as the most vital for the continuity of Hong Kong's 'system and way of life', for turmoil or subversive movements would be a legitimate cause for the PLA's intervention. But the resilience of the Hong Kong people, Tsang remarks in conclusion, will enable the community to bounce back from economic setbacks, to put up with restrictions on democratic development, to steer away from open confrontations with Beijing or, in short, to pragmatically pursue their way of life without overstepping into the sensitivities of the Chinese leadership. Part III concludes on this optimistic note, and the volume moves on to look at matters relating to Hong Kong's economy.

Economy and Society

As we saw in the last essay by Tsang, the British government had little interest in Hong Kong and therefore left the running of the colony largely to the Hong Kong government. Earlier in this volume, Wilding (Chapter 2) in discussing social policy development in colonial Hong Kong, came to the conclusion that the Hong Kong case was a 'very particular model . . . of good governance'. Conversely, Chan's essay (Chapter 4) presents a less favourable portrait of the legacy of the British administration of Hong Kong. Among other criticisms, Chan maintained that the economic policy of free marketism, with its camouflaged bias in favour of

British and big business interests, was carving the way for the political dominance of big business in the post-handover government. Part IV's opening essay by Alex Choi (Chapter 14), however, suggests that the colony was indeed vital to Britain's economic interests, that the Hong Kong case was a stark example of colonial exploitation, and that the economic policy of free marketism has been detrimental to the interests of the local industrial bourgeoisie as well as to the economy and people of Hong Kong.

Choi arrives at the above views through probing the intriguing issue of why, despite its widely recognized importance, industrial upgrading was not pursued in Hong Kong. He begins with a critique of two theoretical explanations. The 'free marketism' explanation, which argues that industrial upgrading will be brought about by free market forces, is, in Choi's view, unconvincing because it was obvious that nothing of that sort had happened even after some three decades of industrial development in Hong Kong. Nor is the 'interventionism' explanation offered by Stephen Chiu (1994) convincing for, Choi points out, this explanation overlooks the fact that, by the late 1960s, Hong Kong's industrial capitalists had already captured so many 'state positions' that they should have been able to overcome the financial and commercial bourgeoisie's opposition to state intervention to upgrade the territory's industrial production. That the state in Hong Kong would refuse to assist in industrial upgrading, Choi goes on to argue, is bizarre because the territory was increasingly facing a disadvantage in competing with its neighbours Singapore, Taiwan and South Korea, where the state's active intervention was putting their industrial structure ahead of Hong Kong's.

The explanation of this anomaly, in Choi's view, lies primarily in the colonial and class nature of the Hong Kong state, which was primarily concerned with serving the interests of Britain and of the British financial and commercial bourgeoisie in Hong Kong. Thus the colony's surplus financial resources, which could have been used for industrial upgrading, were instead deposited as reserves in London to back up Britain's sterling policy. Furthermore, the industrialists' call for measures to protect and advance industrial development was turned down by the state which was bent on upholding the British banking and trading elite's hegemony over the local industrial bourgeoisie. In addition, Choi contends, the industrial capitalists lacked the resolve and internal unity to push for reforms to upgrade industry. They were hesitant about making long-term investments, uncertain about Hong Kong's political future and too occupied with reaping the benefits of the government's 'cheap labour' policy. At the same time, the large industrialists, who were mostly Shanghainese immigrants lacking a solid local political base, were far more concerned about securing positions within the governmental structure than cooperating with the small local Cantonese industrialists in opposing the government's 'non-interventionist' economic policy. Finally, Choi observes, Hong Kong's major labour union federations, the pro-Beijing Federation of Trade Unions and the pro-Taiwan Trades Union Council, were each so preoccupied with pursuing their respective partisan political goals that they largely failed to advance the industrial workers' interests through demanding that the government reform the colony's low-cost, labour-intensive industrial structure. With China's open-door policy and the availability of cheap labour and land on the mainland, Hong Kong's industrialists have, since the 1980s, increasingly relocated their production in the special economic zones across the border. And Hong Kong, Choi states in conclusion, is now suffering from the ill-effects of its 'lost opportunity' for industrial upgrading, which would have enabled the economy to continue with capital- and technology-intensive production in juxtaposition to the low-cost labour-intensive production on the mainland. As it is, the territory's ensuing de-industrialization instead of regional specialization has bred problems of wage stagnation and widespread unemployment among its working population.

The next few essays in Part IV take a detailed look at some of the issues raised in Choi's discussion. The first two examine industrial relations and trade unionism in Hong Kong. These are followed by a study of the impact of Hong Kong's de-industrialization on the workforce, particularly women workers, and finally an account of the problems in Hong Kong's economy and society around the time of the handover.

We now turn to Chiu and Levin's essay (Chapter 15) which seeks an explanation for the low degree of centralization and formalization in Hong Kong's industrial relations system – an anomaly among countries at a comparable stage of economic development. In the Hong Kong case, the authors point out, decisions on pay and other conditions of employment are made largely through individual agreements between employers and employees instead of being determined by collective bargaining between employers and trade unions (hence low degree of centralization) or regulated by an established set of procedural and substantive rules (hence low degree of formalization). There are, Chiu and Levin observe, three possible explanations for this phenomenon. The first, an explanation from the economic perspective, holds that the organization of industrial relations is shaped primarily by market structure and forces. Thus, developments since the early 1950s, such as the preponderance of small-scale industries, the high proportion of young women in the workforce, high labour mobility across firms and industries, and job availability and wage increases are seen as factors impeding the trade union movement. Employers, on the other hand, preferred to bargain individually with employees in order to sustain a flexible labour force, and so had no incentive to recognize trade unions or collective bargaining. But this explanation, Chiu and Levin argue, is incomplete because it cannot account for the absence of collective bargaining even in industries where trade unions were strong, which was obviously due to the unions' lack of interest in collective bargaining.

The second explanation, from the political perspective, sees the organization of industrial relations as framed by collusion between the state and capital. The capitalists' interest in maximizing their freedom and hence power in dealing with employees, according to this explanation, coincided with the colonial state's interest in maintaining political stability through minimizing its intervention into the social and economic affairs of the indigenous community. The outcome was a voluntarist institutional framework which enabled the capitalist employers to set the terms of employment through bargaining individually with employees. But this explanation, Chiu and Levin point out, also predicts that, under such circumstances, trade unions would undertake collective action to defend and advance the interests of employees. Yet, in the case of Hong Kong, trade unions generally refrained from resorting to confrontational collective action against the state and capitalist employers. In short, Chiu and Levin contend that an explanation for the organization of industrial relations in Hong Kong must also satisfactorily account for the trade unions' lack of interest in taking collective action against a voluntarist institutional arrangement which favoured employers rather than employees.

The third explanation, from neo-institutional sociology, in the authors' view offers the most complete and convincing account of the organization of industrial relations in Hong Kong. In applying this explanation, which perceives 'culturally meaningful institutional models' as exerting a significant influence on how actors view and choose possible courses of action, Chiu and Levin focus on the diffusion of British institutional models to Hong Kong, the character of local social, political and educational institutions, as well as the institutional models which circumscribed the founding and operation of Hong Kong's trade unions. Thus, according to the authors, Hong Kong's voluntarist industrial relations system was largely a replication of the British institutional models of economic individualism and voluntarism. They further maintain that the institutional ethos of utilitarian familism, together with the absence of democratic political institutions and an apolitical curriculum in educational institutions, disposed trade unions and workers to adopt an individualist, rather than a collective, political approach to industrial relations. Finally, the authors note that Hong Kong's trade unions were set up on institutional models – such as craft guilds and political parties – which led them to focus on mutual aid functions and Chinese partisan politics rather than on employees' powers and rights. An explanation with due attention to the influence of institutional models and supplemented by insights from the economic and political perspectives, Chiu and Levin remark in conclusion, provides the most thorough and satisfactory account of Hong Kong's industrial relations system.

The next essay by Snape and Chan (Chapter 16) offers further information on Hong Kong's trade unions through an empirically-based examination of their ideologies and practices. In the authors' view, Beijing's interest in the HKSAR government's industrial stability, coupled with the pro-Beijing stance of the territory's largest trade union federation, the Federation of Trade Unions, prompts a concern about the future direction of Hong Kong's trade union movement. But what really are the 'industrial relations' policies and practices of the FTU? And what are the orientations and strategies of its main rival, the Confederation of Trade Unions, and those of other trade union groupings? Questions such as these underlie Snape and Chan's investigation of 'Whither Hong Kong's Unions?'.

Two theoretical models inform Snape and Chan's study. The 'communist' dual model, which the FTU appears to draw on, depicts trade unions as concerned primarily with the dual functions of representing labour's rights and interests and facilitating production through maintaining labour discipline and providing labour education. Conversely, the 'Western capitalist' adversarial model, which the CTU claims to espouse, sees trade unions as pursuing labour interests in confrontation with employers. These two largest union federations, in the authors' view, will probably have the most significant impact on the development of trade unionism and industrial relations in Hong Kong, as the other union federation, the pro-Kuomintang Trades Union Council is rapidly losing its importance with the transfer of sovereignty, and as the independent unions seem to be fragmented and concerned only with the interests of specific, relatively privileged occupational groups.

Snape and Chan's research findings, however, do not entirely support the above conventional portrait of the functions and roles of the territory's union groupings. Thus while the CTU's leaders do see representative unionism as their main concern, the FTU's leaders do not appear to differ from them significantly in this respect and do not seem to be more strongly committed to the production function than other union officials. The independent union leaders, as expected, endorse an inward-looking approach and exhibit a weaker identification with the representative function. The TUC union officials, however, espouse a far greater commitment to both the representative and production functions than the above portrait suggests. But these are only the union officials' *views* of the roles of their respective unions. What does union *behaviour*, as reflected in involvement in labour disputes, tell us?

Turning to their pertinent research findings, Snape and Chan note that the FTU and the CTU have a much higher level of involvement in labour disputes than the independent unions, while the TUC has the lowest and only minimal involvement. The FTU surpasses the CTU in terms of its involvement in disputes, but in terms of an involvement ratio, taking into account the membership size of the two union federations, the CTU outperforms the FTU. Another noteworthy finding is that the FTU has done better than the CTU in dispute settlement, which is contrary to the belief that the FTU is relatively ineffective in representing labour's interests.

Snape and Chan then point out that the future direction of Hong Kong's trade union movement will depend not only on the policies of its two largest and most influential union federations, but also on the territory's political development. The CTU's adversarial approach to industrial relations, which both the Chinese government and local employers view as conducive to instability, renders this union federation liable to repression. Its leaders, the authors observe, have participated actively in the pro-democracy movement for they recognize that democracy is their best safeguard against repression. On the other hand, the FTU, despite their pro-Beijing stance and image of being a moderate union federation, have by no means been compliant and ineffective in labour disputes and labour representation, for its leaders realize that, in Hong Kong's democratizing climate, representative unionism is the way to win labour support. But whether the trade union movement can continue and advance on the current trend, Snape and Chan observe in conclusion, depends on the direction of democratic development in Hong Kong.

We turn now to a study of the negative impact of de-industrialization on Hong Kong workers – a problem which Choi, in the opening essay of Part IV (Chapter 14), attributes to the territory's failure to upgrade its industries when the opportunity came. In the next essay (Chapter 17), Chiu and Lee focus on the plight of women workers, who have fared far worse than their male counterparts under Hong Kong's industrial restructuring.

The pertinent statistics which Chiu and Lee provide testify to the magnitude of the problem. As a result of de-industrialization, the number of workers in manufacturing was almost halved between 1987 and 1994, and this drop dealt a particularly severe blow to women workers, as a substantial proportion of them were employed in the manufacturing sector. Within this period, close to 270 000 women workers lost their jobs in manufacturing. Many of them faced huge difficulties in finding new employment or in adjusting to their new jobs, and the result was a sharp deterioration in their lifestyles and self-esteem. Chiu and Lee's study explains why.

Indeed, Chiu and Lee's research, based on telephone surveys and in-depth interviews, discovers that even those who remained in manufacturing often had to change factories or became underemployed, and felt very much threatened by the prospect of losing their jobs in the tide of plant relocations to mainland China. Many of those who had moved to service jobs similarly felt insecure about their employment, as employers would very likely replace them with younger women when such were available. Others, after working for some time in the service sector, reluctantly chose to exit due to a combination of factors – low pay, physical exhaustion and family responsibilities. Those who chose to remain in service jobs often had to accept lower pay and more work, as well as put up with less job autonomy and an uncongenial work schedule. But the hardest hit, Chiu and Lee observe, were those who, because of their age (40 or above), were unable to find employment even after numerous attempts. The overall effect was a deterioration in these women's lifestyles as they were forced to economize through cutting back on groceries, leisure and social life, and even to borrow from relatives in order to survive through the hard times.

In addition, Chiu and Lee point out, many of the unemployed women suffered a massive fall in self-esteem. While years of searching in vain for a job sapped the confidence and moral strength of the unemployed, their unwilling resort to dependence on welfare support or on their husband's income had a debilitating effect on their self-respect and autonomy. At the same time, after years of gainful employment outside the home, many found full-time housewifery disabling and detrimental to their mental and physical well-being. Hong Kong's industrial restructuring and its discontents, the authors observe in conclusion, testify once again to gender inequality in the society and the vulnerability of women workers in times of economic change and hardship.

We have seen from the last essay and several previous essays (Chapters 4, 11, 12) that Hong Kong, around the time of the handover, was plagued by a host of political and social-economic problems. This raises the question of whether the Hong Kong Dream, so prevalent among the local population during the 1970s and 1980s (see Chapter 1) has now faded and been superseded by less optimistic, or even despondent, social and economic beliefs and attitudes. We have an answer from a series of survey findings, which Lau Siu-kai presents in the concluding essay in Part IV (Chapter 18).

Lau begins with a summary examination of the principal economic and social changes in Hong Kong from the late 1980s to the eve of the handover. Economic growth slowed down; inflation, especially in property prices, soared; unemployment and underemployment rose as a result of de-industrialization; and mobility opportunities for the better educated plunged due to the rapid expansion in higher education. The consequences, Lau argues, were, on the one hand, a disgruntled population driven to making a fast buck through speculative activities and, on the other, a widening gap between the rich and an underclass of unemployed and semi-employed. On top of all these, Lau observes, was the 1997 issue which engendered widespread anxiety and doubts about the government's ability to safeguard the interests of the Hong Kong people. Lau describes the ensuing social milieu as one suffused with greed, disorientation, cynicism, intolerance and nastiness, and he cites survey findings to substantiate his view.

The most obvious change, as revealed in findings from surveys conducted between 1985 and 1994, was the decline in the trust people had in one another. A survey in 1994 also showed that only a tiny minority of Hong Kong people saw their fellow citizens as having a strong sense of social responsibility. Yet rather unexpectedly, surveys in that decade consistently came up with the finding that the overwhelming majority of the local population still saw Hong Kong as a land of opportunity and the rich people as having made their way up through effort and ability. At the same time, Lau notes, survey findings also reveal some ominous trends, such as increasing public unease at the territory's growing economic inequalities, the worsening public image of the wealthy, and the mounting popular belief that people with wealth and status were not fulfilling their responsibility to society. Other findings correspond more directly to the population's economic discontent and political uncertainty. Thus there was a rising public demand for government intervention in the economic sphere, most notably in the provision of unemployment benefits; there were signs that people were increasingly worried about the rising level of social conflict in the society; and there was evidence that people had little confidence in the post-1997 legal system. But amidst all these disheartening messages, there is the encouraging survey finding that Hong Kong people in the 1990s still harboured a strong sense of belonging to the community, which Lau sees as important in reducing the negative impact of the social-economic and political problems faced by the society.

Lau's analysis and the survey findings he presents indeed suggest that some of the difficulties which beset the HKSAR government were not of its own making. While it cannot be gainsaid that the disarticulation of the political system, the occasional but disconcerting blunderings of the civil service (see Chapter 12), and the enlarging mass–elite gap (see Chapter 14) have been crippling flaws of the post-handover regime, one has to recognize that social frustrations and discontents carried over from the colonial era and political distrust engendered by the change of sovereignty have added a substantial challenge to the regime's task of governing under the principle of 'one country, two systems'.

Social Issues and Social Policy

The essays in Part V examine a number of issues which have become the major concerns of the Hong Kong people in recent years. The first essay throws light on some central features of education in Hong Kong through an examination of what the local population perceive as valid and high-status school knowledge and why they have such a perception. The second essay takes us to the domain of the family and, in the light of recent developments in family relationships and ethos, assesses the government's attitudes and actions in this respect. This is followed by two essays on the subject of social movements, both of which seek to understand the nature, as well as the career, of these movements in the context of the larger society. Part V finishes with an essay on emigration from Hong Kong, which demonstrates how personal networks shape considerations to exit Hong Kong. Together, the five essays in this Part add fresh knowledge and insights on how culture, society, politics and the economy bear on the making of the 'Hong Kong experience'.

Part V's opening essay (Chapter 19) by Paul Morris revolves around the intriguing issue of why the curriculum of Hong Kong's secondary schools continues to place a heavy emphasis on abstract remote academic knowledge despite the government's attempts over the past two decades to promote a curriculum which is more down-to-earth and closely related to the students' social milieu and experiences. Morris' main thesis is that, in Hong Kong, a school curriculum that is 'decontextualized' and 'depoliticized' would serve the state's objective in maintaining the territory's political stability, and that this objective was realized, through direct state control of education in the early post-war period and through the operation of the 'educational market' over the past two decades. Morris shows how the market fortuitously relieves the state of the need to control curriculum knowledge and indeed makes it possible for the state to *opportunely* trumpet its rhetoric of progressive curriculum reform.

The educational market operates essentially on the principle of merit: the best schools get the best students and vice versa; and the best schools are those whose students excel in public examinations. Public examinations are based on subjects taught at school and these subjects follow an official syllabus. Since the early 1990s the government has recommended a more broadly-based, 'relevant' and 'utilitarian' curriculum, which shifts the traditional emphasis on academic subjects such as languages, mathematics and sciences to a more balanced encompassing of subjects and cross-curricular activities like cultural studies, social studies, physical education and civic, moral and sex education. But the forces of the educational market, Morris notes with reference to empirical findings, have obstructed the realization of this recommendation. In the first place, students, parents as well as teachers consider the traditional academic subjects to be far more important than the broadly-based subjects as it is the academic subjects that count most in the students' progress to further school education. Second, teachers perceive cross-curricular activities which are expected to be inserted into the teaching of formal subjects as detracting from the time and attention for the official syllabus, with dire implications for their students' performance in public examinations. Further, the government's recommendation for an 'individualized' approach to teaching geared to cultivating the students' 'active thinking' can hardly be implemented because of large class sizes and because a large amount of homework is held to reflect dedicated teaching and high status of a subject. In the race for recognition and prestige, Morris observes in conclusion, Hong Kong schools end up producing students who strive to perform well in the traditional academic subjects in public examinations, but who are poor in physical fitness and deficient in aesthetic sensitivity and social and political awareness.

We saw from the last study a case where the government's recommendations for school curriculum reforms have not been duly implemented because of the operation of the market, which is conditioned by the public's conception of what is valid school knowledge and what makes a high-status school. This case of education furnishes a bearing for understanding the next study, in which Chow investigates the government's reluctance to formulate a family policy against the background of the Hong Kong Chinese people's attitudes towards the family, as well as the problems faced by the family in contemporary Hong Kong.

Chow's essay (Chapter 20) begins with the observation that, in traditional Chinese society, the family performed a wide range of functions and was held to be the best provider of care and support for its members. But in modern Hong Kong, Chow adds, many of the family's traditional functions have been taken over by other social institutions, and the values that bind the family members together into a unit of care and support – filial piety, chastity and faithfulness – have been in decline. The consequences of these developments, Chow points out on the basis of survey findings, are, on the one hand, the waning of the belief that the family is by and large a self-sufficient unit capable of looking after the needs of its members, and, on the other, the increasing severity of family problems such as divorce, child abuse, wife-battering and neglect of the elderly. Chow notes that the government has responded, since the mid-1960s, through providing comprehensive family welfare services aimed at strengthening interpersonal relationships within the family, preventing and remedying family problems, and catering to needs which cannot be met from within the family. But all these do not amount to a family policy which Chow deems necessary for providing the existing family services with coordination and direction.

A family policy, Chow explains through citing authorities in the field, is guided by 'certain objectives regarding the family'. But the Hong Kong government has refrained from setting objectives for its family services, for fear of being accused of using the family as an agent of social control and of imposing goals and responsibilities on a domain which the Chinese people have traditionally held to be self-directing. In short, the government is worried that a family policy would do more harm than good. But the increasing public demands for government intervention in various spheres, Chow argues, suggest that the government's worries are unwarranted. Further, without the benefit of a coordinated and goal-oriented family policy, the existing family services are often contradictory and ineffective in offering a long-term solution to family problems. In conclusion, Chow recommends the formulation of a family policy targeted at preventing abuse and oppression within the family and enhancing the functioning of the family through linking it closely with other social institutions.

The previous two essays gave us an idea of the ethos of the Hong Kong people in respect of the issues of school knowledge and the family. The next essay by Lui and Chiu (Chapter 21) analyses the development of social movements in contemporary Hong Kong in the context of historical changes, and adds substantially to our understanding of the changing social concerns and political ethos of the local population.

Lui and Chiu begin with what many have viewed as an important turning point in Hong Kong's social and political development: the 1966 and the 1967 riots. The 1966 so-called Star Ferry Riots were small-scale, short-lived and sporadic hostile outbursts where the participants were mainly disgruntled youths, but these disturbances were the first clear signal of ominous discontents among the first generation of post-war local-borns. The territory-wide riots a year later, while instigated and led by local left-wing groups under the influence of the Cultural Revolution on the mainland, nevertheless starkly exposed the shortcomings and vulnerability of the colonial government. To Lui and Chiu, these mid-1960s riots had the effect of opening up a new phase in the development of social movements and collective protests in Hong Kong, for the population – especially the younger generation – now began to question the nature of colonial rule and to reflect on the issue of their national and cultural identity. These, according to the authors, were what the student movement of the 1970s was really about.

But the mid-1960s riots had another unintended consequence – the government's subsequent administrative reforms and expanded social service provisions heightened the population's awareness of the government's responsibility in looking after their needs. The 1970s were thus a time of proliferating pressure group politics, through which the grassroots resorted to protest actions to voice their claims and demands for resources from the government. The pressure groups, Lui and Chiu observe, initially took its ideological platform from the student movement, but by the late 1970s replaced the waning student activism as the motor of protest movements. But the protest movements, characterized by a concern for resource redistribution, were soon overshadowed by the pressure groups' involvement in electoral politics in the 1980s, when the community was actively preparing for decolonization and self-government in the aftermath of the Sino-British agreement on Hong Kong's future. With this, pressure group politics gave way to a pro-democracy movement, and to debates among the newly formed political parties about the territory's democratic future in their contention for electoral success. In Lui and Chiu's view, this concern for democratic progress and electoral success since the 1980s has eclipsed the grassroots' movement for a better livelihood, which has now become mainly a subsidiary issue in the politicians' quest for support in electoral and parliamentary politics.

The women's movement, Choi argues in the next essay (Chapter 22), is facing a similar prospect as the grassroots movement. Choi's study charts the career of the women's movement from the setting up of the first feminist organization, the Hong Kong Council of Women, in 1947 to its struggle to retain its cause and identity amidst the political current of decolonization and democratization in the 1980s and 1990s. Drawing her data from printed documents and interviews, Choi observes that it was in the 1980s, when feminism began to catch the attention of local Chinese women, that women's organizations for the *local* Chinese were established. For these women's organizations, being *local* not only means identifying with one's own culture, but also focusing on the grassroots, in contrast to the Council of Women (which disbanded in 1995) whose members were mainly expatriates and middle- and upper-class women. But the grassroots orientation of the women's movement, Choi points out, was also a product of its leaders' participation in the student movement and the pressure groups' protest movements of the 1970s, which led them to believe that resolving the problems of the grassroots women is the first step to looking after the needs of women in general. The strategies of the women's movement were thus similar to those of the pressure groups, relying mainly on protest campaigns, alliance with grassroots groups and media exposure to further its cause.

But the transition to 1997, Choi argues, brought a challenge to the local grassroots identity of the women's movement. In particular, the Beijing-sponsored, well-financed Hong Kong Federation of Women (set up in 1993), with its pro-establishment stance and its subordination to the national objective of building up the strength of China, tended to overshadow the women's groups and truncate their cause and identity. Further, the Federation's top-down approach not only smacked of authoritarianism and intolerance of 'radical' feminist views, but also threatened to dislodge the movement's pressure group 'peripheral politics' with the 'mainstream politics' of networking and negotiating with the influential circles. And after Hong Kong's return to the motherland, the women's movement also faces the prospect of having its identity and cause intermingled with, or even subsumed under, the concerns of women on the mainland. Finally, Choi adds in conclusion, the women's movement also has to struggle against being submerged and drowned in Hong Kong's mainstream politics of democratization. In short, in Choi's view, the politics of identity has remained a principal concern of the women's movement in Hong Kong. It had to carve out a *local* identity under colonial rule, and now has to preserve this identity under the 'one country, two systems' political arrangement, as well as to uphold its *female* identity in its support for the territory's democracy movement.

The final essay (Chapter 23) in Part V (and the volume) is a study of the significance of personal networks in decisions to emigrate from Hong Kong. On the basis of data obtained through survey and in-depth interviews, the authors, Wong and Salaff, show that while personal networks abroad are an important asset for Hong Kong emigrants, different occupational classes make use of these networks in different ways and for different reasons. As such, the study also sheds light on class differences in attitudes to, and the use of, kinship and friendship ties among the Hong Kong population.

Emigration from Hong Kong in the decade or so before 1997 is commonly held to have been a response to anxieties generated by the transfer of sovereignty. But Wong and Salaff discover from their research findings that it was personal connections, not attitudes towards the handover, that determined who planned to exit and where they went. The authors argue that personal networks are important and can be regarded as a form of capital because the holder can use them to attain benefits, especially economic returns. And having such networks in the host country is crucial to the Hong Kong emigrant in finding employment and settling down in a new environment. Wong and Salaff's survey findings show that, on the whole, emigrants had more kinship and friendship ties abroad than non-emigrants. But the authors also discover an exception to this general pattern, as well as variations across occupational classes in their use of overseas personal networks.

Working-class families, Wong and Salaff point out, relied heavily on kin abroad in making their emigration decisions; this was an extension of their close relationship with kin in Hong Kong. Lacking financial resources and information about life abroad, and having to use *family* reunification as the legal basis for emigration, working-class emigrants had little alternative than to count on kin for support. In contrast, lower-middle-class emigrants, who were mostly salaried employees in bureaucracies, seldom sought the help of kin. In the first place, most of them relied on their civil service status to apply for emigration. Second, the kind of jobs they wanted to take in the recipient country depended on credentials rather than kin connections. Finally, they preferred to be independent than rely on their kin. Moreover, the lower-middle-class workers were not keen to emigrate because they did not want to risk a drop in status and living standard in the foreign country. Wong and Salaff then examine the *exceptional* case of the affluent families, among whom the emigrants had fewer kin abroad than the non-emigrants. Indeed, as the authors go on to show, these affluent emigrants, who were mostly businesspeople and professionals, did not require kin assistance economically or in their emigration application. They were also not closely tied to their kin contacts as they could easily make use of their network of friends, colleagues and classmates. And they relied mainly on such friendship networks because they knew one another's skills and talents and because their working relationship with friends would be one of partnership not dependence. In conclusion, Wong and Salaff return to the importance of personal connections as network capital, highlighting that the multitude of personal networks renewed and extended by families in the process of emigration in effect constitute a valuable asset in Hong Kong's globalized economy.

References

Abbas, Ackbar (1996), 'Marginals, Nomads, Hybrids: A Culture of Disappearance', *Critical Studies*, **5**, pp. 9–27.

Chau, L.C. (1989), 'Labour and Labour Market', in Henry C.Y. Ho and Larry L.C. Chau (eds), *The Economic System of Hong Kong*, Hong Kong: Asian Research Service, pp. 169–89.

Chiu, Stephen (1994), 'The Politics of Laissez-faire: Hong Kong's Strategy of Industrialization in Historical Perspective', *Occasional Paper no. 40*, Hong Kong: Hong Kong Institute of Asia-Pacific Studies, The Chinese University of Hong Kong.

Coates, Austin (1975), *Myself as a Mandarin*, Hong Kong: Heinemann.

England, Joe (1989), *Industrial Relations and Law in Hong Kong*, Hong Kong: Oxford University Press.

Lau, Siu-kai (1981), 'Utilitarianistic Familism: The Basis of Political Stability', in Ambrose Y.C. King and Rance P.L. Lee (eds), *Social Life and Development in Hong Kong*, Hong Kong: The Chinese University Press.

Lau, Siu-kai (1982), *Society and Politics in Hong Kong*, Hong Kong: The Chinese University Press.

Leung, Benjamin K.P. (1991), 'Political Process and Industrial Strikes and the Labour Movement in Hong Kong, 1946–1989', *Journal of Oriental Studies*, **29** (2), pp. 172–206.

Leung, Benjamin K.P. (1996), *Perspectives on Hong Kong Society*, Hong Kong: Oxford University Press.

Miners, N.J. (1975), 'Hong Kong: A Case Study in Political Stability', *Journal of Commonwealth and Comparative Politics*, **13** (1), pp. 26–39.

Salaff, Janet (1995), *Working Daughters of Hong Kong* (2nd edn), New York: Cambridge University Press.

Shively, Stan (1972), 'Political Orientations in Hong Kong: A Social-psychological Approach', Hong Kong: Social Research Centre, The Chinese University of Hong Kong.

Siu, Helen (1997), 'Remade in Hong Kong', *Index on Censorship*, **1**, pp. 145–51.

Wong, Siu-lun (1988), *Emigrant Entrepreneurs*, Hong Kong: Oxford University Press.

Wong, Thomas W. P. and Lui Tai-lok (1993), 'Morality, Class and the Hong Kong Way of Life', *Occasional Paper no. 30*, Hong Kong: Hong Kong Institute of Asia-Pacific Studies, The Chinese University of Hong Kong.

Youngson, A.J. (1982), *Hong Kong: Economic Growth and Policy*, Hong Kong: Oxford University Press.

Part I
Profiles of Change and Development

[1]

THE 'HONG KONG EXPERIENCE': CLASS, INEQUALITY AND MORALITY IN POLITICAL TRANSITION

Tai-lok Lui, The Chinese University of Hong Kong,
and Thomas W.P. Wong, University Hong Kong

1.

Talks about Hong Kong's future have turned into a 'political marathon'. Like other marathons, its has been a long, winding and rocky road — new issues keep cropping up and, in most cases, quickly become topics of heated debate either between diplomats representing the governments on the two sides of the negotiation table or among local politicians of different political persuasions. Unlike others, it is not at all clear when and where it will come to an end; 1997 has become a convenient date for political and administrative purposes, telling us very little about how life, social as well as political, would be after the change of sovereignty. But the amazing thing about these future talks is that the contentious issues always revolve mainly, if not exclusively, around the institutional arrangements of the polity and the economy. The picture of the decolonization process conjured up in these discussions seems to suggest that the crux of the matter is simply to look for a blueprint which would guarantee a convergence in the political structures of the pre-1997 British colony and the future Special Administrative Region and the maintenance of a capitalist economy in the process of returning Hong Kong to a socialist regime. The key words are 'restrained democratization' (and thus facilitates political convergence) and 'capitalism unchanged' (assuming that all parties involved, governments as well as the people of Hong Kong, will be happy). In short, the answer to all questions raised in the political debates related to Hong Kong's future is 'one country two systems', the blueprint for preserving Hong Kong's lifestyle in the process of 'decolonization without independence' (Lau 1987).

But can we conceive of the polity, or the economy, in isolation? Is it really sensible to talk about institutional design without some basic understanding of the social structure wherein economic and political institutions are embedded? This essay is an attempt to bring class analysis back into our discussion of Hong Kong's future. By doing so, we probe the social basis of the local political discourse on the future question and discuss, very briefly, the emerging political tensions in the years approaching 1997.

112   TAI-LOK LUI AND THOMAS W.P. WONG

2.

That the promise of preserving Hong Kong's lifestyle within the institutional framework of 'one country two systems' is able to catch the imagination of the people, at least in the 1980s, and more specifically prior to the Tiananmen Incident, has a lot to do with the growth of a local identity (Baker 1983; Lui 1988) among the people of Hong Kong. In this light, the design of making Hong Kong a Special Administrative Region after 1997 is a compromise between on the one side, a straightforward change of sovereignty and, on the other, political independence. The latter has long been ruled out by default, on the grounds of political pragmatism and Chinese nationalism. However, the former arrangement is found unacceptable by the majority of the local population. Doubts and worries concerning life after 1997 are real (as revealed in emigration statistics) and are by no means confined to the middle class. The idea of 'one country two systems' was, in the context of the Sino-British Negotiations, conceived as a 'pact' to ease public anxiety — the people of Hong Kong can continue to enjoy their way of life under Chinese socialism. Whether people really have faith in the 'pact' is not our concern here (but again consult the emigration statistics). What interests us is the strong emotional appeal of the notion of the Hong Kong's way of life to the local people. To the majority of these people, the central concern is more of the freedom and opportunity of making, in their mind, a good (in both economic and cultural senses) life than grand narratives of nationalism and political development. But the definition of what constitutes a good life in Hong Kong has been changing. What we find most interesting is that the notion of 'one country two systems' is socially and ideologically embedded in what we can call the 'Hong Kong experience' — there are opportunities and, equally important, a special way of life can be found in contemporary Hong Kong. The 'Hong Kong system' is in this sense more than the political and economic blueprints of institution-building that are presumably the bones of contention on the negotiation table.

We have argued elsewhere in greater detail (Lui and Wong 1993a; Wong 1992a) that the beliefs in Hong Kong as a place of unmatched openness and opportunities — did not have an auspicious beginning. Structural changes in the society, with expanding 'room at the top', creating opportunities and facilitating upward mobility, have wrought important changes in perceptions and preferences. The 'Hong Kong experience', as generally understood, did not come to fruition until the 1970s. When we compare the relevant and comparable findings from

THE 'HONG KONG EXPERIENCE'   113

survey studies in the late 1960s, with those in the 1980s, we find quite dramatic changes. Table 1 shows some of these changes. (The details of the cited studies can be found in Wong 1992b.)

Table 1: Opportunities and Evaluation of Mobility in Society (%)

	Mitchell 1967	Lau 1977	Lau 1986	SI 1988
Per Cent who opted to stay in Hong Kong despite opportunity elsewhere	23	53		
Per cent who saw themselves as having higher status than parents	31		37	44
Per cent who saw themselves as having lower status than parents	36		11	12

The studies are:
    Mitchell 1967: *The Urban Family Life Survey*; source Mitchell 1969.
    Lau 1977: *Urban Hong Kong Survey*; source Lau 1982.
    Lau 1986: *Pilot Study of Social Indicators Study*; Source Lau and Wan 1987.
    SI 1988: *The Social Indicators Study 1988*; Source Lau *et al.* 1991.
Source: Wong 1992b:246-7

It is clear that in relation to the belief in Hong Kong as a land of opportunities, and to the evaluation of one's betterment as compared with one's parents, there have been significant changes. In particular, the perception of the society as providing the best environment for one's career is probably quite deeply-ingrained. The above table also shows that the proportion of those who opted to stay in Hong Kong, despite the availability of opportunities elsewhere, has more than doubled in the decade following the late 1960s. There have been changes — towards optimism and confidence — in the perception of opportunity and social advancement. But such beliefs in openness and opportunities are no doubt shaped by the changing social environment from the 1960s to the 1980s. We will discuss such changes in relation to specific findings.

    Survey findings of Mitchell's study of Hong Kong families in 1967, the year marked by the violent political riot in the post-war decades, suggest that only 16% of the respondents considered they had some or a lot of opportunities to make a successful career (Mitchell 1969:174). Half of the respondents said that there are fairly good or good chances for a working class child to work hard to become a professional (ibid:175). But

114   TAI-LOK LUI AND THOMAS W.P. WONG

compared with survey findings from Singapore (80%), Taiwan (78%) and Malaysia (68%), the Hong Kong results show a rather strong feeling of pessimism with regard to opportunity of social mobility. This, Mitchell suggested, is connected with the downward mobility experience of a significant proportion of the population. In his rough estimation of mobility experience, Mitchell found that for the male adult population in 1967, 35% of the sons were in positions lower than their fathers (should those of agricultural background be excluded from the calculation, the proportion of sons experienced downward mobility would go up to 44%) (ibid:143-4). Furthermore, it is pointed out that 30% of the respondents had found their jobs through personal introduction, reflecting a rather closed occupational structure and personalized networks in the 1960s.

Meanwhile, it is important to note that traces of optimism were found among the young people growing up in the 1960s. A survey of local secondary school students, also carried out by Mitchell in 1967, shows that 42% of the interviewed students mentioned that personal success was very important (1972:75). Only 27% of the adult respondents in the family life survey (Mitchell 1969:238) gave the same answer. Concerning their expectation of future career, 14% of the student respondents answered 'very successful' and 49% 'above average but not very successful'. The observation of an optimistic sentiment among local youth is confirmed by another study of young people in 1969 (Podmore and Chaney 1973). Of course, we are not suggesting that starting from the early 1970s there has been a sweeping change in people's mood and attitude. As we shall see in subsequent discussion, the optimism of structural openness always goes hand in hand with a personal pragmatism and awareness of inequalities between classes. The point is that given the social structural changes and demographic transitions in post-war Hong Kong (also see Lui 1988; Salaff 1981), the discourse of 'Hong Kong experience' -- the imagery of 'home in Hong Kong' (for a migrant population settled in the colony in the years after the Japanese occupation and the civil war in China) and the popular sayings like 'Hong Kong is a place of opportunities' — was in the process of formation. By the early 1970s, Hong Kong, as people's home and as an ideological construct, was more than a 'life boat in the sea of political turmoil (and so don't rock it)'. A glimpse of the popular mentality at that time can be found in Rosen's ethnographic study of the Chinese middle class families in an affluent private housing estate called Mei Foo Sun Chuen (1976:209):

THE 'HONG KONG EXPERIENCE' 115

'... life in Hong Kong provides the access for individuals and their families to attain financial security, and the residents of Mei Foo represent a model for their Hong Kong brethren of how this security can be achieved. It is not that they are very wealthy, for most of them are not. It is rather the fact that most of them reached this stage of security and affluence via the long route: in flight from native homes in China across the border into Hong Kong, and up the ladder in Hong Kong from factory jobs and low-cost housing to white-collar jobs and a flat in Mei Foo. Their current lifestyle thus represents a greater security than that provided by the many isolated cases of greater financial success achieved in pre-revolutionary China or in the host territory of Singapore or Indonesia or South Viet Nam. The security offered in the Mei Foo model lies in the freedom it permits those who attain it to take some measure of control over the rest of their lives. Many will and already have become immigrants, but none will ever again be refugees.'

It should be noted that the discourse of 'Hong Kong experience' does not develop in a vacuum. The early 1970s was a period of a growing economy, improvement in living standard, increase in opportunity of social advancement, as well as that of rising social conflicts (Lui and Kung 1985). But then the articulation of the 'Hong Kong experience' does not presuppose an affluent society or an end of ideology. In retrospect, the emergence of collective actions and social conflicts in the 1970s marked the beginning of attempts by the Hong Kong Chinese to make their claims at the societal level. The very act of making a claim signified a farewell to the 'refugee mentality' characterized by social and political detachment in the 1950s and 1960s. Whether it was to stage a protest against government policy or to organize a demonstration demanding improvements in the social provision of social welfare, the attempts to act collectively in the public arena revealed the emergence of a new identity — they saw themselves as the Hong Kong Chinese trying to find out their social entitlements and obligations and no longer, like their parents' generation, as the Chinese in a colonial Hong Kong with a sojourner status and identity.

116   TAI-LOK LUI AND THOMAS W.P. WONG

3.

Based upon the findings of our 'Hong Kong social mobility study 1989', Table 2 presents, adopting the class schema put forth by Goldthorpe (1987), the 'class map' of contemporary Hong Kong.[1] Class is here defined, in the light of Weberian sociological theory, in terms of 'market situation' and 'work situation' (also see Lockwood 1958; Goldthorpe 1987:40-3).

Table 2: The Class Structure of Hong Kong

7-folded Class	Brief Description	N	%	3-folded Class
I	Upper Service Class: Higher-grade professionals, administrators and officials, managers in large establishments, larger proprietors	81	8.6	
II	Lower Service Class: Lower-grade professionals, administrators, higher-grade technicians, managers in small business and industrial establishments, supervisors of non-manual employees	107	11.3	Service
III	Routine non-manual employees in commerce and administration, personal service workers and shop sales personnel	90	9.6	
IV	Petty Bourgeoisie: Small proprietors, artisans, contractors, with or without employees	132	14.0	Inter-mediate
V	Lower-grade technicians, supervisors of manual workers	150	15.9	
VI	Skilled manual workers	149	15.8	Working
VII	Semi-skilled and unskilled workers, agricultural workers	234	24.8	

Source: *Hong Kong Social Mobility Study* 1989

---

1   The study was a Hong Kong-wide survey conducted in 1989. It covered 1,000 randomly selected male household heads aged 20 to 64. A structured questionnaire was used to carry out face-to-face interviews. For details of the survey, see WONG and LUI 1992b. Concerning questions of interviewing male household heads in class research, a useful summary of arguments from different perspectives and an interesting discussion can be found in CROMPTON 1993.

THE 'HONG KONG EXPERIENCE' 117

Without going into the technical details of the construction of the 'class map' (see Wong and Lui 1992b), here we would present an intergenerational mobility table constructed in accordance with the adopted class schema (see Table 3). Very briefly, Table 2 shows that about one fifth of the respondents (19.9%) can be categorized as the (both upper and lower) service classes. While the sources of the rapid expansion of the service classes are many (also see Lui and Wong 1994), it is fair to say that the major social force in shaping the rise of the professionals, managers and administrators is the phenomenal growth of the colonial economy in the post-war decades.

Table 3: Intergenerational Mobility Matrix

Father's Class/Son's Class	I	II	III	IV	V	VI	VII	Total Fathers
I	12 (30.8) [15.8]	6 (15.4) [6.4]	3 (7.7) [4.0]	4 (10.3) [4.1]	6 (15.4) [4.9]	2 (5.1) [1.7]	6 (15.4) [3.6]	39 [5.2]
II	7 (15.6) [9.2]	13 (28.9) [13.8]	4 (8.9) [5.3]	5 (11.1) [5.1]	9 (20.0) [7.3]	4 (8.9) [3.4]	3 (6.7) [1.8]	45 [6.0]
III	8 (11.1) [10.5]	15 (20.8) [16.0]	15 (20.8) [20.0]	8 (11.1) [8.2]	10 (13.9) [8.1]	9 (12.5) [7.6]	7 (9.7) [4.2]	72 [9.6]
IV	28 (11.2) [36.8]	27 (10.8) [28.7]	21 (8.4) [28.0]	43 (17.1) [43.9]	37 (14.7) [30.1]	35 (14.0) [29.7]	60 (24.0) [35.7]	251 [33.4]
V	8 (11.9) [10.5]	11 (16.4) [11.7]	7 (10.4) [9.3]	7 (10.4) [7.1]	12 (17.9) [9.8]	16 (23.9) [13.6]	6 (9.0) [3.6]	67 [8.9]
VI	6 (6.7) [7.9]	7 (7.8) [7.4]	12 (13.3) [16.0]	9 (10.0) [9.2]	11 (12.2) [8.9]	20 (22.2) [16.9]	25 (27.8) [14.8]	90 [12.0]
VII	7 (3.7) [9.2]	15 (8.0) [16.0]	13 (7.0) [17.3]	22 (11.7) [22.4]	38 (20.2) [30.9]	32 (17.0) [27.1]	61 (32.4) [36.3]	188 [25.0]
Total Sons	76 (10.1)	94 (12.5)	75 (10.0)	98 (13.0)	123 (16.4)	118 (15.7)	168 (22.3)	752 (100.0) [100.1]

Notes: figures in [ ] are column (inflow) percentages; figures in ( ) are row (outflow) percentages. Source: *Hong Kong Social Mobility Study* 1989

118   TAI-LOK LUI AND THOMAS W.P. WONG

The marginals of Table 3, i.e. the distributions of the class positions of the fathers and the sons, can be taken as some approximate indicator of the social structure of Hong Kong in the respective time periods. We can see that there has been a significant expansion of professional, managerial and administrative positions between the two generations. The increased 'room at the top' (twice as many sons as fathers in Classes I and II) is evident; this structural change probably accounted for much of the upward mobility observed in Table 3. Table 3 shows both the effects of rapid economic development on social mobility (high inflow rates for Classes I and II) and the persistence of social inequality (lower outflow rates for positions of Classes I, II and VII) in the structuration of social class in contemporary Hong Kong. On the one hand, rapid economic development in the post-war decades has created new 'room at the top' and opportunities for people coming from different backgrounds to move into 'service class' positions. Our data suggest that 75% and 79.8% of those in Classes I and II respectively are newcomers from non-service-class background (see Tables 3). Though there are difficulties in long-range upward mobility, especially for crossing the manual/non-manual barrier (see Wong and Lui 1992b), still there are 9.2% of those of working class origin (Class VII) being able to reach Class I. However, on the other hand, social advancement is embedded in inequalities within the class structure. 46.2% and 44.4% of those coming from Class I and Class II origins respectively are able to retain their 'service class' positions. Compared with people from other class origins, they show rather strong retentiveness in retaining their privileged class positions. Meanwhile, at the other end of the class structure, the unskilled production workers show a relatively high homogeneity in both inflow and outflow perspectives. Rarely can we find unskilled manual workers coming from service class origin (5.4%); also they have difficulties in climbing up to the top positions.

Table 4: Odd Ratios (3-folded Class Schema)

Pairs of Origin	Pairs of Destination Classes Competed For		
Classes 'in Competition'	I/II	II/III	I/III
I vs II	2.0	1.8	3.5
II vs III	1.8	1.6	2.9
I vs III	3.7	2.8	10.2

Source: *Hong Kong Social Mobility Study* 1989

THE 'HONG KONG EXPERIENCE'   119

Class inequalities in the process of social advancement are clearly demonstrated in Table 4. It shows the odds of one class as compared with another in arriving at one destination class rather than another. From the 3-folded classification schema, we find that Class 1 (i.e. Classes I and II in the 7-folded schema, the service classes) background men have an advantage of 10.2 times over Class 3 (i.e. Classes VI and VII, the manual workers) background competitors in arriving at Class 1 rather than Class 3 destination. Such advantage, though to a less extent, is also evident in other pairs of destinations. In short, despite that rapid economic development has created new openings at the top and thus opportunities for mobility, class differentials in competition for social advancement are still significant.

4.

In the above section, we have pointed out the two sides of the Hong Kong people's mobility experience. On the one hand, there are ample opportunities brought about by economic development for social mobility. On the other, social advancement is structured by class differentials and inequalities. This explains why in the popular discourse of the 'Hong Kong experience', there are both hopes and frustrations, a belief in social advancement intermingled with a feeling of cynicism.

Our findings concerning the Hong Kong people's perception of their lives in the colony show that most of the respondents share the beliefs of Hong Kong as a place of opportunities (at least better than their parents') and of individual efforts in making success (see Table 5-a and b). There exists, as mentioned in an earlier section, an ideology celebrating Hong Kong as a place of openness and opportunities. However, when we asked our respondents whether the children of a factory worker and a business executive would have the same chance to make successful careers, 65% of them disagreed (see Table 5-c). There is an awareness of class differentials in social advancement across respondents of different classes.

120   TAI-LOK LUI AND THOMAS W.P. WONG

Table 5: Perception of Social Issues by Class (in Percentages)

Statement	Class	Response		
		SA/A	N	D/SD
a) 'In Hong Kong, if one has abilites and tries hard, one will be successful'	I	74	14	12
	II	62	20	18
	III	71	10	19
	IV	81	5	14
	V	73	10	17
	VI	75	8	17
	VII	74	11	15
		Better	No Diff	Poorer
b) 'If you are to compare yourself with your parents generation, what would you say their chance is?'	I	25	6	69
	II	20	7	73
	III	14	7	79
	IV	12	6	82
	V	15	11	74
	VI	12	11	77
	VII	16	13	71
		SA/A	N	D/SD
c) 'Would you agree to the view that in Hong Kong, the child of a factory worker has much the same the chance to get ahead as the child of a business executive?'	I	36	12	52
	II	26	9	65
	III	31	5	64
	IV	31	5	64
	V	28	6	66
	VI	26	4	70
	VII	26	7	67
d) 'If the boss is to make a profit, he has to exploit the worker'	I	15	6	79
	II	19	17	64
	III	38	12	50
	IV	17	10	73
	V	32	16	52
	VI	45	14	41
	VII	39	21	40

Notes: SA 'Strongly agree', A 'Agree', N 'Neutral, D 'Disagree', SD 'Strongly disagree'

Table 5: Continued

Statement	Class	Response		
		SA/A	N	D/SD
e) 'There is bound to be conflict between different classes	I	63	5	32
	II	61	14	25
	III	73	6	21
	IV	67	8	25
	V	79	4	17
	VI	68	9	23
	VII	68	10	22
f) 'The influence of the big corporation on Hong Kong is too great'	I	86	8	6
	II	85	7	8
	III	93	3	4
	IV	88	4	8
	V	90	5	5
	VI	86	7	7
	VII	90	4	6
g) The average wage-earner receives less than he contributes'	I	28	21	51
	II	40	18	42
	III	53	27	20
	IV	36	12	52
	V	51	12	37
	VI	61	11	28
	VII	52	17	27
		Collectivist		Individualist
h) 'Some people say that the Hong Kong workers would be better off if they stick together and work for their common interests. Others say that the average worker would be better off if he makes greater efforts to go ahead on his own. Which view do you agree to?'	I	30		70
	II	33		67
	III	35		65
	IV	31		69
	V	35		65
	VI	38		62
	VII	45		55

Source: *Hong Kong Social Mobility Study* 1989

Indeed, co-existing with the belief in Hong Kong as a land of opportunities, there is a persistent sense of strain and injustice (also see Wong 1992; Wong and Lui 1993). 31.5% of all respondents perceived the employment relation as exploitative (Table 5-d). And when asked of their perception of class relations, 69% of them agreed that 'there is bound to be conflict between different classes' (Table 5-e). In terms of class awareness and class identification, 80.4% of the respondents suggested that they belong to a class and most of them had a clear idea about their positions in the class structure (Wong and Lui 1992a:26-27).

As regards the allocation of economic and political resources, most respondents are rather critical of the existing system. 88.6% of the respondents agreed that 'the influence of the big corporations on Hong Kong is too great' (Table 5-f). Nearly half (48.6%) of them accepted the view that 'the average wage-earner receives less than he contributes' (Table 5-g). Of course, there are class differences in the responses to these statements. For instance, the service classes are found to be less critical of the capitalist system. But this would only support our contention of the importance of class analysis to the understanding of social and political dynamics.

Our findings clearly demonstrate that the 'Hong Kong experience' co-exists with some equally deeply-ingrained perceptions and values with regard to social inequality and social injustice. What we find most intriguing is the chasm between personal experience (the strain) and social ideology (the optimism and economically dynamic spirit). Our tentative explanation of such a phenomenon is that there are structural reasons for the co-existence of both hope and strain. Our findings have, among other things, pointed to the great amount of mobility or fluidity intergenerationally. The expansion of the 'room at the top', largely a result of rapid economic development, undoubtedly contributed to the availability of opportunities and thereby shaping the Hong Kong mobility regime (Wong and Lui 1992b; Chan, Lui and Wong 1993). The social history of that expansion and its effects on the Hong Kong people's values, ethos and morality is yet to be taken as a big agenda, for which studies utilizing different methods and time-frames are obviously needed. But pending such a study, we think it not unreasonable to say that this experience is the structural basis of the social ideology. On the other hand, our findings have also revealed significant differentials in mobility chance. Moreover, structural analysis of the mobility table suggests that there are pockets of greater rigidity in the class structure, with an invidious barrier broadly separating the non-manual and the manual classes (Wong and Lui 1992b:62-70). We cannot enter into a discussion of the implications of these findings for class formation or the demographic and socio-political characteristics of different classes. We hope however that such structural differentials could go some way to illuminating the personal experience or sense of inequality and injustice. Both openness and inequalities are

THE 'HONG KONG EXPERIENCE' 123

revealed in the social structure, and as people enter and benefit differently from its changes, their orientations are likewise moulded.

But more interesting is that in response to barriers and class differences, the people of Hong Kong, instead of getting organized and taking collective action as a strategy for improving their livelihood, tend to adopt an individualist strategy (see Table 5-h). When they were given a choice, they believe that they would be better off by making greater efforts to go ahead on their own than sticking together and working for common interests. Part of the reason for the choice of an individualist strategy is that few attempts have been made by local political organizations and trade unions to articulate class and class interests for collective mobilization (cf. Gallie 1983). But more important is the impact of mobility experience on the Hong Kong people's perception of openings and opportunities in a growing economy. They are not simple-minded, happy-go-lucky Horatio Alger heroes. Yet, they believe in personal efforts and that opportunities are available. The practical answer to the question of survival in economic competition is to find your own way up the social ladder.

5.

This brings us back to the question posed at the beginning of this essay. The 'Hong Kong experience' comes to fruition in the 1970s when economic growth, and the concomitant changes in the social structure and increase in mobility opportunity, bring optimism and confidence. Our comparison of research findings across the post-war decades serves to highlight the change in mentality. Whereas those in the 1960s saw themselves having little control of their own careers and future, the Hong Kong people in the 1980s have hopes and dynamism. The mood is to move on, to something new, and perhaps, to some better life-station. But there are two sides of this 'Hong Kong experience'. On the one side, there are individual efforts of making better lives. In the process of making better lives, the Hong Kong people have developed a certain distinct identity, one which perceives Hong Kong as both the (or their) land of opportunities, and also a society where they have some rightful claim to entitlements. The optimism and confidence are however not something untroubled, for our findings suggest persistent strain and sense of injustice, the objective correlate of which we suggested in the mobility differentials. But that experience itself left an indelible mark on the Hong Kong morality: Hong Kong represents opportunity to make a good living, and it in that sense symbolizes freedom and openness. Whether by luck, by entrepeneurship or by bureaucratic advancement, there are opportunities of mobility (Lui and

Wong 1993b). The diverse channels of mobility and opportunity become something more than economic success; they are part of the actual experience, that actual development or formation of the Hong Kong identity, and in that process shapes the Hong Kong way of life. The freedom to be economically successful, to make a better living, become embedded in personal freedom and societal openness.

When reflected at the plane of morality, such perceptions and values are in the form of individualistic, instrumental, practical, morally-neutral orientations and strategies. This is the other side of the 'Hong Kong experience'. In their perception of the Hong Kong way of life, there is no promise of personal success. What people value is the game itself and not its outcomes. In the Hong Kong people's mind, equality is the equality of opportunity (and not of outcomes), fairness 'the more competent gets more', and competition virtuous (Lui and Wong 1994). Class differentials are recognized; but the more important question is how to move on to make a better living. There is no lack of frustration but personal failure can be forgotten and resentment has rarely turned bitter. But this instrumentalist stance, when it is extended to the larger society (and not just personal concerns and ambitions), takes on a particular meaning in the transition period. It has been remarked that the Hong Kong mood is 'one of masterly expedience and crisis-to-crisis adjustment and recovery. It is partly a gambler's mentality, partly fatalism' (Hughes 1976:129). In relation to greater events which might be politically and morally troubling, Hong Kong perseveres to carve out a small corner for herself. As Coates (1975:4) observed:

> '... [W]hile, on the other side of the border, a civil war of world importance might rage, people in Hong Kong were able to pursue their own small personal wars, undeterred by greater events. To anyone interested in these greater events, life in Hong Kong was lived in two dimensions: a large dimension, in which the individual was, like Hong Kong itself, a dot; and a small dimension, in which ridiculously small local matters seemed very important.'

In the transition period, in a similar approach, but perhaps more alarmingly, 'Hong Kong executives naturally expect to continue running their business and making money from them, while they are going through the citizenship or naturalization process' (Wilson 1990:235). The middle class coping strategy is to buy their 'political insurance' (i.e. foreign citizenship) and to continue their successful career in Hong Kong. On this side of the 'Hong Kong experience', it is an identification of Hong Kong more as a way of life than a place of residence. Emigration or no emigration, the issue is morally neutral. This 'getting on', to make a living and perhaps a better living, shapes the Hong Kong approach to morality. The 'politically correct' line of thinking has no appeal to the public in Hong Kong. Moral rightness or wrongness (say, of emigration) is not irrelevant, but instrumentalism constitutes the framework of popular discourse.

To move on, in the eyes of the rising middle class, the entrepreneurial small business owners, or the working class whose standard of living has been improving in the past two decades, means to do well in Hong Kong. So, the 'Hong Kong experience', hopes as well as cynicism, is where the two sides of the popular mood converges.

The point we intend to make is that the preservation of the Hong Kong way of life requires more than an institutional design of 'one country two systems' in its narrow scope as we find in current political and diplomatic talks. It requires more than some additional seats in the legislature and the promise of maintaining a capitalist economic system. The key to the realization of the 'Hong Kong Dream' lies in a social system allowing individuals to pursue their own goals. While democracy (more precisely the formation of the future government) is the hot issue of political debate, people's concerns are really about liberty and freedom. Although the notion of 'capitalism unchanged' appears to have its popular support, the real issue is actually about a socio-economic system which gives rooms for individuals of different classes to strive for success through diverse channels. As Hong Kong approaches 1997, when debates about democracy must come to an end (convergence or no convergence, there will be a political structure for the Special Administrative Region (SAR) which the Chinese government finds acceptable) and talks about a capitalist SAR become more of a daily business (as the two economies are increasingly integrated), the question of the Hong Kong way of life can hardly be avoided. This is the kind of question that cannot be handled simply by political and/or economic measures, or by fiat. If one has to make a brief statement on the 1997 question, it will be a conflict between two ways of life: the 'Hong Kong experience' vs Chinese socialism.[2]

2   Acknowledgements: The project 'A Benchmark Study of Social Mobility in Hong Kong', from which some data for this paper is taken, was supported by funding from the Strategic Research Grant, University of Hong Kong; Centre for Hong Kong Studies and Hong Kong Institute of Asia-Pacific Studies, The Chinese University of Hong Kong; and Education Eye. The assistance of Tak-wing Chan

126   TAI-LOK LUI AND THOMAS W.P. WONG

REFERENCES

BAKER, Hugh. (1983) 'Life in the cities: the emergence of Hong Kong man', *The China Quarterly*, No.95.

CHAN T.W., T.L. LUI, and T.W.P. WONG. (1993) 'Social mobility in Hong Kong in comparative perspective'. Mimeo.

CHANEY, D., and D. PODMORE. (1973) *Young Adults in Hong Kong*. Hong Kong: Centre of Asian Studies, Hong Kong University.

COATES, A. (1975) *Myself as a Mandarin*. Hong Kong: Heinemann.

CROMPTON, Rosemary. (1993) *Class and Stratification*. Cambridge: Polity Press.

GALLIE, Duncan. (1983) *Social Inequality and Class Radicalism in France and Britain*. Cambridge: Cambridge University Press.

GOLDTHORPE, John H. (1987) *Social Mobility and Class Structure in Modern Britain*, 2nd Edition. Oxford: Clarendon Press.

HUGHES, R. (1976) *Hong Kong: Borrowed Place — Borrowed Time*. London: Deutsch.

LOCKWOOD, David. (1958) *The Blackcoated Worker*. London: Allen & Unwin.

LAU S.K. (1982) *Society and Politics in Hong Kong*. Hong Kong: The Chinese University Press.

——. (1987) 'Decolonization without independence'. Occasional Paper No.19. Hong Kong: Centre for Hong Kong Studies, The Chinese University of Hong Kong.

LAU S.K., and P.S. WAN. (1987) *A Preliminary Report on Social Indicators in Hong Kong*. Hong Kong: Centre for Hong Kong Studies, The Chinese University of Hong Kong. (in Chinese)

LAU S.K., M.K. LEE, P.S. WAN, and S.L. WONG (eds). (1991) *Indicators of Social Development: Hong Kong 1988*. Hong Kong: Hong Kong Institute of Asia-Pacific Studies, The Chinese University of Hong Kong.

LUI Tai-lok. (1988) 'Home in Hong Kong', in C.T. Lee, ed., *Hong Kong Films and Social Change*. Hong Kong: Urban Council.

LUI Tai-lok, and James K.S. KUNG. (1985) *City Unlimited: Housing Protests and Urban Politics in Hong Kong*. Hong Kong: Wide Angle Publication. (in Chinese)

LUI Tai-lok, and Thomas W.P. WONG. (1993a) 'Class, inequality and moral order: a class analysis of Hong Kong in transition', in One Country Two Systems Economic

and S.M. Hsu in carrying out the fieldwork and data processing is gratefully acknowledged. This paper was written while Tai-lok Lui was a visiting scholar at Robinson College, Cambridge University on the CUHK-Robinson College Exchange Programme, 1994. He would like to thank Robinson College for its hospitality and The Chinese University for its support of his sabbatical leave.

Research Institute, ed., *Hong Kong in Transition: 1992*. Hong Kong: One Country Two Systems Economic Research Institute. (in Chinese)

———. (1993b) 'Entrepreneurial strategy in context'. Mimeo.

———. (1994) 'A class in formation: the service class of Hong Kong'. Unpublished report submitted to the Chiang Ching-kuo Foundation for International Scholarly Exchange.

MITCHELL, Robert E. (1969) *Levels of Emotional Strain in Southeast Asian Cities*, Vols.I&II. Taipei: Orient Cultural Service.

———. (1972) *Pupil, Parent, and School*. Taipei: Orient Cultural Service.

ROSEN, Sheey. (1976) *Mei Foo Sun Chuen: Middle Class Chinese Families in Transition*. Taipei: Orient Cultural Service.

SALAFF, Janet. (1981) *The Working Daughters of Hong Kong*. Cambridge: Cambridge University Press.

WILSON, D. (1990) *Hong Kong! Hong Kong!* London: Unwin Hyman.

WONG, Thomas W.P. (1992a) 'Personal experience and social ideology', in S.K. LAU *et al.*, eds, *Indicators of Social Development: Hong Kong 1990*. Hong Kong: Hong Kong Institute of Asia-Pacific Studies, The Chinese University of Hong Kong.

———. (1992b) 'Discourses and dilemmas: 25 years of subjective indicators studies', in S.K. LAU *et al.*, eds, *Indicators of Social Development: Hong Kong 1990*. Hong Kong: Hong Kong Institute of Asia-Pacific Studies, The Chinese University of Hong Kong.

WONG, Thomas W.P., and Tai-lok LUI. (1992a) 'From one brand of politics to one brand of political culture'. Occasional Paper No.10. Hong Kong Institute of Asia-Pacific Studies, The Chinese University of Hong Kong.

———. (1992b) 'Reinstating class: a structural and developmental study of Hong Kong society'. Occasional Paper No.10. Social Sciences Research Centre, Hong Kong University.

———. (1993) 'Morality, class and the Hong Kong way of life'. Occasional Paper No.30. Hong Kong Institute of Asia-Pacific Studies, The Chinese University of Hong Kong.

[2]

SOCIAL POLICY AND SOCIAL DEVELOPMENT IN HONG KONG

PAUL WILDING

Hong Kong is often portrayed as the epitome of a free market economy dominated by an ideology of laissez-faire. This paper explores the development of social policy in Hong Kong and the factors which have shaped that development. It shows that the Hong Kong government plays a major role in promoting social development in the territory and that state involvment is the product of a complex interaction of political, social, and economic forces.

Introduction

In terms of social policy, Hong Kong is a curious mixture. In some areas there is massive state provision - for example in housing. In others, for example, education, there is state funding but most schools are provided by non-state bodies. Most social care is also provided by non-governmental organisations. In health there is a frequently reiterated commitment that no one shall be denied the health care he or she needs because of lack of means, but primary care remains substantially private. In social security, provision is limited to a narrowly residual social assistance scheme.

Official publications frequently assert that Hong Kong is not a welfare state, although it cares deeply about the state of welfare. Raymond Chan, on the other hand, in his recent important study of welfare in Hong Kong, takes a different view:

> Although the state and the majority of policy analysts are reluctant to use the term "welfare state," this is, in fact, a welfare state but with some unique characteristics as compared to others.[1]

Governor Patten, in his last address to the Legislative Council in October 1996, went some way to analyse the unique characteristics of the Hong Kong approach. Hong Kong's welfare system, he insisted:

> does not exist to iron our inequalities. It does not exist to redistribute income. Our welfare programmes have a different purpose. They exist because this community believes that we have a duty to protect the vulnerable and disadvantaged members of society, the unfortunate minority, who through no fault of their own, are left behind by the growing prosperity enjoyed by the rest of Hong Kong.[2]

A key question for the student is why social policy in Hong Kong has developed as it has in this unusual pattern. This is the question which this article addresses.

What has shaped social development in Hong Kong?

Political Factors

"Hong Kong senior civil servants are policy makers," says Lee, "in an executive dominated political system."[3] It is vital, then, if we are to understand the history of social policy in Hong Kong to appreciate how the civil service saw its role and responsibilities, and to grasp its values and beliefs about the nature and responsibilities of government. Four points are particularly important. First, top civil servants, whatever the reality of their position, saw themselves, above all, as administrators, not as policy-makers. Second, their fundamental orientation was conservative and cautious. Third, they would have

Asian Journal of Public Administration

endorsed the judgment of Sir John Bemridge, Financial Secretary in the early 1980s, that less government is better government."[4] Finally they were pragmatic. In the end, they would do what they judged necessary for the preservation of order, stability, and a healthy economy, even if it went against other supposed principles.

There seems to be more than a touch of arrogance about the civil service's ethos. Sir Philip Haddon-Cave expressed this ideology very clearly in 1981 at the end of his ten years as Financial Secretary. "Public opinion," he pronounced, "cannot be the only determinant of government policy; the public interest must be considered too."[5] The assumption is clear. The civil service has been granted insight and wisdom beyond that vouchsafed to public opinion. Lau speaks of the Hong Kong government as regarding political activity "with abhorrence and consternation." It is seen to threaten the hegemony of the bureaucratic regime but "what is equally objectionable is that it will inject irrational criteria into the public decision-making process which would direct resources away from rationally desired collective goals set by professional administrators."[6]

Scott reflects on this conservative ethos in his discussion of why it was difficult to bring about changes in social policy in Hong Kong. He suggests that:

> A major stumbling block, was the civil service itself which was not geared to formulating or implementing social policies and had been traditionally more attuned to the maintenance of law and order, crisis management and balancing the budget.[7]

The problem was self-perpetuating. If government was essentially about administration then the expertise which was most highly prized would be administrative and organisational. To ensure civil servants acquire the requisite knowledge, they move from department to department fairly rapidly because the expertise they need to acquire is knowledge of the government machine rather than of particular policy areas. Lau links this to "generally anti specialist attitudes ... which makes it difficult for specialist expertise to feed into the decision making process at the centre." The result is "to demoralize officials in professional grades and produce a defeatist attitude among them."[8]

The enthusiasm for policy-making, from which expertise in a particular social policy field might be expected to generate, is cooled out by the dominant bureaucratic culture.

The absence of politics has had an important effect on the culture and attitudes of the bureaucracy. So does the fact that it has operated in a colonial setting. Historically, the key senior civil servants have lived secluded from society. Senior bureaucrats always, of course, live a rather rarefied existence but the gulf between the administration and the society is wider and deeper in a colonial situation such as Hong Kong. Top officials are insulated from the continuous unobtrusive pressure of contacts with the society which encourage the rethinking of policies.

The Governor and Financial Secretaries

Until the 1980s political power in Hong Kong was the unchallenged preserve of the Governor and a small group of key civil servants.[9] The powers granted to the Governor, says Miners, "are awesome and may be compared to those once possessed by a King of England before the coming of democracy and the rise of political parties with Ministers responsible to parliament."[10] With the arrival of a totally elected Legislative Council in 1995, the Governor's position and decisions are much more open to challenge but his powers remain unaffected. He can refuse his assent to any legislation passed by the Council and, if his patience runs out, he can dissolve the Council and order fresh elections. The Governor also has "virtually untrammeled" power to control the civil service.[11]

Enormous power also resides in the key senior officials, particularly in the Financial Secretary. The Chief Secretary is deputy to the Governor and head of the civil service but it is interesting that in Miners's classic study, *The Government and Politics of Hong Kong,* there are more entries in the index for Financial Secretary than for Chief Secretary and - even more suggestively - the entry for Financial Secretary refers the reader to entries under the names of the Financial Secretaries who have held office since 1962. These individuals clearly count. There are no such references to particular Chief Secretaries under the "Chief Secretary" entry.

Asian Journal of Public Administration

The Financial Secretary has formal responsibility for the Hong Kong budget and a kind of informal responsibility for the Hong Kong economy. He is subordinate only to the Governor. In a system in which so much depends on the economy the Financial Secretary is immensely powerful. Though subordinate to the Governor, he has, inevitably, a considerable degree of independence because of the technical nature of his work and because of the wide range of other issues to which the Governor has to attend. Responsibility for finance gives the Financial Secretary a broad role in policy-making. Given his position in the power structure his personal views become crucial. Woronoff said of Sir John Cowperthwaite, Financial Secretary from 1961 to 1967, that "a more ardent defender of laissez-faire could hardly be imagined."[12] In 1971, Cowperthwaite announced that free primary education was to be made available to all children regardless of their circumstances. "I cannot say," he added, "that I myself am particularly happy to make this announcement."[13] He was a firm opponent of the provision of free services. He said in 1966 when Hong Kong's public services were little more than embryonic,

> I remain firm in my conviction, first, that our public services are excessively expensive and growing rapidly more so, second, that those who benefit from them should pay directly, according to their means, for what they get.[14]

It was Sir Philip Haddon-Cave who coined the term "positive non-interventionism." He was a firm believer in the free market, in the "compelling need to keep the relative size of the public sector as small as possible," in the principle that "when the economy is enjoying strong growth, the relative size of the public sector should fall," and in charging for public services wherever possible.[15] Summing up the influence of Cowperthwaite and Haddon-Cave, Youngson concludes that:

> Each exercised an enormous influence on the life of Hong Kong during the decade in which he held office.[16]

Sir Hamish Macleod, Financial Secretary from 1991 until 1995, had similar views - that government expenditure should grow no faster than the economy as a whole, that government should strive to leave money where it can do most good, that is, in the pockets of the people, and that freedom for private enterprise and small government were the way to the good society.[17]

What we see in the Hong Kong polity is a concentration of power in a small group of key people. That concentration was normally used to forestall or resist any pressures for change. In particular situations, of course, such a concentration of power could make radical change possible if someone with radical ideas or policies came to power. As regards social policy, this has only happened once in the post-war history of Hong Kong when Sir Murray MacLehose was appointed Governor in 1971. In his key-note address to the Legislative Council on October 18, 1972, MacLehose set out a new agenda for social policy in the territory. Nearly twenty-five years later it is still impressive in its range and vision.

The Concern for Stability

Concern for political and social stability and the perceived need to strengthen and affirm the legitimacy of the government's authority have always been a central thread in Hong Kong's politics. The government was very conscious in the early post-war years that most of the population had no particular loyalty to Hong Kong or to the colonial government. And a few miles away was China, the subject seemingly of almost continuous revolution - communist in 1949, cultural in the 1960s and economic in the 1980s.

Political stability was central to the success of the kind of economy which Hong Kong is and was. Hong Kong had to persuade international companies that it had the stability to ensure the safety of their investment and their staff. Equally, a successful economy was a key factor in ensuring political stability and legitimacy. The feel-good factor generated by economic growth induces and increases satisfaction with government. There is a symbiotic relationship between economic growth and political stability.

The government's concern for political stability was matched by that of the people. Lau writes of "this pervasive fear of conflict among

Asian Journal of Public Administration

the Hong Kong Chinese."[18] Most of them had come to Hong Kong in search of stability and order. In the late 1970s, Lau's research found that 87 per cent of a sample of the Hong Kong population said that social stability was more important to them than economic prosperity. Furthermore, 57 per cent saw the primary role of government as being to maintain social stability.[19]

The riots in Hong Kong in 1966 and 1967 are seen by many as a defining moment in Hong Kong's political and social development. Scott sees the riots as forcing the state elite to look for ways of ensuring social stability without democratic participation.[20] In his view, the riots "showed conclusively that it was no longer possible for the government to be both unrepresentative and slow to improve living and working conditions."[21] The implication of Scott's analysis is that government could ensure legitimacy and stability either by increasing its representativeness or by improving material conditions. If it did neither, then it would be at risk.

The crucial question is how far the riots and the questions they raised affected the development of social policy in Hong Kong in the 1970s. The driving force behind those developments is generally agreed to be Governor MacLehose. How far was MacLehose influenced by the riots and the analysis which they provoked? Castells et al. see the riots as a central element in MacLehose's thinking:

> It is our hypothesis that MacLehose identified social reform as one of the critical elements to fulfilling his fundamental political assignment - to pacify and stabilize the colony after the 1966-67 riots.[22]

MacLehose turned to social policy to promote this stability because it was a tradition and approach with which he was familiar and with which he identified.

There is general consensus about the way in which concern for social and political stability has promoted government involvement in housing. This is so both for the development of public rented housing and for government's more recent initiatives to help the development of owner occupation. The latter concern is to bind the growing middle class - the so called sandwich classes - to Hong Kong by enabling them to buy property which they could not afford without government help. In a recent review of such schemes, Lai's verdict is that "[to] subsidise this class is indefensible except for political reasons."[23] But political reasons have always played a significant role in housing policy in Hong Kong because housing has been seen as an important element in promoting stability.

The argument of this section has been that the Hong Kong government's great concern for political and social stability and legitimacy led it to take initiatives in social policy. Did they work? Scott's judgment is positive. He speaks of developments in social policy in the 1970's as "probably the most important aid to the enhancement of legitimacy."[24]

In the long term, however, developments in social policy had two problematic effects on political and social stability. First, they contributed significantly to the undermining of the undemocratic system of rule by Governor and an administrative elite and stimulated the social mobilisation which led on to District Board elections in 1982 and direct elections to the Legislative Council in the 1990s. In a sense, such democratisation was probably necessary for long-term stability but the government's view in the 1970s was probably that social welfare development would pre-empt, rather than fuel, demands for fundamental political changes. The reality was that developments in social policy were both the product of political change and forces for further development.

Second, developments in social policy both increased satisfaction and dissatisfaction with government. Service provision, in Jones's view, "has come to constitute, if not the *raison d'etre*, then at least Government's strongest claim to popular legitimacy."[25] At the same time, however, as Jones also points out, the problems - or impossibility - of keeping pace with mounting demand for ever more welfare provision makes the government a permanent focus of criticism and dissatisfaction.[26]

Political Attitudes

An important factor in the preservation of undemocratic bureaucratic rule in Hong Kong and hence of the slow development of social policy

Asian Journal of Public Administration

was the attitude of the local Chinese population towards politics. Almost all commentators comment on the political apathy of the people of Hong Kong,[27] and on their "ingrained political passivity."[28] The basis of this passivity is an important issue. Clearly it existed, but whether it was cultural or an adaptation to colonialism and powerlessness is a question worth considering.[29]

People expected little of government in the sense of services. They expected government to maintain stability so that they could pursue the goal of financial independence and prosperity for themselves and their families but they expected little more. Many had come to Hong Kong - in a sense - to avoid government. Their goals were essentially private and familial. They neither looked for, nor expected, government help.

In a sense, too, the Hong Kong government, prior to the mid-1980s, met people's political aspirations by the consultative processes it established - "government by discussion."[30] The series of studies of social development in Hong Kong show that 75 per cent of the population were satisfied with the political system in 1985 and 1988. Even in 1990, 59 per cent agreed that although the Hong Kong political system was not perfect it was the best that was currently possible.[31] What is very interesting, and what helps to explain the high level of satisfaction with the status quo, is people's conception of democracy. The most common conception of democracy - held by 40 per cent of the population - was that a democratic government is one which consults public opinion. Only 28 per cent of the population defined a democratic government as one which is elected by the people.[32] For those who saw consultation as the crucial characteristic of democratic government, Hong Kong was already democratic.

The Power of Capital

Another key factor in Hong Kong is the dominant role of business. The health and development of the economy is crucial to political stability. This gives the business community great influence and power because of its key role in furthering economic growth. However, the business community also needs understanding and help from the political system. It requires assured political stability to encourage international and local investment. It requires low taxes to ensure competitiveness and attractive profit levels. It requires principles of taxation which will offer guarantees about the future tax-take. It needs confidence that no policies will be adapted - for example contributory social insurance or pension schemes - which will push up labour costs and/or fuel demands for wage increases. The business community requires certain government policies just as much as the government needs the business community to deliver the economic growth on which it so heavily depends.

The resulting symbiotic relationship between the political and business elites has been a vital factor shaping the development and non-development of social policy. Chan suggests that the business sector did not have to press for dominant control of the state because "[their] interests were automatically taken care of by state officials, not by force but by a sharing of values and orientation."[33] Business has always been averse to the development of contributory social insurance.[34] Business has always been averse to the extension of democracy because of fears about the pressures for increased welfare spending which it might generate.[35] Business has always opposed any increase in rates of income tax or corporate taxation. Business has always fought any extension of environmental policy which might impose extra costs on business or limit profitable activities on environmental grounds.

Business influence did not mean the non-development of social policy. It meant the subordination of social policy to economic and political goals. When social policy development could be seen as clearly functional to these prior purposes, it would be pursued. Government involvement in housing, or example, was welcome to business interests on four main grounds. First, it helped to free land for profitable industrial and commercial purposes. Second, by reducing rents it helped to reduce pressures for wage increases. Third, it provided continuous work for the construction industry. Finally, it contributed to the political stability on which economic prosperity ultimately rested by increasing popular satisfaction with government. On the other hand, where the development of social policies was seen as an actual or potential threat to the economy then development was less likely to take place.

Asian Journal of Public Administration

The China Factor

Miners observes that:

> Senior officials, have always taken China's possible reaction
> into account when making any policy decision and try, as far
> as they can, to avoid giving her any reasonable grounds for
> complaint. China thus exercises a pervasive influence over
> government policy-making, merely by the fact that she is there
> and potentially capable of causing difficulties.[36]

What has been the nature of that influence? Essentially it has been
both conservative and radical. There has been concern in Hong Kong
that China would be opposed to any radical changes in the territory -
in terms of social policy development or political development. Cheng
postulates that Governor MacLehose in the late 1970s "occasionally
hinted" to political groups in Hong Kong that Beijing would object to
any political reform in the territory. In Cheng's view "this undeniably
constituted an effective deterrent."[37]

Another way in which China exerts an indirect conservative
influence in Hong Kong is by providing the frame of reference for the
majority of the Hong Kong population. It is a frame of reference
which, in Schiffer's words, "affords the Hong Kong government the
benefit of the doubt."[38] It is both encourages satisfaction with Hong
Kong and discourages any radical action which might make the
territory's future problematic.

There are, however, two respects in which the China factor may
well have had a radical impact. The first is in the way in which
discussion in the early 1980s about the future of Hong Kong created
a more politically sensitive middle class. The promise of fifty years of
self-government with a high level of autonomy after 1997 created new
possibilities for politics. The desire to use that freedom to safeguard
the Hong Kong way of life was a spur to political action.

A second, rather unexpected, radical impact of the China factor is
the way the return to Chinese sovereignty on July 1, 1997 has been a
spur to welfare development. Welfare expansion in recent years, says
Leung, "has been seen as a gesture to promote the benevolent image

of the government and the governor in the remaining years of colonial rule."[39] Expansion has certainly been very real. The Hong Kong government committed itself to increasing recurrent expenditure on social welfare (for example, social security and social care services) by 26 per cent in real terms between 1993 and 1997.

Political factors have clearly been central to the development and non-development of social policy in Hong Kong. There have been restraining and limiting factors - the absence of democracy, the orientation and power of the governing bureaucracy and its key members, the power of business interests, local political culture, and the looming presence of China. Equally, there have been factors encouraging policy development - the strong concern of both bureaucratic and business elites for political and social stability, the way policy development fed pressures for democratisation which, in turn, led to demands for increased welfare and service provisions.

Social, Structural, and Cultural Factors

The particular nature of Hong Kong society has clearly affected the development of social policy in the territory. I pick out what seem to me to have been some of the more important influences.

A Transient Population

One of the most striking characteristics of Hong Kong is expressed in the famous description of the territory as "borrowed place, borrowed time." Impermanence and temporariness have always been naggingly present. Sweeting notes that in the early post-World War II days the Director of Education used the transient nature of the population as a reason for not developing education for the local population - though as Sweeting points out, the Hong Kong government had, since 1903, given support to education for the transient and impermanent European population. Hence, transience looks like less than the whole story.[40] Sweeting suggests that "it is not coincidental" that the first signs of genuine planning for Hong Kong education came in the 1960s and 1970s when something approaching an "identification" process was beginning to affect Hong Kong residents.[41] Lau makes his view

Asian Journal of Public Administration

quite clear that the government used the transiency of the Chinese population "as an excuse."[42] Whether it is an excuse or a justification is a matter of interpretation. But it is clear that a transient population does create less sense in government of a need to provide and also makes it more difficult to provide effectively and efficiently.[43]

Hong Kong has always been concerned about immigration. In recent years, however, it has been the emigration of highly trained professional people which has become the important issue. Various social policies have been developed to try to increase the benefits of the middle class in Hong Kong. There were the various schemes for assisting the "sandwich classes" to possess their own properties. There was also the expansion of higher education to increase educational opportunities for young people and hopefully to give them a greater stake in Hong Kong and instil confidence in the future.

The Chinese Family

The family has always been at the heart of the debate about social policy in Hong Kong. It is often argued, for instance, that the family occupies a special position in Chinese society and culture, that it plays a central role in the provision of welfare for family members, that public provision of welfare might have adverse effects on the family, and that the impact of economic and social changes on the family and on its continuing ability to perform its traditional functions is likely to be immense.

Lau has characterised the dominant cultural code of the Hong Kong Chinese as "utilitarianistic familism" by which he means that family concerns and interests override the interests of society and other groups and individuals. The outcomes and implications of such a philosophy are four-fold. First, it leads to "an unusually strong emphasis on the norm of mutual assistance among familial members."[44] Second, it leads to a lack of involvement in groups beyond the family and a lack of involvement, or interest, in politics. Third, it leads to "the depoliticisation of many social and other issues and their deployment and solution within the familial group" which is "extremely functional to the maintenance of the colonial regime."[45] Fourth, this ethic has played a vital role in stabilising Hong Kong's transient, immigrant society. Without the bonding provided by family there would surely have been greater pressure for government action to help bond society together.[46]

The role of the family in the provision of welfare is very plain. The 1990 Social Indicators Study shows a rich web of inter-familial help: 64 per cent of respondents had given financial assistance to parents in the previous six months; 48 per cent had received such assistance from parents; 50 per cent had rendered assistance to siblings; and 39 per cent had received assistance from siblings. There had also been comparable exchange of help during illness.[47] Another area where the centrality of family help is vital is home purchase. Lui's recent work on coping strategies in a housing market with rapidly escalating prices shows how the family "assumes the central position in the facilitation of home ownership."[48]

The major role of the family in welfare both legitimates government inaction - the family already meets social needs - and can be used as an excuse for policies of non-intervention. The fear that the development of collectivist policies would weaken the family and undermine family help networks has been reiterated again and again in official statements. On occasions, no doubt, such anxieties were genuine. On other occasions they were useful excuses for inaction.

In the 1990s, there has been growing concern as to whether the family in Hong Kong will be able to continue to play such a significant role in welfare provision. Chow sees the expansion of social welfare in recent years as "more a result of the diminishing functions of the traditional support networks than of the rising demand for social justice."[49] Leung is equally concerned about the indications that the family's "capacity to provide care and support to its members in need is rapidly eroding."[50] Changes in the family may therefore increase the need and demand for social policy development.

Social Values

What are the central social values of the Hong Kong Chinese population and what has their influence been on the development and non-development of social policy in the territory? In his inaugural lecture in 1976, Professor Peter Hodge described Hong Kong as a society

Asian Journal of Public Administration

"which stresses individualism, pursuit of self-interest, and competitiveness, and which has come to consider inequality of circumstances of living and of rights as a 'natural' order of human existence."[51] Wong and Lui speak of "utilitarian individualism" as "the hallmark of the Hong Kong ethos."[52] Stability is another primary value. Lau and Kuan actually describe it as "the predominant value ... in the ethos of the Hong Kong Chinese."[53] Important also in the culture of the Hong Kong Chinese is short-termism.[54] It is a product of being an immigrant in a society characterised by permanent economic uncertainty with a definite limit to its particular political status. Short-termism, for example, partly explains the absence of popular pressure for contributory social insurance. Such mechanisms depend on confidence in the long-term future and a future orientation.

Hong Kong's social values are clear and powerful but economic values almost always take precedence. In Hodge's view, an equal right to compete in the economic market was more conducive to social consensus than an equal right to share in the social market in Hong Kong.[55] Wong's research confirmed Hodge's supposition.[56] People look to the market for the fulfilment of their hopes rather than to the state.

While the values of the Chinese population in Hong Kong seem to support and legitimate a free market ideology and minimal government, in fact the reality may well be more complex. Lau and Kuan's work shows a generalised support for laissez-faire alongside strong support for government intervention on particular issues.[57]

So too does the most recent Indicators of Social Development Study.[58] What has both confirmed and legitimated the values of independence, self-reliance and competitiveness is the economic success of the Hong Kong economy and the way in which most people have been able to fulfill their economic aspirations. As regards the less successful, then state provision has provided a minimal safety net.

In a way, therefore, the values of the Hong Kong Chinese encourage and support the positive non-intervention approach of the Hong Kong government. Their values parallel the government's commitment to a free market ideology accompanied, when deemed necessary or desirable, by major government programmes and initiatives - by a kind of seemingly unprincipled pragmatism which is in fact guided by the principle that order and stability are the beginning and basis of welfare, that in general free market principles are the surest guide but they should not be seen as precluding major acts of collective provision when that is necessary for the social and economic well-being of the territory.

There has always been a fear in Hong Kong that extended welfare provision could undermine those values on which the territory's economic and social success is believed to depend. In the 1965 Government White Paper, it was emphasised that social welfare services must not be organised in such a way as "to accelerate the breakdown of the natural or traditional sense of responsibility."[59] The latest White Paper expresses anxiety that the expansion of welfare could create "the sort of dependency culture that has emerged in some developed industrialised societies."[60]

That fear has certainly influenced government policy but, if Chow is correct, popular opinion is also not ready to accept the collective meeting of social needs by state action[61] - because of currently dominant values about individual, family, and social responsibility, because of attitudes to government and politics, and because of beliefs about what generates - and might damage - a dynamic economy.

Voluntarism

Another element which has been an influential factor in Hong Kong is the strong ethic of voluntarism and the predominant position of voluntary organisations. Government has preferred to support voluntary agencies rather than to involve itself in direct service provision. It was ideologically more acceptable. It avoided direct government responsibility. It could be seen as stimulating and encouraging self help and community action. It fitted with the government's desire that welfare be seen as charity rather than any kind of right. The result of this approach is that the 1990 White Paper reported that excluding social security, subvented voluntary agencies currently received as grant-aid two-thirds of government expenditure on social welfare services and employed round 80 per cent of all social welfare staff.[62] Some services - for example, non health based services for people with mental illness - are provided completely by voluntary organisations.

Asian Journal of Public Administration

Voluntary provision justifies inaction by the state - apart from the provision of funding. It is also potentially an obstacle to the development of public services because the voluntary sector is already there providing a service - albeit limited and patchy on occasions. But, given the early presence of voluntary bodies in education and the social care field, for example, government action to provide services seemed both inappropriate and unnecessary. A strong voluntary sector has allowed the survival of an ethic of voluntarism and has helped to legitimate state non-intervention particularly in the area of social care.

Social Change

The last thirty years in Hong Kong have been years of immensely rapid social change. Those changes have exerted conflicting pressures on the development of social policies. Five changes are particularly important in relation to the development of social policy - increased affluence, the emergence of a new middle class, democratisation, emigration, and the increased number of women in the formal economy. A new affluence, for example, has clearly increased the population's ability to meet their own basic needs and has helped to cement social stability - an issue to which government might otherwise have had to devote more attention. In this way, therefore, affluence has contributed to the non-development of public policies. On the other hand, the dynamic development of the Hong Kong economy has made the financing of extended welfare services relatively painless. Affluence has also raised aspirations - for example for improved housing and for more education.

Affluence and changes in the structure of the economy have contributed to the development of a new middle class. So and Kwitko postulate that by the late 1970s this had become "a new political force that the Hong Kong colonial government could no longer ignore."[63] The new class exerted pressure on government to develop policies which met its needs and aspirations. The most obvious examples are the development and expansion of higher education in the years after 1989 and the policies designed to extend owner occupation.

Democratisation from the early 1980s was both a product of the development of social policies and a force for their further development. It was a product of their development in that it was in the public housing estates that the new community politicians and pressure groups emerged - facilitated and led by workers from the welfare sector. Democratisation was also a force for further development in that it ate away at the secluded bureaucratic politics which had dominated Hong Kong. It exerted new pressures on the government via District Boards and via the Legislative Council. The government was no longer safely insulated from popular opinion and popular pressures. The extension and expansion of public provision in the 1980s and 1990s clearly owed a great deal to democratisation. The Hong Kong government and the business community had always feared precisely these pressures and they had opposed the extension of democracy largely on the basis of its likely economic implications. Developments proved that their anxieties were justified.

Emigration is another issue with important social policy implications. So and Kwitko speak of the departure of large numbers of the new middle class as having "drained the democracy movement of leadership and resources."[64] On the other hand, this potentially negative effect has been counter-balanced by those policies which the government has developed to try to keep potential middle-class emigrants in Hong Kong.

Another notable trend in Hong Kong - as in all industrial societies in the late twentieth century - was the increase in the number of women in the formal labour market. The proportion of women in the 25-34 age group in paid employment almost doubled in the twenty-five years after 1961 when it reached nearly 65 per cent. This trend created a greatly expanded need for day-care but the government successfully resisted the pressure for expanded public provision on the basis that the care of young children was a family rather than a state responsibility. The family and the market should be expected to provide any care needed.[65]

The government was pragmatically receptive to social change because of its primary and overriding concern with political and social stability and economic growth. Any social change which had, or seemed likely to have, negative impacts on these was a matter of concern even though basic beliefs and values predisposed the government against extending its range of activities.

Asian Journal of Public Administration

What eased the pressure of social changes for the government was the fundamental economic, social, and political orientation of the Hong Kong population. Its apathy and passivity, the "political aloofness" which summed up the local political culture, the low expectations of politics, and the focus on family rather than on society, all reduced the potential impact of social trends.

Economic Factors

What effect have economic factors had on the development and non-development of social policy in Hong Kong? We look first at the nature of the Hong Kong economy. Then we explore the economic beliefs and ideologies which have helped shape policy. Finally, we look at how the nature of the economy, the pattern of economic progress, and economic ideologies have affected social policy developments.

The Fragility of the Economy

What distinguishes the Hong Kong economy is its dependence. No other economy perhaps is as dependent on overseas trade. Hong Kong imports 80 per cent of its food, all its fuel, and almost all the raw materials used by its industries. As much as 70 per cent of its water comes from China. Hong Kong depends for its survival - let alone prosperity - on trade and on its ability to make and sell goods and services in international markets.

Hong Kong also depends heavily on its ability to attract overseas investment. The territory is competing with its investment-hungry neighbours in the region - Taiwan, Singapore, South Korea, China, and the Philippines. Hong Kong has clear disadvantages. It has an uncertain political, and therefore economic, future. Land is expensive and labour costs are higher than in many neighbouring countries. It has, therefore, to try to offer compensation attractions.

There is in Hong Kong a keen sense of this fragility and vulnerability. Sir Hamish Macleod, Financial Secretary from 1991 till 1995, spoke of people in the territory being "particularly keenly aware of the potentially fleeting nature of success."[66] Fragility is a reality but after thirty years of amazing economic development, arguments about fragility must be seen both as reflecting reality and as counters in a political debate. The reality is that Hong Kong does have to ensure it can compete in world markets. It does have to try to offer an economic environment attractive to foot loose international companies and capital. It does have to offer rewards and profits commensurate with the possible risks involved in investing in what is, from July 1, 1997, again part of China.

The sense of economic fragility has always given local business interests great influence. These are the people on whom Hong Kong's future depends - because everything depends on economic success. Business also depends, as was argued earlier, almost equally on government to ensure the political and social stability which are vital to the maintenance of business confidence. Likewise, government depends on business to deliver the economic growth which is so central to political and social stability. What has, over the years, given the Hong Kong government its particular ideology - positive non-interventionism as Haddon Cave styled it[67] - is, partly at least, its relationships with the business and financial world. That relationship has been the product of the particular position of business in Hong Kong life which is, in turn, the product of the particular nature of the Hong Kong economy and polity.

The nature of the economy has encouraged certain approaches and policies. Haddon-Cave's view is that "in Hong Kong there is little scope for managing the economy."[68] Lau sees the open nature of the Hong Kong economy as rendering "any attempts on the part of the government to guide the economy relatively futile." Reality, there-fore, in Lau's view, protects the government from pressures designed to make it play a more interventionist role and also makes it "a less attractive target of control by the dominant economic interests in the colony."[69]

The sense of vulnerability has had certain clear and predictable outcomes. The government has vigorously resisted any policies which would have increased labour costs - such as the development of contributory social insurance. It has vigorously proclaimed Hong Kong to be a low-tax economy with a maximum salaries tax of 15 per cent and more than half the working population paying no tax at all.

Asian Journal of Public Administration

The World Competitiveness Report 1992 estimated that Hong Kong's total tax revenue at 11.25 per cent of the Gross Domestic Product was the lowest of all the thirty-six countries surveyed.[70] The sense of fragility has also encouraged a policy of enormous caution in financial management. Surpluses were predicted by the Financial Secretary in only eight budgets out of twenty in the years 1960-61 to 1979-80 with correspondingly cautious expenditure plans. Surpluses were actually achieved in eighteen of those years.[71] Hong Kong's concern to build up massive financial reserves - which again has been a restraint on expenditure programmes - is partly a product of this same deep anxiety about the future.

Fragility legitimates giving priority to the needs of the economy. It therefore legitimates giving only modest attention to the social costs of economic development. Fragility therefore causes or allows the continuing poverty of long-term recipients of social assistance, the low pay and insecurity of those without skills or whose skills have been superseded, and the continued dreadful living conditions of those living in the worst private and public rented housing.

Economic Beliefs and Ideologies

What are the beliefs and ideologies which have dominated economic and social policy-making in Hong Kong? First, it is important to distinguish between myth and reality. Miners sums up and expresses the myth in his statement that "The Hong Kong government is more committed to nineteenth-century policies of allowing free play to market forces, than is the case anywhere else in the world."[72] Certainly, many in government have sought to portray Hong Kong in this light. The reality, however, is rather more complex - and rather different. "Hong Kong and laissez-faire," says Youngson, "have only an occasional acquaintance."[73]

Certainly, the Hong Kong government is firmly committed to the virtues of the free market. "We have a clear philosophy," said Sir Hamish Macleod, "consistently applied, which is based on a commitment to market forces, free enterprise and free trade. We believe in creating an environment with minimum regulation and interference plus maximum government support in terms both of infrastructure and protection of the needy, leaving business free to flourish."[74] In fact, as we have seen, there are a range of interventions in the economy by the government - provision of housing for almost half the population, careful monitoring of public utilities and control of their prices, and administered prices for a wide range of products.[75]

Haddon-Cave summed up the governing ideology in 1984 in expounding his philosophy of positive non-interventionism. He saw little scope for managing the Hong Kong economy because of its particular nature. He was quite willing to see government control over the enfranchised public utilities in the private sector "to ensure that they enjoy no more than a quasi-competitive return on capital."[76] He asserted his belief in individual ownership and in the acquisition and accumulation of wealth by individuals. He was concerned that the work ethic should not be eroded "by social policies and redistributive fiscal policies."[77] He saw "a compelling need to keep the relative size of the public sector as small as possible."[78] Where public services could be related to individual use, they should be charged for so long as adequate arrangements for remitting charges were available.[79]

More recently, Sir Hamish Macleod summed up his economic philosophy as "consensus capitalism." It has five essential elements - restraint on public expenditure so that it grows no faster than the economy as a whole, low taxation so that money is left to fructify in the pockets of the people, private enterprise and small government, provision of the social, physical and regulatory infrastructure required to support the market, and provision for the disadvantaged and the vulnerable.[80]

The ideology which has dominated Hong Kong stresses individual effort, individual responsibility, self reliance, and self-help. It emphasises competition. It lays great stress on the need to reward success. It assumes the natural superiority of private enterprise and private provision over that of the public. It sees economic growth as the surest way to safeguard the sources of welfare. The ideology may be at odds with reality but it is to the ideology that people tend to attribute Hong Kong's success.

The heart of the Hong Kong ideology is "economy first." It is free market capitalism but when there are things the government can do to enhance economic performance, it acts pragmatically. The major

Asian Journal of Public Administration

programmes of government expenditure have been primarily but not solely driven by economic concerns and benefits. Social and political concerns have also played an important, albeit a subsidiary, role.

Ideology sets the government firmly against social expenditure with no obvious economic benefits - for example, retirement pensions - and expenditure which is believed, even if simplistically, to threaten the values and behaviour which have fuelled Hong Kong's dynamic growth - for example, unemployment benefit, higher rates of social assistance, and contributory social insurance.

Economics and Social Policy Development

How then have the nature of the Hong Kong economy and the dominant economic ideologies affected social policy developments? The real significance of these beliefs and ideologies is without doubt that they put economic concerns at the top of any hierarchy of government concerns though they are modulated by considerations of political and social stability.

Clearly the low tax policy has constrained social policy development. The 15 per cent level of salaries tax and a relatively high-tax threshold seem to have become sacrosanct. Rapid economic growth has meant buoyant tax revenues but the commitment to low tax rates reduces revenue potential and inhibits development in areas which traditionally have been tax-funded and where there are no immediate and obvious economic gains from increased expenditure.

Concern for competitiveness has inhibited developments which were thought likely to increase labour costs or increase the tax-take from the corporate sector. Opposition to the introduction of contributory social insurance has clearly been driven by concern - real or assumed - about its effect on labour costs and on the supply of labour. Until the mid-1990s the normal situation in Hong Kong has been one of shortage of labour. Benefits for the unemployed were therefore (a) unnecessary and (b) likely only to make the situation worse. Again, in a society where labour was short, where independence was so important, where the economic sphere was regarded as the key element, retirement was something alien to the Hong Kong philosophy. This approach is nearly captured in the government's 1977 Green paper on services for elderly people. According to the Paper, "One of the means of preserving the independence of older men and women and at the same time allowing them to contribute towards the wealth of their society is to encourage and facilitate their continued employment."[81]

There is much debate as to the extent to which economic considerations have driven housing development but there are powerful arguments for their central importance. The classic accounts of the development of housing policy in Hong Kong by Castells et al. and Smith all see government's conern for more profitable land use as a key factor. Smith's view is that "the apparently compassionate rehousing of squatters in the colony was primarily aimed at obtaining valuable building land for general urban development."[82] Dennis Bray, a Hong Kong civil servant who retired in 1985 as Secretary for Home Affairs was clear in his statement that "the resettlement programme of the fifties was not a housing programme for the poor. It was a means to clear land for development."[83]

The other question is the extent to which the Hong Kong government's housing policies were driven by consideration of the contribution cheap housing could make to the territory's economic competitiveness. Certainly, low rents helped keep down wage demands which in turn helped to hold down costs. Whether this was an aim or simply a fortunate side-effect it is impossible to say.

It is not only government housing policy which poses these problems of interpretation. Castells and his colleagues make it "a central argument of our analysis that a number of fundamental factors underlying Hong Kong's manufacturing competitiveness are largely the result of deliberate government policies in providing industrial infrastructure, land, housing, social services and subsidised foodstuffs and raw materials."[84] Certainly, government policies were deliberate. The crucial - and really unanswerable - question is whether the government's objectives were specific and particular or whether they were geared to any overarching goal such as enhancing competitiveness or maintaining growth rates.

Economic considerations have clearly been a factor in the development of education in Hong Kong but of a slightly unusual kind. Hong Kong's success in textile production from the 1960s put great pressure on European industry. Allegations were made that Hong

Asian Journal of Public Administration

Kong's success was due to the exploitation of child labour. There were threats of import quotas and restrictions. Hong Kong had to show that the use of child labour was not an element in its success. "The implementation of free and compulsory schooling," says Postiglione, "was used to indicate compliance."[85] Sweeting sees the extension of compulsory education in Hong Kong to fifteen in 1978 as the product of the same concerns.[86] While these specific concerns may have been precipitating factors, there were clearly also broader predisposing factors encouraging education development.

Other educational developments also owed something to economic pressures. The link between industrial and economic development and technical education was a factor in the expansion of technical education from the 1970s. The Hong Kong Polytechnic was founded in 1971, the City Polytechnic in 1984, and the Hong Kong University of Science and Technology in 1988. In 1982 the government created the Department of Technical Education and Industrial Training.

The Governor's announcement in 1989 that the government would double the number of university places by the mid-1990s was driven by both political and economic concerns. The economic element was the need to replace skilled people who were leaving the territory following the Tiananmen Square massacre and to try to anchor educated people to Hong Kong by the virtual guarantee of good jobs which university education had always offered.

Another area of social policy where economic concerns have very clearly dominated policy development is in the environment. The government has created an Environmental Protection Department but there is a widespread sense that this represents symbolic politics rather than a firm commitment to the goals and polices it has set out. Economic concerns, private and corporate interests, and short-term gains triumph over the common good and the safeguarding of the long-term environmental future of the territory and the planet.

The Hong Kong government's primary concerns have always been economic and political rather than social. It has not been able, as have some Western countries, to take economic and political stability for granted. In such a situation of uncertainty they have to come first. Without confidence in present and future political stability, there can be no consistent economic or social development. Without confidence in the future of the economy, there can be no political stability. It is these priorities which have shaped the development of social policy in Hong Kong.

Walker suggests that "Hong Kong might be regard as an extreme case of the dominance of economic policy over social policy."[87] That is true but it is far from being the whole truth. It needs further explanation. In Hong Kong, because of the nature of the economy and of the Hong Kong state, economic policy was social policy. Economic concerns, however, have not always dominated social policy in a crude fashion because political concerns have been a central concern in economic policy.

Conclusion

The aim of this article has been to try to explain the development and non-development of social policy in Hong Kong. The argument has been that development has not simply been the product of a crude application of a laissez-faire ideology but rather of the complex interaction of political, social, and economic factors.

To British eyes what has emerged may look limited and deficient. In many respects it is, but the reality of Hong Kong's situation needs to be kept firmly in mind - the relative recency of affluence, the fact that compulsory primary education and statutory cash assistance, for example, are each only twenty-five years old, that the government has been coping with population increases of around one million in each of the post-war decades, that there have been particular factors such as the permanent uncertainties of the economy and the uncertain political future, the absence of any tradition of voluntary contributory insurance among the working class, and the absence until recently of any sense of belonging or permanence.

Unless the particular nature of Hong Kong is kept clearly in mind it is easy to jump to simplistic conclusions about Hong Kong as an embodiment of laissez-faire capitalism. It is a much more complex society than that. Certainly, social welfare and social development have been seen as subsidiary to other more important and more fundamental goals. The government has put the economy first not simply for ideological reasons, but also because economic success was seen as the most basic element in securing and maintaining social

Asian Journal of Public Administration

and political stability which were regarded as the key elements in social well-being. Economic, political, and social factors dictated this as the priority. There were casualties, but the government's strategy was defensible. If the economy failed to deliver growth, then Hong Kong as a society would be at risk of plunging into chaos. What was at stake was the well-being of the whole society. In the government's view that was better served by putting the economy first.

Economic success also depended on politics - on business confidence in political stability and on government's ability to ensure the active or passive support of the population. There could be no economic miracle without confidence in Hong Kong politics. This sets limits on the influence of capital and ensures the relative autonomy of the state.

Hence, politics and economics were mutually dependent, and both in turn depended on the right mix of social policies. Where such a dependence was detected, the government could, and did, embark on massive exercises in collective provision - in housing, in hospital care, in higher education, for example. There was also an undeniable paternalistic, colonial concern for the public welfare. But policy development was always geared to broader goals - Hong Kong's economic and political needs rather than any abstract concerns about equity, rights, or supposed social needs.

What has shaped the development and non-development of social policy in Hong Kong is a concern for these broader goals. Certainly, there has been a commitment to free market principles but the commitment has not been simply and crudely ideological. It was based on a calculated pragmatic assessment of Hong Kong's particular situation.

The tradition of social policy analysis is to focus on short-comings, limitations, and deficiencies. For someone coming to the study of social policy in Hong Kong with a concept of the European welfare state, there are plenty of those to be found. But that starting point is unsatisfactory. Hong Kong's development has necessarily and inevitably been quite different in character and time-scale. Hong Kong's achievement requires a different yardstick. Constructing a balance sheet is very difficult and inevitably involves value judgements. On many criteria it must be regarded as an economic, political, and social success. What is striking, however, is the high level of public dissat-

Social Policy and Social Development

isfaction with many aspects of life and public policy in Hong Kong. In 1993, more people were dissatisfied or very dissatisfied than were satisfied or very satisfied with the political situation, with public order, with the housing situation, with social welfare, and with employment.[88] In a sense, dissatisfaction is the price which has to be paid for achievement.

This latter point helps to put Hong Kong's achievement in economic and social development in perspective. The achievement is indeed impressive, but there are gaps and shortfalls. It is hard to believe that dealing with them would be likely to imperil the health of the Hong Kong economy or affect the energy to which the development of Hong Kong's economy owes so much. It simply requires the pragmatic interventionism which Hong Kong has used to such good effect in so many areas of its public life.

Is this just an interesting story for students of the development of social policy? I would argue that it is more than that because it challenges many conventional Western conceptions about the nature and role of social policy. The government in Hong Kong has always seen the economy as the main engine of welfare. It has taken the view, in governor Patten's words, that "full employment should ... be the government's single most important welfare objective."[89] It has asserted the centrality of political and social order and stability. It has always been pragmatic rather than principled in its approach. It has used social policy to serve economic purposes. Its central concern has always been the general social good - as defined by government - rather than the needs of individuals or special needs groups. What has emerged from the interplay of political, social, and economic forces is a very particular model of social policy and good governance. It encourages a re-examination of the role of social policy and the nature of the good society.

NOTES

1.   R.H.K. Chan, *Welfare in Newly-Industrialised Society* (Avebury: Aldershot, 1996), p.270.

Asian Journal of Public Administration

2.   Hong Kong Government, *Hong Kong: Our Work Together* (Hong Kong: Government Printer, 1996), para.78.

3.   J.C.Y. Lee, "Civil Servants" in D.H. McMillan and S.W. Man, eds., *The Other Hong Kong Report 1994* (Hong Kong: The Chinese University Press, 1994), p.59.

4.   Quoted in I. Scott, *Political Change and the Crisis of Legitimacy in Hong Kong* (Hong Kong: Oxford University Press, 1989), p.254.

5.   Quoted in N. Miners, *The Government and Politics of Hong Kong* (Hong Kong: Oxford University Press, 1995), p.204.

6.   S.K. Lau, *Society and Politics in Hong Kong* (Hong Kong: The Chinese University Press, fourth printing, 1993), p.36.

7.   Scott, *Political Change and the Crisis of Legitimacy in Hong Kong*, p.106.

8.   S.K. Lau, *Society and Politics in Hong Kong*, pp.51-52.

9.   S.S. Cheng-Lo, *Public Budgeting in Hong Kong* (Hong Kong: Writers and Publishers Cooperative,1990), pp.66-67.

10. Miners, *The Government and Politics of Hong Kong*, p.69.

11. *Ibid.*

12. J. Woronoff, *Hong Kong: Capitalist Paradise* (Hong Kong: Heineman Asia, 1980), p.33.

13. Quoted in A.J. Youngson, *Hong Kong: Economic Growth and Policy* (Hong Kong: Oxford University Press, 1982), p.75.

14. *Ibid.*, p.74.

15. P. Haddon-Cave, "Introduction" in D.G. Lethbridge, ed., *The Business Environment in Hong Kong* (Hong Kong: Oxford University Press, second impression, 1985), pp.15-18.

16. Youngson, *Hong Kong: Economic Growth and Policy*, p.58.

17. S.H. Tang, "A Critical Review of the 1995-1996 Budget" in S.Y.L. Cheung and S.M.H. Sze, *The Other Hong Kong Report 1995* (Hong Kong: The Chinese University Press, 1995), p.164.

18. S.K. Lau, *Society and Politics in Hong Kong*, p.11.

19. S.K. Lau, "Utilitarianistic Familism: The Basis of Political Stability" in A.Y.C. King and R.P.L. Lee, eds., *Social Life and Development in Hong Kong* (Hong Kong: The Chinese University Press, 1981), pp.203 and 207.

20. Scott, *Political Change and the Crisis of Legitimacy in Hong Kong*, p.79.

21. *Ibid.*, pp.81-82.

22. M. Castells et al. *The Shek Kip Mei Syndrome* (London: Pion, 1990), p.140.

23. C.W.C. Lai, "The Property Price Crisis" in McMillan and Man, *The Other Hong Kong Report 1994*, p.207.

24. Scott, *Political Change and the Crisis of Legitimacy in Hong Kong*, p.165.

25. C. Jones, *Promoting Prosperity* (Hong Kong: The Chinese University Press, 1990), p.258.

26. *Ibid.*, pp.195 and 236.

27. Miners, *The Government and Politics of Hong Kong*, p.33.

28. Jones, *Promoting Prosperity*, p.68.

29. See, for example, K.M. Cheng, "Traditional Values and Western Ideas: Hong

Kong's Dilemmas in Education," *Asian Journal of Public Administration* 8(2, 1986): 195-213 and S.K. Lau, "Utilitarianistic Familism," p.198.

30. G.B. Endacott, *Government and People in Hong Kong 1841-1962: A Constitutional History* (Hong Kong: University of Hong Kong, 1964), p.229.

31. S.K. Lau, "Political Attitudes" in S.K. Lau et al., eds., *Indicators of Social Development: Hong Kong 1990* (Hong Kong: The Chinese University Press, 1992), pp.132-3.

32. *Ibid.*, pp.134-50.

33. Chan, *Welfare in Newly-Industrialised Society*, p.232.

34. For example, see M.K.Y. Li, "Interest Groups and the Debate on the Establishment of a Central Provident Fund in Hong Kong," *Asian Journal of Public Administration* 11(2, 1989): 231-52.

35. Tang, "A Critical Review of the 1995-1996 Budget" in Cheung and Sze, *The Other Hong Kong Report 1995*.

36. Miners, *The Government and Politics of Hong Kong*, p.230.

37. J.Y.S. Cheng, "Political Modernization in Hong Kong," *Journal of Commonwealth and Comparative Politics* 27(1989): 294-320.

38. Schiffer, *Anatomy of a Laissez-Faire Government: The Hong Kong Growth Model Reconsidered* (Hong Kong: Centre of Urban Studies and Urban Planning, University of Hong Kong, 1983), p.27.

39. J.C.B. Leung, "Social Welfare" in Cheung and Sze, *The Other Hong Kong Report 1995*, p.378.

40. A.E. Sweeting, "Hong Kong Education within Historical Processes" in G. Postiglione, ed., *Education and Society in Hong Kong* (Hong Kong: The University of Hong Kong Press, 1992), p.67.

41. *Ibid.*, p.68.

42. S.K. Lau, *Society and Politics in Hong Kong*, p.46.

43. D. Bray, "Shaping Up for the Twenty-First Century" in Hong Kong Government, *Hong Kong 1991* (Hong Kong: Government Printer, 1991), p.9.

44. S.K. Lau, *Society and Politics in Hong Kong*, p.75.

45. S.K. Lau, "Utilitarianistic Familism," p.213.

46. S.K. Lau and H.C. Kuan, *The Ethos of the Hong Kong Chinese* (Hong Kong: The Chinese University Press, third printing, 1991), pp.22-23.

47. M.K. Lee "Family and Gender Issues" in S.K. Lau et al., *Indicators of Social Development: Hong Kong 1990*, Table 1-8.

48. Lui Tai-Lok, "Coping Strategies in a Booming Market" in R. Forrest and A. Muric, eds., *Housing and Family Wealth* (London: Routledge, 1995), p.126.

49. N. Chow, "Social Welfare: The Way Ahead" in J.Y.S. Cheng and S.S.H. Lo, eds., *From Colony to SAR* (Hong Kong: The Chinese University Press, 1995), p.408.

50. J.C.B. Leung, "Social Welfare," pp.364-5.

51. Quoted in N. Chow, "The Quest for Human Betterment," *Hong Kong Journal of Social Work* XXVII(2, 1993): 12.

52. T. Wong and Lui Tai-Lok, "Morality and Class Inequality" in B.K.P. Leung and T.Y.C. Wong, eds., *25 Years of Social and Economic Development in Hong Kong* (Hong Kong: Centre of Asian Studies, University of Hong Kong, 1994), p.85.

Asian Journal of Public Administration

53. Lau and Kuan, *The Ethos of the Hong Kong Chinese*, p.56.

54. Lau, *Society and Politics in Hong Kong*, p.118.

55. P. Hodge, "Expectations and Dilemmas of Social Welfare in Hong Kong," *Hong Kong Journal of Social Work* XXVIII(2, 1993): 11.

56. Wong and Lui, "Morality and Class Inequality" in Leung and Wong, *25 Years of Social and Economic Development in Hong Kong*, p.81.

57. Lau and Kuan, *The Ethos of the Hong Kong Chinese*, chapters 2 and 3.

58. S.K. Lau et al., *Indicators of Social Development: Hong Kong 1993* (Hong Kong: The Chinese University Press, 1995), chapter 7.

59. Hong Kong Government, *Aims and Policy for Social Welfare in Hong Kong* (Hong Kong: Government Printer, 1965), p.5.

60. Quoted in I.D. Campbell, "Economic Ideology and Hong Kong's Governance Structure after 1997" in One Country Two Systems Economic Research Unit, *Hong Kong In Transition* (Hong Kong: One Country Two Systems Economic Research Unit, 1993), p.352.

61. N. Chow, "Welfare Development in Hong Kong. An Ideological Appraisal" in Leung and Wong, *25 Years of Social and Economic Development in Hong Kong*, p.323 and N. Chow, "Social Welfare: The Way Ahead," p.405.

62. Hong Kong Government, *Social Welfare into the 1990's and Beyond* (Hong Kong: Government Printer, 1991), p.17.

63. A.Y. So and L. Kwitko, "The Transformation of Urban Movements in Hong Kong, 1970-1990," *Bulletin of Concerned Asian Scholars* 24(4, 1992): 37.

64. *Ibid.*, p.41

65. C.K. Wong, "Economic Growth and Welfare Provision: The Case of Child Day Care in Hong Kong" *International Social Work* 35(1992): 389-404.

66. H. Macleod, "Hong Kong: A Hard Earned Success" in Hong Kong Government, *Hong Kong 1995* (Hong Kong: Government Printer, 1995), p.11.

67. Haddon-Cave, "Introduction," p.xii.

68. *Ibid.*

69. Lau, *Society and Politics in Hong Kong*, pp.173-4.

70. Macleod, "Hong Kong: A Hard Earned Success," p.16.

71. Youngson, *Hong Kong: Economic Growth and Policy*, pp.60-61.

72. Miners, *The Government and Politics of Hong Kong*, p.47.

73. Youngson, *Hong Kong: Economic Growth and Policy*, p.132.

74. Macleod, "Hong Kong: A Hard Earned Success," p.15.

75. Castells et al., *The Shek Kip Mei Syndrome*, chapter 3; Schiffer, *Anatomy of a Laissez-Faire Government*.

76. Haddon-Cave, "Introduction," pp.xii-xiii.

77. *Ibid.*, p.xiii.

78. *Ibid.*, p.xv.

79. *Ibid.*, p.xvi.

80. Tang, "A Critical Review of the 1995-1996 Budget" in Cheung and Sze, *The Other Hong Kong Report 1995*, pp.164-70.

81. Quoted in J. Tao, "Growing Old in Hong Kong: Problems and Programmes" in

Social Policy and Social Development

J.F. Jones, ed., *The Common Welfare: Hong Kong Social Services* (Hong Kong: The Chinese University Press, 1981), p.112.

82. Drakakis Smith, *High Society* (Hong Kong: Centre for Asian Studies, University of Hong Kong, 1979), p.20.

83. Bray, "Shaping Up for the Twenty-First Century," p.9.

84. Castells et al., *The Shek Kip Mei Syndrome*, pp.83-84.

85. G.A. Postiglione, "The Decolonisation of Hong Kong Education" in Postiglione, *Education and Society in Hong Kong*, p.7.

86. Sweeting, "Hong Kong Education within Historial Processes," p.49.

87. Quoted in Cheng-Lo, *Public Budgeting in Hong Kong*, pp.42-43.

88. Lau et al., *Indicators of Social Development*, p.404.

89. Hong Kong Government, *Hong Kong: Our Work Together*, para.26.

Paul Wilding is Visiting Professor in the Department of Public and Social Administration, City University of Hong Kong and Emeritus Professor at the University of Manchester in Britain. This article is a revised version of his paper entitled "Social Policy and Social Development in Hong Kong" published in the Working Paper Series at the Department of Public and Social Administration, City University of Hong Kong in 1996.

[3]

ESCAPE FROM POLITICS:
HONG KONG'S PREDICAMENT OF POLITICAL DEVELOPMENT?

Hsin-chi Kuan
Department of Government and Public Administration
The Chinese University of Hong Kong
Shatin, N.T.
Hong Kong

ABSTRACT

Political development in Hong Kong is analyzed in terms of its existential rationale, non-sovereign autonomy, capability, and democratization. Changes in these aspects have two implications. First, the conditions for social escape from politics before the conclusion of the Sino-British Agreement on the future of Hong Kong in 1984 no longer exist. Secondly, a political society has emerged to mediate between society and the polity. This represents a fundamental structural transformation in state-society relationship away from the "minimally integrated system" whereby the polity was secluded from the society. Can this political society in Hong Kong be eradicated, arrested, or tamed after the transfer of sovereignty to China in 1997? An educated guess is that it can at best be tamed, provided that democracy remains the ultimate objective of the Basic Law and elections continue to be held.

Copyright © 1998 by Marcel Dekker, Inc.   www.dekker.com

INTRODUCTION

Mr. Lu Ping, director of the Hong Kong and Macau Office of the State Council of the Chinese government, once warned the people of Hong Kong not to turn Hong Kong into a "political society." To him, the absence of politics has made Hong Kong what it is today: a world-class trade, communications and financial center. This hard-won success now seems, in his eyes, to be put at risk with governor Chris Patten's package of reforms the tendency of which is to transform the basis of governing from the colonial domination to the consent of the governed. Although Mr. Lu Ping's worry is somewhat overdrawn, it is however quite perceptive. For almost a full century, Hong Kong's society has not been politicized.[1] When it was first ceded in 1841, Hong Kong was nothing but barren rocks, inhabited by a few villagers and occasionally visited by pirates. The whole history of the colonial rule can be characterized by the conspicuous absence of political agitation on the part of the ruled. G.B. Endacott has even argued that the Chinese character of Hong Kong has been one of the main factors in delaying the introduction of essentially Western ideas of political freedom.[2] Neither were the end of the Second World War, nor the seizure of power by the communists in China, nor the huge influx of refugees from China offered any pretexts for politicizing Hong Kong's society. Despite spectacular transformation in society and the economy, there has been no major constitutional change in Hong Kong. It is only now when sovereignty over Hong Kong is about to be handed over to China in 1997, politics suddenly creeps into all walks of life. Is it a good omen? Can it be avoided at all? It is the argument of this paper that the conditions for escape from politics in the past have changed, because the emergence of a political society has ended the seclusion of the polity from the society. This political society is the result of the cumulative effects of several developmental processes that began since the mid-1960s and it is likely to persist if elections will continue to be held after 1997.

DEVELOPMENT AS REALIZATION OF AN EXISTENTIAL RATIONALE

Is there such a goal called "development"? Can developmental processes be consciously directed toward any objectives? If yes, whose objectives? This essay does not purport to provide any general, conclusive answers to these difficult questions. But the experiences of Hong Kong do suggest that at least until very recently, she has developed by default, rather than by design. In contrast, The Sino-British Agreement on the future of Hong Kong concluded in 1984 and the Basic Law for the Special Administrative Region of Hong Kong promulgated in 1990 are both unprecedented in their explicit setting of specific objectives for Hong Kong. Thus, Hong Kong has entered a stage of development by design. As a result, the political agenda of Hong Kong has become unwittingly crowded with a host of old and new issues.

In earlier days, Hong Kong was better known as "a borrowed place, a borrowed time."[3] It was born in 1842 as a military, diplomatic and trade station of Britain in the wake of her Opium War against the egocentric and self-isolated China.[4] There was no British intention to colonialize it, nor was it a popular settlement site for the Chinese. Under these circumstances, both could escape from politics for a protracted period of time, in the sense that there was hardly any domestic issues over which the imperialist master and the local subjects needed to be embroiled in any bitter quarrel. The British administration could afford to govern the least, and preferred to govern by an indirect rule whenever possible. The local Chinese were happy being left alone. As most of them had managed to escape from the turbulent politics in mainland China and wished to return there in due course, there was no incentive to get involved with local politics. Thus, the British administration and the Chinese society remained two different worlds, unconnected to each other. Politics did not exist; it is said to have been "socially accommodated."[5]

Unique circumstances of an unexpected and changing nature have however nourished a precarious political existence for Hong Kong since the Second World War. First of all, Hong Kong defected from the worldwide wave of de-colonialization in the early post-war years. Then as the British empire shrunk, Hong Kong also lost its strategic significance as a British diplomatic and commercial outpost in the Far East. The communist seizure of political power in China in 1949 could have spilled over but finally stopped at the border. Since then, China has officially insisted on her sovereignty over Hong Kong on moral and legal grounds. There were enough agitations to give effect to that insistence during the Cultural Revolution in China from 1965 to 1974 or the riots in Hong Kong in 1966-67. Yet, actual policies of China have been carefully designed to preserve Hong Kong's status quo under British rule.[6] In this sense, the precarious existence of Hong Kong is largely due to the accommodation policy of China, which has been based on pragmatic considerations such as the value of Hong Kong as a point of access to the world and a reliable source of economic benefits.

While the tensions between China and the West created a space for Hong Kong to develop, such an existence gradually acquired a meaning for the local residents different from their forefathers. Immigrants from China no longer expected to return to the mainland and their locally-born descendants grew in numbers and soon made up the majority of the total population since the 1960s. The demographic transformation helped to promote an indigenous society in replacement of the immigrant society. The change in the nature of the society was accompanied by decades of remarkable achievements in the economy. Both contributed to the emergence of a sense of Hong Kong identity among the younger, better educated, relatively better-off generation. On the eve of the Sino-British negotiations in 1982, the status of Hong Kong as a place of transit had definitely been buried, although the structural foundation for a political society was still weak.

The question of an existential rationale for Hong Kong was ushered onto the political agenda by the Sino-British negotiations over the future of Hong Kong in 1982. The intense tug-of-war between the two sovereign powers over the fate of six million people prodded them to reflect on who they were, what they wanted, what had made Hong Kong tick, which of the existing values were preferred, what they could do and whither should Hong Kong go. The reflection has not yet known its end and debates over these questions are still being waged, forming part and parcels of the process of the unsettling transition itself. However, the intermediate impacts of the soul-searching are obvious. The people of Hong Kong have gained on political awareness, the intellectuals' search for answers has brought about an intensive political discourse never known before and contributed to the growth of the public sphere.

NON-SOVEREIGN AUTONOMY AS DEVELOPMENT

There can be no meaningful existential rationale without autonomy. In politics, national independence in the sense of unabridged sovereignty is a crucial criterion of political development for a nation. Both the world system approach[7] and the traditional dependency theories[8] have stressed the adverse impact of dependence on national development. Although positive development is still possible in the context of dependency,[9] the strive for independence in controlling one's own development remains a morally convincing objective for national leaders.[10]

Hong Kong has never experienced any independence movement, although there has been substantial attitudinal support for independence.[11] As a colony in the past and a special administrative region of China in future, Hong Kong's development cannot be measured in terms of sovereign independence, but of non-sovereign autonomy. Non-sovereign autonomy is possible only when the right of sovereignty can be conceived as a bundle of rights which can be dissolved and distributed to sub-state units. While the orthodox conception of sovereignty identifies as fundamental and inalienable those rights over defence matters and foreign relations, in reality no right over whatever matter cannot be delegated, except the residual right to delegate authority and to rescind such a delegation.[12]

Hong Kong's acquisition of the right to and the capacity of autonomous development is based on delegation of authority by the present sovereign state of Great Britain. It has been a long process the analysis of which warrants a full-length book. It suffices here just to summarize the development and stress that this process has been scheduled to continue under the future sovereign state of China, as promised in the Basic Law.

The year of 1958 is a watershed when Hong Kong was given the autonomy to manage its own financial affairs. In less than a decade, the financial autonomy was spilled over to the area of foreign economic relations. At the time when Great Britain joined the European Common Market, Hong Kong had already acquired the independent capacity to directly deal with its trade partners in the world.

The next stage of development was the acquisition of full membership in those international organizations that are not constituted on the sole basis of sovereign states. By virtue of her admission into GATT as a separate customs area in 1986, Hong Kong became officially recognized as a non-sovereign autonomous entity.

Hong Kong as an effective actor in international relations received yet another special recognition when Canada revised its Foreign Missions and International Organizations Act in 1991 and the United States Congress passed the United States-Hong Kong Policy Act in 1992. The Canadian Act of 1991 extends to an office of "a political subdivision of a foreign state" diplomatic and

HONG KONG'S PREDICAMENT 1429

consular privileges and immunities; The U.S. Act of 1992 allows the U.S. government to treat Hong Kong as "a non-sovereign entity distinct from China for the purposes of US domestic law based on the principles in the 1984 Sino-British Joint Declaration."[13]

This status of non-sovereign autonomy is duly recognized in the Sino-British Joint Declaration (Section 3 (1) and Section I of Annex I). It is further provided in Article 12 of the Basic Law that Hong Kong is an administrative region that "enjoys a high degree of autonomy." The kinds of autonomy are further delineated in several other articles. Yet the future can still be regarded as uncertain. First of all, some provisions in the Basic Law are subject to different interpretations and it is likely that in case of a constitutional conflict the central government will prevail over the future Hong Kong government. Secondly, the promises of "one country, two systems" and of "a high degree of autonomy" for Hong Kong is of a transitional nature, as they are valid for fifty years only. Thirdly and more immediately, the "one country, two systems" project has already begun with a rough start. In fact, the advent of the 1997 issue and the Sino-British negotiations over Hong Kong's future have already dealt a severe blow to the autonomy of Hong Kong, as both sovereign powers must "fulfil" their separate as well as joint responsibility over Hong Kong.[14] The reversal of autonomous development is then compounded by the requirement for convergence during the period of political transition through 1997. China has acted on behalf of the future government of Hong Kong to check the otherwise autonomous actions of the present Hong Kong government, by insisting that she must be consulted on all issues straddling 1997, including the launch of infrastructural development projects, the granting of franchises, and so on. The argument that the action of China can be justified under the condition of a lack of trust between the Chinese and the British/Hong Kong government is besides the point. The point is that the conventional right of Hong Kong to autonomous development has partially been suspended and there is a risk that China may habituate herself to asserting central control over local development as a result of cumulative effects of the transition. If on the other hand the setback is only transitional, then the four-decade long legacy of non-sovereign autonomy must provide a fertile ground for better integration between the polity and the society than before 1958.

CAPABILITY-ENHANCEMENT AS POLITICAL DEVELOPMENT

The political development of Hong Kong can be further assessed in terms of capability to achieve results. There are however many analytical difficulties. The first is conceptual: what kind of capability? We may well call a state developed for its capability to provide for order, or to mobilize social-economic resources for modernization, or to overcome the so-called developmental crises such as penetration, integration, participation, distribution, identity and legitimacy.[15] Secondly, we may be more interested in the degree of change in capability, rather than the kinds of results to be achieved. Political development thus boils down to enhancement of political capability. Thirdly, no matter how we conceive of capability, it is no easy task to measure it.

We may first dismiss as inapplicable the concept of mobilization capability, as the regime of Hong Kong has never been mobilizational. In contrast, the "success" of Hong Kong has often been explained by its laissez faire policy.

When we come to the issue of order, be it social or political, there has been broad consensus about the remarkable stability enjoyed by Hong Kong over a long period of time. While tumult is widespread elsewhere, Hong Kong has survived the Second World War, the post-war decolonialization tide, the Korea War, the waves of Chinese refugees, the rise of Chinese nationalism especially during the Cultural Revolution in China, and widening

HONG KONG'S PREDICAMENT   1431

social inequality at home. Explanations for this stability abound. Ambrose Y. C. King has credited the political system's ability to co-opt social-economic elites into the consultative mechanism of the government for the spectacular stability in the past decades.[16] Lau Siu-kai has instead argued that this stability was made possible by the lack of integration between society and the political system.[17] One may further argue that to the extent that capability must be assessed relative to the task, Hong Kong have been fortuitous in the absence of serious challenges. Facing a society of refugees who were happy with Hong Kong as a lifeboat and did not expect much of the government at a time when economy could grow with individual initiatives alone, the political system needed little capability to govern. It follows that with changes in the social, economic and political circumstances, the polity of Hong Kong may face more serious challenges. It is precisely this that has happened.

In the past decades, Hong Kong's society has undergone fundamental changes. As alluded in the above, it is no longer an immigrant society, thus paving the way for the emergence of a Hong Kong identity. It has grown more pluralistic and educated, with implications for the rise of expectation. The economy has also experienced fundamental transformation. With the relocation of manufacturing activities into mainland China and the rise of the service industry closely linked with China trade, Hong Kong has resumed its more traditional role of entrepôt. Gone is the path of development in the 50s through the 70s when seclusion or independence from the Chinese influence was a blessing for success. These social-economic changes and the political agenda of reunion with China jointly presents several developmental problems for the polity of Hong Kong.

First of all, neither economic laissez-faire nor positive non-interventionism serves well to maintain economic competitiveness. In what way should the government henceforth intervene into the economic process? Secondly, governing on the basis of elite consensus alone is no longer adequate, as the mass have become awaken. How should the non-elites be allowed to participate in the political process? Thirdly, impending political reunion and further economic cooperation with China pose a twin crisis of identity and integration. Can Hong Kong find a suitable role to play by pursuing economic integration with China while avoiding political interference therefrom?

Thus, on the eve of reversion to Chinese sovereignty, Hong Kong is no longer as fortuitous as before. The above problems have occurred not only simultaneously but also at a time when the colonial government has begun to suffer from a decline in authority in the face of Sino-British intervention as well as in the minds of the Hong Kong people. As a result of the erosion of colonial authority and the increase of political cynicism on the part of the population, S.K. Lau has predicted an increasing incidence of scattered political actions against the government and more political uncertainties on the eve of transfer of sovereignty.[18] On the other hand, Ian Scott has lamented over the Hong Kong government which is losing legitimacy in the eyes of the Hong Kong people while preparing to transfer power over them to one with even less.[19] Finally, the most alarming forecast comes from a U.S. journalist, Mark Roberti. He has suggested that a new power elite is growing in Hong Kong, consisting of mainland companies, pro-China businesses, and trade unions. Beijing will use this new power elite in Hong Kong to silence its opponents, to obtain an edge in competition, to influence court cases, and to establish centralized control over the economy. The verdict is that within twenty years of 1997, Hong Kong as a financial and commercial center of the world would have been dead.[20]

To sum up, the governability of Hong Kong has become a problem. Although a major factor, i.e. the issue of transfer of sovereignty, is exogenous, another important factor is endogenous and has to do with the mode of governing. Traditionally, Hong Kong was governed by the consent of a rather homogeneous elite co-opted by the government into advisory bodies, who had no mass bases. Hong Kong was thus deprived of the social base for the emergence of a political society, i.e. an ensemble of autonomous groups mediating between the polity and the society. At the same time, the polity had to directly deal with the issues of interest articulation, interest coordination, and legitimation. For the past three decades, the Hong Kong government has tried hard to act as an arbiter of plural interests and provide a tutelage over a variety of social-economic groups. Its capability to coordinate interests has been enhanced by an ever-expanding system of consultation, augmented by the use of opinion polls in more recent years. The system of government by consultation is however inherently limited in its capability. For one thing, consultation must be still selective no matter how widely it is practised. Therefore, the risk of mis-selection always exists. More importantly, as all consultative systems depend on the initiative of the ruler, he must be capable to keep pace with social-economic changes so that the right people, especially elites of the new breed will be consulted and/or co-opted in time. This task is of a tall order, as society grows more rapidly and more complex. In short, there is an institutional limit to the capability of any system of consultation. And there must be a time for a change. Democratization is a possibility. In Hong Kong, however, it was one that was primarily precipitated by a concern about defence against the future sovereign, rather than by a consideration of capability enhancement.

DEMOCRATIZATION AS POLITICAL DEVELOPMENT

Political development can be understood in terms of democratization, or in connection with the above discussion, the building-up of democratic institutions to facilitate interest representation and aggregation, to absorb the expansion of political participation, to promote the development of identity, and to achieve political legitimacy. The author has argued elsewhere that given the level of socio-economic development in Hong Kong, there should have long developed a democratic government. Yet, as a colony dependent on Britain and under the influence of an authoritarian China, political development in Hong Kong has lagged behind its social-economic development.[21]

All political reforms in Hong Kong before 1984 were designed to improve communication between the government and the people, while retaining central power for the colonial executive. The Young Plan of 1946 striving for a new fully representative municipal council, which was stillborn, did not address the issue of representation at the central level of the Legislative Council. Further efforts from 1966 to 1969 to devolve power to local authorities also failed. The absence of political reform had become taken for granted by the majority of the apathetic Hong Kong Chinese who were eager to escape from politics and concentrated on business. This state of affairs could have lasted a little longer had not the issue of 1997 opened up the agenda of democratic development.

The issue of 1997 was settled in principle in 1984 as the British and the Chinese government agreed to have the sovereignty over Hong Kong transferred to China in 1997 in return for the latter's promise to grant a kind of self-rule to Hong Kong and to respect its present social-economic systems for fifty years. The issue of transition ensued. The British/Hong Kong government announced its intention to introduce representative government to Hong Kong, probably as a measure to win over public acceptance of the Sino-British agreement on the transfer of sovereignty of Hong Kong to China in 1997. A small dose of representation was then injected in 1985 with twelve members each returned by an electoral college of District Board members and nine functional constituencies respectively to fill the 56-member Legislative Council. The intended introduction of direct election of a small number of councillors to this body in 1988 was however stalled at the opposition of China who demanded political convergence with the Basic Law which was being drafted. The British and Chinese governments secretly agreed in 1987 to let a small number of seats to be directly elected in 1991.

The reform envisaged by the British/Hong Kong government was never meant to establish a liberal democracy in Hong Kong, but simply to improve on political representation. Consequently, the introduction of electoral elements to the legislature was carefully designed so as not to affect the forming of the government. Such a caution is understandable as Britain was concerned about maintaining effective governing in Hong Kong right up to 1997. In the meantime however, the people of Hong Kong have grown more interested in possibilities of participating in the political process. The increase in interest had in fact been encouraged by both sovereign powers in their promise of "representative government" and "Hong Kong people ruling Hong Kong."

In this context, the drafting of the Basic Law for the future Special Administrative Region of Hong Kong became a convenient battleground for newly inducted politicians to explore their political visions and blueprints. As they were drowned in heated and indeterminate debates about the desirability and modalities of democratization, a critical event erupted in Beijing to change the political landscape in Hong Kong: the suppression of the democracy movement of Beijing students by the regime on June 4, 1989. As the movement built up strength from spring on, emotion in Hong Kong also escalated, with an increasingly large number of people in Hong Kong (around a million) demonstrating against the policies toward the students of the Beijing regime. Came the crackdown, the Hong Kong people were reminded of the nature of the Beijing government to which they would be subject to after 1997. The fear of an unpleasant future not only fuelled the exodus of elites to other countries, but more importantly brought people of different persuasions closer to each other on the issue of democratization. On the one hand, the Hong Kong Alliance in

Support of the Patriotic Democratic Movement of China was born. On the other, consensus models on a faster pace of democratization were forged in and outside the Legislative Council. Most prominent among all was "the UMELCO consensus" of July 26, 1989, whereby all unofficial members of the Legislative Council agreed that twenty members of that Council should be returned by universal suffrage in 1991 and no less than half in 1995.

The Basic Law which was promulgated in 1990 has rejected all consensus models as developed in Hong Kong. Instead, it envisages to elect only 20 members by universal suffrage for the first term of the 60-member Legislative Council, to be increased to 24 and 30 for the second and third terms respectively. For the remaining seats, 30 will be returned by functional constituencies for all three terms through 2007 and ten and six seats to be returned by a selection committee for the first and second term respectively. In other words, it will take a full decade for the legislature in Hong Kong to be half indirectly and half directly elected. Any amendment to this arrangement must be approved by two-thirds of the future legislature, the chief executive, and the Standing Committee of the National People's Congress of China. Apart from these measures to ensure that the directly elected councillors will remain in the minority forever, Annex II of the Basic Law is designed to limit the influence of the legislature by requiring a split vote of each of two groups of directly elected and non-directly elected members for motions proposed by individual members. In sum, the political development of Hong Kong **as planned** must tread a conservative path.

The plan may nevertheless not work, if spontaneous forces released by political reforms running up to 1997 raise enough the consciousness and skills of the people of Hong Kong. Election may be such a candidate. In this light, the elections in 1991 ushered in a new era of political change in Hong Kong. It was the first time that direct elections were introduced. They offered a new channel of political recruitment for aspiring political leaders who had been previously excluded from the appointment system of the Hong Kong government. The elections were held in the legacy of the June 4 event in China. Candidates with pro-China attitudes, records or connections were all defeated, whereas the democratic camp of candidates won a landslide victory.[22] Within a few years after the elections, Hong Kong witnessed not only some changes in the political ethos and alignments but also a transformation of the traditionally compliant Legislative Council from an advisory to a supervisory role.

In midst of these changes arrived the new governor, Chris Patten, in 1992, who gave a further impetus to the democratic development in Hong Kong. His reform package included the adoption of a single-seat single vote system, the lowering of the voting age to eighteen, the abolition of the appointed membership in the municipal councils, and the creation of nine new functional constituencies in which individuals rather than groups as in the old functional constituencies would be eligible to vote. His overall strategy was to broaden the electoral base of the political system and to strengthen the Legislative Council as a representative assembly, without however compromising the executive control. The reform plan of Patten, which was promoted with beats of populism and agitation, crashed head on with the wish of Chinese government to retain the political initiative in safeguarding a smooth transition through 1997.

Upon the failure of seventeen rounds of negotiations over political reforms in Hong Kong, the Sino-British accord of convergence was declared bankrupt. While the Legislative Council passed Patten's reform bills by a narrow margin on 29 June 1994, the Preparatory Working Committee established by the Chinese government a year ago was put on high gear to consider measures for political transitions in Hong Kong, including the replacement of the 1995-elected Legislative Council by a provisional legislature in 1997. The warning issued by Mr. Lu Ping against the birth of a political society must be understood in the context of the breakdown of Sino-British cooperation and Chinese suspicion of a hidden agenda of the British government.

For a while, China was just powerless to halt Patten's reform. The elections of all three levels of assemblies from September 1994 to September 1995 were conducted according to Patten's reforms. It is significant that a wholly elected Legislative Council was returned for the first time in the history of Hong Kong. Although ten political parties participated in the elections, the contest was largely conducted between the Democratic Party and the Democratic Alliance for the Betterment of Hong Kong. The former party which fielded 24 candidates won a total of 19 seats, on a platform that emphasized steadfastness and trustworthiness in their fight for democracy. The latter party obtained only 5 with 11 candidates on a platform that stressed their care for the livelihood of the men on the streets and their ability to communicate with the Beijing government. The China factor remained a dominant factor that shaped the electoral results. The prospect of a democrats-dominated legislature challenging the traditional hegemony of the executive in Hong Kong and the policies of China toward Hong Kong can only reinforce China's fear of a democratic transition in Hong Kong. China immediately reiterated its warning to dissolve the new Legislative Council in 1997 when a "provisional" legislative will be appointed to her liking. How can the Chinese government then contain the aspiration of the Hong Kong people for more democracy?

GROWTH OF THE POLITICAL SOCIETY

We can now return to where we begin, i.e. with the admonition of Mr. Lu Ping against turning Hong Kong into a political society. In fact, a political society has been in the making ever since the 1970s. The question is whether it can be arrested and for how long.

HONG KONG'S PREDICAMENT   1439

A political society is an independent arena mediating between the society at large and the polity. No political society could develop as the colonial government was effectively secluded from the society and when all social-economic elites could be successfully co-opted into the government's machinery of consultation. Given the social-economic changes discussed in the above, the concomitant differentiation of a growing elite strata, the increase in the government's responsibilities, the politics of governing by consultation alone must sooner or later reach its limit of usefulness, no matter how hard the Hong Kong government tries to co-opt the new elites. The process of politicization began in the 1970s when pressure groups emerged as an autonomous force independent of and critical toward the government. They were aided by the prospering and free-wheeling mass media fed by the information-hungry Hong Kong Chinese. What followed was the growth of a public sphere where politics was freely discussed although power could not yet be contested. The mass was gradually socialized to become more interested in politics, more demanding toward the government, and more willing to join actions aimed at influencing the government.

As alluded to in the above, the political society was still small and weak on the eve of the Sino-British negotiations over the future of Hong Kong. It has since grown on strength due to the cumulative pressures of the processes of political development and the specific impacts of the conduct of elections.

The general effect of developmental issues and processes lies in a diffuse **engaging (self-engaging)** of the activist strata from the society into the political process. The search for an existential rationale for a future has been intellectually engaging, thereby contributing to the growth of a political discourse and a public sphere largely sustained by the free mass media. The desire to secure the conditions for autonomous development is by no means confined to the arena of high politics but to all areas of associational life, thus witnessing an array of covert or overt, clientelist or otherwise initiatives to self-strengthening (or self-defense). The need of the government for broadening the basis of political support in order to enhance capability to deal with developmental crises is leading to the expansion of the net of consultative politics, attracting more people of a wider spectrum to the political game. Finally, the introduction of representative government has provided even better conditions for the political activists to play a role well as empowered the mass public.

In an abstract sense, the introduction of elections has made the need for group-based mediation between the polity and the society even more obvious. In response to this need, the political society has been growing fast with a growing public sphere of political discourse, a proliferation of political groupings and a burgeoning of policy-oriented joint meetings and actions. The specific impacts of elections can be understood in the following terms.

First of all, competition between the British and the Chinese government for the support of the people of Hong Kong has opened up new channels for political participation by individuals and groups. On the one hand, there are now four government institutions subject to election, compared to none before the 1980s. The number of members elected to those institutions has gone up from 15 in 1981 through 338 in 1991 to 457 in 1995. On the other hand, hundreds of different types of elites have filled posts created by the Chinese government as members of the Preparatory Working Committee, Advisors on Hong Kong affairs, Advisors on District Affairs, and so on.

Secondly, the introduction of elections by functional constituencies to the Legislative Council in 1985 has turned "state-recognized" interest groups into political establishments with a continued stake in political power. These otherwise commerce- or other-regarding associations have willy-nilly succumbed to a politicized future.

HONG KONG'S PREDICAMENT   1441

Thirdly, the repeated conduct of elections, especially on the basis of popular franchise, has contributed to the growth of political parties. While councillors/members without any party background still constituted the largest group in all levels of assemblies/boards, the proportion of those with party background has been rising. In 1985, members with party background constituted only 6.3 percent of the total membership of the District Boards, the figure nowadays is 49.7 percent. The corresponding figures for the Municipal Councils are 33.3 percent in 1989 and 75 percent in 1995. In 1991, only fifteen out of 60 members of the Legislative Council belonged to political groups (no formal political party existed yet), the proportion of councillors with party background jumped to 73.4 percent in 1995.

Finally, elements of the political society thrived on a growing number of political participants among the general population. While turnout rates in elections remain low, the absolute number of voters has grown significantly, with 6,195 turning out to vote in 1981 to 1,338,205 in 1995.

CONCLUDING DISCUSSIONS

We have seen that there is already a political society in Hong Kong. This represents a fundamental structural transformation away from what was described as a minimally-integrated system between the polity and the society in the past. Is this major transformation only temporary? Can it be eradicated, arrested, or tamed after 1997? This author believes that it can at best be tamed. The most important factor lies in the fact that the polity can no longer be secluded from the society. Not only has a stratum of activists been engaged, but also the people of Hong Kong have been psychologically inducted into the political process. Unlike their forefathers, they no longer rely solely on individual and familial self-help for a living, but entertain increasing expectation from the government and are ready to join collective actions to press demands on the government. The growth in intermediate groups will facilitate the articulation and aggregation of interests. In short, we have definitely entered an era of mass politics.

The government has also changed in their attitudes and policies toward interest group activities from suspicion and distrust to accommodation and cooperation. Besides, the Hong Kong government in general and the top leadership in particular have become increasingly localized, with its members coming from the general society, bringing with them the latter's ethos and aspirations. This will provide a positive backdrop for participation by elements of the political society in the formation of public agenda and the general will. The question for the future therefore does not concern a choice between having a political society or not, but whether the political society will experience a healthy development, so as to better mediate between the polity and the general society.

At a less abstract level, the development of Hong Kong's political society depends on the development of elections, parties and interest groups corporated into functional constituencies. So long elections remain an alternative route to political representation and power, and provided that elections are run in a fair and open manner, there remains a public space for the spontaneous growth of the political society. Interest groups, political alliances and parties will have a life of their own.

In the same vein, the development of the mass media will affect the future health of the political society in Hong Kong. The scope of the public sphere in which political discourse is conducted will shrink to the extent that the freedoms of expression and communication are abridged. There are real risks to those freedoms in Hong Kong in the form of Chinese intervention.[23] But the most immediate and greater risk pertains to self-censorship, as there are limits to governmental intervention, beyond which the costs to the proper function of a mature, industrialized, modern

HONG KONG'S PREDICAMENT   1443

society will become unbearable. It goes beyond doubt that Hong Kong requires an infrastructure of free communication to maintain its competitiveness in the international economy. In addition, the pluralistic structure of mass communication in Hong Kong and especially the accessibility of foreign media and internet forums may provide flexible correctives to occasional encroachment to freedoms of communication. In the ultimate sense, the continued existence of a public sphere depends on the diffuse support of the people of Hong Kong who as attentive spectators may not tolerate a blatant reduction in the availability of free information.

What if the Chinese government is determined to eradicate the political society in Hong Kong? It may do so. But at considerable costs. To the extent a modern society needs a political society to handle the complexity of mass politics, it is imprudent to suppress it. A political society straddles between the voluntary, associational (hence particularistic) sphere of human interaction and the coercive, universal domain of public authority. It simplifies the processes of interest representation and policy-making in peace times and absorb the shocks and pressures from the mass society in crises situations. Eradicate the political society, the government is put into direct confrontation with the society.

One of the first steps taken by totalitarian countries was to abolish the political society. Instead, their rulers governed by means of a system of transmission belts, i.e., government-controlled organization executing party-government orders. Such a system worked for a while, under the conditions of revolutionary mobilization, closeness to the outside world, an uneducated and disciplined mass, and a decent level of performance in material production. In the longer term, however, the utility of such a system has been proven worthless in Eastern Europe in 1989 and the Soviet Union in 1991. It will never fit Hong Kong.

What if the Chinese government resorts to the past practice of the Hong Kong government in consultative authoritarianism? It is quite likely. In this case, the Chinese government does not have to eradicate the political society altogether, but so tame it such that it either becomes inactive or unable to have any impact on public policies. For instance, the Chinese government does not have to eliminate all political parties, but just to make it hard for any undesirable party to develop or in the worst-case scenario to outlaw it. The rest or the most friendly ones can be co-opted into the advisory machinery of government. We have already seen the limits of government by consultation. It is inefficient in meeting the changing needs of a complex, post-industrial society. In addition, a consultative-yet-authoritarian government remains inherently unstable in the modern age of democracy and information. Not only that the expectation of an educated, worldly exposed and political awakening mass sustains considerable pressures on the government to open up, but also the hegemonic idea of democracy in our age subjects an authoritarian regime to a constant crises of legitimacy. That is why no authoritarian government today rejects democracy, although some may claim itself being more democratic than the democracies in the West. An authoritarian regime which has pre-committed itself to democracy is an inherently unstable regime, because it will be taken to task to deliver. After all, democracy has been declared the ultimate objective of political development in the Basic Law. Unless this provision were changed, there will linger on an agenda for further democratic reforms, thereby upholding hope for a political society. At a more specific level, elections often have unintended effects. Unless elections at all levels were abolished altogether after 1997 such that their direct and indirect impacts on political learning and structuring were foreclosed, there will always be room for the political society to persist.

To conclude, it is illusory to ignore the implications of an awakening mass for the political process. The multiplicity and diversity of interests in a rapidly changing society require a flexible and adaptive mechanism to sort out the priorities and to forge consensus. It would be wrong to assume that the past

HONG KONG'S PREDICAMENT 1445

practice of the Hong Kong government in consultative authoritarianism can easily be adapted to provide tutelage over the plurality of fluid cleavages and groups. It may be even less so after the transfer of sovereignty in 1997 since the new government has to accumulate political capital before its tutelage will be accepted as legitimate. The foundation for the past practice in governing through consultation, i.e., elite solidarity and lack of mass mobilization, and so on no longer exists. Alternative institutions are called for to meet the requirements of rational efficiency, interest intermediation and long-term legitimacy in the modern world. A healthy political society will be a blessing rather than a curse, serving as an efficient feedback loop between the general society and the polity. Such a healthy political society requires the institutions of fair elections and free communications. These are the critical areas the development of which is going to shape the future political landscape of Hong Kong.

REFERENCES

1.   As used in the political discourse in Hong Kong, politicization vaguely means instigating or manipulating a socio-economic conflict, forcing it onto the public agenda with a view to seeking a political resolution.

2.   Endacott, G.B. *Government and People in Hong Kong 1841-1962, A Constitutional History*, Hong Kong University Press, Hong Kong, 1964, p. vii.

3.   Hughes, Richard. *Hong Kong, Borrowed Place - Borrowed Time*, André Deutsch, London, 1969.

4.   Victoria island together with the Kowloon peninsula and the remainder of Kowloon were ceded in perpetuity in 1840 and 1864 respectively. The New Territories was leased in

1898 to Britain for a period of 99 years and that lease expires on July 1, 1997.

5.   Lau, Siu-kai. *Society and Politics in Hong Kong*, The Chinese University Press, Hong Kong, 1982.

6.   Lane, Kevin P. *Sovereignty and The Status Quo, The Historical Roots of China's Hong Kong Policy*, Westview Press, San Francisco, 1990.

7.   Wallerstein, Immanuel. *The Modern World-System*, 2 Vols., Academic Press, New York, 1974, 1980.

8.   See for example Furtado, Celso. *Development and Under-development: A Structural View of the Problems of Developed and Underdeveloped Countries*, University of California Press, Berkeley, 1964; Frank, Andre Gundar. *Capitalism and Under-development in Latin America*, Monthly Review Press, New York, 1967.

9.   Evans, Peter B. *Dependent Development: The Alliance of Multinational, State, and Local Capital in Brazil*, Princeton University Press, Princeton, 1979.

10.   Samuel P. Huntington has not included national independence as a goal of development, see his "The Goals of Development," in Weiner, Myron and Huntington, Samuel P. eds., *Understanding Political Development*, Little, Brown & Co., Boston, 1987, pp. 353-390.

11.   In a survey conducted by the author and Lau Siu-kai in 1988, respondents were asked whether they agreed with the statement that Hong Kong should not be returned to China but go independent. 2.9% and 44.3% strongly agreed and agreed, while 6.0% and 46.9% disagreed and strongly disagreed. See Lau, Siu-kai and Kuan, Hsin-chi. *Social Indicators Survey 1988, Module D Data Book,* Centre for

Hong Kong Studies, Chinese University of Hong Kong, Hong Kong, 1988, p. 39.

12.   For a more general discussion of the issue of state autonomy, see Nordinger, Eric A. "Taking the State Seriously," in Weiner, Myron and Huntington, Samuel P. eds., *Understanding Political Development*, Little, Brown & Co., Boston, 1987, pp. 353-390.

13.   See Tang, James T.H. "Hong Kong's International Status." *The Pacific Review* 6 (1993): 205-215. The Act also declares support for democratization as a fundamental principle governing U.S. policy toward Hong Kong and requires the Secretary of State to make periodic reports on the development of democratic institutions in Hong Kong. Prior to this Act, the U.S. Congress enacted in 1990 the Immigration Act of 1990 to establish a separate immigrant visa quota for Hong Kong.

14.   Kuan, Hsin-chi. "Power Dependence and Democratic Transition: The Case of Hong Kong." *The China Quarterly* 128 (December 1991) : 774-793.

15.   See Binder, Leonard, Pye, L. W., Coleman, J.S., Verba, S., Lapalombara, J., Weiner, M. *Crises and Sequences in Political Development*, Princeton University Press, Princeton, 1971.

16.   King, Ambrose Y.C. "Administrative Absorption of Politics in Hong Kong: Emphasis on the Grass Roots Level." *Asian Survey* 15 (May 1975) : 422-439.

17.   Lau, Siu-kai. *ibid.*

18.   Lau, Siu-kai. "Decline of Governmental Authority, Political Cynicism and Political Inefficacy in Hong Kong."

Journal of Northeast Asian Studies 11 (Summer 1992) : 3-20.

19.   Scott, Ian. *Political Change and The Crisis of Legitimacy in Hong Kong*, Oxford University Press, Hong Kong, 1989; Scott, Ian. "Legitimacy and its Discontents: Hong Kong and the Reversion to Chinese Sovereignty." *Asian Journal of Political Science* 1 (June 1993) : 55-75.

20.   Roberti, Mark. *The Fall of Hong Kong, China's Triumph & Britain's Betrayal*, John Wiley & Sons, Inc., New York, 1994.

21.   Kuan, Hsin-chi. *ibid.*

22.   For a comprehensive analysis of the elections in 1991, See Lau, Siu-kai and Louie, Kin-sheun (eds.). *Hong Kong Tried Democracy, The 1991 Elections in Hong Kong*, Hong Kong Institute of Asia- Pacific Studies, The Chinese University of Hong Kong, Hong Kong, 1993; Kwok, Rowena, Leung, Joan and Scott, Ian, eds., *Votes Without Power, The Hong Kong Legislative Council Elections*, Hong Kong University Press, Hong Kong, 1992.

23.   For a pessimist assessment, consult Bonnin, Michel. "The Press in Hong Kong, Flourishing but Under Threat." *China Perspectives* 1 (September/October 1995) : 48-59. For more balanced views, consult Chu, Leonard L. and Lee Paul S.N. "Political Communication in Hong Kong:Transition, Adaption,   and Survival." *Asian Journal of Communication* 5 (1995) : 1-17. According to Chu and Lee, the Hong Kong media are likely to go back to the public relations model in the short run, but they will resume their reformist and public forum roles in the long-run as a result of China's political development. See p. 12.

[4]

The Legacy of the British Administration of Hong Kong: A View from Hong Kong

Ming K. Chan

As the one and a half centuries of British colonial rule draw to a close on 30 June 1997, it is timely to review the true legacy of British administration in Hong Kong. It would be naive to resort to any simplistic blanket judgment or to issue any sweeping endorsement or condemnation on the mixed record of the British administration. It would also be dangerous to look only at the attainments in the final days of the British regime and use them to reconstruct, or even to substitute for, the full span of British rule. Even given a charitable view of this sunset era of the British regime as its finest hour in Hong Kong, a more informed and balanced assessment of its past deeds must be appreciated in the fuller context of the actual inputs and outputs of British officialdom in shaping developments in the territory and the life of Hong Kong people during the entire course of British rule.

Hong Kong today is globally recognized as a remarkable example of a liberal society with a vibrant economy where its population of more than six million enjoy their freedom and opportunity. For this, the British can indeed claim considerable credit. As portrayed by the last British Governor Christopher Patten, four of the major British contributions to Hong Kong's success – the rule of law, the civil service, economic freedom and democratization – can be a useful starting point to articulate the true British legacy from a Hong Kong perspective.[1]

The Rule of Law

A much celebrated British "gift" to Hong Kong has been the rule of law, which included the British-style common law legal system with an independent and impartial judiciary supposedly delivering fair and equal justice to all. To most Hong Kong people, the preservation of the legal system is of crucial importance to the "high degree of autonomy" which the Hong Kong Special Administrative Region (HKSAR) is supposed to enjoy. However, Hong Kong's present legal system has serious defects. Indeed, a good case could be made for reform of the legal system in order to meet the requirements of contemporary Hong Kong as a complex international community and to lay a more solid foundation for the future.

While much of the structure and institutions in Hong Kong's legal system are regarded as basically sound and smoothly functioning, a major flaw has been the legal language of English. Until very recently all court proceedings in both civil and criminal litigations in Hong Kong were held in English. While an official language, English is not the mother tongue

1. Christopher Patten, *The 1996 Policy Address, Hong Kong: Transition* (2 October 1996). This was his last policy address as Governor.

or language of everyday usage for most local people, who are Cantonese speakers. At present, residents must speak English in order to be eligible for jury service. This plays a major part in the composition of the Hong Kong jury and renders it significantly out-of-step with contemporary ideas about the role of trial by jury. Thus, those Hong Kong Chinese who appear on the List of Common Jurors are likely to be better educated, middle-class businessmen or professionals and as such are not representative of the society which they are called upon to serve. Failure to reform the jury franchise would create the risk of the jury system losing legitimacy among the public, even jeopardizing its future under a different sovereign with a Leninist-Stalinist legal system.[2]

A parallel concern is the late start and inadequate progress in bilingual (English and Chinese) codification. Part of the problem stems from the personnel in the legal system. The institutions of justice are still expatriate-dominated in both the Legal Department and the judiciary branch. The government's personnel localization efforts have had only limited results and required more conscientious effort in the recruitment and training of legal personnel through reform of the legal profession and education. Localization of the common law requires long-term and far-sighted reform which could strengthen the rule of law by making the entire legal system more attuned to the demographic and socio-cultural realities as well as the rising democratic and human rights consciousness of the local Chinese populace. This much-needed sinification would help to consolidate the common law legal culture and institutions for HKSAR.[3]

The British legal legacy in Hong Kong also harbours dangerous historical breaches that set unhealthy precedents for the future. The thick piles of discriminatory legislation (mostly racially based, anti-Chinese law passed by an appointed legislature) and draconian, biased (anti-Chinese and anti-grassroots) court rulings are not good examples of decency and fairness.[4]

Moreover, the executive and legislative branches have a rather weak separation of power, with inadequate checks and balances. Until 1985, when indirectly elected seats were introduced into the Legislative Council (Legco), the colonial administration had been able to pass legislation through a compliant legislature constituted by appointed members, with government officials enjoying a majority until 1976. Thus one could say that the colonial regime often operated within the law, as it had always been able to change the law through the appointed legislature. The executive branch could always claim a legal basis for any action that it

2. Peter Duft, Mak Findlay, Carla Howarth and Tsang-fai Chan, *Juries: A Hong Kong Perspective* (Hong Kong: Hong Kong Univeristy Press, 1992), pp. 53, 57, 58. Of the 143,798 names on the 1987 List, two-thirds are Chinese and the others mainly European, Australian and North American.

3. A recent advocate for reform of the legal profession and education is Anthony Dicks "Will the laws converge?" the Hong Kong Lecture, University of Hong Kong, 1996.

4. Peter Wesley-Smith, "Anti-Chinese legislation in Hong Kong," in Ming K. Chan (ed.) *Precarious Balance: Hong Kong Between China and Britain, 1842–1992* (Armonk, NY M. E. Sharpe, 1994).

wished to take. And when the administration did not wish to seek the approval of Legco, it could resort to the Emergency Regulations Ordinance of 1922 which allowed imprisonment without trial and many other breaches of human rights. Yet such acts were considered perfectly legal. Because the Emergency Power Ordinance still remains on the statute books, though all the regulations had been repealed by 1985, the serious risk remains that the HKSAR government may apply it as did the colonial government.[5]

Furthermore, the colonial regime's own illegal official acts should not be forgotten. These included film censorship without proper legal authority,[6] character assassination[7], and extra-legal manoeuvres to undermine or obstruct the course of justice (such as London's executive interference with the Hong Kong legal process in the Chinese government aircraft case in 1950.[8] They amounted to a gross travesty of the rule of law and the pursuit of true justice.

A host of humiliating incidents, ranging from serious criminal activities by legal personnel to dubious conduct and incompetence, have marred effective legal administration, resulting in a tarnished reputation for the legal system.[9] The following recent cases illustrate the perils and predicaments of the Hong Kong legal system: the Reid and Harris cases,[10] the registration of Judge O'Dea,[11] and the alleged interference with Justice Caird.[12] There was also Chief Justice T. L. Yang's controversial 1995 opinion on the Bill of Rights' unsettling impact on the legal system and possible contravention with the Basic Law (in the context of the PRC's intention to resurrect the old version of six local ordinances before

5. The author is fully indebted to Norman Miners on this issue; see his "The use and abuse of emergency powers by the Hong Kong Government," *Hong Kong Law Journal*, Vol. 26 (1996), pp. 47–57.

6. Before the Hong Kong government's enactment of an ordinance on film censorship in 1986, its film censorship had been without proper legal authority; see Michael C. Davis, "Free speech in comparative perspective: the case of Hong Kong," paper presented at the 40th Annual Meeting, Association for Asian Studies, San Francisco, 25 March 1988.

7. The regime's "dirty tricks" targeted pressure groups and politicians; see Robert Adley, *All Change Hong Kong* (Poole, Dorset: Blandford Press, 1984).

8. James T. H. Tang, "World War to Cold War: Hong Kong's future and Anglo-Chinese interactions, 1941–55," in Chan, *Precarious Balance*, pp. 120–21.

9. For a recent expose on the misdeeds of senior expatriate legal officials, see *Hong Kong Economic Journal*, 19 September 1996, p. 15. The relatively low conviction rate in cases handled by the Legal Department, as pointed out in the *Report of the Director of Audit on the Results of Value for Money Audit* (October 1996), ch. 7, reflected partly the legal staff's incompetence.

10. Christopher Harris, a senior Crown Counsel, was convicted of sexual offences. *Ta Kung Pao* (*Dagong bao*), 22 February 1990. Warwick Reid, a Deputy Director of Public Prosecutions, was convicted of corruption and several lawyers were implicated. *South China Morning Post*, 21 June and 2 September 1990; *Far Eastern Economic Review*, 28 June 1990.

11. Supreme Court judge Patrick O'Dea made international news after admitting reading a book while presiding over a robbery trial. Berry F. C. Hsu, *The Common Law System in Chinese Context* (Armonk, NY: M. E. Sharpe, 1992), p. 64.

12. Judge Brian Caird withdrew from the Aaron Nattrass case on 3 September 1996 on medical grounds and retracted his earlier allegations of being improperly pressured. Yet the internal judicial inquiry which found no pressure had been applied was regarded as a massive "cover-up" staged by the judiciary and left a host of questions unanswered. See *South China Morning Post*, 24, 28 August, 6, 26 September, 15 October; *Hong Kong Standard*, 25 August, 4, 11, 25–27 September, 2, 13, 15, 16 October; *Sing Tao Daily*, 4 September 1996.

their amendment by the 1991 Bill of Rights).[13] Public criticism was also levelled at Yang's conflict of roles in his campaign to become HKSAR chief executive while still the titular head of the judiciary.[14]

The rule of law is definitely a foremost British legacy for Hong Kong, and is rightly popularly perceived as such. However, it is necessary to be reminded of the significant lapses and gaps in the common law legal system as it has been practised by the colonial regime. Much effort is still needed to reform and remedy its defects and inadequacies so that it can live up to its own avowed objective: "If the people of Hong Kong are to have confidence in their judicial system, the courts must be seen to be capable of dispensing justice independently, within a reasonable period of time and in a language which the vast majority of people can understand."[15]

Civil Service

The backbone of British rule in Hong Kong has always been the civil service. If the British tradition of an executive-led government is to become even more emphasized in the HKSAR, then the civil service administrative bureaucracy will remain the most powerful political institution. While much has been said of Hong Kong's civil service system as a whole as highly professional, efficient, politically neutral and relatively free from serious corruption, this can only be a partially accurate reflection of recent realities.

Precisely because Hong Kong, at least until the mid-1980s' sovereignty settlement and start of democratization, was often labelled as an "administrative state" or "bureaucratic polity" practising "administrative absorption of politics," local governance and political power were monopolized by civil service administrator-bureaucrats. In this sense, the civil service system, particularly the senior echelon, dominated the process of policy formulation, decision-making, implementation and supervision. With such a concentration of functions and power in its hands, the civil service also nurtured and perpetuated its own bureaucratic culture of elitism and even arrogance at the expense of public accountability and responsiveness. This partly explains why the senior civil service echelon is still ill-adjusted to the growing pressure for open

13. The damage was not only Yang's expression of doubts about the overriding effect of the Bill of Rights. Yang apparently violated judicial independence by bowing to executive pressure and submitted a report explaining his view to Chief Secretary Anson Chan. See Berry Hsu, "Judicial development of Hong Kong on the eve of 1 July 1997," in Ming K. Chan and Gerard A. Postiglione (eds.), *The Hong Kong Reader: Passage to Chinese Sovereignty* (Armonk, NY: M. E. Sharpe, 1996), p. 82.

14. In late August 1996 while on long leave, Yang announced his intention to run for the HKSAR chief executive post. It was not until 5 September that he submitted his resignation as Chief Justice to the Governor to be effective on 4 November. Although Yang did not resume active duty on the bench after his announcement, he still enjoyed his perks as Chief Justice while involved in his campaign until 27 October 1996, his revised day of effective resignation. See *Sing Tao Daily*, 4, 7, 26 September, 29 October 1996.

15. *Hong Kong Judiciary 1994–1995* (Hong Kong: Government Printers, 1996), p. 18. For a more detailed critique of the problems in the British colonial legal system, see Ming K. Chan, "The imperfect legacy: defects in the British legal system in colonial Hong Kong," *University of Pennsylvania Journal of International Economic Law*, Vol. 18, No. 1 (1997), pp. 133–156.

government and public accountability as well as the new inputs from elected politicians and political parties or pressure groups in the policy process.

The 1997-dictated decolonization also accelerated civil service localization. It is ironic that this process, which is needed because of past discrimination, should now be claimed as a British achievement. The fact that civil service localization did not become a high priority until after the signing of the Sino-British Joint Declaration in 1984 raises serious doubts about the British commitment to a genuinely professional and rational approach to personnel policy and management in its own governance. Even now, the administration's localization measures are being successfully challenged in court by some expatriate civil servants under the Bill of Rights.[16] Until the early 1990s, the civil service's top ranks were dominated by British officials, and even though they were few in number (about 3,000 out of a total civil service of over 150,000 in the 1980s), several serious problems were thereby revealed as shown below.

Until the 1980s, these expatriate officials, while small in number, were concentrated at the top and very high levels of the hierarchy and controlled the vital processes of recruitment, assignment, supervision, evaluation and promotion of their subordinates, the majority of whom were local Chinese. The top-heavy strategic concentration of expatriates with their great power, influence, control and policy-making functions distorted this supposedly *Hong Kong* civil service system into a very British state machinery, thus undermining Hong Kong local autonomy.

Secondly, the expatriate domination of the civil service top echelon is not only racially discriminatory but raises serious doubts about the integrity of the entire system as a genuine meritocracy in which recruitment, posting and promotion should be based solely on rational criteria of ability, skills, performance and overall effectiveness and contribution to the local community. With some well-known exceptions, most of the expatriate top officials are illiterate or only semi-literate in Chinese. While few would doubt their professionalism and devotion, their limited literacy and understanding of the Hong Kong Chinese community and of developments in China diminished their effectiveness as Hong Kong administrators and policy-makers.

Such expatriate dominance carried the assumption that local Chinese civil servants somehow, by British official standards, lacked the ability, skills and competence to reach the top. This became the pretext for the non-promotion of local civil servants. Thus deprived of the opportunity to shoulder responsibility and gain greater experience, they of course remained "lesser qualified" or "inadequately prepared" for the top appointments. Local civil servants were also not deemed to be politically trustworthy, with unquestioned loyalty to the British crown. Hence, only

16. *Sing Tao Daily,* 23 November 1996; *South China Morning Post,* 23 November 1996. The recent court rulings found seven areas of the government's localization measures unfair and discriminatory to the affected expatriate officials. The Association of Local Senior Civil Servants regards the rulings as the "perpetuation of discrimination" against local personnel "under the name of human rights." *Sing Tao Daily,* 2 December 1996; *Hong Kong Economic Journal,* 30 November 1996.

British officials were considered loyal enough to be fit for top positions. Such logic makes a mockery of the colonial regime's claim that it has always served the true interests of Hong Kong and has always enjoyed very high autonomy from London. Even though colonial British interests in Hong Kong might at times be at variance with the British government's interests, neither should be regarded as identical with the interests of Hong Kong's Chinese majority.[17]

Another result of the "glass-ceiling" of exclusion from the top ranks (and more favourable terms for expatriates than locals in the other ranks) was that many qualified and gifted local Chinese were discouraged from entering or staying within the civil service, thus depriving it of a full reservoir of candidates of high calibre and great promise. This weakened the overall quality and actual range of skills, capabilities and expertise of the civil service. The delay in full-scale, whole-hearted localization also means a failure to nurture a generation of highly experienced and proven administrative leadership that will be indispensable to the HKSAR.

Because of its elitist bias, institutionalized racism and expatriate domination within its bureaucracy, the Hong Kong government headed by an unelected British Governor could hardly claim to enjoy the legitimacy of a popular mandate. It could only aim to achieve public acquiescence or at best "administrative legality" camouflaged as "functional legitimacy."[18] This has induced a sense of insecurity in the colonial bureaucracy, often resulting in the officialdom's inability to acknowledge mistakes and accept valid public criticism. This has also contributed to the lack of any longer-term vision and overall developmental objectives.

A most glaring failure of the civil service system under expatriate leadership was the senior bureaucrats' lack of any sense of the role of the government in relation to the inevitable issue of 1997 sovereignty retrocession. A lack of guidelines for the political transition rendered Hong Kong civil servants ill-prepared and even confused about how to react to political issues. The situation was exacerbated by the Sino-British discord created in 1992 by Governor Patten's electoral reforms, which politicized the senior civil servants and required them to support policy unacceptable to Beijing. This trend undermined the colonial state's recent promotion of the "political neutrality" of the civil service, probably to pre-empt the public perception in order to ward off future PRC attempts to politicize the HKSAR civil service.[19] The democratization efforts also

---

17. In the 1980s, the UK led the drive in the EEC quota on Hong Kong textile imports, while "foreign students" fees were charged to Hong Kong students in UK universities.

18. Ming K. Chan, "Labor vs Crown: aspects of society–state interactions in the Hong Kong labor movement before World War II," in Elizabeth Sinn (ed.), *Between East and West: Aspect of Social and Political Development in Hong Kong* (Hong Kong: Centre of Asian Studies, University of Hong Kong, 1990).

19. Civil service political neutrality is again jeopardized by Chief Secretary Anson Chan, who upon her decision not to run for the HKSAR chief executive post, issued a political statement on the criteria for the ideal chief executive. She also congratulated C. H. Tung, but not the other two nominated candidates, on his candidacy nomination. *Sing Tao Daily,* 5, 19 November 1996, and *Sing Pao Daily News,* 19 November 1996, headline. All three nominated candidates for chief executive consider the senior civil servants already too politicized and the HKSAR must restore their political neutrality. *Sing Tao Daily,* 30 November 1996.

A Hong Kong View of Hong Kong   573

added pressure on the senior bureaucrats as salespersons of government policies to a Legco with elected members. This new public role moved then Financial Secretary Macleod to acknowledge that "we have increasingly become quasi-politicians."[20]

Finally, the fact that Governor MacLehose found it necessary to establish the Independent Commission Against Corruption in 1974 reflected the existence of widespread corruption among almost all ranks and branches of the civil service. Public criticism and a tarnished international image forced the government to act, but even then MacLehose was compelled to grant a general amnesty to the entire police force in the notorious police mutiny against the Commission in 1977. This is a powerful indictor of how widespread corruption had eroded public confidence in the integrity, fairness and efficiency of the civil service two decades ago. Unfortunately, corruption among civil servants has been increasing recently. The prospect of the serious epidemic of corruption in the PRC becoming part of Hong Kong's post-1997 reality is disturbing.[21]

Another disturbing trend for the future HKSAR polity would be a reconfiguration of Hong Kong's pre-1985 anti-democratic power coalition in which expatriate senior officialdom worked closely with British big business and Anglicized local elites. The new power equation will see local senior bureaucrats join the Beijing-endorsed local elites, such as those on the HKSAR Preparatory Committee and Selection Committee, to preserve the status quo with a conservative, pro-business and anti-democratic orientation. The civil service top echelon's tradition of administrative supremacy and colonial authoritarianism has rendered it ill-adapted to the 1985–97 era of democratization and legislative scrutiny. Perhaps, the localized officialdom should have much common ground with the new HKSAR elites to marginalize electoral politics and the implied popular legitimacy of the democratic camp.[22]

Economic Freedom with Limited Government

Another often trumpeted British contribution to Hong Kong's success has been its promotion of economic freedom with limited government. In the post Second World War era, colonial officials have stressed "positive non-interventionism" as their guiding light in economic policy, helping to raise Hong Kong to its world class economic hub status.[23]

20. Hamish Macleod, "Hong Kong: a hard-earned success," in Government Information Services, *Hong Kong 1995: A Review of 1994* (Hong Kong: Government Printers, 1995), p. 20.

21. John P. Burns, "Civil service systems in transition: Hong Kong and China," in Ming K. Chan (ed.), *The Challenge of Hong Kong's Reintegration with China* (Hong Kong: Hong Kong University Press, 1997), pp. 31–48.

22. Anthony Cheung, "The transition of bureaucratic authority; the political role of the senior civil service in the post-1997 governance of Hong Kong," paper presented at the Conference on Hong Kong in Transition: Political Order, International Relations and Crisis Management, Hong Kong, 18 September 1996.

23. An authoritative view on "minimal government" is then Financial Secretary MacLeod's "Hong Kong: a hard-earned success."

It has often been said that with free trade and a free port as the cornerstones of the economy, the combination of low taxation, a truly open market, the free flow of information/technology/capital, and a minimum of regulation and official red-tape has yielded considerable room for the private sector to maximize profit. In essence, Hong Kong's economic miracle seems to be the direct result of the official minimalist approach.[24] A careful examination of the historical record, however, reveals an interventionalist regime whose actions did not always conform to its projected *laissez-faire* façade.[25]

While free trade remains the hallmark of Hong Kong's economy, the colonial government has continuously played a direct and crucial role as a very significant economic participant. Besides its control of valuable resources, the regime's command of the relevant legal, political and social institutions and processes also indirectly shapes economic behaviour and societal development. It could be argued that the continuous projection of the "positive non-intervention myth," despite a very different reality today, serves several purposes.

This official stance was a useful lure to the international business community. It emphasized free trade, low tax, and unlimited opportunity for free enterprise because of the absence of regulation and official interference in order to attract investments in Hong Kong. Domestically, limited government with a *laissez-faire* façade aimed at minimizing the colonial state's role as an active protector, provider and promoter of many community needs. This stance could help to de-escalate the rising expectations of an increasingly affluent and modern society. In turn, the public's limited claim on governmental services helped to limit expenditure, hence the continuation of low taxation and the slower growth of bureaucracy. Finally, the "minimalist" regime's limited responsibility aimed at encouraging the private sector to carry the lion's share of societal undertakings and also discouraging further influx of immigrants from China by making sure that Hong Kong was no welfare state utopia.

This was the mentality fostered by a colonial regime supposedly practising "positive non-interventionism." However, during the post-War years, three successive waves of the China Factor – the 1949 Communist revolution and the influx of immigrants from China, the 1967 Cultural Revolution-inspired leftist urban terrorism disturbances, and since the 1980s, the 1997 syndrome – all affected the colonial polity and socio-economic order and necessitated considerable reorientation in the actual undertakings and new commitments of the British administration of Hong Kong. Thus, since the late 1960s the China Factor effects, combined with demographic growth and a greater complexity of community requirements, have forced the once uncaring and aloof colonial regime to cater more to grassroots needs. In the last three decades, the Hong Kong government's involvement in everyday life has increased steadily and now reaches into many vital areas of socio-economic development.

24. See Alvin Rabushka, *Hong Kong, A Study in Economic Freedom* (Chicago: University of Chicago Press for the Graduate School of Business, 1979).
25. Ming K. Chan, "Stability and prosperity in Hong Kong: the twilight of laissez-faire colonialism?" *Journal of Asian Studies*, Vol. XLII, No. 3 (May 1983), pp. 589–598.

A Hong Kong View of Hong Kong 575

The government is the monopoly owner of scarce resources (land), and the largest landlord in Hong Kong (with over 40 per cent of the population living in public housing). It also imposes rent control on some private housing. Through its fully-owned public corporations it operates the three railways and also regulates the other major transport services and public utilities as monopoly franchises. Most of the education system is government subsidized. While much of Hong Kong's welfare and charity programmes are provided by private and voluntary organizations, their budgets are heavily subsidized by the government, which also sets basic policy and supervises the delivery of services. All public and subvented hospitals come under the jurisdiction of the Hospital Authority, which is a government appointed and funded public body. In the 1980s, the government took over the ownership and management of several insolvent local banks to prevent a general bank run. These are just a few examples of a government which is actively involved, fully engaged and often interventionist, whether by design or necessity. Taken together they should substantially modify the government's "positive non-intervention" self-characterization. The main focus here is not the merits or demerits of the interventionist or non-interventionist stance of the regime. Rather this discussion aims at clarifying the actual role of the colonial state in societal and economic development vis-à-vis the official claim to, and public perception of, the principle of *laissez-faire* in this articulation of the British legacy in Hong Kong.

One could argue that the myth of *laissez-faire* economic policy has also served the purpose of white-washing dubious, unfair government practices. Contrary to the ideal of free trade and the open market, the colonial authorities resorted to discriminatory and monopolistic measures that constituted interventionist actions. Noted examples include the mandatory use of only British motor vehicles by the franchised public bus companies until 1983, and the monopoly status granted to British university degrees and British-derived professional qualifications for academic, professional and even business purposes.

In this context, the most damaging legacy was the blatantly pro-business bias in the government's decision-making. A notable point is the special status enjoyed by British business in Hong Kong, especially the collusion between the colonial officialdom and the British economic elites, with an almost quarantined Executive and/or Legislative Council seat for the *taipans* of the Hong Kong Bank, and such British hongs as Swire and Jardine. This not only enabled them to enjoy special access to power, information and policy inputs, and in crisis situations (such as the 1922 Seamen's Strike) the support of the colonial state against the local Chinese grassroots,[26] but it also distorted the basic orientation of the state in its larger societal responsibility. Indeed, the colonial regime has been at fault for its subservience to business interests as manifested in its unwillingness until very recently, not because of *laissez-faire* but from its pro-business bias, to legislate against cartels and monopolies and to

26. Chan, "Labor vs crown."

regulate economic activities in the interests of labour, consumers and the environment. Even in the more "enlightened" post-1967 era, business interests received a disproportionally large share of representation in the Executive Council (Exco) and Legco, taking an average of 40 per cent of the seats in the two councils until the introduction of Legco direct elections in 1991.[27]

While no one could deny the remarkable economic growth and rising affluence in Hong Kong during the past three decades under British rule (with GDP per capita increased from US$686 in 1966 to US$23,200 in 1996[28]), the pro-business, positive non-intervention heritage of the regime still left an alarming socio-economic disequilibrium on the eve of British departure. Behind the impressive GDP figures is a widening income gap between the super-rich and the grassroots, with 650,000 people reportedly living below the poverty line.[29] Despite the increase in the budget's allocation for welfare, the British regime has been unable to resolve the survival crisis of the poor and needy effectively.

In other words, free trade and free enterprise with an open market for much of colonial Hong Kong history did not always mean fair trade and equal opportunity: the regime intervened to favour British and big business interests at the expense both of fair play and of a level playing field for all economic players regardless of class or race.[30] The crux of the issue is not whether limited government did promote the economic freedom responsible for Hong Kong's recent prosperity. Rather, the question is when, how, with what motives and in whose interest the colonial state decided to intervene in the social and economic spheres. If the state had to intervene in its roles of public protector, promoter and provider, then it should have been the local Hong Kong majority interest that deserved consideration. Yet, in its very nature as a colonial regime with a British-dominated bureaucracy, it often confused and even blatantly substituted British for local Hong Kong interests. That such British economic giants as the Hong Kong Bank and Jardine had to relocate their corporate domicile overseas before 1997 emphasized the fact that their days of special privileges and unfair advantages would be gone forever.

In unmasking the hypocrisy of *laissez-faire* colonialism with all its past record of unfair and discriminatory socio-economic practices, one may also be apprehensive about the likelihood of mainland Chinese corporations and elites, not unlike the British hongs in the past, striving to claim special privileges and even monopolizing some of the HKSAR's economic sectors. This could be facilitated by the British-groomed local elites who are shifting their loyalty to Beijing, thus providing a powerful network of co-optation and political patronage as in the classical British colonial example.[31]

27. *Hong Kong Economic Journal,* 26 September 1996.
28. The 1966 figure is from MacLeod, "Hong Kong: a hard-earned success," p. 7; the 1996 figure is from Patten, *The 1996 Policy Address, Hong Kong: Transition,* section 36.
29. *Sing Tao Daily,* 18, 25 October, 4 November 1996; *South China Morning Post,* 18 October, 4 November 1996.
30. Recently, the Consumer Council called for the introduction of competition laws to ensure a level playing field for business; see *South China Morning Post,* 29 November 1996.
31. The author wishes to thank Sonny S. H. Lo for his insights on this point.

A Hong Kong View of Hong Kong   577

Perhaps the myth of *laissez-faire* capitalism as projected by the colonial regime has appeared so real that it even prompted PRC officials to criticize the increase in Hong Kong public welfare spending as "runaway welfarism," with the Patten regime "driving recklessly toward a fatal crash."[32] This not only represents Beijing's interference in local policy-making and resource allocation, which should be in the purview of Hong Kong's autonomy; it also reflects a commonly held but erroneous perception of the Hong Kong economic miracle. In this sense, while Hong Kong's *laissez-faire* characterization might have been disproved in academic debates more than a decade ago, the lingering shadow of positive non-intervention still colours the *realpolitik* of sovereignty transition.[33]

Democracy

Perhaps the most regrettable shortcomings of the British colonial presence in Hong Kong have been its inadequate efforts at democratization. The very fact that British-sponsored democratization only emerged after the 1984 Joint Declaration seemed to justify Beijing's criticism, and called into question the motives and potentially troubling consequences of the British eleventh-hour change of heart to democratize Hong Kong right on the eve of their departure. Of course, it is for the British to explain and justify why there was no commitment to democratization during the first 140 years of their administration of Hong Kong. Yet it is still useful briefly to review the many missed opportunities, deliberate non-actions and even anti-democratic manoeuvres of the colonial regime. This should serve as a needed balance to any self-congratulatory portrayal of democratic advancement under British rule. While one might find Mr Patten to be the most energetic among all British Governors in his promotion of democratic reform, the list of democratic reform-minded Governors is indeed rather short – Mark Young (1941, 1946–47), and Edward Youde (1982–86).[34]

32. *South China Morning Post* and *Sing Tao Daily*, 29, 30 November 1995. Also see *Far Eastern Economic Review*, 9 November 1995, p. 36 on Hong Kong's inadequate welfare provisions.
33. See A. J. Youngson, *Hong Kong Economic Growth and Policy* (Hong Kong: Oxford University Press, 1982), ch. 4. A recent input on this issue is the policy platform by Peter K. C. Woo, one of the three nominated candidates for HKSAR chief executive, which calls for a substantial modification of the "positive non-intervention" approach. *Hong Kong Economic Journal*, 13 November 1996. Yet Financial Secretary Donald Tsang resorted to the Basic Law's stipulation on preserving the capitalist system to refute the front running candidate C. H. Tung's call for government assistance to local industry. See *Sing Tao Daily*, 4 December 1996. A scholarly study on industrialization in the context of the colonial regime's *laissez-faire* is Stephen Chiu, *The Politics of Laissez-faire: Hong Kong's Strategy of Industrialization in Historical Perspective* (Hong Kong: Hong Kong Institute of Asia-Pacific Studies, The Chinese University of Hong Kong, 1994).
34. On the MacLehose and Youde eras, see James T. H. Tang and Frank Ching, "The MacLehose-Youde years: balancing the 'three-legged stool,' 1971–86," in Chan, *Precarious Balance*.

578   The China Quarterly

Ironically, it was pressure from both the local grassroots and the revolutionary Kuomintang-Chinese Communist Party United Front in Guangzhou, as manifested in the devastating 1925–26 Guangzhou-Hong Kong General Strike-Boycott, that caused the first Chinese member to be appointed to the Exco in 1926. It was a response to the strikers' demand for the right of local Chinese directly to elect their representatives to the Legco.[35] But the British patronage system to recruit local elites not only perpetuated the elite–officialdom collusion that characterized much of the undemocratic elite politics of colonial Hong Kong, but also sowed the seeds of future miscarriages of democracy.

The post-War British plans for limited democratic reform were shelved by an unsympathetic Governor Grantham and the colonial tycoon elites who, as hand-picked British appointees, were definitely not legitimate representatives of local majority interests. Using the perceived threat of the 1949 Chinese Communist victory and the 1950–53 Korean War as pretexts, the colonialist–tycoon elite axis, with the blessing of Whitehall under the Conservatives, derailed this first attempt at democratization. It could be said that those members of the local elite appointed to the Exco and Legco were more than eager to protect their own privileges and exalted positions against any widening of political participation.[36]

It was not any lack of far-sighted ideas and practicable proposals that underlaid the British administration's deliberate denial, deferment and derailment of democratization until the 1980s, using as pretext circumstances outside Hong Kong, especially the China Factor. Even in the 1980s, there was only lukewarm, half-hearted commitment to democratic reforms and electoral arrangements with a snail's pace approach through highly manipulative and illegitimate means; one notorious scandal was the Wilson administration's deliberate disregard of public opinion (which was mostly in favour of introducing directly elected Legco elements in 1988) in order to appease Beijing.[37] After the historic first direct election of 18 members (out of a total of 60 seats) to the Legco in September 1991, Governor Wilson appointed another 18 Legco members, 16 of them with clear anti-democratic inclination (an exact equivalent to the 16 directly elected democratic camp members).[38] Subsequently many of these 16 Wilson appointees, together with a few indirectly elected Legco members from the business functional constituencies, formed the Co-operative Resources Centre, which in 1992 became the conservative, pro-business Liberal Party. It was this Liberal Party bloc that almost

35. Ming K. Chan, "Hong Kong in Sino-British conflict: mass mobilization and the crisis of legitimacy, 1912–26," in Chan, *Precarious Balance*, pp. 48–51.

36. On the failure of the Young plan for democratic reform under Grantham, see Steve Y. S. Tsang, *Democracy Shelved: Great Britain, China, and Attempts at Constitutional Reform in Hong Kong, 1945–1952* (Hong Kong: Oxford University Press, 1988).

37. Ming K. Chan, "Democracy derailed: realpolitik in the making of the Hong Kong Basic Law, 1985–90," in Ming K. Chan and David J. Clark (eds.), *The Hong Kong Basic Law: Blueprint for "Stability and Prosperity" under Chinese Sovereignty?* (Armonk, NY: M. E. Sharpe, 1991), pp. 9–13.

38. Frank Ching, "Toward colonial sunset: the Wilson Regime, 1987–92," in Chan, *Precarious Balance*.

stopped Governor Patten's electoral reform which was passed by a single vote on 30 June 1994.

Many of these Liberal Party elites had by then already defected to the pro-Beijing anti-democratic coalition which strongly opposed the Patten reform. In the words of PRC National People's Congress member, Dorothy Y. C. Liu, these "ex-colonial old batteries" now under the new sovereign's politics of elite patronage, enthusiastically play their part in Beijing's transition power organs.[39] They fully endorse Beijing's decision to disband the elected 1995 Legco and to install an appointed "provisional legislature" which has no legal or constitutional justification under the Joint Declaration and the Basic Law. In this ironic twist of role transformation from colonial "fire brigade" to PRC hatchetmen, they at least show consistency and continuity in their anti-democratic stance, bowing to the whim of an undemocratic political master. If Governor Patten were indeed sincere, in his democratization push, he should deeply regret the failure of his predecessors in not laying the foundation for Hong Kong democracy with an earlier, fuller-scale and more whole-hearted promotion of democratic political culture, institutions and processes.[40]

If Governor Patten and Whitehall criticize the undemocratic and unrepresentative (hence illegitimate) composition of the HKSAR Preparatory and Selection Committees for being overwhelmingly filled with tycoons and their professional-circle allies, the British should not be too harsh in their condemnation of Beijing's preference for plutocracy as the future HKSAR political leadership for many of their targets of criticism were groomed and nurtured by British colonial elite politics.[41] In this sense, British colonialism is a thoroughly implicated accomplice to Beijing's retardation of democracy in Hong Kong.

Conclusion

In retrospect, it may be fair to say that looking at the full spectrum of the one-and-a-half centuries of British administration of Hong Kong, it is definitely not the best, although far from the worst, record of colonial rule in the world. Even if the intended efforts and actual attainments of the British regime in its sunset era of the last decade were to be accepted as generally positive (as portrayed in Governor Patten's last policy address), that the earlier years were filled with discrimination, unaccountable or even unlawful actions, biases and favouritism cannot be denied. Nor can

39. *South China Morning Post,* 31 March 1997.

40. This author argued, at the end of Wilson's governorship, that his true record could only be measured by the difficulties faced by his successor; see *Hong Kong Economic Journal,* 1 July 1992. When Wilson deferred Legco direct election from 1988 to 1991 to meet Beijing's requirement for "convergence" with the yet to be finalized Basic Law, he could at least eliminate all appointed seats in the Urban/Regional Councils and the 19 District Boards. Wilson's failure doomed Patten's 1994/95 reform of these two tiers as Beijing has promised to reintroduce appointed seats to these councils and boards in the HKSAR.

41. A recent critique of elite politics is sociologist (and HKSAR Preparatory Committee member) Siu-kai Lau's lecture, "Political order and democratization in Hong Kong – the separation of elite and mass politics," the Hong Kong Lecture, University of Hong Kong, 1996.

the damage inflicted on Hong Kong's people because of inaction or the lack of adequate and timely measures by the British authorities be redeemed by later remedial efforts. Of course, such a mixed record of British rule in Hong Kong is also burdened by an acute lack of popular mandate and of genuine legitimacy, inevitable in a colonial regime imposed from London. Indeed, colonialism and imperialism are in themselves, as manifested by the British and other powers' behaviour in Asia and elsewhere in the world, immoral and unjustifiable by today's standards of human rights and democracy.

The British regime in Hong Kong in its final days was still far from truly democratic, fully accountable, uncorrupted, efficient and effective in meeting the basic needs of the populace. The British-style rule of law, always regarded as the single most significant British legacy in Hong Kong, was weakened by the sullied records of its legal system personnel and by a history of legal impropriety and dubious legislation. The dark side of colonialism is also manifested glaringly in what many people still regard as the British betrayal of the long-term interests of the Hong Kong people in the 1982–84 sovereignty retrocession negotiations and in the subsequent British appeasement of PRC infringements of, and non-compliance with, the Joint Declaration, especially on democratization and the formation of the Court of Final Appeal.[42]

After affixing her signature to the 1984 Sino-British Joint Declaration, Prime Minister Margaret Thatcher had repeatedly emphasized the British Government's "moral responsibility" towards Hong Kong's people. However, in her "Foreword" to a recent volume on Hong Kong, Lady Thatcher only asserts that "a free economy, the rule of law, and the excellent administration ... have been Britain's most important contributions to the Territory."[43] It is revealing that she does not include democracy among the British contributions to Hong Kong.

Another still unclosed chapter of unilateral British imposition of a heavy burden under the Thatcher regime is the Vietnamese boat people issue. Since the late 1970s, without the consent of the Hong Kong people, London unilaterally adopted the policy of "first port of asylum" that invited tens of thousands of Vietnamese refugees to the "free port" of Hong Kong. Besides creating problems of law and order and further pressing the very limited space of the territory for camp sites, the British did not even pay the massive expenditures for the boat people's long stay. The United Nations Refugee Commission still owes Hong Kong more than $100 million for this purpose.[44] It is doubtful whether this problem

42. See Mark Roberti, *The Fall of Hong Kong: Britain's Betrayal and China's Triumph* (New York: John Wiley, 1994).

43. Margaret Thatcher, "Foreword," in Sally Blyth and Ian Wotherspoon, *Hong Kong Remembers* (Hong Kong: Oxford University Press, 1996).

44. In his October 1996 policy address, section 17, Governor Patten gave a figure of 12,000 as the total number of Vietnamese boat people still remaining in Hong Kong. The highest figure of 68,748 was recorded in 1979, see Donald H. McMillen and Man Si-wai (eds.), *The Other Hong Kong Report 1994* (Hong Kong: Chinese University Press, 1994), p. 180. The United Nations' non-payment of over $100 million was raised in the HKSAR chief executive candidates' question-and-answer sections; see *Ta Kung Pao*, 30 November 1996.

will be resolved before 1 July 1997. Is this part of the British legacy to be inherited by the HKSAR?

To appraise the legacy of British colonial rule from a proper historical perspective, perhaps one should attempt to divide the history of British administration in Hong Kong into more distinct phases, such as before the Second World War, the early post-War period, initial reforms during the 1960s and 1970s, and post-1984 reforms under the shadow of 1997. Each of these phases, though all underpinned by a fundamental will to impose and sustain colonial rule on the part of the British regime, have been shaped by more time-specific and circumstance-specific factors, especially the inevitable and increasingly powerful China Factor as well as Hong Kong's own growth and complexity. What is to be left behind in Hong Kong as part of the so-called British legacy is therefore often more accidental (in terms of historical circumstances) than institutional or ideological. For example, one cannot say that the British rule of law has throughout the past 150 years been fair to all, Europeans and Chinese. Similarly, the civil service was not really corruption-clean or efficiency- and expertise-driven until perhaps the 1980s. The repeated abortive attempts at experimentation with some degree of local democracy are illustrative of a lack of sincerity on the part of British colonial establish-ment, which pursued mostly an instrumental approach and was ultimately unwilling to give up real power and control. In other words, it would be necessary to place those "institutions of British legacy" both in their historical contexts and in the context of institutional evolution shaped by domestic demands and circumstantial contingencies.[45]

Indeed, Hong Kong has made a name in the world as a successful model of development with few natural resources other than its creative and productive people. This success story of the Hong Kong model bears a significant testimony to the contributions of the British administration, which has been on the whole adaptable in meeting many difficult challenges under fast-changing circumstances. The most recent government can take some credit for the freedom, prosperity and stability enjoyed by this cosmopolitan community of 6.3 million. But it is doubtful that the Hong Kong government of a generation ago (at the time of the 1966 Star Ferry riot and on the eve of the 1967 leftist urban terrorism)[46] even came close to meeting the standards expected of a modern and fair government, even when judged by world standards prevalent at the time. As for the full one-and-a-half-centuries' record, British colonialism in Hong Kong could hardly claim to have been genuinely enlightened and legitimate. While not negating the selective positive aspects of the British presence in Hong Kong, one must be careful not to make virtues out of necessities even if the very crucial moral question of colonialism were to be set aside. In the same light, one must not give too much credit for what supposedly is a fairly effective

---

45. The author is indebted to Anthony B. L. Cheung, City University of Hong Kong, for his insights on these crucial points.

46. On the mid-1960s riots, see John D. Young, "The building years: maintaining a China–Hong Kong–Britain equilibrium, 1950–71," in Chan, *Precarious Balance*.

and even conscientious remedial effort by a would-be rescuer who himself was responsible for the initial problem.

If the post-1997 developments in the HKSAR turn out to be far from ideal, the burden of blame should not rest entirely on misguided policies and actions by Beijing. Rather, the inadequate foundation, unhealthy political culture, flawed legal-administrative framework and questionable bureaucratic practices inherited from the British – together with the inability of the Hong Kong people to stand firmly to defend their much cherished freedom, democracy and high degree of autonomy because of their colonial deprivation – ought to be blamed as well. As such, the true legacy of the British administration in Hong Kong must await the unfolding of HKSAR developments to demonstrate how successful the British-endowed institutions, personnel, procedures and processes as well as the values, skills and mind-set of a free economy and a liberal, pluralist society, have prepared and empowered the Hong Kong people to meet the challenge ahead.

Ming K. Chan is Research Fellow, and Executive Coordinator of the Hong Kong Documentary Archives at The Hoover Institution, Stanford University. He was a member of the History Department, University of Hong Kong during 1980-97. He is General Editor of the multi-volume series 'Hong Kong Becoming China', being published by M.E. Sharpe and HK University Press, nine titles of which have appeared since 1991.

Part II
Culture and Identity

[5]

Eating Metropolitaneity: Hong Kong Identity in *yumcha*

Siumi Maria Tam
Anthropology
The Chinese University of Hong Kong

In Hong Kong, *yumcha* as a pervasive form of eating in the public has become an institution which epitomises the diversity and inclusiveness that Hong Kong people think are quintessential to the civility and economic achievement of the metropolis. It stands out particularly as a practice that has come to represent both locally and overseas a Hong Kong culture which has previously been characterised as residual of 'Chinese tradition' and 'Western customs', and hence is virtually non-existent. This paper looks at the consumption of metropolitaneity and the culturalisation of the *heunggongyan* or Hong Kong personal identity in the form, content and social relations involved in *yumcha*. It points to the significance of *yumcha* as a key to understanding a collective identity which has become more and more apparent as Hong Kong people vehemently try to establish a city with its own culture, first against the colonial grip and then the return of sovereignty to China in 1997.

Introduction

This paper[1] examines the consumption of metropolitaneity and culturalisation of the *heunggongyan* or Hong Kong personal identity through the practice of *yumcha*.[2] It

---

1. Parts of this paper were presented at the International Food Conference on Changing Diet and Foodways in Chinese Culture, Chinese University of Hong Kong, June 1996, and the Anthropology Colloquium, Macquarie University, April 1997. I am grateful for the comments received at these occasions and have incorporated some of them in this current version.

proposes that *yumcha* manifests the value placed upon diversity and inclusiveness, and change and adaptability which are considered by the people of Hong Kong to be integral components of the social ethos and represent their society and its civility as a metropolis. Data are drawn mainly from fieldwork during 1995 and 1996, including participant observation in various Chinese teahouses, restaurants and families, questionnaire interviews done by student assistants on family food consumption patterns during Christmas 1995 and Lunar New Year 1996, as well as interviews with restaurateurs and food columnists.

In the physical consumption of the wide variety of food in *yumcha*, which they do every day, Hong Kong people devour a metropolitaneity that they experience as an almost natural part of their life. *Dimsum*, the small appetisers eaten at *yumcha*, reproduces physically and most vividly the variety of choice and desire for change. It thus carries special meaning and constitutes a powerful symbol of the spirit of the Hong Kong person. Syncretism or inclusion of things foreign are not only a matter of lifestyle in Hong Kong; they exist as part of the self ascription of *heunggongyan*.[3] But *yumcha* is more than an analogy of this identity. It serves as the medium in which the identity is lived out in praxis and reinforced in everyday life.

The subjectivisation of *yumcha* as such contributes to a collective identity in a culture-searching process. In a colonial milieu it on the one hand displays the pride Hong Kong people have in their achievement in the building up of an international city by hard-working individuals who thrive with adaptability yet with roots firmly grounded in 'five thousand years of Chinese civilisation'. But on the other hand the fact that the identification of a modern, metropolitan community has to resort to a remote dynastic Chinese custom itself exhibits a self-consciousness of political impotence as a colony. In a way, the choice of a so-called traditional foodway in a Westernised colony has perfectly suited the 'powerful discourse of "East meets West", and the "modern world meets Chinese tradition"' (Evans and Tam 1997:5) as it is politically unthreatening yet culturally

---

All transliterations in the paper are based on Cantonese (C) unless when pinyin (P) is used. In such cases these words are indicated by (P:). Place names follow the conventional usage, for example, Hong Kong is used rather than Heunggong (C) or Xianggang (P). All names are pseudonyms to ensure informants' privacy.

2. It is in order here to make a distinction between two practices, both of which may be referred to as *yumcha*. *Yumcha*, which is the focus of this paper, is the practice to eat various hot and cold foods, in a restaurant, when Chinese tea is served. The term can also be translated literally as 'drink tea'. Although it has the same phonemes as *yumcha*, 'drink tea' is another custom altogether and is better known as *changai* (pronounced as *cha ngai*) locally. Meaning 'the art of tea', it involves the knowledge of tea categorisation, appreciation of tea according to its look, smell and taste, as well as curating the various kinds of utensils used in containing, preparing and drinking tea. More importantly, eating is not involved in *changai* which I believe is closer to wine tasting in European based culinary cultures. This paper will focus only on the practice known as *yumcha* by the local Hong Kong people, in which eating is a central and necessary component.

3. G. Mathews (1996) has pointed out, for example, that though his informants may use their English names for different purposes in different circumstances such as in identity shifting, they indeed considered their English names part of their 'self' just as their Chinese names were.

assuring to both indigenous Anglophiles and expatriate Sinophiles. The diversity and inclusiveness embodied in *yumcha* would only point to the success of the city. Yet underpinning this sense of pride must be a sense of limitation or even helplessness as a colonised community brought about by the political reality. As a powerful discourse of cross-fertilisation of customs, *yumcha* may well be a discourse of power in praxis.

Interestingly, *yumcha* has been constructed as a representative of Hong Kong culture both locally and overseas in a historical period in which Hong Kong people feel an urgency to reconstruct a politico-economic existence and a cultural identity. In tourist brochures, *yumcha* is always central to the construction of a 'gourmet's paradise', as competitive prices in Hong Kong's Southeast Asian neighbours have snatched away its 'shopper's paradise' title. Hence spring rolls beside a pair of chopsticks and bamboo baskets of shrimp dumplings are constant images in booklets distributed by the Hong Kong Tourist Association, a semi-government organisation. But *yumcha* features just as prominently in local Chinese newspapers, where elaborated reports are written on 'new trends' of *yumcha* and rejuvenation of 'traditional' *dimsum*. Perhaps more obviously using the *yumcha* analogy as a social idiom is the comedian Chow Sing Chi who epitomises a new generation of post-modern popular culture with his *moleitau* (illogical) style. His now famous quotation '*chodai yum bui cha, sik gor bau*' ('Let's sit down, drink a cup of tea, eat a bun')[4] which literally means to talk over things slowly, highlights *yumcha*'s pervasiveness and symbolic power in everyday life. Undoubtedly *yumcha* is *the* way of eating that occupies both the stomach and imagination of the Hong Kong people.

This sense of cultural need for identification was felt all the more keenly as 1997 approached, when the British colonial administration would give way to Chinese rule. The prospective relations with China in the post-colonial period called for the verification of a distinct Hong Kong culture in order to ensure continued survival. The emergence of *gongsik yumcha*, or Hong Kong style *yumcha* as opposed to its prototype *guangsik yumcha*, Guangzhou style *yumcha*, clearly accented the effort in drawing a boundary between mainland Chinese and Hong Kong Chinese. This cultural cleavage was ironically perpetuated by the globalisation of *gongsik yumcha* via the Hong Kong diaspora which was in effect a vote of no confidence for the impending Chinese rule.

Eating as a channel to Hong Kong culture

With and without the actors' conscious design, the ways in which eating is signified, practised and legitimated manifest the social ethos of the society in which they are found. Numerous scholars have discussed the importance of food as a window to understanding culture. Douglas for example discusses meals as systems of codes to be deciphered (Douglas 1972); Chang pioneers the study of Chinese culture via its foodways (Chang 1977); and Appadurai finds cookbooks indicative of Indian national identities (Appadurai

---

4.   This is an expression allegedly started by a comedian duo called *yuen ngang tinsi*, 'the soft and hard celestial masters', whose illogical style brought them immediate fame and popularity among the youth. The expression was further popularised by the comedian Chow Sing Chi in his Cantonese films, for a time the favourite among movie goers.

1988). In a recent book Mintz reminds us that 'food and eating afford us a remarkable arena in which to watch how the human species invests a basic activity with social meaning—indeed, with so much meaning that the activity itself can almost be lost sight of' (Mintz 1996).

In Hong Kong, indeed, as in other cultures, food is heavily laden with social and cultural meanings. With a total of 7,434 eating places registered with the Regional and Urban Councils in 1996 (Regional Services Department October 1996, and Urban Services Department June 1996), the landscape of the industry is constructed and perceived to be international, smart, up-to-date, but without losing local flavour and Chinese tradition. A good example of this kind of thought about eating, is a place called Lan Kwai Fong in Hong Kong's central business district. Within the two blocks congested with office buildings, high brow boutiques, florists, and a shrine for the neighbourhood deity, there are restaurants of all sorts—an American steak house, a Japanese sushi bar, a French cafe, an Indian tandoori place, and a Vietnamese rice noodles stall, comfortably mixing with a Chinese herbal tea shop and numerous pubs, discotheques and expensive European restaurants. The ecology of this cultural microcosm deserves more discussion, but here we shall concentrate on the phenomenon of *yumcha*. For even within Central District where the choice and availability of cuisines are taken for granted, the favourite for the lunch pilgrims is still the Cantonese restaurants where *yumcha* is served. Mok estimates that in 1992 Chinese eateries took up 68% of the total consumption on food, and families on the average spent 56% of their money on food eating out, mainly for lunch (Mok 1992). Although it is impossible to tease out the spending on *yumcha* from the statistics, interviews with both restaurateurs and local families point to the fact that *yumcha* is the most popular choice when local Hong Kong people dine out.

The pervasiveness of *yumcha* in Hong Kong's everyday life is obvious even to the casual observer. Among the thousands of Chinese restaurants, there are regional, temporal and hierarchical variations, but the practice of *yumcha* stands out as the Hong Kong way of eating. Not only has it maintained a hard-core Cantonese identity, but it has also had its influence diffused to non-Cantonese Chinese restaurants, and to food businesses categorised by the government as 'Western' and 'other' (Urban Services Department 1996), which includes such varied enterprises as karaoke bars, pubs, open-air food stalls, fast-food restaurants and convenience stores.

From Guangzhou to Hong Kong

Most local historians agree that old *yumcha* practices in Hong Kong generally followed those of Guangzhou (Canton), historically the political and commercial centre of the Pearl River Delta of which Hong Kong is part. Today, terms such as *yuet choi* (Cantonese cuisine) and *yeungsing mei dim* (delicate *dimsum* of the City of Rams) are still used in everyday speech, reminding us of Guangzhou city's centrality in the food culture of South China. Whether in practices of commerce or in consumption habits, Guangzhou's merchant culture had heavily influenced that of Hong Kong.

EATING METROPOLITANEITY: HONG KONG IDENTITY IN *YUMCHA* 295

According to Kan (1985), in old Guangzhou those who had the leisure to *yumcha* were typically merchants. Guangzhou people used to have meals twice daily around 11 a.m. and 6 p.m., and bosses normally ate with their employees on the shop floor. When business was less busy, the bosses would go to teahouses to drink tea at a leisurely pace. There were different types of teahouses, such as *chasat, chagui,* and *chalau.* These teahouses were perceived as distinct categories according to the food served, the atmosphere, the pricing, and the social status of their patrons, but invariably merchants and traders enjoyed themselves in these teahouses by 'savouring tea at a leisurely pace' or, in Cantonese, *tan cha.* These occasions would never substitute for main meals, and, equally true, they were often not simply occasions of leisure. In a commercial centre such as Guangzhou, these were occasions of information exchange and business negotiations (see for example, Kan 1985, Chen et al. 1990). The teahouses may be better considered a second office and *yumcha* as regular albeit informal business meetings.

The two most important concepts in relation to *yumcha* in old Guangzhou are *yat chung leung kin* and *dimsum. Dimsum* (literally 'touch heart') is a generic term for a variety of small food items which are usually savoury or sweet, and are consumed together with Chinese tea. The phrase *yat chung leung kin* literally means one cup [of tea] and two pieces [of *dimsum*]. In the teahouses, while the customers drank tea, they were served two pieces of delicately made *dimsum.* The purpose of eating *dimsum* was not to fill the stomach. The eating complemented the drinking. It allowed a balance of taste and substance, in accordance with the conventional philosophy of the human body system based on a Taoist worldview. To provide variety in tea drinking, *dimsum* were made in bite-sizes, that were appealing to the senses, and were judged accordingly in three aspects: look, smell, and taste (*sik, heung, mei*). As *dimsum* should complement rather than take precedence over tea, the types of *dimsum* in old teahouses were limited to a few. The most common ones were *hargau* (shrimp dumpling), *siumai* (pork dumpling), and *funguo* (meat and vegetable dumpling). They were commodities of leisure, and sometimes of luxury. It was when competition became fierce that teahouses began to expand the kinds of *dimsum* served, with the more expensive teahouses changing their *dimsum* menu weekly to offer the best seasonal items (Dak 1989:45). Increasingly, the quality and variety of *dimsum* became an important concern, and *dimsum* took a more central role in *yumcha.*

This description is a gross generalisation of the *yumcha* culture in Gunagzhou and is more about up-scale teahouses. Certainly socio-economic factors, significantly class and gender, were important determinants of who went where and how. Among the earlier teahouses, those that catered for the working class were called *yileigun* (roughly translated as 'two *lei* pub') where tea was sold at two *lei* a cup. One *lei* was a bit less than one-third of a cent in today's Hong Kong currency. *Dimsum* sold in the *yileigun* were considerably bigger in size, and rice and noodles were also available. These lower-range teahouses also served lunch. But *dimsum* were not, and to a large extent still are not, considered a proper meal.

The following incident illustrates this point. In 1996, three students and I went to Sheung Hei Lau (Double Happiness Teahouse) for a field excursion. This was the only remaining old-style teahouse (*chalau*) on Hong Kong Island. It has catered for the working class in the older district of Wanchai since 1949, and, according to a food columnist, its business style has remained the same in more than four decades (Tou 1995). When we were there during lunch time, we found that the mostly male staff, as usual, were carrying trays of food around the teahouse yelling out the name of the dish as they walked. There was not much variety. In fact there was little available other than rice with different types of meat on it. We asked one of the staff whether there would be more *dimsum* coming. To this he answered, 'It is lunch time (*ngsi*) now. We don't sell *dimsum*.'

Hence *yumcha* as a form of food consumption can best be understood in contrast to a meal. A meal is considered to be proper, official, elaborate; it involves a fixed schedule with dishes served according to an established program, and it always includes a staple such as rice. *Yumcha* is considered to be casual and unofficial, and staples are not necessarily involved. Whereas meals are served by someone such as a waiter, *yumcha* involves more 'self-help' behaviour. In the past, *dimsum* were placed on trays and carried by older men using a strap which was attached to the tray and went around the back over the shoulder. As the carriers walked around the teahouse they announced the names of the items by calling out loudly. If a customer wanted the *dimsum* he (usually a he) would call the carrier to come to the table. Today the trays have been replaced by carts pushed around the restaurant, often with the names of the *dimsum* written in front of the cart. The customer reads the words and waves his/her hand as a signal. The 'self-help' element is even more obvious now as many restaurants offer *dimsum* 'buffet'. This involves placing a variety of prepared or semi-prepared *dimsum* and other foods on a long table, buffet style. Customers come to the long table, choose the food (to be cooked by an employee if necessary) and bring it back to their own table. Despite these differences, there are commonalities between *yumcha* and a meal, and the most obvious is their use in creating social congeniality, and reinforcing social relations and positions.

Many of the above practices can still be observed in both Hong Kong and Guangzhou today. Yet since the fifties the centre of food culture in South China has shifted from Guangzhou to Hong Kong. As the commercial developments in Guangzhou slowed down after 1949 while those of Hong Kong greatly expanded, Hong Kong assumed for itself the title 'heaven of delicacies' (*meisik tintong*) and took over Guangzhou's central place by saying it is now *sik choi heunggong* (to eat, go to Hong Kong) whereas the old saying was *sik choi guangchow* (to eat, go to Guangzhou). While in the past Guangzhou set the standards for *yumcha,* now Hong Kong does. At the same time in everyday speech the term *guangsik* (Guangzhou style) *yumcha* has been substituted by *gongsik* (Hong Kong style) *yumcha*. In a Taiwan newspaper article entitled 'Irresistible temptation: gourmet food from Hong Kong—*yumcha*', the author describes how restaurants in Taiwan in the past few years have closely imitated Hong Kong style *yumcha*. Targets of imitation include such areas as the use of a small dish for discarding bones, and the expansion of the menu to include real delicacies. She goes on to advise Taiwanese to avoid eating too extravagantly during *yumcha* (such as ordering whole roasted piglets and whole sharks fin), because such behaviour will be considered country-bumpkin by the Cantonese (Hu n.d.). In Sydney's Chinese magazines, too, it is a common sight to see restaurant advertisements that use *gongsik yumcha* to identify themselves as serving authentic Cantonese cuisine. Clearly *gongsik* is the new standard for the food consumption behaviour called *yumcha*.

Hong Kong style *yumcha*

Restaurateurs in Hong Kong are fully aware of these developments overseas and do not hesitate to assume a global leadership role in the Chinese food industry. In the area of *yumcha*, restaurateurs strive to maintain their status by ever-expanding their *dimsum* menu and grafting Hong Kong practices onto their overseas branches. In an interview, Mr. Wong of Heichinrou Restaurant, for example, tells us that their new restaurant in Ginza, Tokyo, will be modelled after their restaurant in Hong Kong, complete 'with *dimsum* carts, . . . there is no precedent in Japan'.

In a restaurant of medium to large size, the average number of kinds of *dimsum* served daily is 150-200. Savoury and sweet, hot and cold, steamed and deep fried, meat, seafood and vegetarian—the sheer variety of *dimsum* indicates the significance of this kind of food and its popularity. There is a consensus that *dimsum* should be eaten in restaurants because only then will it be 'full of the energy of the wok' (*gauwokhei*) which is an essential element in Cantonese cuisine. However, this does not lead to the restaurants' monopoly over the sale of these tidbits. Interestingly, as the mainstay of *yumcha, dimsum* has managed to conquer not only Chinese teahouses and restaurants proper, but also karaoke bars, *taipaitong* (open-air food stalls), and even convenience stores. For instance, Seven-Eleven, the biggest 24-hour convenience store chain in Hong Kong, has as part of its regular shop installation a refrigerated shelf of *dimsum* next to a microwave oven. It is common to see young people nibbling away at such delicacies as *hargau* and chicken feet with plastic fork in one hand, and a giant paper cup of Coca-Cola in the other. *Dimsum* is also served in many cafeteria-style eating places that aim to provide meals for staff and workers. In the Chinese University of Hong Kong where I teach, nine out of the sixteen staff and student cafeterias on campus serve *dimsum* for breakfast, lunch, and/or afternoon tea. Four others have done so in the past, while in another two that serve 'Western-style' food occasionally *dimsum* is found on the buffet table.

There are more examples. Three of the biggest fast-food chains in Hong Kong, Maxim's, Fairwood and Cafe de Coral, all serve *dimsum* in the morning besides the regular Western menu. The choices of *dimsum*, however, are much more limited than those found in teahouses and put a bigger emphasis on staples such as rice and wheat. A local newspaper featured a report comparing the 'Big Three's' breakfast menu, entitled 'Eat and run—Chinese breakfast report' *(Apple Daily,* 22 November 1995). It is obvious that as a meal the breakfast *dimsum* served in all three restaurants have a big proportion of staples. All three restaurants invariably use *dimsum* made of rice: *juk* (congee), *normaigai* (glutinous rice with chicken wrapped in lotus leaves), *lorbakgou* (turnip cake), *cheungfun* (meat rolled in rice sheets) and *maifun* (vermicelli). The two selling points of these *dimsum*, according to this report, are speed ('eat and run') and stomach-filling capability. In great contrast to the original leisurely pace and 'touch on the heart' traditions of *yumcha, dimsum* on its own has become part of everyday meals, and instant meals in particular. In line with this development, it is interesting to speculate how long before fast-food restaurants like McDonald's and Hardee's also find it necessary to follow suit, after teriyaki burger (called 'The Shogun') was served at the former, and sauteed mushroom rice in a styrofoam bowl (interestingly called 'Western country style' rice) at the latter.

Dimsum and *yumcha*, however closely related, are understood to be separate categories. For instance, during a discussion session, Winnie and Jenny who are both college students, were adamant that *dimsum* is not equal to *yumcha*. *Yumcha* has to take place in a commercial eating place, outside of the home. It is not *yumcha* if one purchases frozen *siumai* from a supermarket and consume it at home. Nor will it be considered *yumcha* if one goes to Seven Eleven, pays for a packet of *hargau* at the cashier, microwaves it hot and eats it at the condiments bench. According to these young people, such forms of behaviour are considered 'just eating'. By the same token, Chinese restaurants' take-away service will not allow their customers to *yumcha* at home. *Yumcha*, then, requires a certain content, a certain form, a certain atmosphere, and certain relations—not only among the customers themselves, but also between the consumers and service providers. In short, one needs to eat in a certain context, arousing certain social meanings, for the action to be qualified *gongsik yumcha*, or Hong Kong style *yumcha*.

Gongsik yumcha: globalised identity

Teahouses in Hong Kong have been in business for one and a half centuries. Although *yumcha* is made to conjure up a 'traditional Chinese' image especially by institutions like the Hong Kong Tourist Association, a semi-government organisation, the practice of *yumcha* itself has been redefined and reinvented over time. It has never been short of 'new life' given by innovative chefs, the result of which is 'classiness' and 'delicatisation'. These continuities and changes can be summed up as a process of globalisation. The first aspect of globalisation is seen in the constant change of form and content in *dimsum* and the consumeristic fanfare that surrounds it. The second aspect is the spread of *yumcha* across the globe as part of a neo-Cantonese culture that accompanies the Hongkong diaspora.

Over and over again the words 'diversity', 'syncretism' and 'variety' are used to characterise Hong Kong society and its achievements. *Heunggongyan* are not shy in showing their pride in the inclusiveness and hybridity of their lifestyle. And it is most obvious in *yumcha*. On the restaurant shop floor, imported items such as 'Japanese sashimi', 'Thai-style chicken feet', and 'Western fresh cream cakes' are placed side by side on *dimsum* carts. Whether these are authentic nobody really questions. One also finds on the same *dimsum* cart recent creations like 'mango-sago mix' as well as unmistakably old Canton traditionals such as *charleung* (a savoury long doughnut wrapped in steamed rice sheet).

A *dimsum* order sheet from a local restaurant will illustrate this inclusiveness. There are a total of 61 items on this order form, a conspicuous mixture of food originating from many different places. There are, for example, non-Cantonese Chinese items such as 'Szechuan clams', 'Xiamen vermicelli rolls', 'Beijing pan-fired red-bean cake', and 'fried Shanghai dumpling'. From the Southeast Asian region, there are 'Singapore black glutinous rice sweet soup with coconut milk', and 'Malaysian lotus-seed cake'. East Asia is represented by 'Japanese style salmon salad' and marinated baby octopus; the West by baked puddings and pastries. All these are ranked into categories of 'small', 'medium', 'large', 'extra large' and 'ultra large', not in terms of size but the price, and are carefully

EATING METROPOLITANEITY: HONG KONG IDENTITY IN *YUMCHA* 299

balanced so that every customer may find something that they like or can afford. These items may well be authentic, or may simply be adaptations, combinations or even inventions by the chefs. They are not grouped on this *dimsum* sheet according to their places of origin, but rather each is considered a full-fledged *dimsum* on its own right. What divides them is related to the production method such as steaming, frying and baking, whether they are a cold or hot dish, and whether they are a 'dry' food such as a dumpling, or 'wet' as in congee or soup. The lesson, if you will, is that these items, vastly different in style, taste and form, are all eclectically placed together to achieve a sense of completeness through diversity.

Globalisation in terms of incorporating non-Chinese elements is manifested not only in the content of food items but also in how they are served to the consumers. The more expensive restaurants promote their Western style internal decorations and atmosphere. The Heichinrou Restaurants, which originated in Japan, have a grand piano right in the middle of each of their shops, while gold-plated porcelain plates in the Zen Restaurants remind their patrons of European fine dining. The uniform worn by waiters are predominantly a black Western suit, white shirt and black bow tie. To be classy, or to have good taste (read Western, modern) in terms of one's lifestyle, is a distinguished index of the Hong Kong sense of metropolitaneity. To digress a bit, one only needs to look at the many real estate advertisements that overwhelmingly sell not the apartments (with names such as Royal Ascot, Miami Beach Towers, and California Gardens) but the lifestyle that they supposedly represent. To come back to *yumcha*, classiness is related to delicateness and price. *Hargau* (shrimp dumpling), one of the core *dimsum*, has recently been made in the shape of a cute little rabbit complete with tiny eyes made of carrots, sitting on a bed of shredded lettuce—and sold for three times the normal price. One also finds sharks fin, scallops or crab roe now decorating the top of pork dumplings, whereas it used to be a slab of pork liver.

Another change in *yumcha* is the emphasis placed on efficiency. If *yumcha* has been an occasion of leisure in the past, today it has become part of a globalised instant, fast-food culture. The fast-pace lifestyle that has grown with Hong Kong's economic development has undoubtedly helped to legitimise this 'traditional' form of eating as an instant food. But more so because Hong Kong people believe that to be efficient is one of the requirements to be modern, metropolitan and indeed to be *heunggongyan*. Hence unlike old-style teahouses, Chinese restaurants now serve d*imsum* together with numerous kinds of staples such as rice and noodles. During lunch time, these are prepared in small portions and pushed around in carts, allowing customers to quickly pick something to eat and go. So it is common to see in Chinese restaurants mothers feeding their children with *dimsum* before or after the children's half-day school. Groups of white collar workers pack the Chinese restaurants in business districts, each downs a few dishes of *dimsum* and rice and hurries back to the office.

Despite these changes, however, the core Cantonese authenticity of *yumcha* is never allowed to be lost. Patrons are reminded that after all *yumcha* is a very Cantonese affair as they are greeted at the entrance by the host dressed in ankle-length red *cheungsam* (P: *chipao*), or by the smell of pigs knuckles and ginger stewed in sweet red vinegar, which is normally cooked at home for new mothers but now sold as *dimsum* by the bowl, pushed around in thermos pots on a *dimsum* cart. Over the past decade the increasing use of cheap ingredients such as beef tripe and chicken feet in *dimsum* has been obvious. These relatively cheap dishes are delicatised and promoted in the guise of hometown styles (*garheung fungmei*). Thus 'classiness' in *yumcha* is Janus-faced. As in other rapidly developing economies, it coincides with a simultaneous identification with the modern and a nostalgic yearning for the traditional.

Another aspect of globalisation of *yumcha* is its spread outside of Hong Kong. With its economic success since the Second World War, the Hong Kong lifestyle has become a target of imitation particularly within the South China circle. The practice of *yumcha* as a distinctively Hong Kong practice has been adopted by Chinese in Taiwan and mainland China. The label *gongsik yumcha* commands a disproportionately high price and is indicative of high social status. At the same time, in the past 15 years more and more Hong Kong people have emigrated, motivated by anxieties surrounding the Sino-British talks of 1984 and the Tiananmen incident of 1989. With this new wave of diaspora comes the global spread of a *heunggongyan* identity. Many of these migrants, who used to be highly paid professionals in Hong Kong, find it difficult to settle into a relatively low-pay and low status job, and an unexciting lifestyle in the host countries. Like other *heunggongyan*, they are fiercely proud of their Hongkong identity, and a feeling of forced departure calls for a reproduction of another Hong Kong in another time-space. What better choice than the multi-purpose *yumcha* which is loaded with social meanings and whose familiarity brings a sense of security and friendly understanding in a hostile immigrant life.

Going to *yumcha* Hong Kong style in a foreign country is like going to a church meeting or returning to one's natal home for a family gathering. Uncle Mak, an elderly man who emigrated to Vancouver with his son's nuclear family, religiously went to *yumcha* every Sunday at Maxim's Restaurant in a shopping mall called Tsimshatsui East (the name of a shopping district in Hong Kong) in the middle of Richmond. There he was greeted by Cantonese-speaking waiters, and was able to order his favourite tea just as he used to in Hong Kong. This repertoire of *dimsum* enjoyed by Uncle Mak is the same that other Hong Kong immigrants find in Chinese restaurants in Toronto, Sydney, Brisbane, London, San Francisco and various other cities where the Hong Kong diaspora has mainly spread. Together with a Hong Kong based popular culture involving video-taped TV serial melodrama and sword epic fiction, Canto-pop CDs and concerts, and lately satellite transmitted TV news, *yumcha* contributes to the construction of a neo-Cantonese culture across the surface of the globe. The consumption and subjectivisation of these materials have created a new collective, the transnational *heunggongyan* community.

Culturalising metropolitaneity

Until the late sixties, Hong Kong has been portrayed as an economic outpost of the British Empire, culturally residual of Chinese civilisation, and a political void. After the Second World War, when Hong Kong had become an important entrepot as well as a commercial and industrial centre, administrators and the intelligentsia alike were still of the view that Hong Kong was a city without culture. With a nickname such as 'cultural desert', the city's local traditions whether in the area of performing arts, literature or cuisine, had been subjugated under elite forms of European-American or Chinese origin. In fact history textbooks portrayed Hong Kong as having virtually no history before 1842 when it became a British colony. Likewise, Chinese traditions meant only those that came from the centre of this civilisation—the mainland.

Hong Kong people thus were told that they had had no identity of their own as a collective. The idea that Hong Kong people were purely economic beings who happened to be in the right place at the right time to harvest the gains from the rapid socio-economic development of Hong Kong since the fifties, was a widely circulated and accepted characterisation. Hong Kong was constituted as 'a transient place where one passes through' (Siu 1996:178) and issues of identity were never really raised as everyone living in the colony was thought to be just Chinese, or British, or wherever one's place of origin happened to be. Indeed, Hong Kong was said to have acquired an 'identity by default—as "what it is not"' (Naquin, quoted in Siu 1996:178).

With a sojourner mentality and physical requirements in a colonial society in which the fit survived, Hong Kong society was perceived as a pragmatic collection of refugee labor and refugee capital. These were supposed to explain the lack of public and community participation among Hong Kong Chinese. In a frequently quoted article, Lau (1981) portrayed the Chinese people of Hong Kong as politically apathetic individuals, who remained aloof because of the value they placed upon 'utilitarianistic families'. According to this argument, these family networks provided essential resources for individuals to survive, but at the same time limited them in their willingness and ability to explore outside the family structure. Non-participation contributed to non-identity as a community.

However, it is clear that a Hong Kong identity was beginning to take shape since the early seventies and demanded a revision of how Hong Kong culture was constituted. A series of socio-political disturbances in the late sixties and early seventies, including the 1967 riots, the anti-corruption demonstrations, the movement demanding that Chinese be accepted as an official language, and the protests over Diaoyutai's sovereignty, all pointed to anything but political apathy among Hong Kong people. Given the geo-political location of Hong Kong, a totally Chinese identity could be dangerous. Henceforth the people of Hong Kong were told that Hong Kong was their home and they should be proud of it. Numerous activities such as the 'Clean Hong Kong' campaigns, the various 'Hong Kong Festivals' and 'Hong Kong Art Festivals', the 'Miss Hong Kong Pageants' and the promotion of the Mutual Aid Committees in public housing estates all helped to foster a sense of belonging and communitas. Popular culture played an important role, albeit unintended, with locally produced Cantonese movies taking over the market from Mandarin and English films. Popular songs sung in Cantonese, or Canto-pops as they were called, became the vogue. Novels, tabloids, and even movie subtitles became written in the vernacular. These helped immensely in elevating Cantonese, the lingua franca in Hong Kong, from the status of a dialect to that of a language, and consequently in establishing a Hong Kong culture based on this language.

Like Cantonese, which no longer relegated itself to Mandarin as a dialect of the official language, popular culture moved Hong Kong away from the periphery of Chinese civilisation, to the centre of a neo-Cantonese culture. Indeed, as Chan (1994) points out, popular culture was 'the key dynamic' of this culturalisation project. But, at the same time he argues that, for want of a state and its high culture that would anchor it in the modern world, Hong Kong's popular culture based identity could only be ephemeral. But, unlike generations of Chinese intelligentsia, what the *heunggongyan* identity tries to defy is exactly this subjugation of the local under the central. During the seventies and eighties, this pride of being *heunggongyan* or Hong Kong person was particularly reinforced by the relative wealth that Hong Kong people were able to demonstrate as they undertook their annual *wuiheung* (return to hometown) trip to China. The economic discrepancy between the miracle economy in Hong Kong and the poverty in pre-liberalisation China further bolstered the reversed role in the cultural hierarchy.[5]

The Sino-British talks in the early eighties over the reversion of sovereignty of Hong Kong ushered in serious rethinking about a collective future—continuous economic prosperity, civil liberties and political autonomy—and eventually about a cultural distinctiveness that would allow the metropolis to remain as it was. Fears of being 'swallowed whole by China' were a common undertone of everyday popular discourse, and these became a catalyst for *heunggongyan* to recognise a Hong Kong culture that will allow the city to exist as an entity within but separate from the People's Republic of China. This culture cannot be only Chinese, for it will render Hong Kong just another Chinese city, an identity that will lead Hong Kong into a politico-economic cul de sac that will annul its function as China's door to the international market. And though as a society Hong Kong has adopted some British customs such as the morning tea and bright red double decker buses, and has largely accepted English as the language of government and education, the people themselves have not been proud British subjects. Politically too, it would be unwise to insist on a British linkage after the handover especially since Chris Patten, the last governor of Hong Kong, angered Beijing by drastically changing the political structure soon after he assumed governorship in 1992. To maintain a distinct character, Hong Kong's culture has to be simultaneously Chinese and Western; and more importantly it must go beyond, to be firmly international and cosmopolitan.

The culturisation of the Hong Kong metropolis with such idiosyncrasy and inclusiveness can be seen in many aspects of Hong Kong society. A conspicuous example is the new promotional package produced by the Hong Kong Tourist Association. One of the more wide-reaching tactics in this package is a TV commercial aired several times a day that project Hong Kong as a super metropolis, 'a place above all else'. It boasts of the various world's firsts: (busiest container wharf and longest suspension bridge of its kind etc.), and its quick-flicking images show off the city's sophisticated, high-tech infrastructure, its diverse array of cuisines and consumer goods, and its hard-working and festival-loving people. Hong Kong is defined as a place of (Chinese) traditions and (Western) modernity, a place of contrasts. But it is no longer satisfied with being a place where 'East meets West'; rather it is *the* place where East and West are successfully integrated. Diversity and inclusiveness are definitive traits of the Hong Kong lifestyle, and an obvious sense of pride and confidence emerges from this portrayal. The praxis of this sense of pride and

---

5. For a discussion of a generation who started to identify themselves as *heunggongyan* (Hong Kong persons) see H. Siu's article (1996) on the post WWII baby boomers who became the elite bureaucrats and professionals.

confidence varies in different aspects of daily life. Lilley's discussion of the performing arts in Hong Kong, for example, shows artists submitting to European and American standards while treading precariously to manipulate 'Chinese tradition' to create a sense of local identity though they are never quite able to define such identity (Lilley 1991). In the arena of food, however, Hong Kong chefs show a lot more confidence. In the self-assuredness evident in the organisation of eating, the Hong Kong identity is continuously made sense of and acted out.

Heunggongyan: who am I

The term *heunggongyan*, as Hong Kong people like to call themselves, is quite untranslatable. In general it refers to the people of Hong Kong, their lifestyle, their entrepreneurial spirit. But more importantly it is an identity ascription that highlights a sense of pride of the adaptive self-made person, a feeling of sophistication grown out of metropolitan experience, and a particular brand of regional chauvinism which mocks a centre that is perceived to be backward. It conjures up such 'modern' values as efficiency, rationality, liberal-mindedness and tolerance for difference and change. The identity is readily subscribed to by different regional groups and obviously involves a hybrid culture of mainly Chinese—Cantonese, Shanghainese, Hakka, Chiuchowese, or Fukienese—then add to it a bit of American and Japanese, plus a touch of British/European, Indian and Korean. Yet even the pride in achieving such multiculturalism cannot dispel a sense of liminality that has sprung from the realisation that the local culture is not any of the above per se.

One good example is a recently published and well received anthology of short essays by a group of college students (Liang et al. 1996). These essays are accounts of feelings after a visit to the Hong Kong Museum of History's permanent exhibition titled 'Hong Kong Story'. From this visit the authors talk about their life experience. A strikingly common theme that comes through is the process of acquiring a Hong Kong identity by the authors. And, more importantly, it implies the reality that these young people feel that they need an identity that is not Chinese or Western. Most of the authors were born in Hong Kong, and a small minority in mainland China. But whether they perceive themselves to have originated from Hong Kong or not, their sense of belonging to Hong Kong the cosmopolis is unmistakable. The title of the book is symbolic—*yincha qingin*, or in Cantonese: *yumcha ching chun*, meaning 'to *yumcha*, please come in'. The photograph on the book cover, however, does not show anything related to *yumcha* such as a teahouse or *dimsum* and the like. Rather it shows an old man sitting in a corner outside a Hong Kong style cafe (*chachanteng*) reading newspapers, a common sight in the streets of Hong Kong. A more careful look at the café storefront's glass panel on one side of the entrance, we are led to see loaves of bread, trays of cupcakes and bottles of fresh milk. On the other side is a big board on which a menu of different kinds of rice is written, in Chinese, equally prominently displayed.

To the authors, this picture of the ordinary folk represents a Hong Kong lifestyle that is familiar to them. *Yumcha* is a metaphor for Hong Kong's hybrid culture. To be local is to be both Chinese and Western (how these are defined never really mattered), *as well as* to be non-Chinese and non-Western. Like the *chachanteng* cafe which is so specifically Hong Kong, the college students see themselves as from and of Hong Kong the metropolis, something they are so proud of, yet simultaneously not without regret—that they are neither Chinese nor Western. It is a sense of being helplessly international. The book captures such a mood of being both proud yet suffering from a sense of loss. It opens with a section entitled 'Your presence is welcome' (*foonying guanglam*), indicative of the openness and inclusiveness of a Hong Kong culture built up by migrants, and ends with the section entitled 'Hong Kong Story. I was once here' (*hongkong gusi. doe chi yat yau*) [6] They have come a long way to become *heunggongyan*, but where are they heading now?

For the populace, this existential question carries a realistic twist. What happens after the communists take over? How will the social-economic-political structure achieve the slogan-like objectives of 'stability and prosperity' (*onding fanwing*) and 'smooth transition' (*shunlei guoduo*)? Despite all the promises for civil liberties and non-intervention from Beijing, will Hong Kong be another Shanghai after liberation? Within popular discourse responses are mixed, indicative of the complicated, or even conflicting, feelings *heunggongyan* have as they anticipate the change. Some answer with a patriotic rebuttal that any questions asked of the resumption of Chinese rule contradicts the *oiguok oigong* ('love country, love Hong Kong') principle and should be condemned. On the contrary, some believe the Hong Kong issue must be internationalised on all possible levels in order that Central intervention is minimised and the *yatguok leungchai, gongyan chigong* promise (one country two systems, Hong Kong people govern Hong Kong) can be materialised. Others maintain a nationalistic stand while simultaneously insisting that *heunggongyan* must be not be blindly patriotic, but rather should be assertive in the installation of a democratic political structure. Still others think that if they adhere to a *hosui butfan chengsui* ('river water will not bother well water') attitude, all will be well and life will go on as usual. Some, remembering the post-liberation nightmares of the fifties, 'vote with their feet' and choose to emigrate. Others, having secured a foreign citizenship, 'return with the tide' (*wuilau*).

Whatever the response one chooses or is forced to take, for all the *heunggongyan* it is a time of unprecedented change. Some liken it to a gambling table where different players put in their best bet, calculating the chips they have—you win some, you lose some. To others it is a Sunday family *yumcha* lunch that one cannot refuse to attend, but where each participant is allowed to eat what she/he likes, thus preserving individuality in collectivity. Whatever attitude they hold on to and whichever action they ultimately take, all of them take pride in what they have chosen and show, as one interviewee put it, 'the adaptability and confidence of a *heunggongyan*'. The general sense is a cautious optimism and a superiority complex when *heunggongyan* relate to all other Chinese and may be even the world. In another paper I have discussed this theme in an extremely popular cassette tape/CD production called 'I'm the Best' (*ngo tsi lek*) by Chan Bak Cheung.[7] A brief

---

6. *Doe chi yat yau*, literally translated as '[I] came and visited', is a classic Chinese tourist graffito.

7. I have discussed the identity confusion felt by the younger generation born and raised in the seventies in 'Youth in Hong Kong: re-rooting of an identity', a paper presented at the Association for Asian Studies Annual Meeting, Honolulu, 1996.

reiteration will illustrate the point. This 1995 production starts with the Canto-pop 'I'm the Best' which is basically a series of slogans that projects an almost omnipotent *heunggongyan* image. The great handover then comes on stage as the singer chants 'Horse-racing Will Continue' which is a pun on the promise made by Beijing leaders that everything will remain undisturbed after 1997—you can keep on racing horses, dancing dances (in nightclubs) and speculating in stocks and shares' (*machiupau, mochiutiu, guchiuchau*). Despite all the glamour and optimism, however, there is an inevitable, helpless lament of changing times as Chan sings oldies (in English) like 'Unchained Melody', 'Smoke Gets in Your Eyes', and 'Yesterday Once More'. With this pessimistic note, *heunggongyan* return to what they perceive as the inevitable reality and more and more people are beginning to ascribe themselves as 'Hong Kong Chinese'. Like the cultural grouping it signifies, it is a result of hybridity. On the one hand it cues Hong Kong's 'return to the bosom of the motherland', but at the same time it separates them by distinguishing *heunggongyan* from other Chinese. Obviously *heunggongyan* are aware that uniqueness and autonomy are essential to survival, both as an economy and as a socio-cultural community. The new identity label may be a declaration of regionalism, but undoubtedly a first step towards re-unification. It signifies the shifting identities of Hong Kong people as they add new meanings to post-colonial relationships, and highlights a felt need among Hong Kong people to maintain their cultural autonomy, albeit only a self-sanctioned one, for to the mainland Chinese, *heunggongyan* are at best *heunggong tungbau* (P: *xianggang tongbao*) or (younger, never equal) siblings.

Conclusion

In this paper I have looked at the social and cultural meanings of *yumcha*, a most popular form of eating out in Hong Kong, and how they coincide with the city's ethos of inclusiveness and syncretism. The sense of pride and confidence of *heunggongyan* was dampened by 1997-related uncertainties and anxieties, but in the process of locating themselves in new relations with China, Hong Kong people have recognised a collective identity in the culturalisation of metropolitaneity which is readily consumed in daily life—through *yumcha*. Although the meanings of *yumcha* in the colonial environment tend to romanticise the idea of a Chinese tradition, the celebration of self-made success and the anticipation of post-colonial autonomy usher in new interpretations of being, that of the metropolitan *heunggongyan*. It is in this subjectivisation of a transient material process as food consumption that the identity of a distinct *heunggongyan* community becomes reified and relived every day.

References

Appadurai, A. 1988. How to make a national cuisine: cookbooks in contemporary India. *Comparative Study of Society and History* 30(1):3-24.

Chan, H.M. 1994. Culture and Identity. In H.M. Chan (ed.) *The Other Hong Kong Report.* Hong Kong: The Chinese University Press.

Chang, K.C. (ed.) 1977. *Food in Chinese Culture.* New Haven and London: Yale University Press.

Dak, K.G.D. 1989. *The Roots of Cantonese Cuisine* (in Chinese) Hong Kong: Yumsektindei.

Douglas, M. 1972. Deciphering a meal. *Daedalus* 101:61-82.

Evans, G. and S.M. Tam (eds), 1997. *Hong Kong: The Anthropology of a Chinese Metropolis.* Honolulu: University of Hawaii Press.

Hu, T.L. n.d. Irresistible temptation: gourmet food from Hong Kong—*yumcha* (in Chinese).

Kan, G.Z.1995. Cart Blanch. In *Hong Kong's Best Restaurants 1996.* Hong Kong: Illustrated Magazine Publishing Co.

Lau, S.K. 1981. Utilitarianistic familism: the basis of political stability. In A. King and R.Lee (eds) *Social Life and Development in Hong Kong.* Hong Kong: The Chinese University Press.

Liang, H. et al. (eds), 1996. *Yincha qingin.* Hong Kong: Ciwenhua Tang.

Lilley, R.1991. The double bind: performing arts in Hong Kong. *The Australian Journal of Anthropology* 1991 2(3): 293-306.

Mathews, G. 1996. Names and identities in the Hong Kong cultural supermarket. *Dialectical Anthropology* 21:399-419.

Mintz, S.1996. *Tasting Food, Tasting Freedom: Excursions into Eating, Culture, and the Past.* Boston: Beacon Press.

Mok, C.L. 1992. General analysis of Hong Kong's food industry and shares (in Chinese). *Economic Digest* 21 September 1992:34-35.

Regional Services Department. October 1996. Food business licences in the New Territories and Outlying Islands.

Siu, H. 1996. Remade in Hong Kong: weaving into the Chinese cultural tapestry. In T.T. Liu and D. Faure (eds) *Unity and Diversity: Local Cultures and Identities in China.* Hong Kong: Hong Kong University Press.

Tou, T. 1995. Sheung Hei Lau: home of the working class (in Chinese). *Sing Tao Daily* 1 May 1995.

Urban Services Department. June 1996. Food business licences on Hong Kong Island and Kowloon.

[6]

Hèunggóngyàhn: On the Past, Present, and Future of Hong Kong Identity

On 1 July 1997, Hong Kong was returned to China after a 150-year interlude as a British colony. This transition is not only political in nature, it is also a cultural transition. At present, two broad constructions of Hong Kong cultural identity vie for the allegiance of Hong Kong's people:"Hong Kong as *a part* of China" and "Hong Kong as *apart* from China." These constructions of identity *present* are reflected in vying constructions of identity *past:* British-influenced historians and Chinese-influenced historians offer very different interpretations of Hong Kong's precolonial and colonial history in their arguments over whether or not Hong Kong is truly a "Chinese" city. These different interpretations point to a larger fact: between the competing hegemonies of the British and Chinese empires, Hong Kong people have only lately begun to define themselves as having an autonomous cultural identity. "Hongkongese" as a cultural identity involves a "Chineseness plus" that has three clusters of meaning: "Chineseness plus affluence/cosmopolitanism/capitalism," "Chineseness plus English/colonial education/colonialism," and "Chineseness plus democracy/human rights/the rule of law." This article examines the cultural identity of the people of Hong Kong and argues that the survival of *Heunggongyahn*/Hongkongese as a cultural identity depends on whether or not the voices of democracy will continue to be heard in Hong Kong.

by Gordon Mathews*

On 1 July 1997, Hong Kong was handed over from Great Britain to China, an extraordinary development in the world history of colonialism.[1] Mass media throughout the world have been focusing on the political issues of the handover, the jockeying of Great Britain and China over how Hong Kong is to be ruled. But the issue of Hong Kong's handover is not only political but also cultural: how, in light of their change of rulers, do Hong Kong's people identify themselves? Who do Hong Kong's people think they are? In this paper I examine two dominant conflicting discourses of Hong Kong's cultural identity, Hong Kong as *apart* from China and Hong Kong as *a part* of China; and I examine the past, present, and possible futures of *heunggongyahn* (Hong Kong people), the identity of being Hongkongese.

"The idea of a man without a nation," writes Ernest Gellner, "seems to impose a strain on the modern imagination. A man must have a nationality as he must have a nose and two ears." [2] Indeed, almost all people in the world today are socialized and propagandized to hold a national identity. Some might contest their "identity," but most take it for granted. This, however, has not been the case for those who live in Hong Kong. "Who are the people of Hong Kong?" has been in recent decades a question with no common, taken-for-granted answers. This is true at the most mundane level; as a young Hong Kong resident told me, "Every time I travel to another country, I have to write down my nationality. Because I have a British National Overseas passport, I guess I'm supposed to write 'British,' even though I have no right to live in Britain....What should I write: 'British,' 'British Hong Kong,' 'Hong Kong,' or 'Chinese'?" Her difficulties are

*Oral versions of this paper were given at the Conference on the Social History of Hong Kong, Osaka, 1-2 March 1997, and the Association for Asian Studies annual meeting, Chicago, 13-16 March 1997. The research for this paper was funded by the Research Grants Council of the University Grants Committee of the Hong Kong Government (CUHK 145/96H). I am grateful for comments and criticisms from Grant Evans, Wong Heung Wah, and Alvin Y. So, as well as those of an anonymous referee; Tom Fenton was very helpful in his guidance. Tsang Ching Yi, Suen Man Ping, and especially Wong Ngai Lui, as student research assistants, were of great help in locating articles in Chinese for this paper, as well as in their perceptive criticisms of its arguments. Wong Fung Yu, Eve, Tang Hiu Tung, Daisy, and especially Lo Man Fong were of great help in tracking down graphics and illustrations.

Editor's note: In keeping with the practice of the *Bulletin of Concerned Asian Scholars* diacritics are not used in Asian language terms. An exception is made for the headline only in this article.

1. The handover of Hong Kong from Great Britain to China may be seen in broad scope as representing one of the final chapters in the worldwide history of colonialism and post-colonialism, as societies throughout the world have divested themselves of their European political masters. But Hong Kong's handover is extraordinary in its particulars in at least two respects: (1) Hong Kong boasts a high degree of affluence, cosmopolitan self-consciousness, and personal freedom, and (2) Hong Kong's decolonization is viewed by many in Hong Kong as recolonization, as control over Hong Kong is passed from one empire to another. In this paper, for reasons of space I do not examine Hong Kong's handover in larger, comparative perspective, important though that is, but focus instead on Hong Kong's particulars at this historical moment.

2. Ernest Gellner, *Nations and Nationalism* (Oxford: Blackwell, 1983), p. 6.

Hong Kong children awaiting the arrival of China's leadership at Kai Tak Airport, 1 July 1997. These children now have a *national* identity, but what is their *cultural* identity. This photograph first appeared in the *Ming Pao Daily* on 1 July 1997 and is used with their kind permission.

of course rooted in politics. With the transition in July 1997, Hong Kong people were granted the national identity—Gellner's "nose and two ears"—that has so long been denied them. With Chinese Special Administrative Region passports in hand, residents of Hong Kong should have no trouble in years to come specifying the nation to which they belong.

But while China provided Hong Kong's people with a *national* identity on 1 July 1997, *cultural* identity—in simplest terms, who Hong Kong's people believe themselves to be, rather than who the nation that governs them says they are—is more complex.

For most of Hong Kong's history, its inhabitants thought of themselves as Chinese; it is only since the 1960s that a sense of autonomous Hong Kong identity has emerged, commentators say. That separate identity, in the shadow of China, has been, of late, a matter of intense contestation. "China wants Hong Kong to be the most prosperous Chinese city," Martin Lee, the leader of Hong Kong's most popular political party and vocal critic of China, has emphasized. "They will [only] tolerate Hong Kong as a Chinese city," eliminating Hong Kong's own distinct identity. On the other hand, Tsang Yok Sing, the head of a leading pro-China political party in Hong Kong, argues that "most people in Hong Kong know they are Chinese. If you ask me, I say it is my country." [3] Surveys reflect this division in senses of identity: one recently showed that 35 percent of Hong Kong residents consider themselves Hong Kong people, 28 percent Hong Kong Chinese, and 30 percent Chinese.[4]

Recent writers on ethnic identity have emphasized its negotiable and situational quality. This may particularly be the

case in Hong Kong, given its current precariousness: whether one identifies oneself as "Hongkongese" or "Chinese" may depend upon whether one is talking to a mainland official, to one's Hong Kong compatriots, to immigration officers, or, perhaps, to an American anthropologist. But I argue that senses of cultural identity in Hong Kong are not only situational. Context is important, but identity is not only a matter of context. Although people in Hong Kong, like people everywhere, are full of inconsistencies and contradictions, they are not chameleons. Hong Kong people's self-identification as "Hongkongese" or "Chinese" may be linked to their senses of who they most deeply are in their lives and in their community. The mass media in Hong Kong are engaged in a battle to shape the hearts and minds of Hong Kong's people as "Hongkongese" or as "Chinese." One side of this media battle emphasizes, in both Chinese and English, affluence, capitalism, democracy, human rights, and the rule of law—attributes that Hong Kong now enjoys, they claim, in common with Western nations and in contrast to China. The other side emphasizes nationalism and patriotism: the love of China by Chinese, a love from which other nationalities and ethnicities are excluded.[5] The people of Hong Kong are in the middle: between the exhortations on either side, they wait—often obtaining foreign passports "just in case"—to see what the aftermath of July 1st will be.

The sense of a distinct Hong Kong identity is often said to be particularly the product of Hong Kong's middle class, those

---

3. As quoted in Edward A. Gargan, "China's Cloud Over Hong Kong: Is '97 Here?" *New York Times* International, 5 July 1995.

4. Fung Wai-kong, "Public Softens Stance on Handover but Rights Fears Remain," *South China Morning Post* (Hong Kong newspaper), 17 Feb. 1996.

---

5. Roughly speaking, Hong Kong's two English newspapers and the large majority of its Chinese newspapers support, at least in a weak sense, the first of these positions; Hong Kong's several pro-Beijing Chinese newspapers, of relatively low circulation, support the second position. In the year before the handover, most of Hong Kong's newspapers seem to have become substantially more conciliatory toward China, accepting the *fait accompli* of the handover and censoring themselves for their own future protection.

who form "the backbone of Hong Kong's prosperity," whose "social and emotional ties to China are relatively weak."[6] In this paper, I focus on members of this class—at present the dominant class in Hong Kong, at least in terms of numbers.[7] I have interviewed, primarily in English, thirty-five alumni of the Chinese University of Hong Kong, a key site for the contestation of "Chinese" and "Hongkongese" identities. Now engaged in fields such as business, education, and journalism, these university graduates were asked how they see themselves as "Chinese" and "Hongkongese." To augment what they have told me, and to better gauge its representativeness, I have also extensively analyzed mass media and scholarly reports in English and Chinese on the issue of Hong Kong identity.

A Brief History of Hong Kong Identity

Two broad discourses of Hong Kong identity compete for the allegiance of Hong Kong's people: one can be labeled "Hong Kong as apart from China;" the other "Hong Kong as a part of China," with the center of gravity shifting from the first of these discourses to the second. The shifting nature of Hong Kong identity at present may be seen in terms of the shifting constructions of Hong Kong past.

Lord Palmerston, the British Foreign Secretary in 1841, described Hong Kong as "a barren island, which will never be a mart of trade"; as the historian Chan Kai-cheung notes, "every British official and semi-official narration of Hong Kong history in the past century and a half has repeated one or another version of the 'barren island' remark."[8] Recent archaeological and historical research, however, has led to the presentation of a very different picture of Hong Kong's precolonial past. "Hong Kong," Chan writes, "for most of the past 6,000 years with the exception of recent centuries [has] been a busy crossroads of world trade and cultural intercourse," taking part in the mainstream of Chinese history.[9] Some scholars go further in their conclusions: "The ancient people of Hong Kong already had a strong nationalist consciousness, and a tradition of protecting the family and defending the country," historian Siu Gwok-gihn claims.[10]

This reconstruction of history is due to the discovery of new empirical evidence over the past few years; but the politics of contemporary Hong Kong identity are inescapable. "The British

pretend they created Hong Kong's prosperity from scratch. They say it is all their own work....It is the same in every colony. They want people to forget their history, to forget themselves," comments archaeologist Au Ka-fat.[11] However, what self is to be remembered is disputable. Chan concludes his essay on Hong Kong's history with the observation that Hong Kong is "a *very* Chinese city"; but much of the new archaeological evidence predates the era of Chinese imperial control over Hong Kong. Hong Kong's precolonial history may in this sense support a Hong Kong identity *apart* from China as much as a Hong Kong identity as *a part* of China. The difference lies in what "Chinese" is taken to mean: Is it cultural or political, a matter of ethnicity or of empire? But there is also the ambiguity of the archeological evidence itself. I interviewed an archeological curator at one of Hong Kong's museums, who mulled over the boundaries of fact and interpretation in her work: "I think we should believe in the scientific methods of excavation; but on the other hand, archeologists of Hong Kong each interpret things in their own way; it's impossible to decide who's right....Yes, maybe it is all political. But it's taboo to think about these things; as a museum, we just try to be neutral. We can't give a wrong interpretation to the public; we just try to present what's real." And, then, with a laugh, she finished her thought: "And what's real depends on me!"

If Hong Kong's precolonial history is open to fundamental reinterpretation, so too is its colonial history. The war that led to Hong Kong's founding is known in some history textbooks still in use in Hong Kong schools today as "the first Anglo-Chinese War." A number of Western or Western-influenced historians stress that opium was a minor issue: "The war would not be fought over opium; it would be fought over trade, the urgent desire of a capitalist, industrial, progressive country to force a Confucian, agricultural and stagnant one to trade with it."[12] Mainland Chinese historians, on the other hand, emphasize that the issue was not trade but the British effort to enslave the Chinese people to opium and subjugate them to colonialism. As one historian writes,

> The real reason for the Opium War was that Britain had been selling opium to China on a large scale, and this was forbidden by China, setting off the war. During the war, the Chinese government banned all British merchants from trading in China, not just those engaged in the opium trade, and this became the excuse for the British to twist the truth and claim that the war was a "trade war." However, without a doubt the real nature of the Opium War was the invasion of China by British colonialism. The truth of this part of history should not be changed.[13]

6. Helen Siu, "Cultural Identity and the Politics of Difference in South China," *Dædalus* (spring 1993): 32-33.

7. Hong Kong has a per capita income only slightly lower than that of the United States; the typical Hong Kong worker today is not a street vendor or laborer, as was the case thirty years ago, but a white-collar employee manipulating figures on a computer screen. The views quoted in this paper are those of people within Hong Kong's middle or upper-middle class, in terms of educational and occupational attainment, but they are not unrepresentative of Hong Kong as a whole. See, for example, the polling data of the Hong Kong Transition Project (http://www.hkbu.edu.hk/~hktp) for a quantitative portrait of Hong Kong attitudes at the time of the handover that largely mirrors the qualitative portrait sketched in this article. Since the completion of this paper, my students and I have been interviewing more explicitly working-class Hong Kong people, and have found a range of attitudes not dissimilar to those expressed in this paper.

8. Chan Kai-cheung, "History," in *The Other Hong Kong Report 1993*, ed. Choi Po-king and Ho Lok-sang (Hong Kong: Chinese University Press, 1993), p. 457.

9. Ibid., p. 483.

10. As quoted in Lahm Cheui-fan, "Luhkchinnihnchihn yihyauh toujeuhk geuimahn" (Six thousand years ago, already there were native inhabitants of Hong Kong), *Sing Tao Daily* (Hong Kong newspaper), 2 March 1996. The romanization of books and articles in Chinese published in Hong Kong are given in Cantonese in this paper, using the Yale system, as are their authors' names. For books and articles published in English by Hong Kong Chinese authors, as well as for the names of Hong Kong's Chinese newspapers, I use their own romanization of their names.

11. As quoted in Andrew Higgins, "One Territory but Two Histories," *Eastern Express* (Hong Kong newspaper), 25-26 Nov. 1995.

12. C. Hibbert, quoted in Frank Welsh, *A History of Hong Kong* (London: HarperCollins, 1994), p. 80.

"Hong Kong-Beijing Marathon Celebration" at Wanchai Coliseum. This photograph first appeared in the *Apple Daily* on 24 March 1997 and is used with their kind permission.

Martin Lee and other democracy advocates celebrate their legislative victory in 1995. This photograph first appeared in *The International Chinese Newsweekly* on 1 October 1995 and is used with their kind permission.

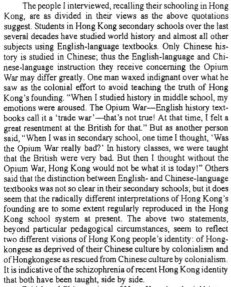

The people I interviewed, recalling their schooling in Hong Kong, are as divided in their views as the above quotations suggest. Students in Hong Kong secondary schools over the last several decades have studied world history and almost all other subjects using English-language textbooks. Only Chinese history is studied in Chinese; thus the English-language and Chinese-language instruction they receive concerning the Opium War may differ greatly. One man waxed indignant over what he saw as the colonial effort to avoid teaching the truth of Hong Kong's founding. "When I studied history in middle school, my emotions were aroused. The Opium War—English history textbooks call it a 'trade war'—that's not true! At that time, I felt a great resentment at the British for that." But as another person said, "When I was in secondary school, one time I thought, 'Was the Opium War really bad?' In history classes, we were taught that the British were very bad. But then I thought without the Opium War, Hong Kong would not be what it is today!" Others said that the distinction between English- and Chinese-language textbooks was not so clear in their secondary schools; but it does seem that the radically different interpretations of Hong Kong's founding are to some extent regularly reproduced in the Hong Kong school system at present. The above two statements, beyond particular pedagogical circumstances, seem to reflect two different visions of Hong Kong people's identity: of Hongkongese as deprived of their Chinese culture by colonialism and of Hongkongese as rescued from Chinese culture by colonialism. It is indicative of the schizophrenia of recent Hong Kong identity that both have been taught, side by side.

British and Chinese views of Hong Kong's colonial history are, as one would expect, drastically different. Recent histories of Hong Kong by British writers portray the cavalcades of British governors and merchants, with the Chinese relegated to no more than a hazy, all-but-forgettable background. As Jan Morris comments in embarrassment about Hong Kong of the 1920s (as well

as, perhaps, about the absence of much Chinese presence in her own book): "For like it or not—ignore it if you could—all around the 4,500 Britons of Hong Kong lived 725,000 Chinese....Very few Chinese names appeared in the history books, because very few Chinese had played public parts in the development of Hong Kong; and the mass of the Chinese population seemed to most observers oblivious to public events, intent only on making a living." [14] In these books, Chinese appear most often as mute victims: Frank Welsh reports Isabella Bird's nineteenth-century comment, "you cannot be two minutes in Hong Kong without seeing Europeans striking coolies with their canes or umbrellas." [15] Morris says that "when I first went to Hong Kong in the 1950s, I noticed that Britons habitually spoke to Chinese in a hectoring or domineering tone of voice." [16] One professor of Western literature in his forties described to me his first meeting with a white person: "Every one of us, in childhood, was afraid of foreigners: because of their size, because they belonged to the ruling class, and because they spoke a language we didn't understand. The first time I met a foreigner was in a music lesson in my primary school; he was an inspector, a big man called Parker. He asked me a question, and I answered 'yes.' He said 'yes what?' and I didn't know how to answer—he shouted 'yes what?' several times and then struck me on my head, very very hard. Do you know what I should have said? 'Yes sir'!" This man attributes the subsequent direction of his studies to the inferiority complex this meeting seared into him—one he struggled thereafter to overcome. [17]

Recent mainland Chinese and Chinese-influenced histories of Hong Kong have similarly emphasized the brutality of the British treatment of Chinese in Hong Kong's history; but the large-scale backdrop of these books, missing from their British

---

13. Louh Faahn-ji, *Ngapin jinjang yuh Heunggong* (The Opium War and Hong Kong) Hong Kong: Jaahpyihnseh, 1983), p. 42.

14. Jan Morris, *Epilogue to an Empire* (London: Penguin Books, 1993), p. 194-195.

15. Welsh, p. 278.

16. Morris, p. 67.

counterparts, is the sense of historical humiliation of China by Britain and other colonial powers, now finally being rectified. These books stress the close relationship between Hong Kong and South China throughout Hong Kong's history. Chapters in one volume, cover such topics as "Hong Kong and the 1911 Revolution" and "The activities of the Chinese Communist Party and other democratic parties in Hong Kong during the Liberation War Era." These minimize all distinctions between Hong Kong and China, thereby shaping a sense of common history and common identity.[18] Great Britain appears in these volumes as no more than a shadowy usurper, robbing China of its territory; British figures and policies, with just a few exceptions, appear in their pages only to be vilified. But, as in their British counterparts, in these books Hong Kong Chinese do not much appear as actors. Hong Kong's people for the most part respond to China, supporting political and social movements on the mainland. This is the primary historical role that they are allowed.[19]

As these Chinese histories of Hong Kong indicate, through most of its colonial past, Hong Kong was indeed closely linked to China. Access to Hong Kong from China was uncontrolled until World War II, and "until recent years, perhaps as late as the 1960s, most Hong Kong Chinese residents considered the mainland to be their 'motherland.' They belonged to it. Hong Kong was only their transitional home."[20] Great Britain was indeed seen as an interloper and usurper by at least some of Hong Kong's people throughout its history, as can be seen in the acts of resistance to British rule that have intermittently taken place, from the poisoned bread case of 1857 (in which a baker spiked his loaves with arsenic for his British customers), to the military struggle of indigenous residents of the New Territories against their British occupiers in 1899, to the General Strike in 1925, to the 1967 riots, which saw "Red Guards in [Hong Kong's]... streets, and the *People's Daily* exhorting... protesters to 'organize a courageous struggle against the British and be ready to respond to the call of the Motherland for smashing the reactionary rule of the British oppressors.' "[21]

By the late 1960s and 1970s, however, a postwar generation, which had only known Hong Kong as a home, reached adulthood,[22] and a sense of Hongkongese as an autonomous

cultural identity began to emerge. For the first time, "Hongkongese" became distinct from "Chinese." The 1967 riots showed the strong ideological influence Communist China held over some of Hong Kong's people; but the Cultural Revolution in all its chaos came to seem for many in Hong Kong less an inspiration than a threat. For the people I interviewed who grew up in the era of the Cultural Revolution, China represented a world that was closed and dark and strange. As one person said, "I remember seeing some of the murdered bodies that floated down from China into Hong Kong waters. I still remember how they stank.... At that time there were so many people risking their lives to escape and come to Hong Kong, so many sad stories. A couple was found tied together, the female was still alive but the male had been eaten by sharks." As another said, "I went to China in 1974, before the Cultural Revolution was over. I still remember the horrifying experience. There were lots of songs everyone had to sing together; everyone was dressed in either gray or blue. I felt I was a Hong Kong Chinese; I had to get out from that place." These words reflect a newly emergent Hong Kong identity of affluent cosmopolitan choice, confronting a communitarian world next door and finding it utterly foreign. And while some critics describe this new sense of Hong Kong identity as one cynically engineered by the colonial government, it seems clear that it was also the fruit of a genuinely new sense of Hong Kong autonomy.

In the decade that followed, Mao Zedong's Cultural Revolution gave way to Deng Xiaoping's economic reforms; the strangeness of the Cultural Revolution to Hong Kong's people gave way to familiarity, as China began to open its doors to the capitalist world that Hong Kong represented. In 1982, negotiations began for the return of Hong Kong to China—Great Britain had a 99-year lease on the New Territories, due to expire in 1997, and without the New Territories, Hong Kong was not a viable entity. The Sino-British Agreement of 1984 guaranteed that although Hong Kong would indeed be returned to China, "the economic, legal and social system in Hong Kong and its citizens' way of life will remain in force for fifty years after 1997"—there will be "one country, two systems." The Tiananmen Square incident, on 4 June 1989, dashed Hong Kong's dreams of a benevolent China. A million people in Hong Kong protested (close to 20 percent of Hong Kong's population)—the first time in Hong Kong's history that Hong Kong people had demonstrated en masse against the Chinese government.

Chris Patten, the last British governor of Hong Kong, initiated electoral reforms during the last five years of British rule, introducing at least a measure of the democracy that Great Britain had denied to Hong Kong throughout the previous 150 years of its rule and enabling the first direct elections to the Legislative Council. China has insistently denounced all such reforms, heaping obloquy on Patten. Many of Hong Kong's

---

17. An Australian brought up within the British public school tradition commented to me that the inspector's behavior in this incident does not necessarily reflect a British sense of racial superiority over Chinese, but rather the typical attitude of teachers within this tradition towards their pupils, no matter what their cultural or racial background.

18. Yuh Sihng-mouh and Lauh Suhk-wihng, ed., *Yihsahp saigei dik Heunggong* (Twentieth-Century Hong Kong), (Hong Kong: Keihleuhn syuyihp yauhhaahn gungsi, 1995).

19. Where, one might ask, between British-influenced and Chinese-influenced histories of Hong Kong, are Hong Kong people's own versions of their history to be found? "There is not yet a Hong Kong history book which is able to use both Chinese and Western materials in giving a complete history of Hong Kong's political, social, and economic changes from different perspectives," writes Fok Kai-cheung, in his *Heunggongsi: Gaauhohk chaamhaau jiliuh* (Hong Kong History: Teaching Reference Materials), Vol. 1. (Hong Kong: Joint Publishing Company, 1995), p. 33—as if Hong Kong has yet to find its own historical identity between the poles of "Chinese" and "Western."

20. Kwok Nai Wang, *Hong Kong Braves 1997* (Hong Kong: Hong Kong Christian Institute, 1994), p. 24.

21. Matthew Turner, "Alterity and Abundance: Early Aspirations to Cosmopolitan Lifestyle and Consumer Culture in Hong Kong," paper presented at the Second Annual Symposium on Cultural Criticism, the Chinese University of Hong Kong, 4-6 Jan. 1996.

22. In 1881, only 3.2 percent of the Chinese population of Hong Kong had been born in Hong Kong rather than in China; by 1931, 32.4 percent had been born in Hong Kong, as Fok (see note 3) tells us (p. 3); by 1981, the Hong Kong census revealed that, for the first time, more than half of Hong Kong's citizens had been born in Hong Kong, as Morris (see note 14) indicates (p. 295).

Hong Kong's chief executive Tung Chee-hwa. Reprinted with persmission from the editors of *Ta Kung Pao*. Hong Kong, 1997

people responded with their feet: half a million of Hong Kong's people have emigrated since the Tiananmen Square crackdown. "We have hard data that a minimum of 2.2 million people will leave or try to leave if something goes wrong" after 1 July 1997, notes the Hong Kong academic Michael DeGolyer.[23] There is a widespread sense among Hong Kong's people over the last several years of having been cast aside to fend for themselves by Great Britain, just as there is the widespread apprehension about China and its intentions. Mark Roberti argues in great detail, neither Great Britain nor China negotiated the handover in good faith, and Hong Kong's people were the losers.[24] Many prominent business people over the past several years have increasingly cooperated with China, seeing their profit as lying with their coming master. Two staunch advocates of Hong Kong's autonomy and democracy, Martin Lee and Emily Lau, contend (in Lau's words) that "China has no respect for freedom, especially freedom of the mind." [25] "Eventually, I expect to be arrested," Lau states. [26]

In late 1996 China appointed an interim provisional legislature to take the place of the elected Legislative Council after July 1; Tung Chee-hwa, a shipping magnate with close links to China, was selected as the future chief executive of Hong Kong by a committee chosen by Beijing. He has advocated the repeal of Hong Kong's Bill of Rights ordinances, and has spoken of the need for a return to "Chinese values" in Hong Kong— "a belief in order and stability; an emphasis on obligations to the commu-

nity rather than the rights of individuals." He links these values to the Chinese nation: "We have to turn pro-China into a very positive definition. It's a good thing to love our country." [27] Tung's charge that the Democratic Party—the lopsided winner in Hong Kong's 1995 Legislative Council elections—is "anti-Chinese" provoked the party's spokesman, Martin Lee, to reply: "Many polls have found the Democrats to be the most popular party in Hong Kong. Is [Mr. Tung] therefore suggesting all Hong Kong people are anti-Chinese?...We will support what is good for Hong Kong while opposing what is bad, no matter whether it is [done by] the Chinese, British, or Hong Kong governments." [28] A similar point of view was expressed by Albert Cheng, a popular Hong Kong radio talk show host: "Hong Kong people are very different from mainland Chinese.... [Mr. Tung] should stay in China." [29]

These two different versions of Hong Kong's cultural identity were reflected in the two largest popular protests in Hong Kong in 1996 (each drawing 20,000 to 40,000 people): the demonstration on the anniversary of the Tiananmen Square incident and the candlelight vigil that followed the death of Hong Kong activist David Chan, who drowned while protesting Japan's occupation of the Diaoyu/Senkaku Islands. At the Tiananmen demonstration, the speeches echoed these sentiments: "We will not go away, despite what the Chinese government says. We will remember the atrocities China has committed, and we will continue in future years to gather and speak and mourn, in the fight for freedom and democracy." At the memorial for Chan, speaker after speaker shouted, "As Chinese, we will not tolerate any resurgence of Japanese militarism. The Diaoyu Islands are Chinese, and David Chan died protecting them for China. Let us stand up for our country!" At the Tiananmen demonstration, Hong Kong as *apart* from China was emphasized: Hong Kong as a free and democratic place that will resist the Chinese government's tyranny. At the tribute to David Chan, Hong Kong as *a part* of China was emphasized: Hong Kong and China unified in their Chineseness against a "militaristic" Japan. The different emphasis at these two demonstrations exemplifies in a nutshell the conflicting currents within contemporary Hong Kong identity at present.

As the foregoing summary indicates, the sense of a separate Hong Kong identity—of *heunggongyahn*—has existed only in the short period between the late 1960s and today, with tomorrow very much an unknown. It is as if, between the competing hegemonies of the British and Chinese empires and the competing historical meanings of Hong Kong they set forth, a tiny fissure opened up for a few decades, and Hong Kong people began to define themselves. Hong Kong's people took the conception of their British rulers of "Hong Kong as apart from China" and reworked it on their own terms, to define themselves as not British, not Chinese, but Hongkongese. This self-definition is now increasingly in doubt; but before speculating as to the future, let us focus on the present. What is the nature of this Hong Kong identity?

23. As quoted in Anthony Spaeth, "One Year and Counting," *Time International*, 1 July 1996.

24. Mark Roberti, *The Fall of Hong Kong: China's Triumph and Britain's Betrayal* (New York: John Wiley & Sons, 1996).

25. As quoted in Keith B. Richburg, "Fearful Press in Hong Kong Toes China's Hard Line," *International Herald Tribune*, 30 March 1995.

26. As quoted in *The Economist*, "Turning Back the Clock in 1997," 6 Jan. 1996.

27. As quoted in *Asiaweek*, "Tung Chee-hwa: In His Own Words," 20 Dec. 1996.

28. As quoted in No Kwai-yan, "Tung Sees Role for Democrats in 1998 Polls," *South China Morning Post*, 7 Jan. 1997.

29. As quoted in Edward A. Gargan, "Airwave Thunder, Hong Kong Style," *International Herald Tribune*, 10 Feb. 1997.

Hong Kong Identity in the Shadow of 1997

Hong Kong identity, as I have indicated, emerged in the 1970s; by 1986, a major survey on Hong Kong identity found that 59 percent of respondents thought of themselves as "Hongkongese" and 36 percent as "Chinese." [30] Recent survey data show that 49 percent of respondents identify themselves as "Hongkongese," and 36 percent as "Chinese," [31] a considerable degree of agreement over a ten-year span, although the percentage of "Hongkongese" has dropped, perhaps as a consequence of the approach of 1997. Other surveys show that the percentage of people identifying themselves as "Chinese" has been increasing, at least by a few percentage points, over the past several years. [32] Surveys have found that the more educated one is, the more likely one is to claim Hong Kong identity, although at the same time, the more likely one is to seek to emigrate. [33] The more educated one is, the more likely one is to be worried about what happens to Hong Kong after 1997. [34] This is reflected in the people I interviewed, the majority of whom saw themselves as "Hongkongese," a minority as "Chinese." But survey numbers are limited in their meaning, in that they do not tell us what the terms "Hongkongese" and "Chinese" mean to those who identify themselves as such. [35]

Most of the people I interviewed, as well as scholars of Hong Kong whose works I have read, take pains to distinguish Hong Kong identity from Chinese identity. "Hong Kong is Chinese in many ways.... Yet it is also evident that Hong Kong... has developed its unique identity and culture," writes Choi Po-king [36] "Hong Kong is not a Chinese city, although more than 97 percent of its population are ethnic Chinese," writes Kwok Nai Wang. [37] One person told me, "It's ridiculous to say that Hong Kong is not a Chinese city. But from a cultural point of view, we Hong Kong people identify ourselves with a kind of special Hongkongness." Mass media frequently report on this sense of difference. Describing Hong Kong business people, one newspaper article said: "The more they have contact with China, the more they realize that, although they themselves are Chinese, their ways of thinking, their characters, their styles, are completely different from those of mainland Chinese." [38] The Deputy Director of the New China News Agency in Hong Kong (China's representative in Hong Kong before 1 July 1997) has commented that "most Hong Kong people do not understand Chinese culture, history, and national events....They do not have deep feelings toward the country and the race....They have received a different education from the mainland Chinese, which leads them to biased ideas about things in China....The Chinese government must undergo great effort to make Hong Kong citizens understand the truth." [39] Despite their coming unity, China and Hong Kong are presented in these accounts—from very different points of the Hong Kong political spectrum—as being fundamentally different.

I asked people who identified themselves as Hongkongese to explain to me at length what they meant by this designation. All seemed to formulate "Hongkongese" as what might be termed "Chineseness plus": "Chineseness plus affluence/cosmopolitanism/capitalism" or "Chineseness plus English/colonial education/colonialism" or "Chineseness plus democracy/human rights/the rule of law." Placed geographically, this "plus" was thought of as "Chineseness plus Westernness" or "Chineseness plus internationalization."

These formulations have a ring of cliché to them ("Hong Kong is the meeting ground of East and West," as the tourist brochures proclaim), but they were not uttered merely as clichés by the people I interviewed. Rather, they were given a distinct emotional edge. For some, this "plus" was a matter of salvation, of escape from a Chinese cultural identity that they detested. One person I interviewed said, "To speak bluntly, mainland Chinese people aren't civilized.... Yes, because Hong Kong was colonized it is civilized!" A letter to a local newspaper stated: "All the bad traits in my personality are from my Chinese side, and the better characteristics, if any, are the products of Westernization." [40] For others, this "plus" was experienced as a profound minus: the loss, through colonialism and "the colonization of the mind" they had undergone, of their true Chinese cultural identity. Another letter-writer speaks of "a latent bigotry...that some in Hong Kong have towards fellow Chinese. Perhaps our colonial heritage has indoctrinated in many of our residents the assumption that what is foreign is superior and what is Chinese is inferior." [41] One person told me, "We Hong Kong Chinese had

30. Lau Siu-kai and Kuan Hsin-chi, *The Ethos of the Hong Kong Chinese* (Hong Kong: Chinese University Press, 1988), p. 178.

31. As cited in Edward A. Gargan, "Will the Virtues of Tea Be Enough to Calm Hong Kong's Jitters?" *International Herald Tribune*, 11 June 1996.

32. Alex Lo, "Growing Number Prefer to Be Identified as 'Chinese,'" *Eastern Express*, 17-18 Feb. 1996.

33. Lau and Kuan, p. 181-184.

34. Michael E. DeGolyer, "A Collision of Cultures: Systemic Conflict in Hong Kong's Future with China," in *One Culture, Many Systems: Politics in the Reunification of China*, ed. D. McMillen and M. DeGolyer (Hong Kong: Chinese University Press, 1993), pp. 280-281.

35. This paper deals with "Hongkongese" and "Chinese" as cultural identities, but pays scant attention to the middle category of "Hong Kong Chinese," an identity adhered to by many respondents in some surveys and by a few in others. Few people I interviewed identified themselves as "Hong Kong Chinese," although this label may increase in importance in the future, as a neutral identity marker between what some may regard as the more politicized poles of "Hongkongese" and "Chinese." Another middle category is that of Cantonese, the Guangdong cultural identity shared to a degree by Hong Kong and its hinterland, Guangdong Province. Certainly the economic integration of Hong Kong and Guangdong over the past decade has been extraordinary, as is emphasized in Reginald Yin-wang Kwok and Alvin Y. So, eds., *The Hong Kong-Guangdong Link* (Hong Kong University Press, 1995). However, few people I have spoken with in Hong Kong seem to feel much shared sense of cultural identity with the people of Guangdong; at present, although perhaps not in future, cultural difference rather than cultural similarity between Hong Kong and Guangdong is what is stressed in Hong Kong.

36. Choi Po-king, "Introduction," in *The Other Hong Kong Report 1993*, ed. Choi Po-king and Ho Lok-sang (Hong Kong: Chinese University Press, 1993), p. xxxiii.

37. Kwok Nai Wang, p. 111.

38. Jeh Wai-jeun, "Dou maht si yahn" (Remembrance from things left behind), Faat Fuh Chihng Column, *Ming Pao* (Hong Kong newspaper), 6 March 1996.

39. Lauh Bing, "Jeung Jeun-sang: Ngoh batnahng chong huhngdang" (Zhang Zhun-shen: I cannot walk when the light is red), *Daaihhohksin* (magazine published at the Chinese University of Hong Kong), Feb. 1996.

colonization. We were brought up in a very strange way. Because of that, we don't have our original [Chinese] culture; we have no home."

Indeed, China as a lost home—or, in an even more emotionally loaded metaphor, a lost mother—was an image in the minds of many of those I interviewed. As one woman said, "Even though I'm ignorant about China, I feel like an abandoned child. I don't know who my mom is, but there's the longing to return to her. No, my country's not China, but there is that dream." Yet for this person the metaphor lost much of its power once she actually beheld that "mother." "All these years I've only visited China once, in 1988. I had this ideal about China, an emotional tie to China, until that visit. I only went to Guangzhou, joined a local tour, but there was a sense of alienation....One section of the hotel where we stayed had a Chinese restaurant, where the locals, at that time, were allowed to eat. The other parts of the hotel, locals couldn't go in. I was shocked. The glorious image I had of China was shattered." As a Hong Kong Chinese, this person had the run of what she called the "Western" section of the hotel, unlike the local Chinese; the rules of segregation, as well as her own sense of difference born of affluence, made her sense that she was not "Chinese" but "Hongkongese."

Southern China has of course become far more affluent in the past decade: the segregation this Hong Kong resident observed is now no longer practiced; many Chinese now have enough money to enjoy the luxuries once reserved for foreigners. However, some Hong Kong people continue to mock mainland Chinese for their lack of sophistication, as shown by their dress. As some of my students tell me, "visitors from China can afford brand names now, but they still don't know how to wear them." "If you see women in the streets wearing Chanel from head to toe, chances are they're from the mainland," says the Hong Kong newspaper columnist David Ho. "They know the brands, but do not have real taste or style." [42] A recent newspaper article describes a symptom of what it terms "the downfall of Hong Kong": the waitresses in a restaurant in the Mandarin Hotel (one of the fanciest hotels in Hong Kong) have become informal and sloppy. "Are we in a government hotel in mainland China?" the writer of the article asks sarcastically. [43]

These comments indicate what is perhaps the dominant discourse of difference among those I interviewed: "Hong Kong is wealthy but China is not"; "Hongkongese are cosmopolitan, Chinese are not"; "Hongkongese have the freedom to consume what they want, Chinese do not"; and underlying these, "Hong Kong is capitalist, China is not." Hong Kong is popularly thought of as a city of capitalist swagger. (A recent popular song—*ngoh ji lek* [I'm the smartest]—boasts of Hong Kong people's capitalist prowess: "We are...flexible and adaptable, conscientious, quick and shrewd....Money-grabbing over the border, knowing at least Chinese and English.... Quick and skilled at constructing

buildings and bridges, foreigners look on stunned.... Our horse-racing pools are greatest, enough to buy up an army." [44])

Hong Kong is also popularly thought of as a city of flashy consumption. (Hong Kong, the guidebooks note, has more Rolls-Royces per capita than any other city on earth; one occasionally sees nouveau-riche businessmen, in their pink Rolls-Royces, flaunting their wealth for all to see.) Many of the people I interviewed showed a high regard for money and the freedom it brings. As a young businesswoman commented: "The best thing about Hong Kong is we can make lots of money here. Money is important in that it gives you choices as to how you want to live, where you want to travel. We have these choices." Hong Kong people have the means to choose the material bases for their lives and identities, but, she is saying, Chinese do not. Another person described the difference between his mode of life and that of his cousins in a southern Chinese city: "My cousins spend their leisure time mostly at home, just drinking tea and watching television. In Hong Kong we spend and consume most of the time. We go out, to the cinema, to restaurants. Freedom of choice is much less in China than in Hong Kong." Hongkongese, these people are saying, have maximum freedom to consume; this, they seem to feel, is the essence of Hong Kong identity as opposed to that of China: the ability to consume and become whatever one likes. [45]

The affluence of Hong Kong, and the freedom of consumptive choice it has led to, was taken for granted as a good by most of those I interviewed; but several also emphasized the "greed" and "money-hunger" of Hong Kong people as indicating their lack of higher civic ideals, whether an ideal of "greater China" or an ideal of "Hong Kong democracy." *Newsweek* magazine set off a storm in Hong Kong in May 1996 with a cover story on "The Betrayal of Hong Kong." "The tycoons have gained the most from a freewheeling society. So why are they working with China to impose communist-style control?" the article asks. It quoted the governor of Hong Kong, Chris Patten, as saying of the tycoons, "they wouldn't be doing [Beijing's bidding] if most of them didn't have foreign passports in their back pockets." [46] But to several of those I interviewed, this attitude is entirely understandable. One said, "Of course Hong Kong people want lots of money; with all the uncertainty about the future, money and a passport are all you can be sure of." In a place of transition like Hong Kong is at present, roots, a sense of home, are difficult to adhere to; but money and passport can, after a fashion, make any place home.

Affluence and the freedom that money can provide are one perceived mark of difference between Hong Kong and Chinese identities; another key marker is that of language. A scholar has noted that "although Cantonese is the vernacular of Hong Kong ... as a communicative medium it is not as socially prestigious as *putonghua* [Mandarin Chinese]." [47] Still several recent newspa-

---

40. Liu Kin-ming, "Chinese and Far from Proud of It," Letter to the Editor, *Eastern Express*, 16 April 1996.

41. David Chu, "Taunting the PLA Manifests Bigotry," Letter to the Editor, *Eastern Express*, 1 Feb. 1996.

42. Angelica Cheung, "Cultural Conflict," *Asiaweek*, 28 June 1996.

43. Arai Hifumi, "Heunggong dik muhtlohk" (The downfall of Hong Kong), *Seun Bou* (Hong Kong Economic Daily), 15 April 1996.

44. The translation of the lyrics of this song is that of Szeto Mirana May, in her paper "Rethinking Community through 'Hong Kong Identity' as a Commodity," presented at the Second Annual Symposium on Cultural Criticism, the Chinese University of Hong Kong, 4-6 Jan. 1996.

45. I explore the issue of consumptive choice and Hong Kong identity in my article "Names and Identities in the Hong Kong Cultural Supermarket," *Dialectical Anthropology* 21 (nos. 3, 4): 399-419.

46. Dorinda Elliot, "Betraying Hong Kong," *Newsweek* International, 13 May 1996.

per articles discuss the vitality of Cantonese and imply its superiority to *putonghua* because of Hong Kong's affluence.[46] Cantonese, the people I interviewed say, marks Hong Kong, and Guangdong Province, as distinct from the rest of China. It also marks, albeit to a lesser degree, the Chinese diaspora as opposed to those who stayed in mainland China: "In Vancouver, you can live your entire life in Cantonese," one person said. Of late, a large demand for *putonghua* lessons has emerged in Hong Kong, as more and more people begin to feel that they will need *putonghua* in the future in order to prosper. But a backlash against *putonghua* is also evident. As one Hong Kong artist says, "Language is one of the prime things to separate yourself from others....In my drawings, I write in Cantonese. I don't care if the [mainland] Chinese don't understand."[49]

Even more than Cantonese, English for the people I interviewed serves as the dominant linguistic marker of Hong Kong's difference from China. If affluence points to capitalism as the root of the difference between China and Hong Kong, English points to colonialism, a colonialism directly experienced by Hong Kong people through, among other forms, their schooling. English has served as the primary medium of instruction in most secondary schools, in textbooks, and often in oral instruction. "The fact that students are being forced to learn in a non-native tongue in Hong Kong is an extremely rare practice. I wonder if you could find it anywhere else on earth," the critic of Hong Kong education Tso Kai-lok has said.[50] Some of those I interviewed were outraged at their enforced English training: "My high school...was the essence of colonial education. We had...to forget our past, forget our culture, what actually Chinese or China means to us....For a long time we wrote letters in English, even to loved ones; we said 'I love you' in English." Others were grateful for this training; as one teacher and activist said, "My feminism has a lot to do with my English literature training. In secondary school I read Catherine Mansfield and Virginia Woolf. Before I read them, I had the very traditional idea of 'men should be at work, women at home'....If I hadn't received a Western education, if I couldn't speak English and read English books, I'd be bearing children, working as a secretary or a saleslady selling clothes; I wouldn't be the person I am now." In these divergent views, we see once again that same stark split in Hong Kong identity: Colonial education is seen as either stealing the linguistic heritage of Hong Kong people or saving

Cartoon dramatizing fears that "China may take away Hong Kong's nascent democracy...its prosperity, and so too the autonomous identity of Hong Kong's people." *The Free China Journal,* 19, April 1996. Artist Tsai Hai-chin.

them from "Chineseness." But whether they resented or appreciated their colonial education, the people I interviewed acknowledged that they were who they were, as "Hongkongese," because of that education, placing them at a remove from "Chineseness." And just as money can make the world beyond China home, so too can the English language: it is English, as well as the presence of a large diaspora community, that makes Canada, Australia, and the United States the leading destinations of Hong Kong emigrants.

A third mark of Hong Kong identity—fundamentally different from the first two, as I will argue—involves political ideals such as democracy, human rights, and the rule of law. A social worker I interviewed stated: "'Hongkongese' can be used as an identity to distinguish ourselves from the Chinese government. The Chinese government acts as if they are parents; the leaders try to command us as if we are children. To claim one's identity as Hongkongese is a political statement....When I visit my relatives in China, I feel Chinese in day-to-day social life. But when I see the Chinese government spokesman on television, I think 'I am Hongkongese; you are Chinese.'"

This, however, seems a fragile basis for Hong Kong identity. First, Hong Kong's partially democratic elections are of very recent vintage, only existing in the past five years. (Hong Kong's colonial governor was never elected by the people of Hong Kong.) In the decades previous, Hong Kong was ruled by a more or less benignly autocratic British hand, leading to the view, set forth by Hong Kong social scientists in the 1960s and 1970s, that Hong Kong people were apolitical. Great Britain's slowness in allowing for democracy in Hong Kong is the stuff of considerable cynicism in Hong Kong. As one recent writer put it, Hong Kong's present-day democracy advocates "deprecate China... because they say it is totalitarian....They should have stood up long ago to defy colonialism, under which the Hong Kong Chinese have been living as second-class citizens until lately."[51]

Second, Hong Kong democracy, and perhaps human rights and rule of law, may shortly be no more. Chinese government officials make claims about the coming of human rights and democracy to Hong Kong: Chinese Foreign Ministry spokesman Cui Tian-kai has said that "the return of Hong Kong to China

47. Herbert Pierson, "Cantonese, English, or Putonghua—Unresolved Communicative Issues in Hong Kong's Future," in *Education and Society in Hong Kong: Toward One Country and Two Systems,* ed. G. Postiglione, (Hong Kong: Hong Kong University Press, 1992), p. 185.

48. For example, *Sing Tao Daily* (Hong Kong newspaper), "Yiu Gwong-dungyahn mgong baahkwa?" (Don't Cantonese people need to speak Cantonese?), Editorial, 17 Aug. 1995.

49. As quoted in Edward A. Gargan, "Will Mandarin Leave Cantonese Speakers in Hong Kong Tongue-Tied?" *International Herald Tribune,* 12 Dec. 1996. Cantonese and *putonghua* (Mandarin) use the same basic writing system, but *putonghua* in China is written in simplified characters. Cantonese in Hong Kong is written using traditional Chinese characters and often uses expressions that are incomprehensible to mainland readers of *putonghua.*

50. As quoted in Felix Lo, "Mother's Tongue Is Best," *Eastern Express,* 21 March 1995.

marks the beginning of the time when Hong Kong people enjoy human rights. From July 1st, there will be no more foreign governor or foreign flag raised in Hong Kong." [52] Beijing's spokesman on Hong Kong affairs, Lu Ping, claims that "July 1, 1997 will mark the real beginning of democracy [in Hong Kong]. There's been no democracy in the territory during the more than a century of British rule." [53] But in fact the Chinese government has, as earlier noted, dissolved the elected Legislative Council in Hong Kong in favor of its own appointed Provisional Legislature (although it claims that democratic elections will be held in 1998); Hong Kong's chief executive, Tung Chee-hwa, has recommended rescinding several laws concerning human rights in Hong Kong. Opinion polls in Hong Kong show that over two-thirds of those polled believe that the method of choosing the Provisional Legislature was "less acceptable" than that of the Legislative Council[54]; 55 percent oppose Tung Chee-hwa's recommendations to scrap human rights legislation, as opposed to only 18 percent who support them.[55]

The Chinese government labels electoral democracy and human rights as Western "cultural imperialism" (Western nations are "assuming superiority and imposing their own... standards on other countries without considering these countries' different history, cultural background, and social conditions," claims Xinhua, the New China News Agency).[56] Hong Kong advocates of "Western" democracy and human rights for China are merely suffering from "colonization of the mind"—a Chinese phrase used facetiously by several of those I interviewed. What the Chinese government sees as baneful foreign influence, these people saw as a universal good. As one person said: "Human rights don't differ in different cultures. The idea may have originated in the West; but human rights are true everywhere, for all people.... The same is true for democracy." To many of those I interviewed, it appears that democracy in Hong Kong will have a lifespan of some five years; and as goes democracy, they felt, so too may go any autonomous Hong Kong identity.

The Fall of *Heunggongyahn*/The Rise of *Junggwokyahn*

What, then, of *heunggongyahn?* "The naked truth about Hong Kong's future can be summed up in two words: It's over,"

---

51. J. Wong, "Happy Medium Seen for Future," Letter to the Editor, *Eastern Express*, 4 Apr. 1996.

52. As quoted in Jasper Becker, "'Don't Pressure Us,' Christopher Told," *South China Morning Post*, 21 Nov. 1996.

53. As quoted in Chris Yeung, "Actions, Not Words, 'To Be the Test of Mainland Promises,'" *South China Morning Post*, 6 June 1996.

54. Andy Ho, "Provisional Legislature Fails to Inspire Public Support," *South China Morning Post*, 24 Dec. 1996.

55. Jonathan Braude, "New Breath of Optimism for Hong Kong's Future, Survey Reveals," *South China Morning Post*, 3 Jan. 1997. That these pessimistic numbers were presented beneath a glowing headline I see as indicative of the efforts of Hong Kong newspapers to instill optimism in their readers about the control of China over Hong Kong. In April 1997 the *South China Morning Post* hired a mainland Chinese "senior consultant"—a founding editor of the *China Daily*, an official English-language newspaper in China. This move was criticized in Hong Kong as a possible indication of the coming muzzling of Hong Kong's English-language free press.

56. "'Human Rights' Mask Racism" (Report from Xinhua, The New China News Agency), *Eastern Express*, 8 Feb. 1995.

---

the business magazine *Fortune* declared in 1995.[57] At present, anyway, this judgment seems far off the mark: the Hong Kong stock market is close to record highs and the local property market is soaring.

If economic confidence remains high, however, political and cultural confidence seems more fragile. *Heunggongyahn*, as I have discussed, only emerged over the past few decades, within a small period of cultural autonomy between the overpowering cultural influences of Great Britain and China. There has been a brief period during which, almost despite Great Britain's political rule, Hong Kong's people became affluent and cosmopolitan, and a still briefer period during which a departing Great Britain gave Hong Kong a degree of political freedom. China may take away Hong Kong's nascent democracy, then, inadvertently, its prosperity. If these go, the autonomous identity of Hong Kong's people will also be lost. As one person told me: "Hong Kong Chinese are unique, but maybe they will become just a name in the history book. At the moment we are still protected; but when the British go, the Chinese influence will come in.... What you see as unique in Hong Kong will not last very long. Hong Kong identity will die out"—not immediately, but gradually and inevitably over the decades to come, he argued.

Apart from the weight of China's political domination, the Hong Kong identity bears the seeds of its own extinction. Affluence and language, money and English: these markers of Hong Kong identity may mark the tickets out of Hong Kong. Those who adhere to these markers of identity often strongly emphasized that they were "Hongkongese," but, these very markers of identity may serve to bring about the dissolution of Hong Kong identity in that those who possess them will find it easiest to leave Hong Kong and make their lives and find their identities elsewhere. Democracy, on the other hand, was not seen as portable by its advocates: it was not just anywhere, but *Hong Kong* that had to be democratic. If two of these "Western" markers of identity enable one more easily to flee China for "the West," the third may involve a potential and highly threatening penetration of China by "the West." And indeed, a number of those who adhere most strongly to "democracy" and "human rights" as values saw themselves not as Hongkongese but as Chinese.

A majority of the people I interviewed thought of themselves as Hongkongese; a minority thought of themselves as Chinese. No one I interviewed was a direct supporter of the Chinese government, but several expressed optimism toward 1997 and its aftermath, due to the cultural and ethnic background they felt they held in common with the mainland Chinese. In the words of one activist who took part in the Diaoyu/Senkaku Islands protests, "We love our country and we want our country to be united again: China, Taiwan, Hong Kong—we're all Chinese—we are all members of the same family; we're all one race." The common ethnic and cultural background the speaker attributes to Chinese overrides all political division: China will behave well toward Hong Kong because both are one and the same Chinese people.

Most of those I interviewed made no such claims; but even some of those who were quite pessimistic about the Chinese government and its future actions toward Hong Kong nonetheless called themselves not "Hongkongese" but "Chinese." The most vociferous advocates of democracy that I interviewed

---

57. Louis Kraar, "The Death of Hong Kong," *Fortune*, 26 June 1995.

dreamed of the emergence of a "true China," as opposed to what they saw as the stilted "Chineseness" of the Chinese state today. Their dreams were not so much of preserving *heunggongyahn* —Hong Kong identity—but of reshaping *junggwokyahn*—Chinese identity, in the mold of the intellectuals of China's 1919 May Fourth Movement and the dissidents imprisoned after the Tiananmen Square incident. One told me of how, when living abroad, she saw on television the events in Tiananmen Square and felt an urge to return to Hong Kong rather than live overseas; her mission, she felt, was to help to create in Hong Kong the "China" that the Tiananmen Square protesters envisioned, a "China" that could perhaps from its base in Hong Kong eventually spread throughout China.

These views parallel those of a number of Hong Kong political figures: "In the long term, China will have to change and be more democratic," Martin Lee has said.[58] Anson Chan, the Chief Secretary of the last colonial Hong Kong government and of the first post-handover government as well, has stated that "in one hundred years, Hong Kong and China will merge into one system—the Hong Kong system."[59] Definitions of "China" are, as we saw earlier, hotly disputed in Hong Kong at present. Whereas Chief Executive Tung Chee-hwa warns Hong Kong people "to be alert to 'international forces' trying to use the territory in a campaign to isolate China,"[60] newspaper columnist Raymond Wong Yuk-man disavows any such narrow political definition of China: "I identify with China culturally and am proud of our national heritage.... [But] if you say to be a Chinese I must support either China or Taiwan, I would rather not be a Chinese."[61]

Cultural China

These people identify themselves not with political China —the present Chinese nation—but with "cultural China": with Chinese ethnicity, history, and cultural traditions, and including China, Taiwan, Hong Kong, and the Chinese diaspora. Over and over again, the people I interviewed would make statements criticizing the Chinese government, after which some would say things like, "you'd better not put that in print. That might be dangerous after 1997." Their loyalty was to a larger China. In one person's words: "I am a Chinese nationalist... but of the land and the people, not the Chinese constitution. If you respect and obey the constitution of the PRC, does it mean you consent to be a Chinese?... To me the most important factor to be Chinese is Chinese culture, traditional Chinese culture. It is easier to be Chinese in Hong Kong than in China because the traditional values are much more affected by the government in China. In China, they understand and interpret the ancient scholars by one approach, the party's approach." Of course, Chinese "traditional culture" is vast and variegated; but for this man, the very multiplicity of readings of "traditional culture" is a mark of authentic

"Chineseness," as opposed to the Chinese government's inauthentic denial of all multiplicity.

Heunggongyahn may soon be dead; but from the ashes of that identity...a resurgent, independent junggwokyahn may emerge.

The revulsion that the people I interviewed expressed toward the Chinese government is widespread in Hong Kong. Indeed it seems difficult to justify the likelihood of going from a free press to a muzzled press (even if through self-censorship); it seems difficult to justify the potential erosion of an ideal of human rights (although I wrestle with this: to what extent is my belief in human rights no more than my American ethnocentrism?).[62] "Despite Hong Kong's return to its homeland in 1997, an overwhelming majority of the working class as well as the business people and intellectuals, if offered a choice, would not elect to identify themselves as citizens of the People's Republic of China. Hong Kong is, at least in spirit, part of the Chinese diaspora," writes Tu Wei-ming.[63] But at the same time, there is very clearly an increasing acceptance of "Chineseness" in Hong Kong at present. This can be seen in the earlier-discussed revising of Hong Kong's archeology and history to make it "Chinese"; and this can be seen too in such commercial ventures as *Shanghai Tang*, a large new store devoted wholly to traditional Chinese clothing that has been attracting considerable attention in Hong Kong and throughout Asia.[64] These trends imply that for all the sense of apprehension felt toward the Chinese government, there is indeed a sense of "returning to 'Chineseness'" in Hong Kong —a turning away from a separate *heunggongyahn*, to become *junggwokyahn*. Cultural identity may in this sense be moving in tandem with national identity. As Hong Kong reverts to China, Hong Kong's people become once again Chinese, and *heunggongyahn* fades into history.

And yet it may be that the very spirit of confident independence that led to the emergence of an autonomous Hong Kong identity may lead, with that identity's gradual dissolution, to the emergence of a new identity: an autonomous, critical, independent Chinese identity, uncontrollable by any government. This is what the current Chinese government most fears from Hong Kong, perhaps with good reason. *Heunggongyahn* may soon be dead; but from the ashes of that identity, if Martin Lee and his ilk are correct, a resurgent, independent *junggwokyahn* may emerge. If this comes to pass, the demise of *heunggongyahn* will have been a worthy cultural sacrifice indeed.

---

58. As quoted in Alan Friedman and Jonathan Gage, "A View from Hong Kong: China Must Change," *International Herald Tribune*, 11 Nov. 1996.

59. As quoted in Cheung (see note 42).

60. Chris Yeung, "Anti-China 'Forces' Out to Use Hong Kong, Warns Tung," *South China Morning Post*, 28 Oct. 1996.

61. As quoted in Mariana Wan and Genevieve Ku, "Beyond Patriot Games," *South China Morning Post*, 29 Dec. 1996.

62. The recent volume *Human Rights and Chinese Values: Legal, Philosophical, and Political Perspectives*, ed. Michael Davis (Hong Kong: Oxford University Press, 1995), offers little relativism. Human rights are universal, the contributors maintain, and current Chinese government policy violates human rights. See also Linda Butenhoff's article "East Meets West: Human Rights in Hong Kong," *Bulletin of Concerned Asian Scholars* 28, no. 2 (April-June 1996), pp. 4-15.

63. Tu Wei-ming, "Cultural China," in *The Living Tree: The Changing Meaning of Being Chinese Today*, ed. Tu Wei-ming (Stanford: Stanford University Press, 1994), p. 15.

64. Dahng Dak-sahng, ed. "Geuihyauh junggwok dahksikdik seungbanchiuh" (A new trend of commercial goods with Chinese character), *Sing Tao Daily*, 22 Feb. 1996.

[7]

Hong Kong Families:
At the Crossroads of Modernism and Traditionalism*

HOIMAN CHAN**
and
RANCE P.L. LEE***

Hong Kong has a total population of 5.7 million, of which about 98 per cent are Chinese in origin. In this demographic sense, Hong Kong is essentially a Chinese society. In the cultural sense, Hong Kong also appears as a Chinese society. Most of the Chinese people, especially the adults, were immigrants from China. They have carried with them the cultural tradition from various parts of China, especially the Guangdong Province. About 90 per cent of the Hong Kong Chinese are Cantonese.

Nevertheless, to emphasize Hong Kong as a Chinese society is only partially correct. Hong Kong has been a British Colony for one and a half centuries. It is the British who have provided the administrative-legal framework within which the Chinese conduct their daily lives and make their living. Furthermore, Hong Kong is a society caught in the throes of change. Recent decades have seen rapid economic growth and urbanization in Hong Kong, marking its emergence as a highly modernized industrial-commercial center in Asia (Lin, et al., 1979). Its Chinese residents have been widely exposed to cultural influences from other parts of the world, particularly the West. To understand the beliefs and realities of Hong Kong society, therefore, we should recognize not only its Chinese cultural heritage, but also how its character has evolved and transformed beyond its original cultural context.

DOMINANT IDEOLOGICAL ORIENTATIONS

As Hong Kong is only partially a Chinese society, a more adequate vantage point into its dominant beliefs or ideologies would be its shifting proximity to and distance from Chinese cultural tradition. The belief system of Hong Kong obviously cannot be divorced from its cultural genesis, but its subsequent divergence can hardly be avoided. It is our basic proposition that what the belief system of Hong Kong owes to Chinese culture is essentially a sense of Chinese traditionalism, whereas divergence from the latter stems from the ideological and pragmatic quests of modernization and the modern way of life. Chinese cultural tradition per se can offer no satisfactory guidance for modernization, let alone binding moral or behavioral codes. Hong Kong is then a society caught in a macro-cultural

---

* We wish to acknowledge the assistance of Miss Michele Kiang in compiling some of the research material.

** Lecturer of Sociology, Department of Sociology, The Chinese University of Hong Kong, Shatin, New Territories, Hong Kong.

*** Dean of the Faculty of Social Science, Professor of Sociology, The Chinese University of Hong Kong, Shatin, New Territories, Hong Kong.

balance. Chinese heritage sets the elementary behavioral code and moral fabric of everyday interaction, while the imperatives of modernity stretch, break, or add on to the plasticity of this original scaffolding. The result is a novel and idiosyncratic reconfiguration of traditional and modernistic values. Moreover, this change has to take place within a given socio-political framework of colonialism, which posits political and practical demands that are contradistinctive from both Chinese autocracy and modern Western democracy.

How the crisis-crossing currents of traditionalism and modernism have contributed to the formation of the belief system of Hong Kong is a question that merits closer examination. It is well known that Chinese tradition harboured a strong Confucian moral overtone. While this observation may be generally accurate, it is not necessarily the sole important key to the cultural landscape of Hong Kong. For one thing, Confucianism has been heavily shaken during its encounter with the modern age. This enigmatic experience went far beyond the compass of Hong Kong society, but figured prominently througut the tortuous history of late and post-Imperial China. Another thing is, while Confucianism may be the ideal culture or 'great tradition' in the Chinese past, yet even then the state sponsorship of Confucianism never displaced the mass culture or 'small traditions' that took on a strong facade of religious syncretism, especially where folk beliefs and cults were concerned.

It would therefore appear that the best way to underline the essential character of a Chinese-inspired belief system — as in the case of Hong Kong — does not simply lie in reiterating at face value the main philosophical and moral doctrines of Confucianism, but to probe into its more deep-rooted ideological bearing and thrust. Further, such attempts would do well to go beyond the terrain of Confucianism as narrowly defined, and into the wider fabric of syncretism that juxtaposed Confucianism with other constituting facets of folk beliefs, including popular versions of Taoism, Buddhism, and others. This would be the context of belief that underlines the Hong Kong scenario, a context that directly built upon Hong Kong's embeddedness in the Chinese cultural heritage.

In this connection, it should be pointed out that when Weber (1951) depicted the religion of China as largely a spiritual attempt to attain a 'rational adaptation to the world', he was mainly dealing with the Confucian side of the picture. In this respect, his characterization is not as encompassing as, for example, what Arthur Smith (1890) has put forth: the Chinese people is simultaneously pantheistic, polytheistic, as well as atheistic. The importance of Weber's conceptualization cannot be doubted, especially when applied to the situation of Imperial China. Its pertinence to the contemporary Chinese societies, however, is another matter. As far as Hong Kong is concerned, the strive for 'rational adaptation' cannot be the whole story in a society thoroughly infused with achievement motivation and spirit of progress. Rather, as a result of the significant loosening of the grips of Confucianism in post-imperial China, two contradistinctive trends took place simultaneously. On the one hand, with the eclipse of Confucianism, the small traditions of folk cults and beliefs were liberated from the repressive stronghold of a hegemonic orthodoxy. On the other hand, the onset of modernity brought on a stronger sense of enlightenment and rationality, over and above the grips of religious ethics and beliefs. The combined impact of these two trends is a paradoxical one. There is a revival of folk beliefs and even superstitions in the post-Confucian Chinese societies, and at the same time there is a progressive extension of the sphere of scientific reason that comes with the onset of

Hong Kong Families: At the Crossroads of Modernism and Traditionalism   85

modernity. Hence, the complex cultural consequences of traditionalism and modernism result in a system of belief that appears enlightened and secular, yet at heart only incompletely so. Hong Kong is situated precisely at this juncture. Confucian hegemony is out, yet survivals of quasi-sacred superstitions stand hand in hand with the progress of modernity and instrumental reason. This explains why, when secularization is the trend in most parts of the world, yet Hong Kong has seen not only the survival but verily the widening of the sacred domain, especially in the spheres of folk religion and ritual practices.

In his study of religion in China, Yang (1967) made a distinction between institutional religion and diffused religion. Institutional religion has an independent theology, an independent system of symbols and rituals, and an independent organization of personnel. In traditional China, it was represented by major universal religions such as Buddhism and Taoism and by religious or sectarian societies. On the other hand, diffused religion has its theology, cults, and personnel so intimately diffused into one or more secular social institutions that they become part of the latter. Ancestor worship, the worship of community deities, and the ethico-political cults were some of the examples.

To speak of religious ideas or belief system prevalent in Hong Kong society today is in the first place not to emphasize institutional religion as such. The politico-institutional framework of Confucianism is of course long obsolete, while other forms of world religion like Buddhism, Catholicism and Protestantism are far from reoccupying the lacunae. It is, in our opinion, more appropriate to focus on the diffused, noninstitutional forms of belief system in the case of Hong Kong, which seeks to weave together strands of both the sacred presence and the tides of secularisation. As matter stands in the present day Hong Kong, this prevailing belief system mainly encompasses three distinctive levels or facets: (1) the sacred presence, (2) the Confucian, ethical imperatives, and (3) the pragmatic, instrumental outlook.

At the level of sacred presence and even mystical cult, one witnesses the popularity once again of many traditional superstitions and beliefs. Yet most of them have been assimilated into quasi-rational or scientific outlook of sort. This applies to all kinds of astrological practices based on the I-Ching, star movement, land form, sacred numbers, terrestrial elements, or the circulation of 'qi'(inner energy). Practitioners of such cults have mostly shed off their traditionalist appearance, and adopt postures of professionalism and scienticity. The title 'Xuan xuejia', which can be loosely rendered as 'mystic-specialist', has now by and large replaced the traditional epithet of 'fortuneteller'. As for the general populace, their connection with the revival of sacredness does not stop at good-natured acceptance of - or sometimes actual consultation with— the new breed of mystic-specialists. Ritualistic practices still constitute a great part of the spiritual life of the populace. These range from the age-old custom of ancestral-worship, the observing of festivals and the lunar year cycle, all the way to mundane prohibitions or taboos that articulate superstition within the framework of everyday life (Berkowitz, 1969; Freedman, 1979; Hui, 1991). Even in the absence of any hegemonic religious system as such, the presence of the sacred has reasserted itself in ways that are more diffused but by no means less forceful on that account. The sphere of syncretic folk beliefs and superstition has acquired a significant boost on its newly found vitality and legitimacy. The dimension of the sacred presence must be viewed as among the core facets in the belief system of contemporary Hong Kong.

This is not to imply that Hong Kong society at large subscribes unquestioningly to the bondage of supernaturalism. In fact, the Confucian tradition itself resisted the influences of the spiritual world. One of Confucian's famous mottos prescribed clearly that the social order must come before the supernatural order. The reality of the latter is not denied, but rather viewed as secondary to the arduous task of building a harmonious human society. It is often asserted by Confucian scholars that "If one cannot manage the human world, how can one manage the spiritual world?", and that "Without understanding life, how can one understand the after-life?" A well-known Confucian advice is that "Pay homage to the spirits but stay away from them."

The persistence of the Confucian heritage in Hong Kong assumes the form of ethical imperatives. These are moral norms that have obtained their authority not from any transcendental forces, but from the systematic ideology of Confucianism itself. The fact that the latter has been well accepted and long persistent in Chinese history has lent it an aura of being immutable and self-evident. Even though the practice of Confucianism can no longer be consciously and wholesalely abided by, yet many of its ethical tenets have become part of the social norms in everyday life. Most prominent of these would be a sense of order, of propriety, in shaping the dynamics of daily conducts. Thus, in the belief system of Hong Kong, one is still looking upon a society where in principle elderlies will be respected, where seniority in general remains a factor in social interaction, and where strengths of kinship ties can still affect one's behaviour toward others. Aside from social order conceived as a well-patterned process of differentiation, propriety and order are also articulated in terms of harmony. Thus, the principles of generalized benevolence ('jen') and loyalty ('yi') are taken to be cardinal rules of conduct even in informal social interaction. These principles provide a degree of humanity and piety to balance potentially inflexible social encounters as based on patterns of differentiation. In brief, the ethical norms of Confucianism must still be seen as of fundamental importance for the society of Hong Kong, in their provision of both a framework of differentiation and of harmony in the structuring of social life.

The complex dynamics of responding to traditional heritages, however, does not stop here. Sacred presence and ethical imperatives aside, Hong Kong also assumes the outlook of a pragmatic, effective society operating along the principle of instrumental reason. This is a characteristic that Hong Kong shared with modern industrial-commercial societies at large. The question for Hong Kong is how this pragmatic-instrumental streak can coexist with, even reinforce the competing facets of religious sentiment on the one hand, and ethical ideals on the other. It is in this light that the distinctiveness of the pragmatic outlook in the Hong Kong society can be fully appreciated. In a very real sense, the strive for profit, for achievement, and for other forms of socio-cultural rewards can become the prime motivating force for Hong Kong (Lee R., 1985b). The concern for individual well being often takes primacy over concerns for collective good. In these respects, the outlook of Hong Kong is quite thoroughly Westernized and modern. Yet to the extent that this pragmatic streak must accommodate itself alongside other sacred or moral frameworks, it remains essentially self-limiting. Unlike most other advanced, modern societies, pragmatism in Hong Kong cannot incarnate to become itself ethical frameworks of some form — to become an amoral moral framework, so to speak. Instrumental reason or profit motive, in and of themselves, are insufficient to account for the validity or acceptability of actions, and excessive individualism becomes self-defeating once beyond the threshold of sense of righteousness. These are the schistic parameters of pragmatism in the belief system of Hong Kong — both its inevitability and its self-limitation must be equally confronted.

This is then a highly schematic picture of the belief system as currently prevalent in Hong Kong. This is a belief system which situates upon the broad conjuncture of traditionalism and modernism, while mitigated by the specific substance of traditional Chinese culture and present Hong Kong conditions. Such a depiction necessarily implies inner tension and strains of some kinds. Yet when looked at as a whole, one must conclude that this belief system functions remarkably well, both in terms of its own persistence, as well as in providing the ideological groundwork for economic advancement and social stability. Apparently any tension or conflict of beliefs there might has often been creatively channelled to become part of the motive force of progress. How this happens is not an easy puzzle to unravel. Yet especially for the purpose of the present paper, at least one important variable in this mutation of idea and society is the role played by the family. As one of the most fundamental arenas in the social structure of Hong Kong, families provide the empirical, middle-ranged setting within the concrete set up of everyday life, where divergent cultural orientations can become articulated and resolved into a more or less coherent framework of life. This is the wider socio-cultural significance of the family system in Hong Kong, which indicates that the relation between families and belief system can be a crucial theoretical link in understanding the nature of Hong Kong society at large. The next natural step to take is therefore to examine the family structure of Hong Kong in this light

FAMILY AND SOCIETY

Following the lines of discussion established in the above section, the situation of the family system can be examined by way of beliefs and ideologies regarding family life in Hong Kong. With these family ideologies in place, the discussion will then take up more detailed analyses of the structure and functioning of the family system in Hong Kong. The advantage of this approach is to maintain a sustained focus on the relation between the family and the belief system, which is the key rationale of the present volume.

Family Ideology and Norms

The first question to be addressed is about the presence, if any, of certain underlying attitude, posture, or beliefs that shape the dynamics of family life in Hong Kong society. While this may seem an unduly abstract vantage point, it is nonetheless one of the prominent themes in studies of Chinese families. The premise here has to do with the traditional preeminence of familism in Chinese culture. For many centuries, the social functions of family life have been so numerous and so significant that families have become indispensable units of social integration. They bridged the vast distance separating the individual from the overarching socio-political system. The family — and its extension into clans, lineages, even communal village and manors — has long been seen as the corner stone of Chinese society and polity. Fundamental ideologies regarding the nature and operation of the family system thus acquired impetus beyond the family as narrowly conceived. Family, in this light, can be regarded as the social institution par excellence in Hong Kong and other Chinese societies.

The best known endeavour in mapping the attitude and beliefs regarding families in Hong Kong is the thesis of utilitarianistic familism (Lau, 1982). In formulating the thesis, Lau's purpose is precisely to address the issue of social integration by firstly identifying the ideological form that families in Hong Kong subscribe to. In his view, the precepts and guidelines that govern familial dynamics are essentially utilitarianistic in nature. Such down to earth, calculative attitude shapes not only the structure and process of the family institution, but even more importantly underlines the interrelation between family and the society as a whole.

Lau's thesis may be analyzed in terms of a series of interrelated propositions: (1) Utilitarianistic familism can be defined as the normative and behavioral tendency of an individual to place his familial interests above the interests of society and of other individuals and groups, and to structure his relationships with other individuals and groups in such a manner that the furtherance of his familial interests is the overriding concern. (2) Among the familial interests, material interests take priority over nonmaterial interests. (3) Utilitarian considerations within the familial group manifest themselves usually in an emphasis on economic interdependence among familial members, and in the criteria used in recruiting peripheral members into the familial group. (4) Structured in such a manner, the familial group can be used in an elastic and flexible manner by the Hong Kong Chinese to organize their own primary groups for coping with what they see as a far-from-benign social environment (Lau, 1982: 72-93).

Self-interest and economic gains are here unambiguously recognized as the prime concern for people of Hong Kong. Yet such interests and gains are conceived not so much at the level of individuals as at the familial group one belongs to. It is hence a kind of 'collective egocentricism', with the family constituting the elementary framework of calculation and consideration.

This is the picture of a family system suspended — just as the wider society is — between two worlds. Modernistic instrumental attitudes have taken root at the expense of traditional order, yet not to the extent where individualism can overturn familism. Familism stands its ground against the onset of Westernization, yet it must concede some of its core organizing premises. In this way, the precept of utilitarianistic familism encapsulates in a remarkably cogent manner the resolution of schistic forces underlying Hong Kong society at the level of the family, which remains the elementary units of social life for the great majority of population. This line of conceptualization, it must be said, is not without its critics. Giving allowance for further fine-tuning, the precept of utilitarianistic familism remains a useful key in uncovering the organizing principle of family life in Hong Kong.

More can be said regarding family ideologies in Hong Kong. Yet the above should suffice in indicating the traditional-modernist bind that haunts family ideologies in Hong Kong, as well as how this bind is resolved and articulated in the form of utilitarianistic familism. For the wider purpose of this essay, the following will turn to examine the more concrete family norms that are accepted in Hong Kong today. Such norms are of a more specific, practical nature; they prescribe how different aspects of family life should be conducted.

In this regard, the combined findings of a number of researches have again shown that prevailing family norms in Hong Kong are derived from both traditional Chinese and modern Western influences. The trend that has been taking place since the 70's is succinctly summarised in one study:

> With regard to relationships between husband and wife, and parents and children, the normative view of the respondents are more consonant with the conjugal family than the traditional Chinese family. Where relationships with the older generation are concerned, a pattern quite consonant with traditional norms is revealed. (Chaney & Podmore, 1974: 404)

More recently, the emerging familial norms have also been detailed along the following lines: (1) nuclear family is the most widely accepted familial form; (2) some traditional and ethical familial norms are still in practice, e.g., care for aging parents, and economic assistance among siblings; (3) while propinquity is not emphasized, frequent social contacts are still maintained with other members of familial group; and (4) conjugalism is augmented by mutual aid among members of the wider familial group (Lee, M., 1991a). To put things in perspective, the following observation is worth noting:

> The picture emerging from these findings is that although the nuclear family is very much an ongoing reality, people have not entirely given up all traditional family norms and ideals. They expect siblings to perform obligations, but are less insistent about supporting their parents. Many believe that sons and daughters should not be treated differently, but would think twice when asked to depart from traditional sex-roles. Under the label 'nuclear family' there is therefore a complex mix of values and norms which do not entirely accord with one another. (Lee, M., 1991b: 44)

From notion of utilitarianistic familism to emerging familial norms, these constitute the general foundation of principles, beliefs, and outlook of family organization in Hong Kong. From this vantage point, we can turn to examine the more immediate issues of family structure and processes in Hong Kong society.

Family Structure

In approaching the family structure prevalent in Hong Kong society, the first thing to note is the relative predominance of nuclear families. Aside from geographical and demographic explanations, two sociological factors seem to contribute to this situation, factors that involve both general and specific considerations. Thus, it is widely accepted that, in industrialized societies, the proportion of nuclear families is likely to increase at the expense of other extended, traditional forms. Moreover, in the case of Hong Kong, the high proportion of immigrants from Chinese mainland also predisposed the structure of society to a large number of nuclear families. This is not to suggest that these factors can be accepted straight-forwardly as explanation for the case of Hong Kong. There have in fact been controversies — at both theoretical and methodological levels — surrounding their relative validity. Be that as it may, however, the bottom line is that, in strict formal and quantitative terms, nuclear family does vastly outnumber any other familial form in Hong Kong. The situation may be readily seen in Table 1.

Table 1	DOMESTIC HOUSEHOLDS BY HOUSEHOLD COMPOSITION 1981, 1986 AND 1991		
	1981	1986	1991
	%	%	%
Nuclear Family	54.4	59.2	61.6
Stem Family	13.6	11.9	10.7
Extended Family	8.7	8.3	6.6
Single Person	15.2	12.9	14.8
Others	8.1'	7.7	6.3
Total No.	1,244,738	1,452,576	1,582,215

Note: Data computed from the 1981 and 1991 Population Census, and the 1986 By-Census.

It is clear from this table that nuclear family is by far the most widely adopted familial form in Hong Kong during the past decade. This pattern in fact also stretches back into the 70's as well. As the study by Wong (1975) has shown, the industrialization of Hong Kong during the 1960's is associated with a corresponding increase in the proportion of nuclear families. While the sequence of causation implied in the studies concerned has been questioned, at least the prevalence of nuclear family cannot be denied.

The rise of nuclear family in the process of industrialization is presumed to carry distinct socio-cultural consequences. These have been analysed in the case of Hong Kong especially in the works of Wong (1972; 1975), who essentially followed the analytical line of Parsonian functionalism. From this position, the progressive increase of nuclear family signified a process of functional changes in the family system. What was in particular emphasized was the structural isolation of the nuclear family. It was argued that on the one hand, the nuclear form allows for greater flexibility and adaptive power for the family units, thus satisfying more readily the needs and demands of an industrializing society. On the other hand, however, the structural isolation of the family units suggests a certain functional 'breakdown' of the family, to the extent that the familial units no longer fulfilled its hitherto multi-functional role. In analogous with proclamations of community in crisis, therefore, is a parallel thesis of 'family in crisis'. This appears to be the main socio-cultural consequence of the dominance of nuclear family in Hong Kong.

This scenario, first rigorously formulated by Wong in the 1970's, remains the baseline for investigation into the character óf modern Hong Kong family. It has also been subjected to criticism, along with the functionalist assumptions it founded upon. At stake is not only the implied causation between industrializations and the rise of nuclear family, but also the validity of the alleged structural isolatedness of this familial form. While the scenario presented by Wong may seem more clear cut than is warranted by the findings obtained, yet in fairness he does try to qualify his basic proposition in three important ways: (1) there have not been conclusive findings concerning the relation between industrialization and the nuclear family form; (2) other factors may also lead to the rise of nuclear family; and (3) the nuclear family is not as a rule structurally isolated (Wong, 1972: 140). These are important qualifications, though it is not clear how they have been incorporated within Wong's own

Hong Kong Families: At the Crossroads of Modernism and Traditionalism   91

studies. This explains why rooms are left for critical reexamination of the applicability of functionalist propositions in the study of Hong Kong family system. And more to the point for the present paper, rooms are also left for focusing attention not only on the overall structure of the family system, but also on the more subtle, complex processes taking place within the family. This is to cast the outward facade of nuclear family in more realistic lights.

Family Processes

What is important to note under the rubric of family processes is how the formal structure of nuclear family is mitigated through the use of a variety of familial networks, norms, strategies, and functions. The outcome is a unique family system, which adheres to the form of nuclear family, yet circumventing much of the perceived limitations and simplicity of the nuclear structure. This family system may be called a system of 'modified nuclear family', in contrast to the notion of 'modified extended family' previously advanced (Litwak, 1965; Lee, M., 1991b). In this light, the nuclear family remains as the basic form of familial organization. Yet familial processes are often not confined to its strict compass. For each familial unit, therefore, one can distinguish between a core and its fringe. The core, as in present-day Hong Kong, is usually a nuclear family. Yet its fringe often diffuses outward to merge with the fringe of other kinsmen's nuclear families, thus forming a familial network or cluster that can provide assistance and affection for each nuclear family unit involved. Presumably the flexibility and efficiency of the nuclear form can be retained, while side-stepping possible danger of isolation and crisis at the same time. How this image of 'modified nuclear family' operates in practice is the subject of this section.

Internal processes As noted, the functionalist vantage point of researches has led to the conclusion that the nuclear form of the majority of Hong Kong families underwent a process of defunctionalization. The ideal-typical image of Chinese extended families is a multi-functional complex, providing shelter for individual members in economic, political, cultural, educational, and even religious terms. The transition towards nuclear family system implies that most of these familial functions will have to be delegated to other social agents and institutions, such as schools, business enterprises, and social service agencies. The case of Hong Kong, however, suggests that this defunctionalization does not automatically entail a decline in the meaning and significance of the family institution. The situation is simply that, together with the rest of society, the family system also undergoes a process of specialization (Lee, R., 1985a). Instead of retaining a multi-functional, all-purpose outlook, the family has become more specialized in cultivating ties of intimacy and affection among its members, and in preparing its members for other social responsibilities and roles. This is a crucial internal process of the nuclear family in Hong Kong. It has far-reaching implications on other facets of the familial operation.

This emphasis on affective function of the family system is well documented by researches conducted in Hong Kong. Mitchell (1969), for instance, reported that less than 10 per cent of 3966 adults surveyed were unhappy with their marriages, whereas about 80 per cent indicated their spouses as a major source of emotional support. For the married adults, family life clearly performs important affective function. This is the case for the younger family members too, as reported in a study of 3917 Hong Kong youths, conducted by the Family Planning Association of Hong Kong (1983). In this study, 41.4 per cent of the respondents reported they have a very good or good relationship with their fathers, with only 8.1 per cent claiming that the relationship is poor or very poor. The same pattern is found in the relationship with mothers, with 58.2 per cent reporting a very good or good relationship, and only 3.7 per cent reporting a poor or very poor relationship. And in general, most respondents felt they were beloved by their parents. More to the point, 43.6 per cent of the respondents thought their families to be sweet and warm, and less than 10 per cent considered their relations with their own families to be either "not good but tolerable" or "intolerable and would like to leave". On the whole, then, the family in Hong Kong appears to be the same sphere of loving care and emotional support that traditional image depicts, both for the parents and for the children. This is the case even when conflicts and strains take place within the family, in intra and inter-generational terms. Examples of such familial strain would include high expectation of education performance placed on the children, or in less well-off families the need for teenage daughters to start work in support of the family (Salaff, 1981; Lee, R. 1985a).

That the family system, nuclear or otherwise, is perceived to be an arena of emotional and affectionate support, says a lot about the relative stability of the family system in Hong Kong. Such a stable environment would be especially important for the socialization function of the family. As shown in a questionnaire survey conducted by Lau (1984), parents were deemed most trustworthy and understanding by adolescents in Hong Kong, scoring much higher than friends, teachers, or mass media. The family was also seen as the most important source of influence for these young adolescents, scoring higher than friends, mass media, or schools. Parents and families are clearly in good standing as the most important agents in shaping the socialization and emotional growth of youngsters in Hong Kong.

In more ways than one, the increased affective and socialization function of the family have important ramifications on the mode of interaction among members of the family. As is well known, Chinese families are traditionally patriarchal in character, the family being more or less a miniature of the autocratic state. Such a familial system would make for a good training ground for obligation and compliance, but not necessarily a source of affective warmth and emotional support (Ho, 1986). The prevalence of nuclear family, in both ideal and reality, however, has stimulated change in a different direction. In general, the internal processes of the family have become more open, consensual, and egalitarian. While the traditional image of a patriarchal family continues to be a prevalent frame of reference, family life is no longer conducted along the same absolutist lines. The husband-father is still the espoused head of the family, yet his authority is now far from being autocratic or encompassing as used to be. The mode of interaction being put in practice is not so much a strict hierarchical structure as a division of labour among family members, especially between husband and wife (Wong, 1981; Lee, M., 1991b).

This situation is well reflected in investigations into the decision-making process of the family. The recent survey by M.K. Lee (1991b), for instance, revealed a pattern of allocation of family responsibilities as shown in Table 2. This pattern is rather similar to the pattern found by Wong (1981) in an earlier study. Such a pattern, it can be suggested, is closely associated with the character of nuclear family in modern society. Because the family is limited in size, therefore each member attains higher relative importance. And the

Hong Kong Families: At the Crossroads of Modernism and Traditionalism   93

Table 2	ALLOCATION OF RESPONSIBILITIES AMONG NUCLEAR FAMILY MEMBERS IN FOUR FAMILY TASKS			
Persons principally responsible	Family tasks			
	Household chores	Making of important decision	Supervision of children	Budgeting daily expenses
	%	%	%	%
Wife	76.8	13.8	39.5	52.5
Husband	3.3	31.1	4.7	14.9
Husband & Wife	8.8	36.4	48.8	24.8
Children	2.0	2.6	1.2	2.0
Whole family	4.6	13.8	3.1	3.0
Friends, maids or other relatives	4.6	2.3	2.7	2.9
Total No.	306	305	258	303

Note: Data were compiled from the report by M.K. Lee (1991b: 48) of a random sample survey of 393 households in Hong Kong in 1988. A total of 308 households were found to be nuclear families, based on which the above percentages were computed.

importance of a member is measured not according to ascriptive status such as fatherhood or manhood, but according to the actual contribution — usually economic and materialistic ones — that this member may bring to the family. In this connection, the progressive participation of women in the labour force earned for them — whether as wives or as daughters — a greater say in decisions affecting the family as a whole (Wong, 1981; Salaff, 1981). In Hong Kong, it is not unusual for daughters from lower income families to enter into the work force while still in their teenages. And it is even more widely accepted practice for women to continue with their work after marriage and childbirth. All these would tend to tilt the patriarchal balance in an opposite direction, with the result that negotiation and consultation progressively become the widely accepted strategy employed in family decision-making.

External processes. The case of Hong Kong has confirmed the post-functionalist insight that structural isolatedness of nuclear family can be much of a myth. Especially in the field of community studies, western sociologists working in a framework of social network analysis have countered the notion of community in crisis with the new notion of a liberated community. The crux of the debate is that geographical or territorial separation is not the sole ground in determining the strengths and frequencies of social ties, whether among friends or members of the same familial group. It is rather the entire network of ties and interaction that must be looked at, in order to determine the actual extent of isolation and embeddedness. Nuclear family may turn out to be mainly a form of residential accommodation befitting the heavily crowded urban space of Hong Kong, but which does not necessarily undermine the level of structural embeddedess of the family units concerned. Granted that physical proximity and availability can remain a significant factor, yet other modes of communication and interaction may still countervail the gap of territorial separateness.

Here, Chinese traditionalism steps in to mitigate the alleged prevalence of nuclear family under industrialization. It can be suggested that, in a real sense, the nuclear family units in Hong Kong are seldom taken to be autonomous units, even by members of the units concerned. While a high degree of social and economic independence is part and parcel of the nuclear family form, yet it needs not render these units overly introversive on that account. In the case of Hong Kong, more often than not each nuclear unit would recognise its close ties with larger, overarching "home base", constituted by parental homes of the husbands and wives of the nuclear household, and to which all their other siblings and their own nuclear units would belong. This scenario, in other words, is the distinction between core and fringe already alluded to. In this way, familial unit does not coincide with the boundary of familial identification. At its extreme, the familial unit can be merely nuclear in form, but extended in operation. The majority of cases however converge in the middle, retaining their relative independence as nuclear units, yet cultivating ties and emotional attachment with their broader home bases or familial groups.

In practice, this system operates both on a regular and an occasional scheme. On regular terms, the close relation between nuclear units and parental home base is maintained through schedules of regular reunion and visitation. Moreover, such family gatherings are conducted not only as social events, but they also have ritualistic and moral forces. Major festivals throughout the lunar year cycle are largely obligatory dates of family gatherings, notwithstanding geographical dispersal of the nuclear units. Tradition is here lending its force in the maintenance of extended family ties. And for family units that live in closer proximity, it is not surprising to have visitation to the parental home on a weekly or even more frequent basis. As suggested by M.K. Lee (1991a), relatives and kinsmen are the most frequent partners in social occasions and in leisure activities. All these signify that cultivation and maintenance of regular ties with the broader familial group can be an important facet of family life in Hong Kong, even when nuclear family is ostensibly the dominant form.

On a less than regular basis, the significance of extended familial ties is demonstrated also by the great emphasis on mutual assistance among members of the familial group. It is well known that nuclear family, because of its limitation in size and membership, can be especially vulnerable at times of social and economic needs. Studies have shown that the wider support bases of the nuclear unit are the other nuclear units belonging to the same familial group (Mitchell, 1969; Lau, 1982; Lee M., 1991b). Siblings not living together in an extended household are still expected to help one another should the needs arise. And when one's own resources fail, relatives are the first persons that one might turn to for help, before friends, voluntary associations, or government agencies. At times of need or crisis, therefore, one falls back on the support of the home base.

This description of the external processes of the family can be further highlighted in more conceptual terms. Dynamics of these processes has been unravelled by R. Lee (1992) using the concept of gradated network, based on the work of the famous Chinese sociologist Fei Hsiao-tung. In this depiction, a gradated network suggests that in Chinese society,

> each individual places himself in the centre and organizes his relation with others in a gradated manner. Normally, the closer to the centre, the stronger are the moral and instrumental obligation. (Lee, R., 1992: 84)

Hence, the intimacy and strength of kinship ties become the criteria for separating different layers of gradated network. Such a pattern of gradation exists with regard to the relationship with other family members and relatives. This would be a three-circle gradation pattern centered around the focal person,

> the inner circle consists of family members, the middle circle of next of kin; the other circle of distant relatives. The strength of the focal person's moral and instrumental obligation towards other people tends to decline from the inner circle to the outer circle. (Lee, R., 1992:88)

This is clearly a more complex image of family structure and processes, distinguishing different degrees of relatedness and sense of belonging. Even if it is likely that nuclear family will remain the dominant family form in the foreseeable future, it would not be the only rubric for the conduct of family lives. The metaphor of a gradated network illustrates the inter-connnectedness of different levels of familial obligation and involvement, to the extent that the prevalence of nuclear familial forms generates new social and ethical meanings. It is in such a manner that the notion of a modified nuclear family is applied to the case of Hong Kong.

BELIEF SYSTEM AND FAMILIAL REALITIES

At this point, it is possible to look back at the terrain surveyed so far, and elicit in more concrete terms the intertwining of the family system and belief system in Hong Kong society. As noted before, this intertwining is part and parcel of the juxtaposition between modernity and tradition in Hong Kong. This juxtaposition is by no means simply the displacement of one mode of societal process by another. Rather, the best of both worlds is to be retained and resolved into a coherent social order. Modernity brings with it material and economic progress, whereas traditionalism sets the foundation of moral and cultural stability. The familial system, in this light, is one of the key mediatory arena for the articulation and integration of these contradistinctive forces. In other words, to what extent the "best of both worlds" can indeed be realized in social life per se would hinge significantly on its extent of assimilation within the operation of the family system.

Part One of this paper has already delineated in broad strokes the belief system of Hong Kong, with reference to the encounter of modernity and tradition. The consequent belief system is mapped out at three ideological levels, i.e., sacred presence, ethical imperatives, and instrumental concern. Along the vein of analyses in the above, this concluding section will illustrate more closely their articulation with the family system reconstrued in Part Two.

At the level of sacred presence then, family units must be seen as still the important center for maintaining a sense of religious or ritualistic pieties in society. Ancestral worship is still a prevalent practice; it serves as important signification of the common origin of the wider familial group. Besides ancestral worship, observing festivals and other rituals of the lunar cycle are also by and large conducted around both independent nuclear units, as well as the larger familial groups. Often the more important festivals, like the Chinese New Year's Eve, the mid-Autumn Festival, or the Winter Equinox, are celebrated together by nuclear units that belong to the same familial group, at the parental household. The less important festivals or ritualistic dates, like the Dragon Boat Festival or the Double-Yang Festival, are often observed with less festivities by the nuclear units themselves. In these ways, the conduct of family life is converged with the passage of the natural and even supernatural cycle, thus confirming the meaning of the latter as part and parcel of social life as a whole.

Even other more eclectic practices of a supernatural or ritualistic sort often bear close relation with the familial system. The belief in geomancy (feng shui) or spatial harmony, for example, is often practiced with regard to the familial living space. It is generally assumed that inappropriate location or arrangement of familial space can instigate misfortune or even tragedy for family members (Freedman, 1979). While this belief does extend to other spheres such as commerce or architecture in general, yet for most people, the concern is mainly tied in with how this spatial harmony can affect the economic success, health or even longevity of members of the family.

Another example of the familial motif in religious belief is the notion of 'predestined affinity' (yuan fen). It is generally believed by the Hong Kong Chinese that the success or failure of social relationship is affected by predestined affinity (Lee, R., 1992). This affinity is not necessarily a deliberate design by God or fate, but is nevertheless something mystical, unknowable, and outside of human control. While this notion has a wide range of applicability in explaining all sorts of mishaps or happy coincidences in social life, it is particularly favoured in accounting for courtship and marriage. To be blessed with predestined affinity is the single crucial condition in the successful marital ties of husband and wife to form a family, all other objective considerations would be secondary. The religious psychology behind this precept is of course a complicated one. Here, it would suffice to note simply that, for those who subscribe to this psychology, the fulfillment of the predestined affinity is the key foundation for families to even exist at all.

At the level of ethical imperatives, the articulations of familial and belief system can also be clearly seen. Loyalty and benevolence towards members of the family are still part of social expectation. Pieties towards parents and seniors are still emphasized, so is mutual assistance and obligation among kinsmen in times of need. All these may be seen as in commensurate with the traditionalism of Confucian ethical imperatives, concerning social order as a well-patterned process of both social differentiation and harmony. At a more macroscopic level, the persistence of such ethical imperative goes a long way in explaining the modified nuclear family form that has become prevalent in Hong Kong. While social and economic independence can account for the existence of the nuclear unit, its embeddedness within a gradated network of familial group can be more adequately explained in ethical terms, in the suffusion of a sentiment of obligation and a sense of belonging that the family can inspire in its members. Such ethical imperatives would persist even after members have left to form nuclear units of their own. In this way, the ethical level ensures a state of mental and spiritual lineage, even in the face of economic independence and physical separation.

Hong Kong Families: At the Crossroads of Modernism and Traditionalism   97

Last but not least, at the level of pragmatism and instrumental reason, the familial system also serves as important ground of articulation. The very form of nuclear family, after all, is the form that can cope more effectively with a social order that places great emphasis on pragmatic efficiency and flexibility. At a more down to earth level, the diffusion of instrumental reason in the family results in such familial ideology as utilitarianistic familism. In this familial ideology, as noted before, economic and material considerations often attain primacy over other concerns. Family processes, such as the division of labour within the household, the attitude towards more peripheral kinsmen and relatives, the redistribution of authority within the household, or the creation of quasi-kinship ties like "sworn-brothers" or even "sworn-parents", cannot be divorced from the progressive primacy of a utilitarianistic climate.

A far-reaching ramification of this transformation is that the very nature of Chinese family in Hong Kong has changed. The formal and functional attributes of the family, of course, can no longer be the same. Even the cultural meaning of the family has also undergone significant metamorphosis. Family ceases to be the hegemonic focal framework for one's own identification and strives for advancement. As the family is perceived in largely instrumental and utilitarianstic terms, the traditional notion of personal advancement as filial and familial duty is severely undermined. The structure of the family, whether qua nuclear units or wider familial groups, is hence based more or less equally on affective and instrumental terms. While this dual basis appears to serve well the stability of the Hong Kong familial system, what is undeniable is that the very ideal and nature of Chinese family have undergone substantive metamorphosis and disenchantment on their own. This being the case, the modified nuclear family form must then be seen not simply as an adaptation of the traditional extended family ideal, because such ideal ceases to be the acknowledged goal towards which nuclear units are but transitory way-stations. This is the reason why, in this paper, the concept of modified nuclear family is preferred to the concept of modified extended family.

The juxtaposition of modernism and traditionalism in the familial system of Hong Kong, as depicted in this paper, is an ongoing, dynamic process. Their resolution in the prevalent balance of form, ideology, and process will not stay unchanged. Yet the future direction of change is inevitably engrossed in the overall trajectory of Hong Kong society. In the case of Hong Kong, this is more than general rhetorical observation. The reclaim of the sovereignty over Hong Kong by China in 1997 is only a few years away. Rarely can a society foretell with certainty the oncoming of massive and far-reaching systemic change. The more precise paths of change, however, can only be the subject of educated guess or of wild speculation. The familial system cannot be exempted from the predicament instigated by this general scenario. To suggest that family system of Hong Kong is standing at the crossroads of change, therefore, has more enigmatic truth to it than is ostensibly implied in this otherwise unexceptionable observation.

98   *Journal of Comparative Family Studies*

REFERENCE

Berkowitz, Morris I.
    1969   Folk Religion in An Urban Setting. Hong Kong: Christian Study Centre on Chinese Religion and Culture.

Chaney, D.C., and David Podmore
    1974   "Family Norms in a Rapidly Industrializing Society: Hong Kong." Journal of Marriage and the Family 36 (May): 400-407.

Family Planning Association of Hong Kong
    1983   A Study of Hong Kong School Youth. Hong Kong: Family Planning Association.

Freedman, Maurice
    1979   Chinese Geomancy: Some Observations in Hong Kong. Pp. 189-333 in G.W. Skinner (ed.), The Study of Chinese Society. Stanford: Stanford University Press.

Ho, David
    1986   "Chinese Pattern of Socialization: A Critical Review." Pp. 1-37 in Michael H. Bond (ed.), The Psychology of the Chinese People. Hong Kong: Oxford University Press.

Hui, C.A.
    1991   "Religious and Supernaturalistic Beliefs." Pp. 103-144 in S.K. Lau, et al. (eds.), Indicators of Social Development: Hong Kong 1988. Hong Kong: Hong Kong Institute of Asia-Pacific Studies, The Chinese University of Hong Kong.

Lau, S.K.
    1982   Society and Politics in Hong Kong. Hong Kong: Chinese University Press.

    1984   "Perception of Authority by Chinese Adolescents: The Case of Hong Kong." Youth and Society 15: 259-284.

Lee, M.K.
    1991a   "Organization and Change of Families in Hong Kong." Pp. 161-170 in Chien Chiao, et al. (eds.), Chinese Family and its Change. The Chinese University of Hong Kong. (In Chinese)

    1991b   "Family and Social Life." Pp. 41-66 in S.K. Lau, et al. (eds.), Indicators of Social Development: Hong Kong 1988. Hong Kong: Hong Kong Institute of Asia-Pacific Studies, The Chinese University of Hong Kong.

Lee, Rance P.L.
    1985a   "Changes in the Family and Kinship Structure in Hong Kong." Pp. 133-157 in K. Aoi, K. Morioka and J. Suginohara (eds.), Family and Community Changes in East Asia. Japan Sociological Society.

    1985b   "Social Stress and Coping Behavior in Hong Kong." Pp. 193-214 in W.S. Tseng and David Wu (eds.), Chinese Culture and Mental Health. New York: Academic Press.

    1992   "Formulation of Relevant Concepts and Propositions for Sociological Research in Chinese Society." Pp.81-98 in Chie Nakane and Chien Chiao, (eds.), Home Bound: Studies in East Asian Societies. Tokyo: The Center for East Asian Cultural Studies.

Lin, T.B., Rance P.L. Lee and Udo-Ernst Simonis
    1979   Hong Kong: Economic, Social and Political Studies in Development. New York: Sharpe.

Litwak, E.
    1965   "Extended Kin Relations in an Industrial Democratic Society." Pp.290-323 in F. Shanas and C.F. Streib (eds.), Social Structure and the Family. Englewood Cliffs: Prentice Hall.

Hong Kong Families: At the Crossroads of Modernism and Traditionalism   99

Mitchell, Robert E.
    1969   Family Life in Urban Hong Kong. Survey Research Centre, The Chinese University of Hong Kong.

Salaff, Janet W.
    1981   Working Daughters of Hong Kong. Cambridge: Cambridge University Press.

Smith, Arthur
    1890   Chinese Character. China: Shanghai.

Weber, Max
    1951   The Religion of China. New York: Free Press.

Wong, F.M.
    1972   "Modern Ideology, Industrialization, and Conjugalism: The Case of Hong Kong."International Journal of Sociology of the Family 2 (September): 139-150.

    1975   "Industrialization and Family Structure in Hong Kong." Journal of Marriage and the Family 37 (November): 9581 000.

    1981   "Effects of the Employment of Mothers on Marital Role and Power Differentiation in Hong Kong." Pp.217-233 in Ambrose Y.C. King and Rance P.L. Lee (eds.), Social Life and Development in Hong Kong. Hong Kong: The Chinese University Press.

Yang, C.K.
    1967   Religion in Chinese Society. Berkeley and Los Angeles: University of California Press.

[8]

Reinventing Hong Kong
Memory, identity and television

● Eric Kit-wai Ma

Chinese University of Hong Kong

ABSTRACT ● This paper probes the processes by which commercial television has become an important site for reinventing Hong Kong identity at a key moment of political transition. Analysis foregrounds the complex interplay of local, national and global forces that are transforming Hong Kong media and the popular histories they produce. Of primary importance is how these forces shape collective memories and refigure popular imagination of membership in the Chinese nation-state. The concepts of 'overdetermination' and 'fuzzy' production process are applied to examine the political economy of mediated history. A specific television program, 'Hong Kong Legend', is used to chart the contours of these interactions and to interrogate the role that television is playing in the resinicization of Hong Kong. ●

KEYWORDS ● history ● identity ● memory ● overdetermination ● television

In 1996, I was invited by TVB, a Hong Kong television broadcaster, to be the presenter of a heavily promoted evening program. The program, entitled *Hong Kong Legend*,[1] featured Hong Kong's social history and aimed at fostering a sense of belonging among the people of Hong Kong before the sovereignty change in mid-1997. I took the opportunity to write an ethnographic diary on the production routines and do textual analysis of the program. This participant approach enabled me to probe the processes by which popular memory is constructed by commercial television at a key moment of political transition. The following analysis foregrounds the complex interplay of local, national and global forces which constitute commercial *television*, collective *memory* and cultural *identity* in Hong Kong.

Identity, memory and television

Issues of identity and memory have been of grave scholarly and public concern in recent years. The destabilizing consequences of industrialization in high modernity have heightened the psychological need to revive the past. In face of a highly mobile present and a rapidly changing future, the past has been used as an identity resource to foster a sense of belonging and security. In spite of the popular notion which perceives the past as a stable and secure refuge, recent scholarship has significantly problematized the past, deconstructing all its forms into discursive imprints of the present.[2] While governments tend to essentialize identity and popularize collective memory for political legitimacy, scholars tend to denaturalize identity and problematize memory, some to the extent of postmodern nomadity and idiosyncratic subjectivity. However, as Morley (1996) argues, nomadic identity has a geography and is largely an experience of the industrial west. In Asian contexts, where nation-states and traditional discourses have a great power over identity politics, identities are less nomadic and more subjected to the power geometry at both the local and global level. Although Hong Kong has been described as a hybrid culture without formidable cultural barriers (Abbas, 1997), the present case shows that the hybrid Hong Kong identity is shaped by global capitalism and increasingly by nationalistic imperative. It illustrates the constructionistic nature of cultural identity, but at the same time analyses the power dynamics behind the attempts of fixing and essentializing Hong Kong identity through the media.

The media have increasingly been one of the major social institutions in which the past is stored and categorized (Gillis, 1994). Mediated past has become a resource for legitimizing power, fostering collectivity, and reinforcing identity. Among the few media scholars who explore the relation between mass media and the past, Thompson (1995, 1990) has given us a sophisticated theorization of what he called the 'mediazation of modern culture' and the 're-mooring of tradition'. He argues that modern culture is largely colonized and redefined by electronic media. Traditions, histories and heritage are deritualized and recontextualized in mediated and thus more flexible and accessible forms. Besides Thompson's theoretical works, Dayan and Katz (1992) have pioneered in the empirical study of media events, which relate global media, television in particular, to issues of history, tradition, heritage, memory and identity. They demonstrate powerfully how electronic media produce 'instant histories', which are remembered by a large body of audiences through massive participation in media events.

As a complex organization, commercial television produces televisual history, heritage and memories in organizational processes which inevitably involve commercial and political calculations. Television's capacity for local, regional and global articulation invites contextual retouches of the collective memories that television produces. This paper uses the specific case of *Hong Kong Legend* (*HKL*) to interrogate the roles that popular television is playing in resituating Hong Kong inhabitants as members of the Chinese nation-state. It charts the complex contour of collective memory as produced by commercial television in the particular socio-political context of Hong Kong in the 1990s. Specifically, I ask, first, how does popular television represent Hong Kong identity during the transition period? Does it represent the past by selective remembering and forgetting? Second, what are the political, economic and institutional practices that are shaping the remapping of Sino-Hong Kong identity through televisual construction of memory and amnesia? Thus this paper comprises two parts which deal respectively with the textual ideologies and the production processes of the television program *HKL*.

I. Texts

Hong Kong Legend was part of the upsurge in nostalgic materials in the popular media of Hong Kong before the sovereignty reversion in July 1997. The program, comprised of 39 hour-long episodes, is a significant case because it was not broadcast on the fringes of the Hong Kong television schedule, but was produced and heavily promoted by TVB, the most powerful commercial broadcaster, which commands the dominant share of local and regional television markets. The program was not a project of TVB's news and public affairs department, but of the entertainment division, which produces the majority of the station's programming. It was initiated by and produced under the guidance of TVB's top level executives. Thus it is an appropriate case to study how the institutional logic of TVB plays out in the construction of collective memories. After unusually long preproduction planning of almost one year, the program was launched in August 1996, and ended in May 1997. Contents ranged from significant historical events to legendary stories, from Chinese traditions to western practices, and from city planning to public transport. The textual analysis below is based mostly on the first round of 13 episodes which received most public attention.

Hong Kong re-members

Although Hong Kong had been under British control and therefore politically separated from China for more than a century, the formation of a distinctive local identity only took place after the emergence of the Cold War

332   INTERNATIONAL journal of CULTURAL studies *1(3)*

era. In the 1970s, the new-found Hong Kong identity was largely constructed by foregrounding cultural differences between Hong Kongers and
mainland Chinese. In the mass media, mainlanders were stigmatized as
'uncivilized' outsiders against which modern, cosmopolitan Hong Kongers
could define themselves. In the 1970s and 1980s, mainland immigrants were
given a collective name, 'Ah Chian', which was a label with a derogatory
sting. It was the most popular name for the newcomers from mainland
China for more than a decade. The name originated from a television melodrama in which a character, nicknamed Ah Chian, came to Hong Kong from
China to rejoin his family (Ma, 1999). Since most Hong Kongers are ethnic
Chinese, the Sino-Hong Kong cultural differentiation, or the 'othering' of
mainland Chinese, was a significant process from which a distinctive local
Hong Kong identity has emerged. This desinicization produced an ambivalent and sometimes contradictory Sino-Hong Kong identity. On the one
hand, Hong Kong people identify with traditional Chinese culture in an
abstract and detached sense, but on the other hand, they discriminate
against the particular cultural practices which are affiliated with the Communist regime on the mainland.

After the news of the inevitable return of Hong Kong to China broke in
the mid-1980s, the Sino-Hong Kong identity boundary in the popular media
underwent a complex resinicization process. Here resinicization refers to the
recollections, reinvention and rediscovery of historical and cultural ties
between Hong Kong and China. The realignment of political power in the
1990s entailed the reconstruction of identity categories to legitimize new
social relations. No longer could one see the unrestrained stigmatization of
mainlanders in the vivid social imaginations of the popular media in the
1970s and the 1980s. An interesting case is the award-winning *Comrades,
Almost a Love Story* (produced in 1996), which romanticizes the story of
mainland immigrants of the 1980s and represents Hong Kong as a city built
by mainlanders rather than a city apart from China. While some popular
movies featured mainland heroes who were modern, sophisticated and
superior, some soap operas emphasized the virtuous characters of mainlanders associated with grand Chinese traditions. This remapping of the Hong
Kong identity boundary involved complex and dynamic struggles among
institutions and actors, which engendered selective processes of remembering and forgetting. It is in this context that *Hong Kong Legend* reconnects
Hong Kong to China.

The remembering of cultural roots contrasts sharply with the active forgetting of the ties with China in the 1970s, when Hong Kongers in general
did not have a historical bearing and were living in the eternal present. The
sharp Sino-Hong Kong identity border, which was unabashedly clear in the
media of the 1970s and still lingers in social discourse, is now remapped
and made blurry by placing Hong Kong within the tides of continuous immigration from China (episode 1). The previously invisible ethnic tie is made

visible again, while the previously visible differentiation between Hong Kongers and mainlanders is made invisible. The motif of a 'melting pot' is deployed (episode 8) to portray a harmonious mix of different ethnic groups of Chinese immigrants in Hong Kong. The history of emigration from China to the territory is rediscovered as the lifeline of Hong Kong, providing the city with hardworking immigrants and the entrepreneurial skills of mainland merchants. The program also features traditional legends, festivals, and long-forgotten ritual practices (episodes 9, 10 and 11), which are now replayed and highlighted to enhance the membership of Hong Kong within a reimagined cultural China. The once desinicized Hong Kong is now resinicized televisually as a member of the Chinese nation-state.

The politically correct past

The maiden episode of *HKL* provides an overview of the social crises during the previous 50 years. This ambitious grand opening was heavily promoted and attracted wide attention. After this first episode, the program returned to contents such as lifeways, festivals and transportation, which are politically less sensitive and more familiar to the production team. The first episode deserves close examination since it stretches to TVB's institutional limit and can indicate the ideological boundary which TVB allows.

The one-hour episode, entitled 'Turbulent Days', covers the Japanese occupation between 1941 and 1945 and the riots in 1956, 1966 and 1967. These riots and unrest are used to draw a lesson for Hong Kongers, who faced the uncertainty and change before and after 1997. The program stresses that Hong Kong people have an incredible ability to recover from blow after blow to their prosperity and stability. It constructs, within a historical narrative, the identity of Hong Kongers as tough, with a formidable survival spirit, and capable of turning curses into blessings. It is a process of concretizing Hong Kong identity through selective historical reconstruction.

This 'remembering' is accompanied by active 'forgetting'. Rather than dealing comprehensively with the major historical moments of post-war Hong Kong, the first episode of *HKL* elides significant moments of political turmoil. Popular street demonstrations in 1967, a watershed event in Hong Kong often seen as marking the emergence of a local political consciousness, are characterized in this documentary as a rebellion against colonial rule. The program makes no mention of the consequences of the spillover of China's Cultural Revolution into the territory. The riot in 1967 was a multifaceted incident which had political, social and industrial dimensions. It was *industrial* because it was triggered by the conflicts between Hong Kong citizens, factory workers and industrialists; riots and industrial conflicts hampered local industries and brought about grave economic loss. The incident was *social* because it had its origins in the frustration of the

334   INTERNATIONAL journal of CULTURAL studies *1(3)*

population over the unjust colonial administration. From a *cultural* perspective, the riot signalled the rise of a local consciousness and an emerging
indigenous cultural identity. The riot also had a very obvious *political*
dimension. It is widely remembered that the Cultural Revolution was the
cause of the resulting violence. Local leftists seized the opportunity to activate nationalistic and anti-colonial sentiments. Widely remembered also
were protesters holding 'the little red book' of the quotations of Mao
Zedong. In the program, archive film clips clearly show the protesters with
the little red book waving it frantically outside Government House. Some
extremists, who were labelled 'left guys', started to plant bombs as a challenge to the colonial rule. However, both the narrator and interviewees in
HKL fail to mention anything about the connection between local riots and
the political movement on the mainland. The audience can hear loudly and
clearly the industrial, cultural, and social aspects of the riots, with the anti-
colonial argument at the forefront. As Britain was losing its power, the colonial government became an easy target, while for China the Cultural
Revolution is still perceived as a politically problematic issue and was suppressed in televisual discourse.

Unfailing capitalism

Hong Kong Legend celebrates the territory as a land of opportunities and
attributes its success to the unique characteristics of its people. The economic miracle becomes a major component of the program. In fact, the discourse of unfailing capitalism has a grain of truth and cannot simply be
dismissed as purely ideological and illusory. Hong Kong's post-war economic take-off provided unprecedented opportunities of upward mobility for
the local generation. Economic transformation and sudden population
increase led to a truncation of whatever class structure existed before
(Leung, 1994). Despite the ups and downs of the world economy, Hong
Kong experienced a persistent economic growth all through the post-war
decades. Contrary to popular prediction, political transition did not hamper
economic growth and the Hong Kong economy was very strong in the mid-
1990s until the Asian economic crisis in the last quarter of 1997.

Backed up with such an economic achievement, *HKL* expresses a fully
blown capitalistic discourse. Often repeated is the phrase 'Hong Kong has
evolved from a small fishing port to a leading financial center in the world'.
This discursive statement is either made explicit or becomes the underlining
logic of many episodes. In episode 2, real estate agents and entrepreneurs
talk about a bright future for the real estate market despite the fact that
unreasonable real estate prices have been widely condemned and have gone
far higher than the affordable limit of an average middle-income family.
There are many stories of successful entrepreneurs who stress their personal
ability to capture the opportunities given to them. In episode 3, a young

billionaire is interviewed.[3] The selling point of his story is that 'a thirty-nine year old Hong Kong man is able to make two hundred millions' (episode 3). This line is consciously used by the programmers in program teasers and promos. There has been virtually no sad story of the social underclass or of the poor. Even members of subordinate social groups, if they appear on the program, seem satisfied with their living conditions. Those who participate and work hard can eventually make a success here. For Hong Kongers, economic success is the source of pride and constant reassurance in a time of uncertainty. Personal life stories of successful businessmen tend to decontextualize success by highlighting personal endeavour and downplaying family background (Peneff, 1990). In the program, these stories are further set in the ideological context of capitalism and grounded within a historical narrative. This positioning further essentializes capitalistic ideology as an inseparable part of Hong Kong identity. This 'excessive' capitalistic discourse coincides with the discursive position offered to Hong Kong by the international community. It is also politically correct since China continuously defines Hong Kong as a commercial and apolitical city. In the 1990s, Hong Kong is politically subordinate to China. Yet, Hong Kong's brand of capitalism has become a dominant ideology underpinning the open-door politics of mainland China. Hong Kong, as a political periphery, has become a role model for China's economic reform. The excessive capitalistic discourse in *HKL* should be seen in this discursive position on national reintegration. Hong Kong's social and political activities are masked and restricted under these unfailing and excessive capitalistic discourses (Chiu, 1996).

Hong Kong the cosmopolitan

Besides remembering ethnic, cultural and historical ties with China and forgetting politically incorrect history, the program also mobilizes stereotypical motifs such as 'the meeting point of the East and the West', 'the melting pot of Chinese and western culture', and 'an internationalized society embracing ethnic Chinese and foreigners'. The program fosters a strong sense of locality for the people of Hong Kong by foregrounding the city's distinctive historical mix of peoples and experiences, while at the same time stressing Hong Kong's international ties and its willingness to be an active member of the global community.

In the program, Hong Kong is frequently given labels such as the world's leading trade centre and the world's financial centre. Hong Kong is located in a dominant position in global and regional capitalism. *HKL* tells the stories of how new innovations, such as airplanes and movie cameras, were imported to Hong Kong within days and months after their invention in the west. From the large quantity of archive footage used in the program, the audience observe a harmonious mix of Chinese and foreigners at the horse races, in streets and markets, trams and rickshaws. As visual dressing, these

336 INTERNATIONAL journal *of* CULTURAL studies *1(3)*

images are not used in a way which is sensitive to the implicit exploitative nature of colonial practices. In some instances, anti-colonial discourse is obvious but not militant. Here and there are comments such as 'westerners were more powerful in the past, but local people have gained status and security as they have been making lots of money'. Foreign culture is treated as a natural and modern component of the Hong Kong experience. The mixing of western and Chinese practices – Christmas, Valentine day, Halloween, Mid-Autumn festival, Chinese and western wedding and funeral rituals – is not conceived as a dilution of indigenous culture but rather a unique Hong Kong characteristic which is to be celebrated. This insensitivity to cultural barriers and differentiation with the west is quite astonishing if compared with some xenophobic Asian countries. For instance, in Japan, popular intellectuals have been promoting a unique Japanese culture with an ontological difference from western culture (Yoshino, 1992). In contrast, the locality of Hong Kong culture is not essentialized in the expense of excluding the global. Rather, the uniqueness of Hong Kong lies in the 'ontological' connection and hybridization of the local and the global, the Chinese and the western, and the traditional and the modern.

The above textual analysis has shown that *HKL* constructs Hong Kong as a resinicized, apolitical, capitalistic, and international city. These textual descriptions are inevitably schematic for a long-running series such as *HKL*, but the descriptions are based on dominant textual features which are by and large clear and obvious. In the next section, I shall map the complex articulation between these textual configurations and the socio-political context of *HKL*.

II. Contexts

The textual representations spelled out in the previous section are densely interrelated with the social contexts, not in a simple and direct way, but mediated through organizational processes which involve conscious and unconscious political and commercial calculations. Through participant observation, I shall examine the program history and the production routine of the documentary and suggest some of the contextual processes at play.

HKL is a very specific case, but it is also a localized manifestation of a global trend of nostalgic consumption. As Lowenthal (1994) points out, people the world over, even when exalting unique heroes and virtues, celebrate success, stability, progress in much the same way. Part of the reason for this is that the pursuit of popular history worldwide can be related to the destabilizing effects of modernization and the pessimistic mood associated with the 'declining' industrial and imperialist west. The past becomes a decontextualized identity resource for coping with the rapidly changing present. However, the past is also deployed for political and commercial ends. In the past few decades, major western countries have organized public rituals and grand anniversaries to foster political commitment by popularizing national histories. Towards the end of the 20th century, there is a strong current of commercialization of the past by popular media, global tourism and local heritage industries. Historical narratives, myths, icons, stories and artefacts are framed and repackaged in museums, festivals, parades, heritage sites, music videos and nostalgic infotainment shows.

The production of *HKL* reflects this global need to strengthen collective identity in times of change. Yet this global trend manifests itself in the sociopolitical particularities of the Hong Kong context. In the year before the sovereignty change in July 1997, magazines and newspapers produced specials and supplements on histories and collective memories of post-war Hong Kong; there were popular books, memoirs, photo-collections in book stores and newspaper stalls. In all the major electronic media there were programs on historical and nostalgic themes. Among similar television programs, *HKL* received more public attention and was the longest running of all. In this section, I shall first talk about the political economy of the program. In the last part, I shall single out my experience of the first day of studio recording to illustrate the ideological contestation embedded in the production process of *HKL*.

The political economy of electronic memories

The political economy approach to the media stresses the determination of political and economic powers in the textual production process (Murdock, 1982). In a free capitalist society like Hong Kong, the media do not have a direct link with the polity. However, editorial influence can be achieved through political and commercial networking within and beyond media corporations. As Hong Kong's first commercial broadcast television station, TVB dominated a rapidly growing broadcast and advertising market throughout the 1970s and 1980s. In the 1990s, increasing competition from other media producers and technologies, along with a stagnant advertising market, caused TVB's profit rates to fall. Pressured to develop new revenue streams, TVB executives began aggressively to pursue expansion into mainland and overseas Chinese markets. This corporate strategy requires TVB to maintain good relations with the Chinese government. Besides, the local media market has also been sensitized to Chinese politics. The power shift in political transition has privileged the Chinese government and its affiliates. From the early 1990s, the proportion of local advertising money connected with Chinese interests steadily increased. In the early 1990s, news broke that the Chinese government had blacklisted 'hostile' newspapers and asked all Chinese business affiliates not to place advertisements in these papers. Political influence, which sinks through in economic terms, sensitizes media executives to the imagined ideological boundary that they should not trespass (Chan et al., 1997).

The textual analysis has shown that *HKL* suppresses politically incorrect contents and produces an excessively capitalistic discourse. This textual ideology is congruent with Hong Kong's socio-political situation, in which Hong Kong has increasingly been under the influence of Chinese politics and served, in economic terms, as a gateway between China and the West. Participant observation reveals that this resonance between the text of *HKL* and the power geometry of Hong Kong is articulated in complex organizational process.

Sponsor

At first Hong Kong and Shanghai Bank, the quasi-central bank of the colony, was quite enthusiastic in sponsoring the program. The bank had launched a series of advertisements playing on collective memories of postwar Hong Kong. These TV ads aimed at boosting the bank's image of standing by Hong Kong in all the previous years and the years to come. *HKL* has a strong intertextual link with this corporate image. However, the bank suddenly pulled out. Outsiders will never know the real reason for the last-minute withdrawal. But according to Ng Ho, the program's consultant, the bank might have thought sponsoring a program on local history politically too risky. Finally, the program came under the sponsorship and guidance of Goldlion, a very powerful Hong Kong financial interest with strong ties to China. Tsang Hin-chi, a politically active businessman who frequently expressed a strong pro-China stand in public forums, runs Goldlion.[4]

Although members of the production team claimed that they were not subject to intervention by the marketing department, there were widespread perceptions of implicit boundaries and constraints. For instance, when I discovered that Tsang, the sponsor, was on the list of the program's interviewees, I asked a senior researcher whether TVB took the initiative or Tsang made the request. His answer was that Tsang did not ask for it, but the team thought it was 'appropriate' to interview the sponsor, since Tsang is a rather famous and successful Hong Kong businessman. The program ended up having an interview of Tsang retelling how he succeeded after years of hard work. Another incident was the choice of a presenter to replace me after the first set of 13 episodes.[5] The production team invited Wong Yuk-man, a famous talk-show host known for his critical stand towards Communist China. At a farewell barbecue, the executive producer told me that they had contacted Wong but the sponsor was not happy with the choice for political reasons. He stressed that they would make the decision independently. When the choice was made, it was not Wong but a movie actor/director, Mr Cheung Kin Ting, who replaced me.

Self-censorship

This informal feedback maps out and strengthens the imaginary ideological boundary for producers and executives. Since *HKL* was a project of the entertainment division, the production was less bound by journalistic norms and more susceptible to ideological pressure. Deserving closer examination is the revision of the first episode, which says a lot about the logic of self-censorship. In July 1996, a rough cut of the first episode was made by the producer and was approved by the executive producer and senior creative director. The rough cut was then circulated and screened to unit managers and executives. It did not get through. TVB took the trouble to reshoot the first episode which required, among other things, a dozen additional expensive studio hours. The reason for this, according to the team, was that the draft cut stressed the hardship and ordeals of previous social crises too much and the management wanted to shift the focus to how people recovered from those crises. I was able to get the unreleased version from reluctant producers several months after I left the program. Comparing the two, it becomes obvious how malleable 'electronic memory' can be. Basically the two versions are using similar historical 'facts' (audio-visual materials, interviews, voice-over narration, etc.), but different compilations construct quite different ideologies. The differences can be summarized as follows.

Offering a more optimistic outlook

The extensive and vivid descriptions of social turmoil were edited out. In both versions there are about 14 minutes of program time covering the 1967 riot. Within the 14 minutes in the new version, there are $3\frac{1}{2}$ minutes of new interviews which include how factory workers and owners worked together to meet deadlines. The message is clear; the new version interprets the crisis along the industrial dimension and emphasizes the ability of Hong Kong to regain its economic vitality. How local and left-wing activists 'recovered' and 'revised' their course of action is completely invisible.

Suppressing the politically incorrect

The unreleased version includes a studio interview involving two interviewees and myself. One is a veteran actor Mr Lam, who helped the police clear up street bombs during the riot. Mr Cheng Kai Nan, the other guest, who participated in the demonstration in 1967, had become the vice president of a pro-China political party.[6] In the interview, I asked about his role in the 1967 demonstration, why the demonstration turned violent, and whether he took part in the violence. I also briefly related the local riot of the 1960s to situations in France and China. Cheng coined the term Cultural Revolution and implicitly related it to the riot in Hong Kong. The

340   INTERNATIONAL journal *of* CULTURAL studies *1(3)*

interview was politically very tame and descriptive. However, the whole studio interview was edited out and replaced by a new interview of Mr Lam alone. This time another presenter, Miss Winnie Yeung (Miss Hong Kong 1995), went to Lam's home and repeated the interview about his heroic and dramatic deeds in 1967. Lam told the same entertaining human-interest story that he told me in the studio session. The reason for the retouch is obvious. The well-known pro-China identity of Cheng could easily trigger the widely remembered leftist involvement in the riot. In contrast, Lam's comments are very show biz and can dilute the political aspect of the riot.

Legitimizing censorship

The new version tries to legitimize the show and build up its credibility. First, the new version plays up the involvement of dozens of consultants by introducing them and putting their names, titles and photos on screen. All of them are professors, historians and heads of museums and research institutes. Second, the new version superimposes Chinese characters on the documentary footage during voice-over narration; this audio-visual treatment resembles serious documentary and current affairs programs. Third, it adds new maps and location shootings as visual paddles to retell the riots in a more detailed and visually attractive form. All these changes, according to the executive producer, reflect the grave concern and goodwill of the management to present history in a clear and orderly manner, so as to inform the audience of what had happened in those significant moments of Hong Kong history. The intriguing thing is that TVB has the goodwill to tell a 'factual story' but at the same time deletes a significant political aspect of the riot. The political censorship is taken for granted and does not trouble the good conscience of these corporate executives. The ideological is dressed up by a technical, factual, current affairs mode of discourse which naturalizes, obscures and legitimizes the uneven distribution of textual meanings in service of the perceived political power.

It should be noted that this streamlining of the first episode was not done in a commandist fashion which links political powers directly to media executives and program producers. It was done in a series of negotiations with perceived and imaginary political boundaries which are subjective and ambiguous. Through negotiative cooperation of a group of corporate employees, the imaginary ideological boundary was mapped into the text. This was especially true for a large production such as *HKL* which had a high-profile corporate involvement. I am not suggesting a direct link between textual ideology, the production team, the sponsor and political powers in the larger context, but it is quite clear that sponsors' publicly declared political and economic interests in mainland China are matched by TVB's own corporate strategy of maintaining good relations with the Chinese government.

The streamlining of episode 1 suggests a process of overdetermination on textual ideology. What I mean by overdetermination is that multiple contextual imperatives (sponsor's preference, commercial pressure, corporate strategy, development of regional market, perceived political pressure, etc.) are not connected but nevertheless work towards similar directions and thus articulate political and economic powers in the production of textual ideology.

The medium is the message

My analysis so far is a combination of a political economy and Gramscian approach to chart the relation between the textual and the contextual. The political economy approach has the strength of highlighting the determining power of political and economic factors, yet it fails to describe the fine grain articulation of the textual production process. In the above, I have supplemented political economy analysis with the Gramscian approach, which pays more attention to textual analysis and focuses more on the overdetermination of textual ideology by multiple discursive powers (Hall, 1986). Here I would like to add an organizational dimension to the analysis, which I think can enrich the Gramscian approach by grounding the articulation process in the context of the medium of television. Besides contextual overdetermination, the very nature of commercial television, to a certain extent, can also define the messages it is able to carry. My participation in the production provides a chance for me to 'experience' the production process and the 'discursive force' of overdetermination. My emotive and analytic encounters become, metaphorically, the 'scale' and 'measurement' against which I can conceptualize the characteristics of commercial television. By characterizing commercial broadcast television as a 'fuzzy' process which presses for the production of 'fluent' and 'consensual' televisual texts, I try to rethink McLuhan's media determinism (1964) and examine how far televisual ideologies are constituted by the televisual form. The following ethnographic experience of my first day of studio recording will provide this abstract argument with concrete illustrations.

Production pressure

The process of television production seems to have a built-in inhibition of ideological negotiation. Despite TVB's promise to give me the studio script in advance, I was only able to get the script when I arrived at TVB five hours prior to the studio recording of the first episode. I wished to make some revisions but could not do so because those five hours were fully occupied by activities which involved dozens of TVB staff. I had to get through the routines of make-up, hair-styling, costume, rehearsal, having lunch with senior executives and attending press conferences. During the studio

342   INTERNATIONAL journal *of* CULTURAL studies *1(3)*

recording there was a team of no less than 20 staff: cameramen, lighting and other studio technicians, researchers, producers and assistants, all worked together to orchestrate the images and soundbites on tape. The stereotypical image of a lone artist pondering over a creative story is completely different from the industrial production process of television. The production comprises a team of creative workers, not lone artists. All working staff and studio facilities are costly. Since overtime rates are high, the producer works under great pressure to finish the recording on time.

Textual fluency

During recording sessions, most attention was paid to the way the presenters looked and talked. We worried about our appearance and whether our lines were spoken smoothly and interestingly. Researchers on the spot were very conscious of what should and should not be said, what details were missing and how the individual segments could be bridged together by the correct tags and intros. We worked together to sweeten the dialogue, iron out the wrinkles and simplify complications. In television there is a tremendous emphasis on the 'textual surface' of the show. In contrast, little effort is paid to negotiating the complexity and subtlety behind the textual surface. The team couldn't care less about the wrinkles on the back of my shirt as long as they were covered by my coat. It is fine if the 'surface' looks good. During the recording, my identity as an academic asked me to be critical while my identity as a presenter asked me to look friendly, and talk interestingly and fluently. The production environment tended to suppress the former and bring out the latter. The team did not want a critical professor, but they did want my academic identity to serve the preferred identity as a presenter. They asked the other two presenters to address me as 'Dr Ma,' in English, because it sounded more authoritative. My academic title is instrumental, it helps enhance credibility and cover up ideological wrinkles. My inability to shape the text was as much a product of my designed role as it was of the organizational routines. Negotiations require time for research and development which television broadcasters often cannot afford. Since researchers and writers have to produce 'fluent' scripts within a tight time frame, they try the most convenient information sources such as popular books, prominent experts and easy informants. They reproduce the most popular ideas which are circulated and readily available in social discourses. They play the role of cultural brokers who popularize mainstream ideology in an audio-visually attractive and fluent format.

Manufacturing consent

The first episode inhibits politically incorrect interpretation of the 1967 riot. Since I took part in the production, I have to ask myself how and why I cooperated in the manufacture of consent. With the benefit of hindsight, I have realized that one needs to be very sensitive and determined to work against the discursive force of commercial television. First, although I have written a book (Ma, 1996) in which I state explicitly the connection between local demonstrations in 1967 and the Cultural Revolution in China, I was not alert enough to point out the missing element in the script in the hectic production process. Part of the reason for this was I was only eight years old during the riot and can recall very few details of what had happened. My knowledge about the incident is mostly acquired from books. I did not have the personal memory and direct experience to compare with the reconstruction in *HKL*. Second, my collaboration had something to do with my expectation of maintaining good relations with TVB and my awareness of the imagined boundary of commercial television of not saying things that are too politically provocative. On top of these factors, I was only able to get a skeletal script for the studio recording. The suppression of the political aspect of the 1967 riot was less discernible given the incomplete nature of the script. In the unreleased version of episode 1, I was able to introduce a very tame political interpretation of the riot in the unscripted interview with politician Cheng Kai Nan and actor Lam, but the interview was replaced. In fact, I was not aware of the censorship until I read the critical reviews in newspapers. The act of censorship was semi-conscious and camouflaged in the hectic and 'fuzzy' process of television production.

Fuzzy process

As a compilation program, *HKL* is multivocal and polysemic. It contains both conservative and liberal ideologies expressed in various information sources. The production process is 'fuzzy' in the sense that it involves too many actors and contingencies which allow the articulation of polysemic and contradictory elements in the text. Since political pressure is not direct, *HKL* has not reduced histories to an official history of Hong Kong. During the production, I was able to revise and delete, with ease, some celebratory phrases in the scripts and dilute the excessively economic interpretation of Hong Kong success. There were also occasional instances of researchers bringing in interviews with ordinary people. However, alternative views were usually reduced to very short fragments compared with the mainstream ideology embodied in long and extensive interviews with successful businessmen in glamorous settings. *HKL* carries mainstream ideologies with occasional polysemic alternatives.

These observations suggest that the production of commercial broadcast television can be characterized as a fuzzy process which presses for 'fluent' and 'consensual' televisual text. Ideological contestation is sometimes disguised or recast in terms of production pressures. Discourse theories talk about subject positions which constitute discursive relations (e.g.

Fairclough, 1995). Likewise, the concept of discourse can be applied to the production process, conceptualizing it as an organizationally embodied discourse. TV production is a complex process with 'positions' offered to lots of people: scriptwriters, presenters, artists. For those who have internalized the production norms, these positions become their institutional habitats (Ma, 1995). They feel comfortable playing a cooperative role in textual production, a process in which they themselves become structural parts. Thus, in *HKL*, we have willing researchers and writers relaying popular histories from popular sources. This popular history tends to confirm but not examine given and existing knowledge. Of course, the production team are · not merely the unconscious agents of institutionalized ideological processes. As Bennett and Woollacott (1987) argue in their study of James Bond films, deliberations and calculations which actually inform the making of a text have a direct and discernible bearing on the processes through which ideologies are worked over and transformed into a specific textual form. However, ideological deliberations are sometimes disguised or recast in terms of production pressure.

The specifics of the medium which I have spelled out combine to become a strong discursive force which presses for the mainstream. I should qualify my argument and restrict it to the case of commercial broadcast television, specifically, the entertainment division. Since there are institutional variances within and between different types of television organization, my analysis does not apply to the medium of television as a whole. In fact, there were other competing history programs produced by TVB's news department and the quasi-government television RTHK.[7] The point to make here is that televisual discourse generated by commercial broadcast television has a unique kind of 'organizational texture' which predisposes towards mainstream ideology. There is no production-line content control. Creative control is achieved through unobtrusive means such as cost consciousness, mutual expectation, internalized agreement between the production team and executives, the constant need for programs to fill up airtime, the tremendous emphasis on 'textual surface', all these blend together to become a unique production process which I describe as a fuzzy televisual process. Mainstream ideologies, stereotypes, prominent myths, or in Hall's words, 'pick-up' meanings (Hall, 1980), fit more easily into this televisual process. Alternative and oppositional discourses, which require negotiation, and thus time and resources, are less likely to be expressed in commercial television. The fuzzy process can absorb polysemic elements and contradictions, but as a whole commerical television still predisposes textual meanings within the imagined ideological boundary overdetermined by contextual powers. Of course this predisposition does not seal off alternative perspectives. This same fuzzy process might make it easier at certain moments for members of the production team to advance alternative meanings. Here resistance and opposition come into play, but these moments of ideological competition should be put into the context of the production process as a whole.

Concluding remarks

Recent scholarship has denaturalized identity and problematized the past by highlighting their subjective and constructionistic nature. Consequently, researchers now pay more attention to little histories of small communities. They privilege individual memories which play a large role in local histories. However, thick descriptions of particular cases might easily lose sight of the larger social structure at work (Furedi, 1992). In this article, I have described a very particular case of televised collective memory in Hong Kong, but at the same time related the particular textual patterns to the overdetermination of multiple contextual forces.

Like other popular histories, mediated memories in *HKL* are celebratory and consensual. But the consensual surface is in fact the product of complicated contestation. Twisting Chamber's stylish prose (1994: 3), Hong Kong history is harvested and collected, to be assembled, made to speak, remember, reread and rewrite, and televisual language comes alive in transit, in interpretation, and transforms the collective beyond the boundary of individual memory. The production of televisual past is subjected to political and economic constraints within and beyond the television organization. For commercial television, not all pasts are viable. In the case of *HKL*, the past of Hong Kong involves global, national and local positioning. These contextual forces restrict the kind of 'memory work' which TVB is able to produce.

In the past few decades under colonial rule, Hong Kong had fostered an indigenous culture without direct national imperative. In the Cold War era, the colony, with mostly Chinese inhabitants, did not affiliate with China, but considered itself an affiliate of the west. Located in a key satellite position of global capitalism, the colony was desinicized and internationalized with relative ease. It embraced the west without sensing any intrinsic contradiction and cultural barrier. This easy positioning was not only an indigenous choice, but also a space offered to Hong Kong by global and colonial powers. The cosmopolitan city described in *HKL* is a representation of this global and colonial positioning.

In the run-up to 1997, there was no strong inhibition of Hong Kong being an international city, yet Hong Kong was increasingly under pressure to nationalize its culture. Nationalizing Hong Kong culture involved changing the spaces of public culture so as to limit the realm of what might constitute meaningful choice and position (Chun, 1996). It involves the inhibition of politically incorrect ideologies and the installation of nationalistic identity. However, this nationalizing has to negotiate with a very localized Hong Kong culture which is characterized by a near paranoid alienation from

346   INTERNATIONAL journal *of* CULTURAL ⬭studies⬭ *1(3)*

communism. Hong Kongers identify with China in a cultural sense, which is affiliated with the ambiguous 'Great Tradition' of Chinese civilization (Wang, 1991). However, this cultural affiliation has a missing link with Communist China. *HKL*, and the recent upsurge of nostalgic materials in the media, somehow serves as an agent to rehistoricize and resinicize Hong Kongers into a continuous Chinese history. This resinicization strengthened the historical identity of Hong Kong, yet the national identity, which often implies implicit political commitment, was still incompatible with the indigenous Hong Kong culture. Although the Chinese government voiced the wish to install a nationalistic identity in the territory, the locally produced *HKL* did not carry strong nationalistic sentiments. It fostered a receptive cultural space for the sovereignty change but did not celebrate or promote a patriotic national identity.

In these global, national and local contexts, *HKL* materializes the past of Hong Kong as a cultural product of commercial broadcast television. These contexts provide the available cultural spaces for *HKL* to construct a viable past and a set of identity possibilities. The articulation of the textual and the contextual is achieved by the overdetermination of multiple imperatives, which include global and national positioning, sponsor's preferences, market pressure, corporate strategy and perceived political boundary. It is a semi-conscious process of willing compromise and cooperation, of selective remembering and forgetting, orchestrated by a fuzzy production process which is medium specific. The medium, in part, constitutes the message.

The collective memories reproduced in *HKL* are not context free. It is in the above contexts, articulated by the electronic medium of commercial television, that *HKL* discursively constructs Hong Kong as a resinicized, apolitical, capitalistic, and international city with an aestheticized past. 'Electronic memories' recollected by *HKL* are characterized by the repression of the politically incorrect, the exaggeration of economic success, the cover-up of depravities and the sublimation of the extraordinary. The past is complexly mediated and transformed by memory, fantasy, desire and, increasingly, by the media. This case study has mapped out, only partially at best, the complexity of these mediated transformations.

Notes

Earlier versions of the paper were presented at the 5th International Symposium on Film, Television and Video, Taiwan, May 1997 and the 47th Annual Conference of the International Communication Association, Montreal, Canada, May 1997.

1 It is literally titled 'Hong Kong Legend' in Chinese. TVB's official English title for the program is 'Hong Kong Epic Heritage'.

2 Memory (Butler, 1989; Lass, 1994), heritage (Brett, 1996; Walsh, 1992),

tradition (Hobsbawm and Ranger, 1983) and history (Lowenthal, 1994), in the same order of relative stability, are said to be embedded in highly problematic processes of retrieval and revision by actors of the present.

3 Lee Siu Fung, head of a corporation which specialized in pottery.

4 In 1995 he was named by the media as one of the prominent candidates for the chief of the Special Administrative Region (SAR) of Hong Kong. Tsang was latter involved in a scandal, being accused of hiding his criminal record from some decades ago. In 1996 his name was not on the list of the candidates who ran for the seat of the SAR chief.

5 I withdrew voluntarily because of the unresolved contradiction between my identity as an academic and my designed role as a presenter. TVB shared the same sense of contradiction (I had commented on the issue of censorship in the newspapers after the first episode was released) and did not ask me to stay when I told them my decision not to host another 13 episodes.

6 The Democratic Alliance for the Betterment of Hong Kong, formed in 1992, represents the major pro-China party in Hong Kong.

7 This paper focuses on the textual and the contextual. The negotiation between competing texts and audiences is dealt with in a forthcoming paper in *Social Text*, issue 58.

References

Abbas, A. (1997) *Hong Kong: Culture and the Politics of Disappearance*. Minneapolis: University of Minnesota Press.

Anderson, B. (1983) *Imagined Communities: Reflections on the Origin and Spread of Nationalism*. London: Verso.

Bennett, T. and J. Woollacott (1987) *Bond and Beyond: The Political Career of a Popular Hero*. London: Macmillan Education.

Brett, D. (1996) *The Construction of Heritage*. Cork: Cork University Press.

Butler, T., ed. (1989) *Memory: History, Culture and the Mind*. Oxford: Blackwell.

Chamber, I. (1994) *Migrancy, Culture, Identity*. London: Routledge.

Chan, J., E. Ma and C. So (1997) 'Back to the Future: A Retrospect and Prospects for the Hong Kong Mass Media', in J. Cheng (ed.) *The Other Hong Kong Report 1997*. Hong Kong: Hong Kong Chinese University Press.

Chiu, F. (1996) 'Politics and the Body Social in Colonial Hong Kong', *Positions* 4(2): 187–215.

Chun, A. (1996) 'Discourses of Identity in the Changing Spaces of Public Culture in Taiwan, Hong Kong and Singapore', *Theory, Culture & Society* 13(1): 51–75.

Cohen, A.P. (1985) *The Symbolic Construction of Community*. London: Routledge.

Connerton, P. (1989) *How Societies Remember*. Cambridge: Cambridge University Press.

348   INTERNATIONAL journal of CULTURAL studies *1(3)*

Dayan, D. and E. Katz (1992) *Media Events: The Live Broadcasting of History.* Cambridge, MA: Harvard University Press.

Fairclough, N. (1995) *Media Discourse.* London: Edward Arnold.

Friedman, J. (1994) *Cultural Identity and Global Process.* London: Sage.

Furedi, F. (1992) *Mythical Past, Elusive Future: History and Society in an Anxious Age.* London: Pluto Press.

Gillis, J. (1994) 'Memory and Identity: The History of a Relationship', in J. Gillis (ed.) *Commemorations: The Politics of National Identity.* Princeton, NJ: Princeton University Press.

Halbwachs, M. (1992) *On Collective Memory.* Chicago, IL: University of Chicago Press.

Hobsbawm, E. and T. Ranger, eds (1983) *The Invention of Tradition.* Cambridge: Cambridge University Press.

Hall, S. (1980) 'Recent Developments in Theories of Language and Ideology: A Critical Note', in S. Hall, D. Hobson, A. Lowe and P. Willis (eds) *Culture, Media, Language.* London: Hutchison.

Hall, S. (1986) 'The Problem of Ideology: Marxism without Guarantees', *Journal of Communication Inquiry* 10(2): 28–44.

Lass, A. (1994) 'From Memory to History', in R. Watson (ed.) *Memory, History and Opposition under State Socialism.* Santa Fe, NM: School of American Research Press.

Leung, B.K.P. (1994) 'Class and Class Formation in Hong Kong Studies', in S.K. Lau et al. (eds) *Inequalities and Development: Social Stratification in Chinese Societies.* Hong Kong: HK Institute of Asia-Pacific Studies, CUHK.

Lowenthal, D. (1994) 'Identity, Heritage, and History', in J. Gillis (ed.) *Commemorations: The Politics of National Identity.* Princeton, NJ: Princeton University Press.

Ma, E. (1995) 'The Production of Television Ideologies', *Gazette* 55: 39–54.

Ma, E. (1996) *TV and Identity.* Hong Kong: Breakthrough. In Chinese.

Ma, E. (1999) *Culture, Politics and Television in Hong Kong.* London: Routledge.

McLuhan, M. (1964) *Understanding Media.* London: Routledge & Kegan Paul.

Morley, D. (1993) *Television, Audiences, and Cultural Studies.* London: Routledge.

Morley, D. (1996) 'EurAM, Modernity, Reason and Alterity or, Postmodernism, the Highest Stage of Cultural Imperialism?', in D. Morley and K.H. Chen (eds) *Stuart Hall: Critical Dialogues in Cultural Studies.* London: Routledge.

Morley, D. and K. Robins (1995) *Spaces of Identity: Global Media, Electronic Landscapes, and Cultural Boundaries.* London: Routledge.

Murdock, G. (1982) 'Large Corporations and the Control of the Communications Industries', in M. Gurevitch et al. (eds) *Mass Media and Society.* London: Edward Arnold.

Peneff, J. (1990) 'Myths in Life Stories', in R. Samuel and P. Thompson (eds) *The Myths We Live By.* London: Routledge.

Rappaport, J. (1990) *The Politics of Memory.* Cambridge: Cambridge University Press.

Thompson, J.B. (1990) *Ideology and Modern Culture.* Cambridge: Polity Press.

Thompson, J.B. (1995). *The Media and Modernity: A Social Theory of the Media.* Oxford: Polity Press.

Walsh, K. (1992) *The Representation of the Past: Museums and Heritage in the Post-modern World.* London: Routledge.

Wang, G.W. (1991) *China and the Chinese Overseas.* Singapore: Times Academic Press.

Watson, R. (1994) 'Memory, History and Opposition under State Socialism: An Introduction', in R. Watson (ed.) *Memory, History and Opposition under State Socialism.* Santa Fe, NM: School of American Research Press.

Yoshino, K. (1992) *Cultural Nationalism in Contemporary Japan.* London: Routledge.

● **ERIC KIT-WAI MA** is Assistant Professor of Communication at the Chinese University of Hong Kong. He is the author of several books written in Chinese and a forthcoming book entitled *Culture, Politics and Television in Hong Kong. Address*: Department of Journalism and Communication, Chinese University of Hong Kong, Shatin, Hong Kong. [email: B682790@mailserv.cuhk.edu.hk] ●

Part III
Recent Trends in Political Development

[9]

Hong Kong's Embattled Democracy: Perspectives from East Asian NIEs

Alvin Y. So

This article examines the origins and development of Hong Kong's embattled democracy. It seeks to explain why, although Hong Kong has already acquired most of the "prerequisites" for democratization, its democracy has been so restricted and contested compared to those in the East Asian newly industrialized economies (NIEs). It also argues that Hong Kong possesses much less favorable structural conditions for democratization than its East Asian neighbors because it has a stronger conservative alliance of state and Big Business on the one hand, and a weaker populist alliance of service professionals and grass-roots population on the other. However, the Tiananmen incident of June 1989 revitalized the Hong Kong democracy project almost overnight, and now that the democrats have been strengthened, they have been able to contest the rules of restricted democracy as imposed by the conservative alliance of Beijing and Big Business.

Keywords: democracy; comparative politics; Hong Kong; Taiwan; South Korea

* * *

The "third wave" of democratization finally reached East Asia in the 1980s,[1] as democratization was carried out from above in Taiwan and below in South Korea. In Taiwan, the Kuomintang (KMT, Nationalist Party) lifted martial law, allowed opposition parties to compete in the electoral arena, revised the constitution, and opened up key legislative posts for elections. In South Korea, prolonged workers' strikes and urban protests first forced the military Chun Doo Hwan government to step down, then helped to institute

---

Alvin Y. So is a Professor of Sociology at the University of Hawaii. His recent publications include *East Asia and the World Economy* (co-author, 1995) and *Hong Kong-Guangdong Link: Partnership in Flux* (co-editor, 1995).

[1]Samuel P. Huntington, *The Third Wave: Democratization in the Late Twentieth Century* (Norman: University of Oklahoma Press, 1991).

ISSUES & STUDIES

competitive elections in the South Korean polity. Consequently, literature posits democratic breakthroughs in Taiwan and South Korea in the 1990s.[2] Not only has there been a high level of civil liberty, political participation, and competition for key positions of government power, but election rules have also been institutionalized in both countries.

At first glance, it seems that Hong Kong should follow the paths of Taiwan and South Korea toward democratization as it has attained most of the necessary "prerequisites."[3] Similar to Taiwan and South Korea, Hong Kong has experienced a rapid rate of economic growth, a high level of wealth, a rising middle class, and the lack of extreme and intolerable inequalities. Following Samuel P. Huntington's wealth explanation,[4] Hong Kong's robust economy should encourage high levels of industrialization, education, literacy, and mass media exposure, all of which are conducive to democracy. In addition, unlike Taiwan and South Korea, the Hong Kong colonial government was liberal; its civil society was well-developed; and it had not undergone authoritarian rule, a military regime, or press censorship. Following Larry Diamond's argument,[5] such a strong civil society should lay the foundation for Hong Kong's democratic transition.

Historical development over the past two decades, however, reveals that Hong Kong has experienced a much more restricted and contested democratization than Taiwan and South Korea:

1. As a *restricted democracy*, Hong Kong has a narrow scope of electoral competition. Instead of a one-person, one-vote direct election, Beijing's Basic Law (mini-constitution) for the post-1997 Hong Kong Special Administrative Region (SAR) favors indirect elections. Thus, after 1997, the Chief Executive of Hong Kong will be indirectly selected by a 600-people electoral commission. In the 60-seat legislature, only one-third of the seats will be directly elected, while the remainder will be indirectly elected through functional (i.e.,

---

[2]Tun-jen Cheng and Eun Mee Kim, "Making Democracy: Generalizing the South Korean Case," in *The Politics of Democratization: Generalizing East Asian Experience*, ed. Edward Friedman (Boulder, Colo.: Westview Press, 1994), 125-47; Tien Hung-mao, "Taiwan's Evolution Toward Democracy: A Historical Perspective," in *Taiwan: Beyond the Economic Miracle*, ed. Denis F. Simon and Michael Y. M. Kau (Armonk, N.Y.: M. E. Sharpe, 1992), 3-23.

[3]Seymour Martin Lipset, "The Social Requisites of Democracy Revisited," *American Sociological Review* 59, no. 1 (1994): 1-22.

[4]Samuel P. Huntington, "Will More Countries Become Democratic?" *Political Science Quarterly* 99, no. 2 (1984): 193-218.

[5]Larry Diamond, "Introduction: Persistence, Erosion, Breakdown, and Renewal," in *Democracy in Developing Countries: Asia*, ed. Larry Diamond, Juan J. Linz, and Seymour Martin Lipset (Boulder, Colo.: Lynne Rienner, 1989), 1-52.

occupation and industry) constituents and an electoral commission.[6]

2. In regard to *contested democracy*, there is a lack of consensus on what the electoral rules in Hong Kong exactly are. The Basic Law, which was passed in 1990, was plagued by resignations and the purging of key members of the Basic Law Drafting Committee (BLDC). In 1992, Chris Patten, the last British governor of colonial Hong Kong, further delegitimized the electoral rules by introducing his own version of democratic reforms which deviated from the constitutional framework set by the Basic Law. In response, the Beijing government made it clear that it will invalidate Patten's democratic reforms, ask the present legislators to step down, and impose its own provisional Legislative Council when it resumes sovereignty over Hong Kong in 1997.[7] In short, Hong Kong's democratic transition has been highly problematic, marred by narrow electoral competition and feuding over electoral rules.

The aim of this article, therefore, is to trace the origins and development of Hong Kong's embattled democracy. In particular, it seeks to explain why, although Hong Kong has already acquired most of the "prerequisites" for democratization, its democracy has been so restricted and contested compared to those in the East Asian newly industrialized economies (NIEs).

A Sketch of Hong Kong's Democracy Project

The Genesis of the Democracy Project

As Hong Kong became a newly industrialized economy, it gave rise to a new generation of college students. In the early 1970s, college students initiated a nationalist movement calling for identification with the Chinese motherland and campaigning for Chinese as Hong Kong's official language. Many student activists also entered and radicalized service professions after graduation, initiating a community movement to criticize the colonial government's policies and forming "pressure groups" to address the grievances of the grass-roots population. Nevertheless, the nationalist and community movements failed to articulate a democratic discourse because the Hong Kong government was still a nondemocracy jointly ruled by an expatriate al-

---

[6]Ming K. Chan, "Democracy Derailed," in *The Hong Kong Basic Law*, ed. Ming K. Chan and David J. Clark (Armonk, N.Y.: M. E. Sharpe, 1991), 3-33.

[7]Suzanne Pepper, "Hong Kong in 1994," *Asian Survey* 35, no. 1 (January 1995): 48-60.

liance of British colonial officials and pro-British Big Business elites.

In the early 1980s, when London negotiated with Beijing over Hong Kong's future, Big Business and the mass media endorsed the continuation of colonial rule. To counteract this pro-British movement, Beijing proposed a package of "Hong Kong people ruling Hong Kong" so as to craft a strategic alliance with service professionals.[8] In response, service professionals were politicized, forming such "political groups" as the Meeting Point, the Hong Kong Forum, the Hong Kong People's Association, etc. These new political groups shared the following common traits. First, their members were former student activists in the 1970s, and were linked to one another through friendship networks formed in their school days. Second, members of the political groups belonged to the service sector (i.e., social workers, teachers, journalists, and doctors), and they had developed links with Hong Kong's grass-roots population through their professional practices. Third, the policy statements of these political groups revealed their common commitment to nationalism, democracy, and welfare capitalism. The Meeting Point, for instance, affirmed that Hong Kong is a part of China, Hong Kong should move toward democratization, and there should be more rationalization of social resource distribution. Finally, the political groups relied on legal channels to achieve their goals. Their common tactics included organizing public seminars, holding news conferences, and presenting position papers on important issues. Consequently, this article argues that the nationalist and community movements of the 1970s served as a prelude to the democracy movement a decade later because there was a continuity of values and beliefs, leadership and organizational forms, and strategies of protest among the three movements.[9]

After it decided to accept Beijing's demands, London pushed for the insertion of democratic clauses such as "election" and "accountability" into the Joint Declaration in order to sell the package to Hong Kong service professionals and the British Parliament. In short, this was the genesis of the democracy project in Hong Kong. However, it resembled more a case of "muddling through" rather than a grand inauguration because the democratic terms were sneaked into the Joint Declaration without any clarification of what these terms meant. Moreover, London covered up its disagreements with Beijing on this critical issue, thus presenting a misleading impression

---

[8]Alvin Y. So, "Hong Kong People Ruling Hong Kong! The Rise of the New Middle Class in Negotiation Politics, 1982-1984," *Asian Affairs* 20, no. 2 (1993): 67-87.

[9]Alvin Y. So and Ludmilla Kwitko, "The Transformation of Urban Movements in Hong Kong, 1970-90," *Bulletin of Concerned Asian Scholars* 24, no. 4 (1992): 32-44.

that Beijing, too, endorsed a Western-style system of full democracy in Hong Kong after 1997.

The Formation of a Restricted Democracy

Once London let the democracy genie out of the bottle, it proposed modest democratic reform in 1984. Alarmed by the prospect of losing their political hegemony in the Hong Kong government, Big Business elites in the Executive Council (Exco) and the Legislative Council (Legco) managed to slow the democratization process and restrict it as a quasi-"corporatist democracy," thus guaranteeing their monopolistic representation in the Legco through functional constituencies and indirect elections. Nevertheless, the limited democratic openings in the mid-1980s had already empowered service professionals. They enthusiastically participated in elections, mobilized community support, articulated a pro-welfare electoral platform to appeal to the grass-roots population, and won a landslide victory against the traditional business community in local District Board elections. Some were even selected into the Legco through indirect elections. The entry of a small number of service professionals thus greatly transformed legislative politics; "consensus politics" was replaced by "oppositional politics," and government and business policies were increasingly challenged by service professionals in the Legco.[10]

Rising democratic expectations in the mid-1980s, however, gave way to rising democratic frustrations in the late 1980s. The growing power of the service professionals triggered an "unholy" alliance between Big Business elites and Beijing; the Basic Law drafting process was then used by Beijing and members of the Big Business community to imprint the restricted, corporatist democracy system on Hong Kong's mini-constitution.[11] Service professionals put up a fight for their populist democracy system; they wanted the Hong Kong government to fulfill its promise of installing direct elections in 1988, and protested against the restricted democracy model both in the BLDC and on the street. Nevertheless, they were overwhelmed by the conservative Beijing-Hong Kong Big Business-London alliance, failed to mobilize the grass-roots population on issues concerning the Basic Law, and lacked unity and strategic planning. As a result, the service professionals' populist democ-

---

[10]Alvin Y. So, "The New Middle Class and the Democratic Movement in Hong Kong," *Journal of Contemporary Asia* 20, no. 3 (1990): 384-98.

[11]Ian Scott, *Political Change and the Crisis of Legitimacy in Hong Kong* (Honolulu: University of Hawaii Press, 1989), 298-305.

ISSUES & STUDIES

racy package was defeated. As the restricted democracy system was written into the mini-constitution, many service professionals became disillusioned about the democracy project, with some even considering emigration and setting up a democratic front overseas to oppose the conservative alliance.[12]

Impetus Toward a Contested Democracy

The conservative alliance of Beijing, Hong Kong Big Business, and London was torn apart by the Tiananmen incident in 1989. At the height of the incident, some Big Business and pro-Beijing forces voiced their opposition to the Beijing regime, and London defected by asking for a faster pace of democratization in Hong Kong.[13] In addition, service professionals emerged as popular leaders during the Tiananmen protests; they adopted the label of "democrats," and were solidified into the "United Democrats." They also deepened their community network and articulated a pro-welfare platform to appeal to the grass-roots population. Through this populist alliance, service professionals won a landslide victory in the first direct elections to the Legco in 1991, capturing two-thirds of the popular votes and sixteen of the eighteen directly-elected seats. Since service professionals also won five of the twenty-one functional seats elected indirectly along occupational and sectoral lines, they formed a fairly solid democratic camp in the Legco together with a few other liberal-minded independents. The symbolic significance of the Tiananmen incident was that it imposed a democratic discourse on Hong Kong politics. Every political group, including Hong Kong Big Business' conservative organizations, put on a democratic label in order to appeal to democratic sentiments in Hong Kong in the early 1990s.[14]

In the mid-1990s, London abruptly shifted its decolonization policy toward Hong Kong, dropping its cooperative policy and adopting an antagonistic policy toward Beijing.[15] In 1992, Governor Patten crafted a strategic alliance with service professionals by appointing them to the Exco and promoting a pro-welfare policy. He subsequently reinterpreted the gray areas of the Basic Law to fit the populist democracy system of the service profes-

---

[12]John D. Young, "Red Colony: Hong Kong 1997" (Paper presented at the conference on Hong Kong: Towards 1997 and Beyond, Honolulu, January 1989).

[13]Mark Roberti, *The Fall of Hong Kong: China's Triumph and Britain's Betrayal* (New York: John Wiley & Sons, 1994), 249-54.

[14]See note 6 above.

[15]John Burns, "Hong Kong in 1993: The Struggle for Authority Intensifies," *Asian Survey* 34, no. 1 (January 1994): 55-63.

sionals. With the support of London and the Hong Kong government, service professionals were thus further empowered, formed a Democratic Party, and won another victory in the 1995 elections in which they took nineteen seats out of a total of sixty seats. With support from like-minded independents, the democratic camp commanded as many as thirty-one votes in the legislature.[16]

Alarmed by the democratic alliance of service professionals and London, the Big Business elites were pushed back to Beijing. In response, Beijing consolidated the alliance by appointing them as "Hong Kong affairs advisors," and as members of the Preliminary Working Committee (PWC) and the Preparatory Committee which handled transition affairs. Furthermore, Beijing announced that it would dismantle the popularly-elected legislature and replace it with a provisional legislature in July 1997 because Patten's reforms had violated the Basic Law.[17] In this respect, Hong Kong society was polarized into two opposing camps: an alliance between Beijing and Big Business versus a populist alliance between service professionals and the grass-roots population. Instead of focusing on winning elections and working through electoral rules, these two camps debated, reinterpreted, and challenged the Basic Law, laying the foundation for a contested democracy in 1997.

Toward a Societal Explanation

Bringing the Societal Forces Back In

In the literature on Hong Kong's democracy,[18] the power dependence explanation highlights the crucial role of Beijing and London; thus the Hong Kong government is characterized as a dependent polity controlled by London and Beijing. In this view, it was London and Beijing that set the rules of democratization in Hong Kong, while the Hong Kong people were denied the right to participate in the shaping of their own future. This power dependence explanation tends to overlook Hong Kong's societal forces because the local elites are seen as power-seeking and preoccupied with "pure" political issues; they have thus been divided and manipulated by Beijing and London, due to

---

[16]*Asiaweek*, September 29, 1995, 34; *Far Eastern Economic Review*, November 29, 1995, 36.

[17]Ngai-ling Sum, "More Than a War of Words: Identity, Politics, and the Struggle for Dominance during the Recent Political Reform Period in Hong Kong," *Economy and Society* 24, no. 1 (1995): 67-100.

[18]Hsin-chi Kuan, "Power Dependence and Democratic Transition: The Case of Hong Kong," *The China Quarterly*, no. 128 (December 1991): 775-93; Lau Siu-kai, "Hong Kong's Path of Democratization," *Asiatische Studien Etudes Asiatiques* 49, no. 1 (1995): 71-90.

political alienation among the masses. Consequently, in focusing on the structural outcome of democratization, proponents of the power dependence theory have a pessimistic view of Hong Kong's prospects for democracy.

However, using the insights of "third wave" democratization research, this article points to the crucial role of societal forces and their shifting alliances in Hong Kong's democratization. Instead of being manipulated by Beijing and London or playing the role of spectator, Hong Kong's societal forces are thus seen as agencies that have made strategic decisions affecting the course of Hong Kong's democratization. Rather than characterizing the elites as power-seeking and preoccupied with "pure" political issues, this article traces their economic interests to examine why certain actors were more prone to raise livelihood issues than others in the quest for democratization. Finally, instead of focusing on structural outcomes, this article examines how the societal forces and their shifting alliances have complicated the genesis and transformation of the democracy project in Hong Kong.

In particular, the critical role of Big Business in blocking Hong Kong's democratization should be emphasized. Big Business had a monopoly on interests in the colonial government before the 1980s; this is why Hong Kong was well known as a "capitalist paradise." In the mid-1980s, Big Business elites, through their control of both the Exco and Legco, were able to restrict the scope of democratization so their interests would still be guaranteed majority representation through functional constituencies and indirect elections. In the late 1980s, they were successful in integrating their restricted democracy model into the mini-constitution through their alliance with Beijing. By the mid-1990s, the Big Business community was openly criticizing Patten's electoral reforms and laying the groundwork for a contested democracy in Hong Kong in 1997. Except for a brief moment at the height of the 1989 Tiananmen incident, the Big Business community has consistently provided strong opposition to the democracy project. In this respect, it is business hegemony rather than power dependency that explains Hong Kong's problematic democratic transition.

On the other hand, service professionals have been instrumental in Hong Kong's democratization attempts. In the late 1970s, they emerged as a new political force in the community, challenging the colonial government's welfare policies. During negotiation politics in the early 1980s, they articulated the concept of a "democratic national reunification" and pushed London and Beijing to include democracy terms in the Joint Declaration. By the mid-1980s, they had grown into an opposition force inside the Hong Kong government after being elected into the Legco. In the late 1980s, they tried unsuccessfully to influence the Hong Kong government to install direct elections in 1988 and Beijing to include their populist democracy model in the mini-constitution. Nevertheless, the democracy project was revitalized by the Tiananmen incident, as service professionals emerged as popular leaders speaking for Hong Kong's interests; cultivated a populist alliance; and won landslide Legco elections in 1991 and 1995. Since service professionals have been highly committed to the democracy project and receive support from the grass-roots population, they will be a pivotal political force to be reckoned with during the 1997 transition.

Even the grass-roots population has not been as passive as has been described in the power dependence explanation. While it has failed to emerge as a class because of historical reasons, it was politically active in presenting urban grievances through the community movement in the late 1970s. It was attracted to the pro-welfare livelihood issues of the service professionals in the 1985 elections, and chose the latter to represent it in the legislature. Although the grass roots was "bored" by the technical issues of the Basic Law debates in the late 1980s, it has again lent its strong support to the pro-welfare, anti-Beijing platform of the service professionals in the 1990s, giving them a popular mandate to push for Hong Kong's democratization.

State-Societal Alliances

To be sure, Beijing and London have been two powerful state actors in Hong Kong's democratic politics; yet, they have not dictated the evolution of democratic politics as they have pleased. They have required the support of Hong Kong's societal forces, and have frequently entered into alliances with the latter in order to carry out their policies.

London, for instance, ruled Hong Kong through an expatriate alliance with the pro-British business elites. London and the British corporations shared common interests in maintaining political stability and their dominance of the Hong Kong economy, and this expatriate alliance was institutionalized through the appointment of British (or pro-British) business elites to the Legco and Exco, thus guaranteeing business' monopolistic representation in the Hong Kong government. This institutional alliance was highly stable and existed in the colony for over a century; however, it began to crack during the negotiation politics in the early 1980s, as London decided to hand over the sovereignty of Hong Kong to Beijing in 1997.

In the late 1980s, the new alliance between Beijing and Hong Kong Chinese business elites gradually replaced the traditional expatriate alliance of London and pro-British business elites. Beijing and the Big Business elites

ISSUES & STUDIES

shared common interests in mainland investments as well as in maintaining Hong Kong's economic prosperity and political stability.[19] This alliance was institutionalized through the appointment of Big Business elites into the BLDC in the late 1980s and into the PWC in the mid-1990s, thus ensuring business' monopolistic representation in the SAR government. The present alliance, like the former expatriate alliance, is highly stable; even though it was shaken by the Tiananmen incident, the rising power of the service professionals and Patten's political reforms in the 1990s have quickly restored and consolidated it.

Apart from the expatriate and "unholy" alliance, a populist alliance emerged in Hong Kong society in the early 1990s. Service professionals and the grass-roots population have shared common interests and commitments to promoting a pro-welfare policy, especially when the livelihood of the urban masses was threatened by the restructuring from a labor-intensive manufacturing economy to a service economy, and by the prospect of the importation of mainland Chinese and foreign workers into Hong Kong. This populist alliance was institutionalized through direct elections to the Legco, guaranteeing that the voice of the populist sector would at least be heard in the Hong Kong government. This populist alliance has begun to achieve consolidation in the 1990s, as shown by the repeated landslide victory of the democrats at the ballot box in the 1991 and 1995 elections.

The above analysis therefore shows that there are deep-rooted institutional bases for both pro- and anti-democracy forces. Historical events over the past two decades, however, have been on the side of the pro-democracy forces, helping them to usher a democratic discourse into Hong Kong politics. Yet, both the Tiananmen event and the 1997 negotiations failed to galvanize the sufficient democratic force necessary to push full democratic transition and consolidation through in Hong Kong. What, then, explains Hong Kong's democratic frustrations in light of the fact that its East Asian neighbors (Taiwan and South Korea) have achieved democratic breakthroughs since the late 1980s?

Hong Kong's Democratic Transition
from the East Asian NIEs' Perspective

Hong Kong, like the other East Asian NIEs, has acquired most of the

---

[19]Alvin Y. So and Reginald Kwok, "Socioeconomic Core, Political Periphery: Hong Kong's Un-

"prerequisites" for democratization; in addition, Hong Kong, Taiwan, and South Korea have shared the same time frame of democratization, with all three regions experiencing the genesis of their democracy projects and democratic transitions in the 1980s.

Despite their similarities, however, there are also profound differences among South Korea, Taiwan, and Hong Kong in regard to state autonomy, the state-business relationship, labor activism, class alliances, and the role of historical events in democratic development. This article argues that it is especially their differences in state, class, and historical events that explain the East Asian NIEs' separate paths of democratic development.

Taiwan: Party-State and Democracy from Above

Among the three East Asian states, Taiwan was the strongest and had the most autonomy. It was often described as a party-state because it possessed a powerful party which penetrated throughout the state structure, the economy, and society. When the KMT imposed its rule on the native Taiwanese after it was defeated by the Chinese Communists on the mainland in 1949, it feared Taiwanese ambitions for independence. As a result, the KMT regime maintained a large state sector (about 16 percent of Taiwan's gross domestic product [GDP] in the early 1970s) in order to have more control over the economy. Aside from economic dominance, the KMT also deeply penetrated civil society so as to curb the development of any organized opposition forces. Due to martial law, civil liberties (such as freedom of assembly) were severely limited and there was strict censorship of the mass media. Moreover, government surveillance and the strong threat of repression were central mechanisms for the KMT's labor control.[20]

Although Taiwan was development-oriented, it was not state enterprises but small and medium-sized (hereafter abbreviated as S&M) firms in Taiwan that filled the export sector. Of the 260,000 business enterprises in Taiwan, 98 percent were considered S&M firms, which employed 70 percent of all employees and accounted for 65 percent of total exports. But the small business sector failed to link itself to the Taiwan state, as the KMT neither paid serious attention to the trade associations organized by S&M firms nor sought

---

certain Transition Toward the Twenty-First Century," in *The Hong Kong-Guangdong Link: Partnership in Flux*, ed. Reginald Kwok and Alvin Y. So (Armonk, N.Y.: M. E. Sharpe, 1995), 251-58.

[20]Thomas B. Gold, *State and Society in the Taiwan Miracle* (Armonk, N.Y.: M. E. Sharpe, 1986), 62-63, 89.

ISSUES & STUDIES

S&M firms' advice in formulating economic policies. Receiving no support from the state, S&M firms in Taiwan had to rely on self-financing or informal money markets for credit, and also complained that the state failed to supply them with information on international trade trends.[21]

Although the Taiwan "economic miracle" resulted in a rapid rate of capital accumulation, it did not completely eliminate such social problems as environmental pollution, industrial safety, traffic congestion, consumer fraud, and human rights violations. By the early 1970s, opposition to the KMT regime began to emerge. Instead of waging a class struggle, however, the opposition took the form of an ethnic struggle against mainlanders' domination over the native Taiwanese.[22]

In the 1970s, a new generation of indigenous Taiwanese intellectuals began to demand their rights of political participation and representation in the state, freedom of speech, and the lifting of martial law. They complained about the KMT's violations of human rights; used new magazines such as *Xiachao* (China Tide) to resurrect struggles against despotism which began during the Japanese occupation of Taiwan; started a "native soil" literature focusing on the lives of indigenous Taiwanese people; and organized a *dangwai* (non-KMT) political group to challenge the KMT in elections. United through their native Taiwanese language and culture, this new generation of intellectuals forged an ethnic alliance among native Taiwanese S&M businesses, the new middle class, and labor against the mainlanders' KMT in the late 1970s and 1980s. The *dangwai* utilized street protests, squabbles on the legislative floor, and literary offenses to express their claims. In response, the KMT shut down their magazines, suppressed their street demonstrations, and placed their leaders in prisons.[23]

However, Taiwan experienced a crisis of legitimacy in the mid-1980s due to Western states breaking off their diplomatic relationships with Taiwan to recognize China; a series of economic and political scandals; and the aging of its strong leader Chiang Ching-kuo. It was at this historical juncture that Chiang persuaded the KMT conservatives to promote democratic reforms from above. The KMT lifted martial law, released political prisoners, legal-

---

[21]Hsin-huang Michael Hsiao, "Formation and Transformation of Taiwan's State-Business Relations: A Critical Analysis," *Bulletin of the Institute of Ethnology*, no. 74 (1993): 1-32.

[22]Walden Bello and Stephanie Rosenfeld, *Dragons in Distress: Asia's Miracle Economies in Crisis* (San Francisco: The Institute for Food and Development Policy, first revised printing, 1992), 175-285.

[23]Chen Guying, "The Reform Movement among Intellectuals in Taiwan since 1970," *Bulletin of Concerned Asian Scholars* 14, no. 3 (1982): 32-47.

ized a multiparty system, granted de facto recognition of the Democratic Progressive Party (DPP), and liberalized the mass media. It was argued that democratic reforms would polish up the KMT's international image and increase its ruling legitimacy by incorporating opposition forces into the electoral game. Moreover, since the KMT was much more powerful than the opposition forces (as it still monopolized the mass media, controlled enormous resources, and received credit for economic development), it would be guaranteed victory in elections and remain in power.[24]

Since the KMT took the initiative in democratic transition when it was still strong, it has managed to dominate the transition–conducting reform where and when it has seen fit–and kept the pace and direction of reform relatively well controlled. In the early 1990s, the KMT further promoted the "Taiwanization" of the party apparatus, increased government spending on some welfare programs, incorporated Big Business into the state's decisionmaking process, and held strategic negotiations with opposition parties so as to dilute the ethnic divide between mainlanders and Taiwanese and gain support from the new middle class, the business community, and moderate elements in the opposition parties.[25]

As a result, democratic consolidation has been taking place in Taiwan. The KMT has been able to achieve constitutional reform and open up the parliamentary bodies, two major cities' mayors, the Taiwan provincial governor, and the presidency for election. Most importantly, the KMT won the highly competitive presidential election in March 1996 at the ballot box. Since the KMT allowed competition in an open electoral system and opened strategic negotiations with opposition parties, it seems that democratic reforms have been firmly institutionalized in Taiwan society. In fact, researchers have already begun to shift their attention from Taiwan's democratic breakthrough and consolidation to the economic consequences of its democratization.[26]

South Korea: Working Class Struggles
and Democracy from Below

Like Taiwan, South Korea has also been development-oriented, with

---

[24]Alvin Y. So, "Democratization in East Asia in the Late 1980s: Taiwan Breakthrough, Hong Kong Frustration," *Studies in Comparative International Development* 28, no. 2 (1993): 60-79.

[25]Hsin-huang Michael Hsiao and Alvin Y. So, "The Taiwan-Mainland Economic Nexus: Sociopolitical Origins, State-Society Impacts, and Future Prospects," *Bulletin of Concerned Asian Scholars* 28, no. 1 (1996): 3-21.

[26]Tun-jen Cheng, "Economic Consequences of Democratization in Taiwan and South Korea" (American Political Science Association Working Papers in Taiwan Studies, 1996).

ISSUES & STUDIES

economic development (in terms of growth, productivity, and competitiveness) being the state's foremost priority.[27] Nevertheless, in contrast to Taiwan, South Korea did not have a strong political party that could secure grass-roots support at the community level, nor did it have a strong leader that could unify the state and party in democratization. Thus, South Korea relied on a strong dose of authoritarianism to maintain its control over civil society. For example, in Park Chung Hee's regime, strikes were banned, existing unions were outlawed, dissenting newspapers and magazines were closed down, protesters were arrested, and political activists were jailed.[28]

The state-business relationship in Taiwan and South Korea was also different. While in Taiwan there was a large state sector and S&M businesses were excluded from participating in state decisionmaking, in South Korea the state cultivated a small number of conglomerate business groups (*chaebol*) as its clientele. The state supported *chaebols'* economic monopolies and granted them credits and export licenses in exchange for a steady supply of political funding. As Carter J. Eckert remarks, despite its wealth, Korean Big Business remained a decidedly unhegemonic class, estranged from the very state and society in which it continued to grow.[29] On the one hand, *chaebols* were frequently "disciplined" by state officials through withdrawal of loans and licenses if they failed to meet the specified stringent performance requirements for exports. On the other hand, *chaebols* were the focus of intense popular resentment because of the aggressive and selfish manner in which *chaebol* owners had accumulated wealth.[30] As Kim Woo Choong, the chairman of the Daewoo Group, complained: "Our country's businessmen have not been able to acquire public esteem. On the contrary, [they] have been denounced or kept at a safe distance with feigned respect."[31] This "*chaebol* bashing" phenomenon emerged, Hagen Koo explains,[32] because *chaebols* pursued maximum labor exploitation under the protection of an authoritarian capitalist state. It is impressive that although they were subjected to a high level of exploitation

---

[27]Alice Amsden, *Asia's Next Giant: South Korea and Late Industrialization* (New York: Oxford University Press, 1989), 79-113.

[28]Choi Jang Jip, *Labor and the Authoritarian State: Labor Unions in South Korean Manufacturing Industries, 1961-1980* (Seoul: Korea University Press, 1989).

[29]Carter J. Eckert, "The South Korean Bourgeoisie: A Class in Search of Hegemony," in *State and Society in Contemporary Korea*, ed. Hagen Koo (Ithaca, N.Y.: Cornell University Press, 1993), 95-130.

[30]Hagen Koo, "Introduction: Beyond State-Market Relations," ibid., 1-21.

[31]Quoted in Eckert, "The South Korean Bourgeoisie," 95.

[32]Hagen Koo, "The State, *Minjung*, and the Working Class in South Korea," in Koo, *State and Society in Contemporary Korea*, 131-62.

without much cultural or ideological preconditioning, Korean workers developed class consciousness and organization.

Subsequently, despite the admiration of outsiders for the "Korean miracle," the Korean state was never successful at buying popular support with economic achievement. The Park Chung Hee regime (1961-79) and the Chun Doo Hwan regime (1980-87) were continuously plagued by student demonstrations, dissident movements, and grass-roots labor protests. Tun-jen Cheng and Eun Mee Kim point out that throughout the 1960s and 1970s, university campuses were annually shut down for at least a month by student demonstrations against the authoritarian Park regime. In the 1980s, student demonstrations became better-organized and more violent. The frequency of violence particularly increased as the suppression of the Chun regime intensified, producing "a vicious cycle of opposition and suppression."[33]

However, unlike the Taiwanese opposition movement, which focused on the ethnic conflict between mainlanders and Taiwanese, Korean opposition took place along the class divide of the state/*chaebols* versus the middle class/working class. Despite labor repression, the high concentration of workers in *chaebols'* large-scale heavy industries provided favorable conditions for a militant labor movement, which became explosive when it developed close links with political conflicts outside industry and was supported organizationally and ideologically by the new middle class *minjung* (the people's or the masses') movement. *Minjung* became a powerful opposition discourse and a political symbol, and provided a new identity for all who participated in political, social, and cultural movements in opposition to the authoritarian state.[34]

In the mid-1980s, the merging of the labor movement and the *minjung* movement led to South Korea's democratic breakthrough. The torture and killing of a Seoul National University student, which was revealed in January 1987, intensified the anti-regime struggles. Labor strikes grew to 3,749 in 1987. Chun's announcement to postpone presidential elections until his term expired brought massive demonstrations. News reports said that more than 300,000 people demonstrated, and street demonstrations escalated until they had grown too large and widespread to be controlled by the police alone. Finally, by June 1987, Roh Tae Woo, the chairperson of the ruling party, admitted defeat, accepting practically all the demands of the opposition movement and agreeing to hold a direct presidential election at the end of 1987. While

---

[33]Cheng and Kim, "Making Democracy," 134.
[34]See note 32 above.

ISSUES & STUDIES

Taiwan's democratic reforms were initiated from above by a strong party leader, the Korean experience typifies the case of promoting democracy from the bottom; the transition from authoritarian rule occurred because of unstoppable working class struggles and street demonstrations by civil society.[35]

These events were followed by strategic negotiations among the opposition elites to merge their parties into one dominant party–the Democratic Liberal Party–in 1990, the election of a civilian government with Kim Young Sam as the president in 1993, and a radical shakeup of the military establishment. In 1996, two former presidents–Chun Doo Hwan and Roh Tae Woo– were even put in jail.

Democratic transition, however, has been accompanied by the deradicalization of middle-class politics. Hsin-huang Michael Hsiao and Hagen Koo observe that some of the middle class felt threatened by labor's militant attitudes after the democratic transition. As the *minjung* movement waned, the middle class began to promote its own interests through new social movements rather than joining labor's class struggles. Thus, a new Citizen Coalition for Economic Justice was formed to promote economic justice, protect the environment, ensure clean elections, fight against gender inequality, and raise civil consciousness. In this respect, the middle class' new social movements and regional loyalty became prominent issues in Korean electoral politics. Currently, as the military is under civilian government control and electoral rules have become institutionalized, there is little doubt that South Korea has achieved not just democratic transition but also democratic consolidation.[36]

Hong Kong: Business Hegemony and
Democratic Frustration

Although the governments in South Korea and Taiwan intervened extensively in the marketplace to direct industrialization according to state priorities, the colonial state in Hong Kong left investment decisions to the private sector. Certainly the Hong Kong state has assisted private capital accumulation in a variety of ways, most notably in infrastructure provisions (such as the maintenance of a low-tax business environment, the expansion of the public education system, and a massive public housing program), but the Hong Kong

---

[35]Hsin-huang Michael Hsiao and Hagen Koo, "The Middle Class and Democratization in East Asia: Taiwan and South Korea Compared," in *Consolidating the Third Wave Democracies*, ed. Larry Diamond (Newbury Park: Sage Publications, forthcoming).

[36]Ibid.

state is still a far cry from the developmental states portrayed in statist literature. Sir Philip Haddon-Cave thus describes Hong Kong's economic policy as
"positive non-interventionism."[37]

Furthermore, Hong Kong has had a state-business relationship pattern
different from its counterparts in Taiwan and South Korea. While Taiwan's
Big Business was excluded from the state altogether and the Korean *chaebols*
were merely the clientele of the authoritarian state, Hong Kong's Big Business
achieved an expatriate alliance with the colonial state for over a century.

From a comparative perspective, Hong Kong's community movement
has been much weaker than the opposition movements in Taiwan and South
Korea. The size of the participants, the strength of the movement organization, and the violence and confrontations involved have been on a much
smaller scale in Hong Kong than those in Taiwan and South Korea. While an
ethnic bond united the native Taiwanese and a *minjung* discourse united the
Korean middle and working class against their authoritarian states, there was
no such ethnic bond or radical discourse uniting Hong Kong service professionals with the grass-roots population. Although the Hong Kong workers
received low wages, they were relatively well-off compared to their Korean
counterparts in large *chaebol* enterprises under strict, repressive labor laws.
Thus, Hong Kong's labor movement was much weaker than the Korean labor
movement. Moreover, Hong Kong protesters had less grievances than their
counterparts in Taiwan and South Korea. Although the Hong Kong state was
a nondemocracy, it was not authoritarian; there was little repression and
bloodshed, and few political prisoners in Hong Kong. Although Hong Kong
was not a welfare state, the Crawford M. MacLehose era in the 1970s did
provide some welfare measures such as public housing to mitigate urban
grievances. These factors may explain why the Taiwanese and South Korean
opposition movements articulated a democratic discourse against their authoritarian states (and against bourgeois hegemony in South Korea) much earlier
than Hong Kong's community movement.

In sum, Hong Kong has possessed much less favorable structural conditions for democratization than its East Asian neighbors because Hong Kong
has had a stronger conservative alliance of state and Big Business on the one
hand, and a weaker populist alliance of service professionals and the grass-
roots population on the other. However, the negotiations regarding Hong
Kong's future in the early 1980s served as a catalyst to inject the democracy

---

[37]Philip Haddon-Cave, "Introduction," in *The Business Environment in Hong Kong*, ed. David
Lethbridge (Hong Kong: Oxford University Press, 1984), xv-xx.

project into Hong Kong society. The talks simultaneously politicized service professionals and created a crack in the expatriate alliance. Nevertheless, Hong Kong's democratic transition has neither been initiated from above (like in Taiwan) nor from below (like in South Korea); rather, it is a case of "muddling through." London merely used the democracy project as a tactical means to sell the Joint Declaration to Hong Kong service professionals and the British Parliament, and had no intention to carry the democracy project through with the Hong Kong government, which was already a lame duck after the signing of the Joint Declaration, and felt no need (unlike the KMT state in Taiwan) to manipulate the course of democratization in order to remain in power.

Had there been no economic integration between Hong Kong and mainland China, Hong Kong Big Business might have formed an ethnic alliance across class lines (like that in Taiwan) to safeguard Hong Kong's interests. However, the very rapid economic integration of Hong Kong and mainland China has prevented the formation of such an ethnic alliance; instead, an "unholy" alliance between Hong Kong Chinese Big Business and Beijing has been cultivated, greatly strengthening the economic power and social status of Hong Kong Chinese businesses at the expense of British corporations. Since this alliance was formed, bourgeois hegemony has been restored in Hong Kong. When London forsook its democracy promise and joined the conservative camp in the late 1980s, the Beijing-Hong Kong Big Business-London conservative alliance left little room for the democratization of Hong Kong. During this period, Hong Kong service professionals forged a populist alliance with the grass-roots population, but were in no position to challenge the all-mighty conservative alliance. The product of these unbalanced pro- and anti-democracy forces was a restricted democratic system written into the mini-constitution.

However, Hong Kong has been on the road toward a contested democracy because the Tiananmen incident revitalized the democracy project almost overnight, demoralizing the Beijing regime; injecting a democratic discourse into Hong Kong politics; causing London to defect to the democracy camp; uniting service professionals into a democratic party; and enabling the democrats to forge a populist alliance with the grass-roots population on an anti-Beijing, pro-welfare platform.

What will happen after 1997? Only time can tell whether Hong Kong will pursue the Korean, Taiwanese, or other paths of democratization, but it seems that Hong Kong still has a long way to go before reaching a full democratic transition.

[10]

REBUREAUCRATIZATION OF POLITICS IN HONG KONG

Prospects after 1997

———————— Anthony B. L. Cheung

The introduction of "representative government" in Hong Kong after the signing of the Sino-British Joint Declaration in 1984 brought hope that by 1997 the classical bureaucratic authoritarian governance that had characterized Hong Kong during the colonial era would be replaced by a democratic and accountable government. Such hope for a full democracy now seems futile, given China's hostility to democratic politics in the Hong Kong Special Administrative Region (HKSAR) and its preference for a new hybrid of elite politics. This article reviews the politics of Hong Kong's transition and examines the future role of civil service bureaucratic power amid the conflicting interests and demands of various key players in the larger political arena.

Foundations of the Hong Kong "Bureaucratic Polity"

Colonial Hong Kong, often called "an administrative state" or a "bureaucratic polity," was governed by administrator-bureaucrats under an appointed British governor; they were not held accountable to any local representative institutions and were not challenged by local politics in any meaningful sense. Until 1985 all the unofficial members of the Legislative Council (Legco) had been appointed by the governor. The Executive Council, which assists the governor in policy-making, has remained wholly appointed. Indeed for most of the time during Hong Kong's colonial history (until the late 1980s) party politics were not tolerated. Social groups such as student organizations and

———————— Professor Anthony B.L. Cheung, City University of Hong Kong

© 1997 by The Regents of the University of California

ANTHONY B. L. CHEUNG 721

trade unions that opposed the government were perceived as "pressure groups" that might threaten the stability of colonial rule.[1]

Colonial absolutism had not excluded local participation, but such participation was confined to established elites—mainly business and professional elites—who were co-opted into the system of governance by "administrative absorption of politics."[2] Through formal and informal interactions with local Chinese leaders, the colonial administrators were able to evolve a form of elite integration that provided the necessary elite legitimation of the regime. People acquiesed to bureaucratic rule partly because of sociocultural factors (e.g., a political culture dominated by features of "utilitarianistic familism"[3] in which the family social resource networks were looked upon to satisfy needs and redress grievances) and partly because of the Chinese population's sense of political impotence within a colonial context.

The backbone of the government mandarinate came from the administrative class, which was instituted locally only in 1960 to replace a previous cadet scheme started in 1861 through which young officers recruited from Britain by competitive examination were groomed for high administrative posts within the colonial service. Its officers formed a nascent administrative elite, operating as "a minuscule band of officials with the same values and from the same social backgrounds."[4] Following the tradition of the British civil service, administrative officers occupied all the senior posts within the administration, headed most government departments, and constituted the policy center of the government. As the administrative class (and its cadet scheme precursor) was essentially staffed by British recruits, it ensured the security of colonial rule.[5] Administrative officers within the colonial context were expected to act as "political officers" having a clear task to keep full control over the indigenous population with the assistance of co-opted local leaders.

1. A Special Committee on Pressure Groups existed during the 1970s to monitor the activities of certain pressure groups critical of government polic*y*. Group leaders were under surveillance by the Police Special Branch, which also tried to infiltrate the groups. See N. Miners, *The Government and Politics of Hong Kong*, 5th ed. (Hong Kong: Oxford University Press, 1991), p. 194.

2. A. Y. C. King, "Administrative Absorption of Politics in Hong Kong: Emphasis on the Grass Roots Level," in *Social Life and Development in Hong Kong*, ed. A. Y. C. King and R. P. L. Lee (Hong Kong: Chinese University Press, 1981).

3. S. K. Lau, *Society and Politics in Hong Kong* (Hong Kong: Chinese University Press, 1982), pp. 26–29.

4. H. J. Lethbridge, *Hong Kong: Stability and Change* (Hong Kong: Oxford University Press, 1978), p. 32.

5. Local officers began to be recruited into the administrative class after the Second World War under the newly introduced policy of localization for all British colonies. However, until the signing of the 1984 Sino-British Joint Declaration, only those holding British passports and thus claiming British nationality of some form were allowed to join the administrative class.

722   ASIAN SURVEY, VOL. XXXVII, NO. 8, AUGUST 1997

Bureaucratic rule was, however, not without challenge. As I. Scott notes, Hong Kong's political history had been punctuated by many crises, most notably the 1967 riots started by pro–Chinese Communist elements that had almost sabotaged the territory and were only suppressed with the help of the British garrison.[6] These were crises over consent in which the colonial government's claim to the right to rule was questioned. The legitimacy crisis of 1967 had forced the colonial government to rethink its governing philosophy and strategy. A series of reforms followed, including the introduction of the City District Officer scheme in 1968 as the first sign of reaching out to the ordinary people, the McKinsey modernization of the colonial government machinery in 1974 to strengthen program planning and policy coordination,[7] and the localization of the civil service and the anticorruption reform during the 1970s. To secure public support, the colonial government also undertook bold attempts to expand public services (such as housing, education, welfare, and labor legislation) and extend the state's interventionist role.

The 1970s under Governor Murray MacLehose were considered to be "the golden age" of social services. As a response to the challenge to its "right to rule" from an indigenous population that had seen rising demands from its younger members who did not identify with their parents' apathetic refugee mentality, the colonial administrators had to reorient their political management tactics. For the first time, the government had to address the people's needs to play a role in governance. As the deputy secretary for home affairs put it in 1969:

> We have no general elections for the central government and yet the general trends of government policy conform to the wishes of the mass of the people. . . . The Government here through formal councils, committees and boards, through reading the press, through informal contacts with individuals and groups, in high station and low, has its antennae turned constantly to public wishes in a thousand fields of our administration. . . . Our methods can certainly be improved, our thoughts thrown wider open, but we do have the essential ingredients of a democracy which has produced a general understanding of the people by the government and the government by the people.[8]

The 1960s and 1970s also saw the transfer into the Hong Kong administrative class of senior British colonial service officers who had served in other British colonies that had become independent. Some of these "new" expatriates, known as "African retreads," having gone through the experience of

---

6. I. Scott, *Political Change and the Crisis of Legitimacy in Hong Kong* (Hong Kong: Oxford University Press, 1989).

7. McKinsey & Co., *The Machinery of Government: A New Framework for Expanding Services* (Hong Kong: Government Printer, 1973).

8. Dennis Bray, quoted in J. Rear, "One Brand of Politics," in *Hong Kong: The Industrial Colony*, ed. K. Hopkins (Hong Kong: Oxford University Press, 1971).

ANTHONY B. L. CHEUNG 723

decolonization elsewhere, were enthusiastic about progress toward reforms and greater public participation. Although their enthusiasm for devolving powers to the indigenous population was not shared by the "locals" within the administration,[9] their new thinking may have reinforced the local momentum for reforms triggered by the 1967 crisis.

Indeed, many of the administrative and social reforms of the 1970s would not have been possible without an active search by top administrators, under the leadership of the reformist governor, MacLehose, for innovative solutions to the political crisis. Such solutions were all administrative in nature mainly because more fundamental political (or constitutional) means were ruled out, allegedly because of China's objection to introducing into Hong Kong party politics and elections along Western lines.[10] But the traumatic experience of the 1967 riots may have convinced the colonial administrators that political reforms could easily lead to an ungovernable local polity dominated by indigenous Chinese leaders who would be susceptible to infiltration by both Communist and Kuomintang elements, hence turning the colony into a place of political turmoil.[11] Bureaucratic reformism was an important driving force, apart from the social protest movements organized by opposition pressure groups during the 1970s, behind the transformation of colonial rule into a relatively more responsive administration that claimed to govern by public opinion.[12]

---

9. Brian Hook, "From Repossession to Retrocession: British Policy towards Hong Kong 1945–1997," paper presented at the International Conference on Political Order and Power Transition in Hong Kong, September 18–19, 1996, Lingnan College, Hong Kong. Hook had served in the Hong Kong government in the 1950s.

10. This was according to the British government's *White Paper: Representative Government in Hong Kong*, presented to the Parliament by the Secretary of State for Foreign and Commonwealth Affairs, February 24, 1994 (Hong Kong: Government Printer, 1994).

11. The local government reforms of the late 1960s, centered on the Urban Council, which was then the only local body with some elected representative elements, were hastily brought to an end in 1971 with the publication of the *White Paper: The Urban Council* (Hong Kong: Government Printer, 1971) that gave greater financial autonomy to the council and withdrew all official members (i.e., civil servants) from its membership in exchange for clawing back some important functions and powers, notably in public housing. The idea of turning the Urban Council into some kind of "municipal government" authority was rejected.

12. Despite the fact that Hong Kong had always been characterized as having the most classical minimalist form of government, which left the market to private business unrestrained by administrative interventions (e.g., A. Rabushka, *Hong Kong: A Study in Economic Freedom* [Chicago: University of Chicago Press, 1979], p. 83), such a description was not entirely accurate. Being an absolutist administrative regime, the Hong Kong government had not found itself inhibited from engaging in what could be regarded as "state interventions" whenever it saw the political or administrative necessity to do so. To that extent government decision makers were more guided by their bureaucratic rationale than by any particular ideological or philosophical fiat. The scope of government intervention and expansion that accumulated throughout the years, especially since the late 1960s, was credited by some as being instrumental in the strength-

724   ASIAN SURVEY, VOL. XXXVII, NO. 8, AUGUST 1997

The Politics of Transition: Uncertainties over the Bureaucracy's Political Role

Despite these changes as noted above, the government remained a "government by administrative officers." Political change of a more fundamental nature only began to take place in the 1980s when the question of Hong Kong's future after 1997 surfaced on the agenda. Before that, however, the role of administrative officers had undergone some subtle change since the 1974 McKinsey reforms of the central government machinery. The reforms based on the McKinsey consultants report resulted in the restructuring of the government secretariat with the establishment of high-level "secretaries" directly below the chief secretary and financial secretary to oversee departments. The policy role of senior administrative officers has since been played up. This change could, to some extent, be conceived as some form of "ministerialization" à la Wettenhall, who had examined how the independence of colonies had brought about the need to introduce new institutions such as ministries and the cabinet.[13] The evolution of the colonial government structure took various "modes" of ministerization whereby the former Secretariat divisions simply became new ministries or ministerial secretariats. The full integration of departments within the ministries was not always an early feature of independence, though eventually ministers achieved control over departments that came within their portfolios. In the case of Hong Kong, independence was not on the agenda; however, the administrative reforms of the 1970s had a very strong connotation of modernization representing an attempt to "re-invent" the government administration to remove its colonial wrappings as far as practicable. The McKinsey reorganization thus facilitated the evolution of a cabinet of some kind—made up of "political" administrative officers rather than politicians. These administrative officers were increasingly expected to act and operate as ministers and ministerial staff and to be politically more sensitive and responsive to external political challenges and turbulences.

The British administration only started to consider seriously a decolonization plan for Hong Kong in the 1980s when the future of Hong Kong was to be negotiated with China. The most significant attempt at decolonization was initiated on the eve of the signing of the Sino-British Joint Declaration in

---

ening of Hong Kong's capitalist economy and having constituted a unique model of growth (see J. R. Schiffer, *Anatomy of a Laissez-faire Government: The Hong Kong Growth Model Reconsidered* [Hong Kong: Centre of Urban Studies and Urban Planning, University of Hong Kong, 1983]). Bureaucratic reformism could be very efficient once the will for change was obtained.

13. R. L. Wettenhall, "Modes of Ministerialization, Part I: Towards a Typology—The Australian Experience," *Public Administration* 54:1 (Spring 1976), pp. 1–20 and "Modes of Ministerialization, Part II: From Colony to State in the Twentieth Century," *Public Administration* 54:4 (Winter 1976), pp. 425–51.

ANTHONY B. L. CHEUNG   725

1984 with the introduction of "representative government," with an aim to install "a system of government the authority for which is firmly rooted in Hong Kong, which is able to represent authoritatively the views of the people of Hong Kong, and which is more directly accountable to the people of Hong Kong."[14] The political reform proposals first set out in the July 1984 Green Paper on representative government,[15] apart from calling for direct and indirect elections to the Legco also advocated that the majority of appointed unofficial members of the Executive Council be replaced progressively by members elected by the Legco, a clear sign of moving toward some kind of Westminster system. Both the direct election component for the Legco and the partially elected Executive Council proposal were abruptly dropped in the White Paper of November 1984. After the 1987 review of the development of representative government, the government decided, against strong popular demands, that direct election to the Legco would not take place in 1988 as previously hinted but had to be postponed to 1991 after the Basic Law for the future HKSAR was promulgated by China.

The year 1988 marked the end of the first attempt by the British government to decolonize Hong Kong in the wake of the Sino-British Joint Declaration. Afterward, the British government concentrated on cooperating with and influencing the Chinese government in the drafting of the Basic Law in what can be termed as a "convergence" strategy.[16] As the former chief of the New China News Agency Hong Kong Branch (China's de facto representative office in Hong Kong) Xu Jiatun recalled in his memoirs, every letter in the Basic Law had British input.[17] A second round of decolonization, more with an eye to shore up Hong Kong's political system against future Chinese central interference, was to be launched in 1992 when Chris Patten took up the governorship of Hong Kong. Apart from any possible undisclosed British

---

14. Hong Kong government, *White Paper: The Further Development of Representative Government in Hong Kong* (Hong Kong: Government Printer, 1984), para. 2.

15. Hong Kong government, *Green Paper: The Further Development of Representative Government in Hong Kong* (Hong Kong: Government Printer, 1984).

16. "Convergence" was first emphasized by the Chinese government in 1985–86 to denote the obligation on the part of the British government not to preempt the Basic Law by causing fundamental changes to be made to Hong Kong's political system in the name of representative government. After compromising with the Chinese government on "convergence," the British government began to promote actively the notion of "through train" under which political institutions and personalities (including members of the Executive and Legislative Councils and principal officials of the government) in place before 1997 could straddle beyond the changeover. Such a strategy allowed the British government to play a role in shaping the post-1997 governance of Hong Kong and was accepted by China on the premise of full Sino-British cooperation over the political transition. The "through train" arrangements came to an end as a result of serious dispute between the two governments over the electoral reforms introduced by Governor Chris Patten in late 1992.

17. Xu Jiatun, *Xu Jiatun's Hong Kong Memoirs* (Hong Kong: United Daily Press, 1993).

strategic considerations, Hong Kong's political environment since the late 1980s had undergone drastic changes that may have contributed partly to the British shift in policy toward China over the Hong Kong question. In the aftermath of the 1989 Tiananmen crackdown on the pro-democracy movement in China and the resultant restlessness among the local population, the Hong Kong government in 1990 (while still under former governor David Wilson, who was considered to be reconciliatory toward China) saw the need to introduce a Bill of Rights that for the first time recognized the civil and political rights of Hong Kong residents. Political parties began to emerge in the early 1990s and took the lead at various levels of local elections.

Patten launched new initiatives on administrative and political fronts. His political reform proposals sought to strengthen the directly elected elements within the legislature without openly exceeding the perimeters set by the letter of the Basic Law. The election to the nine new functional-constituency seats was to be based on universal franchise of the working population, and the election committee—which selected ten seats—was to be formed by elected district board members. In administrative reforms, the governor introduced performance pledges for all government departments and public agencies, along the lines of the United Kingdom Citizen's Charter[18]; annual policy progress reports and policy commitments by all policy secretaries who now had to face public monitoring of their policy performance; and a code of access to information. Senior civil servants, particularly those in the policy branches of the Government Secretariat, also had to explain government policies and address matters of public concern at open meetings of the Legislative Council panels. Patten set an example in this form of public accountability by introducing the Governor's Question Time in the Legco.

Routes of Change

Two clear developments under Chris Patten played up the instrumentality of the public bureaucracy. First, the civil service continued to become more professionalized and modernized, or to put it in the latest *reformspeak*, "managerialized." The whole of the civil service and its associated public sector had undergone changes in structure and processes under the 1989 Public Sector Reform, the aim of which was to transform civil servants from administrators into better and more efficient managers. However, as I have argued elsewhere,[19] Hong Kong's public sector reforms were not motivated so much by the standard global claims about suppressing Big Government

---

18. C. Patten, *Our Next Five Years: The Agenda for Hong Kong*, address at the Opening of the 1992–93 Session of the Legislative Council, October 7, 1992, Hong Kong.

19. A. B. L. Cheung, "Efficiency as the Rhetoric: Public-Sector Reform in Hong Kong Explained," *International Review of Administrative Sciences* 62:1 (March 1996), pp. 31–47.

ANTHONY B. L. CHEUNG 727

and improving efficiency or coping with the fiscal crisis as by institutional needs to manage macropolitical changes. Public sector reform was a means to "remanage" the changing realities of the external and internal environments of the administrative elite to restore legitimacy for the public services concerned. A shift toward the microeconomic notion of efficiency in service as advocated in the new public management orientation helped to depoliticize performance evaluation of public service, thus reducing the pressure for greater political accountability from elected politicians and the population at large. Managerialization would also have helped to replace politics with efficiency as the main if not the sole criterion to evaluate institutional performance. The new managerialist logic was also to be a useful tactical means to keep away politics from China during the political transition.

Second, public sector reform operated as a new political management strategy to manage intrabureaucratic conflict between policy branches and executive departments and between the administrative class and professional civil servants. In exchange for giving departmental managers greater managerial autonomy and microbudgetary powers, the administrative class was assured paramountcy in policy and resource-control functions. The new civil service configuration consisted of two distinct layers: the policy management layer dominated by administrative power and the policy execution layer gradually dominated by departmental professional power. The policy officials were to become more "ministerialized" (in line with the logic first triggered by the McKinsey reforms) and to behave more like ministers and junior ministers, lobbying elected legislators and major interest groups and defending government policies. As the first and only former cabinet-minister appointed governor of Hong Kong, Patten showed his senior administrative civil servants how they should perform their political role and paved the way for that change to take shape. The professional civil servants and departmental managers would settle into a more conventional type of neutral civil service role.

The need to prepare for a more accountable government as prescribed by the 1984 Sino-British Joint Declaration led to a more representative and demanding Legco as well as the advent of electoral politics and political parties. In theory, a possible route of institutional change for the civil service was for it to become a neutral service constitutionally subservient to an elected political leadership. A Westminster system as originally envisaged in the 1984 Green Paper whereby an Executive Council ultimately elected by a representative Legco would have functioned as a full-fledged cabinet could have facilitated such a change. Top administrative civil servants could then have surrendered their policy-making role to elected politicians. However, both British reluctance to dispense with effective rule prior to the change of sover-

728   ASIAN SURVEY, VOL. XXXVII, NO. 8, AUGUST 1997

eignty in 1997 and China's rejection of any Westminster system[20] meant that top policy officials became more politicized while political neutrality was advocated for the civil service as a whole. Public sector reforms also indirectly contributed toward strengthening the powers of policy officials through a process of policy centralization in exchange for managerial decentralization.

On the eve of the handover, the Hong Kong government administration remained firmly in the hands of the team of central policy mandarins headed by the senior trio—the chief secretary, financial secretary, and attorney general—and the 15 policy secretaries. This team oversees more than 70 government departments, agencies, and related statutory bodies. As a result of localization, particularly since the 1984 Sino-British Joint Declaration that requires that occupants of all "principal official" posts after 1997 must be of Chinese nationality, local Chinese officers now make up some 86.7% of the senior management/professional ranks and 72.4% of the directorate ranks of the civil service bureaucracy (as of October 1, 1996). All 23 principal posts (secretary-rank posts plus the headships of sensitive departments like the Police, Customs & Excise, Immigration, Audit, and the Independent Commission against Corruption) have also been localized in preparation for the transition, except for Attorney General Jeremy Matthews.[21] So essentially there is an indigenous center of power within the Hong Kong polity dominated by career bureaucrats who carry much political clout and may not be entirely willing to share power, much less give it up.

The Transition of Bureaucratic Power: Mixed Motives and Agendas

Although Patten was instrumental in making both the political and administrative processes more transparent and accountable in the final days of British rule, his controversial political reform package led to a serious confrontation with the Chinese government, which insisted that the reforms were in contravention with the Basic Law and previous Sino-British agreements. China declared that the original "through train" arrangements to facilitate continuity of Hong Kong's political institutions after the change of sovereignty in 1997 would come to an end. A provisional legislature selected by an exclusive group of 400 members—the same selection committee that had earlier chosen Tung Chee-hwa to be the first chief executive of the HKSAR—was installed

---

20. The Chinese government's desire for a civil service-governed Hong Kong was underlined by its rejection of a party-political government and its preference for a civil servant or former civil servant to head the future HKSAR government, as explained below.

21. British government, *White Paper on the Annual Report on Hong Kong 1996 to Parliament* (reproduced version) (Hong Kong: Government Printer, 1997), para. 127.

ANTHONY B. L. CHEUNG   729

in late December 1996; it was to draw up new election rules based on the decisions of the HKSAR Preparatory Committee appointed by China in December 1995 to oversee the takeover.[22] The question of continuity in the senior civil service has yet to be settled, given China's demand for their political loyalty and the rising exodus of senior officials. However, the appointment by Tung in February 1997 of all serving secretary-rank officials (except Attorney General Jeremy Matthews, who cannot meet the Chinese-nationality criterion) to the new HKSAR government with the same portfolios helped to restore some confidence among senior civil servants who had been nervous about being politically victimized for their past association with the Patten administration.

It became clear that the Chinese government would take active steps to halt, if not reverse outright, the whole process of political democratization and administrative opening-up belatedly started by the British government. The Chinese were suspicious about a British conspiracy on the transfer of Hong Kong, a local democratic politics had emerged[23] that was very often critical of Chinese policies, and most importantly the Chinese leaders preferred that Hong Kong remain bureaucratically controlled and governed in much the same way as it had been during the more typical earlier British colonial days. Two endogenous factors might also have served to reinforce China's determination to derail most of Patten's reforms. First, the major business interests would like to seize the opportunity of the British departure to gain a stronger say over public policies. They regard elected local politicians as mostly antibusiness and too prowelfare and prograssroots. A return to politics by "appointment" would safeguard the domination of the SAR government by business and professional elites, a feature that used to characterize the colonial administration. Second, although senior civil servants

---

22. As of April 1997 the Preparatory Committee was still considering views and submissions on how the first HKSAR Legislative Council should be elected in 1998. The newspaper *Hong Kong Economic Journal*, however, reported on March 25, 1997, that an internal document issued by the New China News Agency Hong Kong Branch advocated the adoption of a "multiseat, single-vote" system for the 20 seats to be elected by district constituencies, as opposed to the "single-seat, single-vote" method used in the 1995 Legislative Council election. The said document encouraged pro-China organizations to reflect the "mainstream" opinion to the Preparatory Committee. It is widely thought that a "multiseat, single-vote" system will reduce the number of directly elected seats that the democrats would be able to secure in an election.

23. Some local observers believed that China had a fundamental ideological opposition to democracy in Hong Kong. King commented in the mid-1980s that "the PRC Government was indeed of the belief that the development of representative government was a conspiracy of *min-chu k'ang-kung* [Using democracy to resist communism]; it was not intended to preserve the characteristics of capitalism, instead, it was to transform Hong Kong into a 'separate political entity', a city-state of its own, subject to no constraints from the Central People's Government." See A. Y. C. King, "The Hong Kong Talks and Hong Kong Politics," *Issues & Studies* 22:6 (June 1986), pp. 52–75.

730   ASIAN SURVEY, VOL. XXXVII, NO. 8, AUGUST 1997

have yet to pass the political loyalty test imposed by Beijing, 1997 represents an opportunity for them also to strengthen their hold on government administration since China obviously intends another form of bureaucratic polity in post-1997 Hong Kong either in the classical Hong Kong model or as a variant of the Singapore model. Both the business and senior civil service interests would gain in power over the HKSAR by displacing newly emergent politicians and political parties, and the two would converge on the need to marginalize electoral politics.

Major Political Actors and Their Agendas

A change of sovereignty brings about a process of reconstitution of political order and a reconfiguration of institutional power. In the case of Hong Kong, colonial administrative authority has been repeatedly challenged by newly emerging local forces and rising public demands ever since the 1970s. From 1980s onward China has entered the scene of political turbulence in Hong Kong. Because of the political transition, business interests that used to coalesce around the colonial authority have gradually sought to realign themselves with the successor regime—i.e., China and its agents in Hong Kong. A China-centered governing coalition has displaced the previous Britain-centered governing coalition. The senior civil servants, while still in name working for the British administration, have been forced to switch loyalty to their incoming political masters.

Though Hong Kong political actors both old and new have their respective agendas, they have one objective in common: to subdue bureaucratic power for their own interest. The British government tried to continue with bureaucratic rule until the last days of colonial rule, but in the process and in view of changed political realities it had to play up the political neutrality of the civil service to prevent other powers from capturing the civil servants. Besides, a localized and seemingly independent civil service could have been portrayed as proof of the British government's returning the power of government to the local population. The Chinese government, on the other hand, has been working to take over the civil service and to turn it into the loyal instrument of HKSAR rule under Beijing's oversight. What China seeks is a governing agent, and the civil service that had done its job so efficiently and loyally under the British political masters is considered most suitable for the new but similar role under the Chinese sovereign. Essentially, China's mind is set on continued bureaucratic rule with only a change of flag and object of political loyalty.

Locally Hong Kong's big-business elites have seen the weakening of British rule and the departure of the British as a great historical opportunity to further assert their power and influence. While most had advocated retaining British administration and presence during the Sino-British negotiations of

ANTHONY B. L. CHEUNG   731

the early 1980s, once the 1984 Sino-British Joint Declaration was signed, they immediately turned their attention to forging a new alliance with the Chinese government, which was happy to co-opt them through "political absorption of economics"[24]—i.e., through appointing big business notables to China's national and lower-level People's Congresses, the People's Political Consultative Conferences, the Basic Law Drafting and Consultative Committees, the HKSAR Preliminary Working Committee and Preparatory Committee, and as China's Hong Kong affairs advisers. On the other hand, newly emerging elected politicians, particularly prodemocracy forces, have demanded a greater say in the way the government is being run on the basis of their popular mandate gained through elections. To them, political reform is not just constitutional change per se to suit Hong Kong's new post-1997 status but also an important process of redistribution of political power from the business and professional elite sectors to the less-endowed middle class and working class. In the course of political reform, these elected politicians demanded the same form of accountability from the civil servants that exists in established democracies. Even though elected legislators do not govern, the fact that there is no longer any appointed seat in the Legco means that the administration will sometimes have to do the bidding of the legislature if government bills are to be passed and financial appropriations are to be obtained.

In face of the offensive from China, the business and professional elites and elected politicians, and the senior civil service, while not being free entirely from the oversight of the British governor, were working under various political constraints. Bureaucratic authority was not as unbounded on the eve of the handover as in the old colonial days. The pressure was particularly felt by the administrative class that had formed the "government" with the governor and governed the territory by co-optation and consultation. The senior professional civil servants, though anxious about their professional power, have not felt their pride shattered to the same degree, partly because they were often not at the forefront of the conflict and partly because they were used to work under the administrative class and had not accumulated as much institutional power.

However, the senior civil servants have profited from the competing bids to share power with the administrative bureaucracy and the disagreements among these competing forces. It has been pointed out that China has always preferred Hong Kong's "executive-led system," which implies government by civil servants, at least as long as these civil servants are politically loyal to Beijing. The New China News Agency Hong Kong Branch Director Zhou Nan made the point, when interviewed by *Time* magazine in June 1996, that

---

24. Ibid.

732 ASIAN SURVEY, VOL. XXXVII, NO. 8, AUGUST 1997

even if a political party secured a majority of seats in the legislature, it had no right to form a government.[25] China's stance is based not so much on the constitutional nicety of the Basic Law—which provides for the separation of powers between the executive and legislature—as on a fundamental distrust of political parties, which could gain entry to the organs of political power only through legislative elections. Such antipolitics sentiments were fully articulated within the China-appointed HKSAR Preparatory Committee when it decided in October 1996 that a candidate for the post of chief executive could not be a member of a political party.[26]

The Effect of Sino-British Conflict

To strengthen the executive-led system, partly on the ground that it had been British practice to appoint a civil servant as the governor of Hong Kong but more importantly to keep the chief executive from being embroiled in party politics or business interests, China had always intended to appoint a civil servant (or ex-civil servant) to be the first HKSAR chief executive. In the name of holding the chief executive above party and commercial interests, China's considerations served to continue the Hong Kong brand of politics so prevalent in its colonial days. Speaking to U.S.-based ABC News in May 1996, Lu Ping, director of the Hong Kong and Macau Affairs Office of China's State Council, confirmed rumors that the first chief executive could have been chosen as early as 1994 had it not been for the Sino-British conflict over political reforms. "We also even thought of having a vice-governor, a Chinese vice-governor before July 1, 1997, so that by July 1, this vice-governor could be Chief Executive," he was quoted as saying.[27] Both the Chinese and British governments were apparently looking to the top echelon of the civil service for a suitable candidate for the chief executive post. Apart from providing continuity and stability, such a choice would have served to emphasize the independence from and impartiality toward the sectoral interests of the post-holder.

The Sino-British conflict started after Patten's arrival spoiled the original plan. China was determined to prevent the British government from having any say over the appointment of the chief executive. Because Chinese lead-

---

25. Zhou Nan, "The Pearl Will Shine Brighter," interview by *Time*, July 1, 1996, pp. 28–29.

26. Article 4 of *The Procedure for Selecting the Candidates for the First Chief Executive of the Hong Kong Special Administrative Region of the People's Republic of China*, passed by the Preparatory Committee at its 5th Plenary Meeting on October 5, 1996, requires that candidates can accept nomination only in a personal capacity and must give up their capacities in any political party or other political organization when declaring their intention to stand for election (*Ta Kung Pao* [Hong Kong], October 6, 1996).

27. C. Yeung, "Beijing's Search for a Face that Fits," *Sunday Morning Post* (Hong Kong), September 8, 1996.

ers were unable or unwilling to pick someone from among top officials within the existing senior civil service (such as Chief Secretary Anson Chan), who would be suspected of having been "Pattenized," they were forced by circumstances to consider potential candidates from within the ranks of business leaders in Hong Kong; hence the rumor that Tung Chee-hwa, a shipping magnate well connected to leaders in China and Taiwan, was the handpicked choice of the Beijing leadership. However, intrabusiness rivalries and conflicts of interest also meant that it was not easy for business leaders to agree on a candidate, not to mention the restlessness of nonbusiness interests about a probusiness candidate. Consequently, some local pro-China leaders supported Sir T. L. Yang, the former chief justice, who nominally could be argued as belonging to the public service system and independent from various sectoral interests. The loss of an opportunity for a top civil servant to be selected and groomed to take over from the British governor as the chief executive of the HKSAR is one of the main casualties of the Sino-British conflict.

Political Window for Bureaucratic Polity of Another Kind

Without delving into the issue of selecting the chief executive, suffice it to say that within China's thinking, the strategy for taking back Hong Kong has always been premised on the assumption that the senior civil service would remain the most important pillar of HKSAR governance and, given the present disarray on the local political scene and the suppression of bottom-up electoral politics, probably the only pillar left. Hence, the senior civil service as a power group has not fallen into political disgrace but has in fact come out from the Sino-British conflict relatively untainted by its close association with Patten's controversial policies. On the contrary, it can be argued that the escalation of Sino-British differences and the polarization of Hong Kong society (e.g., between pro-China and prodemocracy forces) have left the civil service institution as the only force that is still acceptable to all sides and whose continuity is valued by most. Even the democrats have given up any hope for an elected government but have instead come to the rescue of top civil servants by demanding that Beijing should not impose any "political vetting" of present principal officials and should allow the latter to stay in the same posts after 1997.[28]

The difficult position faced by the Chinese government and the limited range of political choices available to it means that the Hong Kong adminis-

---

28. See Democratic Party (Hong Kong), "Civil Service System" (Section 1.5) (in Chinese), *Challenges to a New Hong Kong*, political platform of Szeto Wah, nominee of the Chief Executive Mock-Nomination Campaign (November 1996).

734   ASIAN SURVEY, VOL. XXXVII, NO. 8, AUGUST 1997

trative elite would have a considerable amount of bargaining power vis-à-vis the future sovereign government over the running of the HKSAR. Indeed, it is an open secret that China has accepted that the HKSAR chief executive must be acceptable, inter alia, to the civil service—in practice, the administrative elite. Thus, although the administrative elite does not appear to be able to grasp the highest office of the HKSAR government for its member, it stands certain to secure a second-best solution, that is, a leadership team on which it should have an institutional voice. During the 1980s, when the future of Hong Kong after 1997 was still being negotiated between the two sovereign governments, administrative officers in the Hong Kong government were already tossing out the notion of "Hong Kong bureaucrats governing Hong Kong" (*gang guan zhi gang*). The ultimate scenario seems not too far off from what they had set out to achieve.

After all the hiccups and debates about decolonization, representative government, and democratization over the past decade, the governance of Hong Kong has come back to square one—sustaining a form of bureaucratic polity of another kind with the support of China and the connivance of the business and professional elites. However, this prospect does not necessarily mean that local politics will accordingly be assigned to the wilderness. For one thing, elections and legislative politics have already become part of the political system and will remain so. The senior civil service and the chief executive, thus, will not be able to ignore elected politicians and their political parties entirely. The local population will also try to demand political accountability from the HKSAR government through "opposition" politicians and parties. Furthermore, the administrative elite would not benefit by being left on its own to deal with the central government in Beijing. To maintain its unbeatable position in HKSAR governance, the administrative bureaucrats need the central government's support to suppress local challenges to their position; at the same time, they also need to be able to defer to the need to consult and obtain consent from local business, professional, party, and political elites when confronted with demands from the central government that they find it difficult to accept.

China's "dependence" on the administrative civil servants to govern the HKSAR, however, needs to be properly qualified. Despite repeated remarks by senior Chinese officials praising the high quality of Hong Kong's civil service, the Chinese government is apparently still obsessed with the fear that senior local bureaucrats might try to defy its political diktat. China has looked for a bureaucracy that can act as the central government's governing agent in much the same way as it did in the British colonial days. While on the one hand trying to protect the civil service from challenges by local politics through suppressing political democratization and sidelining of elected politicians in the name of preserving Hong Kong's "executive-led" features,

ANTHONY B. L. CHEUNG 735

China has on the other hand attempted to tighten its political control of the civil service. A kind of China-centered politicization is to displace gradually local politicization. Such a prospect generates all the more a need for the senior civil service not to cut itself off entirely from the local politicians and their "support."

Bureaucratic Power under Tung Chee-hwa

Despite the change of sovereignty and government, top civil servants are still keen to keep their policy-making powers largely intact. Some resented the high-profile leadership of Chris Patten, who was blamed for unduly politicizing the administration. The new chief executive, Tung Chee-hwa, seemed to be seeing eye to eye with these bureaucrats when he campaigned on a platform of depoliticization.[29] However, Tung also claimed that he would exercise "strong leadership" and suggested that the noncivil service members of his executive council would play a more prominent political role by promoting government policies. A vocal and assertive executive council team would certainly provide an important bulwark to counterbalance the power and influence of the top civil service team headed by Anson Chan, the chief secretary who became the HKSAR secretary of administration. The first battle over who should be in charge of government policy formulation was fought in late March 1997 when top civil servants expressed great displeasure over Tung's appointing three of his executive councilors to lead policy teams comprising policy secretaries to formulate policy proposals on housing, education, and elderly welfare. In the face of this bureaucratic opposition and a media uproar over the potential role conflict, Tung was forced to concede by emphasizing that those three executive councilors would only perform a research role and not convene any formal policy teams.

Despite Tung's denial of any intention to move toward a ministerial or quasiministerial system, the nightmare of top mandarins has been that Tung would reduce or take away their policy-making powers and render them simply civil servants. Two competing "executive-led" models will thus be interacting—the bureaucrats favoring a system with civil servants in charge and Tung probably preferring a more presidential kind of executive government. However, given that Tung does not enjoy a clear popular mandate from within Hong Kong or any firm local political power base (not being linked to any political party), he has to strive hard in his early years of administration

---

29. Tung asked, for example, "Is our civil service too bogged down in the politics of our legislative process? Should they be devoting more time and energy to the formulation and efficient implementation of policies?" (Tung Chee-hwa, *Building a 21st Century Hong Kong Together* [Hong Kong, October 22, 1996]).

to deliver effective government to build up his authority. In so doing he must rely on the support and cooperation of his top civil servants; he cannot afford to alienate them. A strong leadership by Tung would not therefore necessarily mean the demise of public bureaucratic power in any substantive sense.

Conclusion

This article explores the changing features of bureaucratic politics in Hong Kong amid the transfer of sovereignty and the conflicts of transition. Historically the administrative elite had enjoyed unchallenged authority in the governance of Hong Kong as a British colony, although it had to share some of its powers with the local elites and secure the latter's consent and support through an elaborate system of appointments and administrative co-optation. With the erosion of colonial authority, the administrative elite was subject to more and more challenges to its power, both locally and from China. Administrative modernization of the civil service and the gradual opening up of the government since the 1970s represented attempts to restore some form of legitimacy through efficient and responsive institutional performance.

The post-1984 political transition has, however, brought about uncertainties with respect to both the institutional rules of the SAR political structure and to the role of the senior civil service. For a while, particularly under the British rhetoric of developing representative government fueled by strong local demands for democratization, it seemed that the civil service might be relegated to being a politically neutral administrative arm of an elected government. Such a possibility, considered remote even when it was contemplated in the less constrained climate of the 1980s, has now definitely proved to be unattainable in view of China's clear preference for an executive-led system with the civil service in charge and of the underdevelopment of party politics. Although big business has enjoyed much political clout in influencing the thinking of both sovereign governments and will likely continue to sustain such clout after 1997, the intrabusiness rivalries and popular distrust of the self-interest of business elites mean that China would be reluctant to lean too much to the side of the business elites. In the tactics the central government takes to strike a delicate balance among various competing local forces and ensure top-down control, the most desirable mode of HKSAR governance is probably a new bureaucratic polity. This polity would be regulated to some degree by the checks and balances of a local legislature representing major interest groups with stakes in Hong Kong, but it would also have an unquestioned, executive-led ethos. Ironically, it is its lack of any substantial social power base within the local scene (unlike the business elites or political parties) that gives the administrative elite of the civil service its most appealing "right to govern" in the eyes of Chinese leaders.

ANTHONY B. L. CHEUNG 737

A new mode of bureaucratic politics can thus be anticipated. The dominance of the civil service bureaucracy after 1997 will be supported ideologically by the "executive-led" ethos and facilitated institutionally by the restraining of political parties and electoral politics. The impossibility of party government in the HKSAR and the likely adoption of election rules preventing a single political party from gaining majority control of the Legco would make the legislature too fragmented to provide an effective countervailing force to the power of the executive branch. Without a strong mandate that can be obtained through popular elections and a political party machinery to provide backup support (as in Singapore, the "authoritarian" rule of which has attracted admiration from Chinese leaders), the HKSAR chief executive has unavoidably to rely almost exclusively on the civil service bureaucracy for policy advice and organizational strength. Thus, the chief executive is susceptible to bureaucratic capture. The influence of the Chinese central government and its agents in the HKSAR will remain a critical factor affecting the prospect of bureaucratic governance. Some form of amalgamation of two sources of bureaucratic authority—the Chinese Communist party-state bureaucracy in mainland China and the HKSAR civil service bureaucracy—could emerge. Like its colonial predecessor, the HKSAR bureaucracy is likely to have to share some power with local power elites, including a new stratum of elected politicians and their party supporters. Though still being checked in its power by endogenous and exogenous forces, the administrator-bureaucrats would continue as the institution that ultimately aggregates and brokers major interests in the HKSAR society. Any change to that predominant role will come only if the HKSAR is allowed to move toward a wholly directly elected legislature and popularly elected chief executive in 2007 when, under the Basic Law, changes can be made to the electoral systems if certain conditions are satisfied.[30]

---

30. Under Basic Law Annexes I and II, amendments to the method for selecting the chief executive and forming the Legislative Council after 2007 require the endorsement of a two-thirds majority of all the members of the Legislative Council and the consent of the chief executive, and have to be reported to the Standing Committee of China's National People's Congress for approval.

[11]

Lau Siu-kai

The Rise and Decline of Political Support for the Hong Kong Special Administrative Region Government*

ON 1 JULY 1997, THE END OF COLONIAL RULE USHERED IN A NEW government for the 6½ million people of Hong Kong. Ironically, in stark contrast to other new regimes which took over from colonial rulers, the Hong Kong Special Administrative Region (HKSAR) government was greeted with only a moderate amount of enthusiasm by the governed. Nevertheless, public support for the new regime mounted in the first four months of its existence. Since October 1997, however, it has declined continuously and has now reached a low level. Evidently, the governing strategy crafted by the HKSAR government had achieved a certain degree of success in the early months of its existence. Since then, though the HKSAR government obstinately follows this governing strategy, changes in the conditions in Hong Kong have in any case rendered this strategy obsolete. Low public support is bound to undermine effective governance in post-colonial Hong Kong.

Though the HKSAR government assumed office only on 1 July 1997, in fact it was in existence as early as December 1996, when Tung Chee-hwa, a shipping magnate, was elected as Chief Executive of the HKSAR by a 400-member Selection Committee, which was drawn from the social and economic elites of Hong Kong. The

*This article is based on a research project entitled 'Indicators of Social Development: Hong Kong 1997', which was generously funded by the Research Grants Council of the Universities Grants Committee. It is a joint project of the Centre of Asian Studies at the University of Hong Kong, the Hong Kong Institute of Asia-Pacific Studies at the Chinese University of Hong Kong, and the Department of Applied Social Studies at the Hong Kong Polytechnic University. I am grateful to Ms Wan Po-san, Research Officer of the Institute, for rendering assistance to the project in many respects. Special thanks are due to my research assistants, Mr Shum Kwok-cheung and Mr Yiu Chuen-lai, for their help in organizing the questionnaire survey and in data analysis.

composition of the Selection Committee was greatly influenced by Beijing, which was determined to see the highest and most powerful office in Hong Kong go to a person who was mindful of China's vital interests there. By virtue of Tung's personal image of integrity, honesty, sincerity, modesty and kindness and his advocacy of Confucian verities, he was also capable of commanding a fair level of acceptance by the Hong Kong people. In the six months before the handover, Tung and his team had managed to restore a certain degree of public confidence. Their efforts were greatly helped by the normalization of Sino-British relations from late 1996 and the practice of pragmatism and self-restraint by the Chinese government on the eve of Hong Kong's reversion.

None the less, at the time of the handover, public nostalgia toward colonial rule and anxieties about a 'pro-Beijing' and pro-business government still haunted the new regime of Hong Kong. The fact that Tung was a political newcomer without a strong and organized base of political support exacerbated the political difficulties of the HKSAR government. Nevertheless, the Tung administration had managed to 'borrow' some political capital from the departing colonial regime by maintaining intact its civil service, by appointing a number of 'pro-British' politicians to his governing team, by vowing to retain the basic policies of his predecessor, and by promising performance in the economic and social spheres. At the same time, the narrow base on which he was 'elected' and his political conservatism did not endear him to the liberal elements in the community.

Since the end of colonial rule in Hong Kong is not the result of public demand and as the colonial regime still enjoyed solid, though declining, popularity in its last days,[1] the HKSAR government is constantly compared to its predecessor by the people. In fact, Tung's regime is acutely aware of and sensitive to the fact that the population constantly measures its performance against the not uncommonly 'romanticized' performance of the colonial regime. In a highly diverse society such as Hong Kong, it is impossible for any government to command respect from all the people. In view of the differences between the colonial regime and the Tung administration, it is unavoidable that the new regime has to cultivate a support base which overlaps but does not coincide with that of its

---

[1] See Lau Siu-kai, 'Democratization and Decline of Trust in Public Institutions in Hong Kong', *Democratization*, 3:2 (Summer 1996), pp. 158–80.

predecessor in order to govern competently. It will have to tailor its governing philosophy and policy stands to appeal to those sections of the public whose political attitudes are likely to make them its supporters. Accordingly, an analysis of public attitudes toward the old and new regimes in Hong Kong is critical to the understanding of the support base of the HKSAR government, the governing strategy it has crafted, and its subsequent loss of political support. It was with these purposes in mind that a questionnaire survey of a territory-wide sample of Hong Kong Chinese was conducted immediately before the handover in 1997.[2] Unless otherwise specified, the data used come from that survey.

LINGERING NOSTALGIA FOR COLONIAL RULE

Hong Kong stands out from other colonial societies in that throughout its history colonial rule was never challenged by the colonized. On the contrary, colonial rule has often been credited with the achievements of Hong Kong's economic miracle, its social stability, the rule of law, personal freedom and escape from the political turmoil in China. Even after Hong Kong's political fate was sealed by the Sino-British deal to return the territory to China, there remained a strong, though weakening, sentiment in favour of continued colonial governance. In a survey of 868 Hong Kong Chinese in the summer of 1992, a plurality (38.9 per cent) agreed with the suggestion to retain Hong Kong as a British colony. Slightly fewer (32.1 per cent) were opposed to it. When we take into consideration the finding that 15.8 per cent were indifferent, it seems reasonable to argue that the retention of Hong Kong as a British colony would have been acceptable to the people.[3]

---

[2] The sample (N = 1,410) used in the questionnaire survey was drawn by means of a multi-stage design. The target population of the survey were the Chinese inhabitants of Hong Kong aged 18 years or over. Fieldwork was conducted mostly from 7 May to 30 June 1997. By then 90.5 per cent of the successful interviews were completed, while the rest of the interviews were carried out from 2 July to 9 September 1997. At the end of the survey, 701 interviews were successfully completed, yielding a response rate of 49.7 per cent.

[3] See Lau Siu-kai and Kuan Hsin-chi, 'Public Attitudes toward Political Authorities and Colonial Legitimacy in Hong Kong', *The Journal of Commonwealth and Comparative Politics*, 33:1 (March 1995), p. 81.

Public nostalgia for colonial rule was gradually diluted as people became increasingly resigned to the unalterable reality; and public confidence in Hong Kong's future has meanwhile strengthened.[4] Still, in the 1997 survey, 29.3 per cent of the respondents favoured retaining Hong Kong as a British colony, as against 47.5 disagreeing and 10.8 per cent indifferent. Declining interest in preserving Hong Kong's colonial status is accompanied by an expression of nationalist pride. As many as 46.5 per cent of respondents were happy at the termination of colonial rule and the return of Hong Kong to China, while a smaller proportion (34.2 per cent) felt otherwise.

Nevertheless, what is most pertinent to political support for the HKSAR government is the fact that people still harbour a high regard for the performance of the colonial rulers: 66.5 per cent of respondents said that in the past 150 years Britain had done a good job in running Hong Kong. Even though many Hong Kong people were dissatisfied by Britain's placing its interest above Hong Kong's when it handed the territory's reversion to China, a substantial minority (32.8 per cent) of respondents still considered that Britain had fulfilled its moral obligation to the Hong Kong people before its withdrawal from the colony (as against 39.2 per cent who held the opposite view). Overall, it can be said that despite the sense of nationalist gratification springing from the termination of colonialism, the people of Hong Kong retained fond memories of British rule. The nostalgia for the old regime unavoidably hampers the efforts at cultivating political support by the HKSAR government, for in the mind of the public the change of regime in 1997 was not something to really enthuse about.

LIMITED SUPPORT FOR THE POLITICAL SYSTEM

In comparison with the colonial political system, the political system of the HKSAR, of which the HKSAR government is an integral part, represents both advance and regression. The Provisional Legislative Council, which formally took office when the HKSAR was established, was 'elected' through a much less democratic

---

[4] Slightly over half (53.4 per cent) of the respondents expressed confidence in Hong Kong's future, as against 11.1 per cent who had no confidence. About one-third (32.1 per cent) were betwixt and between.

method than the colonial legislature it had displaced.[5] On the other hand, the Chief Executive of the HKSAR, though elected only by a small elitist electoral college, is in a better position to stake a claim to a democratic mandate than the last governor, Chris Patten, who was appointed by London without consultation with the people of Hong Kong.

Nevertheless, as a result of minimal public trust in the Chinese government, which had been influential in the selection of the Chief Executive and the setting up of the provisional legislature, overall the people of Hong Kong preferred the pre-1997 political system to the post-1997 one as stipulated in the Basic Law — the mini-constitution of the HKSAR promulgated by China on 1 April 1990. A plurality of respondents (40.9 per cent) thought the pre-1997 political system better than the projected post-1997 one. Only 12.8 per cent preferred the post-1997 system to the pre-1997 one. Judging from previous studies, apparently Hong Kong people thought the pre-1997 political system more open and 'democratic'. There is also a trend of declining public acceptance of the political system of Hong Kong. In the 1995 survey, 63 per cent of respondents agreed with this statement: 'Although the political system of Hong Kong is not perfect, it is still the best under the realistic circumstances of Hong Kong.' Only 20.1 per cent disapproved of it. Two years later, the proportion of supporters of the view that the political system stipulated by the Basic Law was still the best under present circum-stances despite its imperfections was 49.2 per cent, while the opponents amounted to 26.5 per cent.

While public support for the post-1997 political system was on a downward trajectory, still the method for electing the Chief Executive was more popular than that for the provisional legislature. The method for electing the Chief Executive was endorsed by 28.7 per cent of the respondents, but rejected by 34.9 per cent. On the other hand, only 17.3 per cent of the respondents found the way the provisional legislature was elected acceptable, in contrast to a plurality (40.8 per cent) who were displeased with the undemocratic character of the electoral method.

Despite public displeasure with the political system of the HKSAR, Tung's administration is however severely handicapped by its

[5] The 400-member Election Committee, which elected the Chief Executive, was also the body which elected the Provisional Legislative Council.

conservative nature, its narrow elite base, and the slow pace of democratization laid down by the Basic Law. The establishment of the HKSAR has also seen a certain degree of curtailment of the rights of demonstrations, parades and organizing societies on the ground of 'national security'. Consequently, in its quest for political support, the HKSAR government cannot seek to emulate, let alone to improve upon, the colonial regime in promoting democracy and human rights. Aside from its minimal democratic foundation, the support for the new political system of the HKSAR, and as a corollary that of the Tung regime, is adversely affected by public perception that the political system is designed by China in order to produce a pro-Beijing regime in Hong Kong. Throughout the past decade, people tended to perceive the relation between China and Hong Kong as conflictual, though with time the proportion of people holding this belief fell gradually. In 1995, 60 per cent of the respondents saw conflicts of interests between China and Hong Kong. In 1997, the pragmatic efforts of China to woo the Hong Kong people had apparently borne fruit, for a smaller proportion of 44.8 per cent considered the interests of China and Hong Kong as in conflict, while an unprecedentedly large 40.2 per cent denied the existence of conflict of interest.

Nevertheless, public trust in the Chinese government remained at a low level at the time of the handover. More respondents (30.3 per cent) mistrusted than trusted (25.7 per cent) the Chinese government. At the same time, more respondents (39.9 per cent) were inclined to believe than disbelieve (32 per cent) that the HKSAR government would place the interests of the Chinese government above those of the Hong Kong people. Furthermore, a large majority (70 per cent) thought that the Chinese government had great influence over the HKSAR government. In all, public belief in the pro-Beijing bias of the HKSAR government poses an obstacle to the build-up of its political support base.

THE QUEST FOR POLITICAL SUPPORT BY THE NEW REGIME

In face of the difficulties and constraints encountered in the process of legitimizing itself, the new regime has adopted a three-pronged approach to seek public support, drawing upon its strengths. The foremost strategy is to demonstrate to the people that the new regime enjoys the complete trust of the central government and hence is capable not only of averting Chinese interference in local affairs but also of enlisting China's support in resolving Hong Kong's problems. The old colonial regime in its last years was embroiled in serious conflicts with China, and Hong Kong suffered huge losses as a result. Since the new regime is largely groomed and very much trusted by Beijing, the establishment of the HKSAR is widely expected to bring not only peace, but cordial relations between Hong Kong and the mainland. China did cooperate actively with the Tung administration before the handover, and major Chinese leaders openly applauded Tung with an eye to building up his prestige and authority in Hong Kong. To a certain extent, Tung's intimate relation with China had served him well. More people (24 per cent) rated Tung and the HKSAR government as having done a good job before the handover than those who failed to do so (17.3 per cent). This testifies to the importance of China to the political support for the new regime.

Another tactic of the new regime in seeking political support is to appeal strenuously to Chinese tradition. While it is clear that Tung's close political allies are not 'Confucianists', Tung himself however is definitely an ardent advocate of traditional virtues. The political philosophy advocated by Tung is manifestly traditionalist in tone, though Tung has not been able to articulate a coherent political doctrine. There is extensive resort to values such as collectivism, the family, harmony, peace, filial piety, respect for the elderly, benevolence, obligations to the community, modesty and integrity in Tung's speeches.[6] Tung's ideal government is one practising paternalistic and benign rule. The ideal society is one devoid of conflicts, particularly political strife. In that society, people are respectful and deferential to political authorities. To all appearances, Tung's governing philosophy should be anachronistic and at odds with a society which is highly modernized and Westernized. This however is not the case. Survey findings in the past have repeatedly shown that traditional beliefs remain strong in the ethos

[6] See for example speech by Tung Chee Hwa, 'Building a 21st Century Hong Kong Together' (22 October 1996), pp. 5 and 10; and C. H. Tung, 'A Future of Excellence and Prosperity for All', speech by the Chief Executive at the Ceremony to celebrate the Establishment of the Hong Kong Special Administrative Region of the People's Republic of China, 1 July 1997, Hong Kong, HKSAR Government, 1997, p. 18.

THE HONG KONG SAR GOVERNMENT   359

of the Hong Kong Chinese.[7] At the time of the handover, Tung's way of thinking apparently still had broad public appeal. A large majority of respondents (69.7 per cent) agreed with the view that 'a good government should treat the people as though they are its own children'. The idea of paternalistic government was only opposed by 17.3 per cent of respondents. Similarly, most (68.2 per cent) respondents were of the view that the government should enact legislation to punish those who refused to care for their parents, as against 21.5 per cent who disagreed. People also enthusiastically expected the government to be a moral exemplar and educator. Many more (75.6 per cent) respondents agreed than disagreed (12.4 per cent) that a good government was one that taught the people how to conduct themselves.

When asked to choose between Chinese and Western values, the former were preferred more often than the latter. When asked whether traditional Chinese values such as loyalty, filial piety, benevolence and righteousness, or Western values of freedom, democracy and human rights, were more appropriate for Hong Kong, a plurality (32.1 per cent) chose traditional Chinese values, followed by 31.1 per cent choosing both and 27.4 per cent picking Western values. More specifically, social order was deemed to be more important than individual freedom by 59.1 per cent of respondents. Only a tiny 14.1 per cent placed greater importance on individual freedom. And 23.7 per cent saw both as of equal importance. The emphasis on collectivity by the Hong Kong Chinese is also unmistakable. Just under half (44.7 per cent) of respondents said that public interest took priority over human rights. The opposite view was endorsed by 23.7 per cent of respondents, whereas 25.7 per cent saw them as equal in importance. These findings show that Tung's traditionalist philosophy is in many aspects in harmony with the majority opinion in Hong Kong, and it provides the new regime with a potential source of support which can be tapped.

However, in one significant aspect Hong Kong Chinese are far from traditionalist. As many as 63.7 per cent of respondents were opposed to the view that equated the head of government to the head of a large family and demanded everyone to abide by whatever decisions he made. Less than one-third (27.9 per cent) of respondents

[7] Lau Siu-kai and Kuan Hsin-chi, *The Ethos of the Hong Kong Chinese*, Hong Kong, Chinese University Press, 1988.

would willingly submit themselves to the whim and caprice of such a person. Therefore, Tung cannot easily bank on public submission to legitimize his government. He has to deliver on his promises to his supporters who apparently will only exchange political support for benevolent governance. In addition, he has to be able to communicate with the people and persuade them to support his policies. Needless to say, at the time of handover, people had yet to convince themselves that the new regime was able to put its governing philosophy into practice, but Tung's favourable personal image in no small measure increased people's confidence that he would run a benevolent government.

The crucial tactic in the new regime's quest for support lies in its unequivocal promise of governmental performance in the economic, social and employment spheres. Many long-standing economic, social and employment problems, which had not been seriously dealt with before the handover,[8] furnished the Tung administration with the opportunity to win public support. As a matter of fact, even though the old colonial regime did introduce democratic reform in the last one-and-a-half decades of its rule with much fanfare, it was its competence and achievements in Hong Kong's economic development and social stability that formed the base of its political support.[9] The HKSAR in its early years could only offer a legislature less democratic in nature than that under colonial rule, so the new regime's reliance on performance as the principal means to seek support was crucial. Before assuming office, Tung explicitly gave top priority to the problems of housing, education and the elderly in the programme of his administration. He also vowed to strengthen Hong Kong's economic competitiveness and improve the well-being of the people. His promises certainly were credible as the economy was vibrant and the HKSAR government was about to inherit large fiscal reserves from the colonial regime. Tung's emphasis on economic and social issues was in accord with public preferences. When asked about the top priority of a good government, 34.5 per cent chose social stability, 33.4 per cent people's livelihood, 20.5 per cent economic development, 5.4 per cent human rights and freedom, and 1.7 per cent democratic

[8] Lau Siu-kai, 'The Fraying of the Socio-economic Fabric of Hong Kong', *The Pacific Review*, 10:3 (1997), pp. 426–41.
[9] See Lau and Kuan, 'Public Attitudes toward Political Authorities'.

development. By the same token, among economic prosperity, social stability, personal freedom and democratic government, 52.6 per cent deemed social stability most important, 23.3 per cent economic prosperity, 10.4 per cent democratic government and 7.7 per cent personal freedom. In view of the pragmatic disposition of the Hong Kong Chinese, the attempt by the new regime to build its political support on performance should have had a reasonable chance of success.

Political appeals based on traditional values and performance are in fact interrelated. Implicit in the traditional conception of benevolent governance is a government that takes care of the material and spiritual needs of the people. As Muthiah Alagappa has put it: '[E]ffective performance can be deployed to generate moral authority. The enormous concentration of power in the state cannot be justified except in terms of its use in the pursuit of the collective interests of the political community. And success or failure in this endeavor affects the legitimization of government and the regime in which it is rooted.'[10]

BASES OF POLITICAL SUPPORT FOR THE OLD AND NEW REGIMES

Notwithstanding China's intention and promise to preserve Hong Kong's institutional structure and way of life after the end of colonial rule, the different governing doctrines of the old and new regimes must mean that their bases of political support are not the same.

Immediately before the handover, the colonial Hong Kong government was trusted by 52.8 per cent of the respondents, as compared with 31.7 per cent who trusted the new HKSAR government. This finding is no surprise at all in view of the lingering nostalgia for colonial rule and pervasive public ambivalence toward the new regime. It is important and interesting to note that trust in the Hong Kong government and trust in the HKSAR government are positively correlated. That is to say, people who trust the Hong Kong government are likely also to trust the HKSAR government. Furthermore, in a socio-demographic sense, the kind of people who trust the old regime

[10] Muthiah Alagappa, 'Introduction', in M. Alagappa (ed.), *Political Legitimacy in Southeast Asia: The Quest for Moral Authority*, Stanford, Stanford University Press, 1995, p. 22.

are very likely to be also the kind of people who trust the new regime. Hong Kong Chinese tend to calibrate their trust for the Hong Kong government and the HKSAR government chiefly in accordance with what they see as respectively its past performance and its expected performance. Thus, respondents who rated the performance of the Hong Kong government as good or very good in improving the living standards of the people, creating an incorrupt government, promoting economic development, maintaining social stability, safeguarding the rule of law, promoting democratic development, protecting human rights and freedom, delivering social welfare, promoting moral education, and establishing good relations with China were more likely to trust the Hong Kong government. Similarly, respondents who expected the HKSAR government to perform well in these areas were also more inclined to trust it.

Despite the many similarities between people who trust the old regime and people who trust the new regime, there are still some differences. Respondents who trusted the HKSAR government were comparatively more confident about the future of Hong Kong, more likely to believe that the objective of 'Hong Kong people ruling Hong Kong' and 'a high degree of autonomy' could be realized, more optimistic about the future of China, more trustful of the Chinese government, more trustful of Tung Chee-hwa, and more likely to identify themselves as 'Chinese' rather than 'Hongkongese'. They were also more likely to have a stronger sense of belonging to Hong Kong, be opposed to retaining Hong Kong as a British colony, feel happy about the return of Hong Kong to China, have a sense of obligation to the motherland, feel proud to be Chinese, see no conflict of interest between Hong Kong and China, believe that the HKSAR government would run Hong Kong better than the British, believe that the HKSAR government would place the interest of the common people above that of the rich, support the post-1997 political system and leaders, and be sanguine about Hong Kong's democratic future. Respondents who trusted the old regime, on the other hand, were more likely to be those who trusted the British government and had fond memories of colonial rule. Hence, the differences between the supporters of the old and the new regimes lay primarily in their political attitudes, and it is obvious that those attitudes underlying support for the old regime were held by a majority of the Hong Kong Chinese. No wonder the old regime is more widely trusted than the new regime.

In contrast with the finding that the old regime was more trusted than the new regime, what is quite unexpected is that Tung, a political novice, received a slightly higher level of public trust (26.4 per cent) than Chris Patten (21.5 per cent), the last colonial governor and a seasoned politician. The higher level of public trust in Tung is also reflected in the finding that slightly more respondents (24.7 per cent) supported Tung's governing strategy than Patten's (22.8 per cent). In addition, unlike the positive correlation between trust in the old regime and trust in the new regime, trust in Patten and trust in Tung are negatively correlated. In other words, people who trusted Patten were more likely to distrust Tung, and vice versa. It is pretty clear that the bases of political support for Patten and Tung were very dissimilar. Statistical analyses show that the characteristics of respondents who trusted the HKSAR government are largely shared by those who trusted Tung. However, respondents who trusted Tung were ardent believers in traditional Chinese values. Tung's supporters were more likely to define the relation between government and people as that of father and son, agree that a good government should teach people how to conduct themselves, consider traditional Chinese values more appropriate for Hong Kong than Western values, place public interest above human rights, and see economic development, people's livelihood and social stability as higher priorities to a good government than democratic development as well as human rights and freedoms.

On the contrary, respondents who trusted Patten were more likely to be those who had little confidence in Hong Kong's future, were trustful of the British government, believed that Britain had fulfilled its moral obligation to the Hong Kong people, wanted to retain Hong Kong as a British colony, had fond memories of colonial rule, supported the democratic reform introduced by Patten, believed that the new regime would perform less well than the old regime, thought that the pre-1997 political system was better than the post-1997 political system, trusted the pre-1997 political leaders more, preferred democratic government to other forms of government, imputed more importance to democratic government in comparison with economic prosperity and social stability, rated human rights more important than the public interest, and were convinced that Western values rather than traditional values were suitable for Hong Kong.

Evidently, Patten's supporters and Tung's supporters differed sharply on political values. People who were Westernized and democratically oriented were more likely to trust Patten. Tung, on the other hand, had support largely from people who espoused traditional values, and who at the same time had strong nationalistic sentiments. Since traditional values are held by a majority of the Hong Kong Chinese, it is no surprise that Tung, despite his association with the new regime, still enjoyed slightly higher popularity than Patten.

Aside from minority support for Western values, Patten's personal popularity was also adversely affected by the public's belief that material interests were hurt by Patten's conflict with China during his governorship. Less than one-third (29.4 per cent) of respondents were of the opinion that Patten's political reform was to the good of Hong Kong, whereas 19.4 per cent thought it was detrimental to the territory's interests. The impact of public anxiety about material interests on public trust of Patten can be seen in Table 1, where in general respondents who were satisfied with the conditions in Hong Kong were more likely to trust Tung and less likely to trust Patten. Very clearly, people who had vested interests in the status quo were more likely to blame Patten for undermining their interests.

Table 1

Relationship between Satisfaction with Conditions in Hong Kong and Trust in Tung and Patten (percentage)

	Tung		Patten	
Satisfaction with:	Distrust	Trust	Distrust	Trust
1. Economic Conditions	22.5	34.2	31.7	28.9
2. Law and order	18.4	38.8	–	–
3. Political conditions	14.0	44.7	33.1	32.2
4. Transport	21.2	38.7	33.2	29.8
5. Housing	19.2	44.4	29.2	32.9
6. Medical services	18.1	37.6	34.8	29.2
7. Education	20.9	36.8	33.1	30.7
8. Social welfare	16.1	42.5	30.7	34.2
9. Employment	16.5	42.9	29.3	31.3
10. Culture and recreation	20.6	34.6	32.0	27.7

Note: only figures associated with relationships between variables significant at .05 level, using the chi square test, are shown in the Table.

Significantly, in the mind of the respondents there was no complete correlation between the old regime and Patten, even though it is true that those who tended to trust the Hong Kong government also trusted Patten. In Table 2, it can be seen that seeing good governmental performance in several areas — creating an uncorrupt government, promoting economic development, maintaining social stability and safeguarding the rule of law — had not led to trust in Patten. By contrast, Table 3 shows that if the respondents anticipated that Tung would do well in the future, they would trust him now.

Table 2

Relationship between Assessment of Performance of the Hong Kong Government and Trust in Patten (percentage)

		Patten	
HK government's performance good in:		Distrust	Trust
1.	Improving the living standard of the people	27.1	31.6
2.	Creating an incorrupt government	29.6	27.0
3.	Promoting economic development	30.0	27.1
4.	Maintaining social stability	30.5	27.9
5.	Safeguarding the rule of law	30.4	27.6
6.	Promoting democratic development	28.6	30.3
7.	Protecting human rights and freedom	27.0	31.5
8.	Delivering social welfare	26.3	35.9
9.	Promoting moral education	29.6	38.0
10.	Establishing good relations with China	22.0	44.7

Note: only figures associated with relationships between variables significant at .05 level, using the chi square test, are shown in the Table.

Table 3

Relationship between Estimation of Performance of the HKSAR Government and Trust in Tung (percentage)

		Tung	
HKSAR government's performance good in:		Distrust	Trust
1.	Improving the living standard of the people	12.7	45.2
2.	Creating an incorrupt government	9.2	57.5
3.	Promoting economic development	17.0	38.0
4.	Maintaining social stability	16.3	40.5
5.	Safeguarding the rule of law	10.5	49.0
6.	Promoting democratic development	9.5	62.2
7.	Protecting human rights and freedom	9.6	61.7
8.	Delivering social welfare	12.8	49.2
9.	Promoting moral education	18.3	43.7
10.	Establishing good relations with China	24.2	32.3

Note: only figures associated with relationships between variables significant at .05 level, using the chi square test, are shown in the Table.

These findings apparently show that unlike the Hong Kong government, Patten was regarded by the people with a degree of ambivalence. Patten's democratic rhetoric and populist style might have endeared him to a section of the public, but at the same time a substantial proportion of Hong Kong people (including many who shared his political views) might have looked askance at the way he dealt with China. Many people were worried that political confrontation with China would not be in Hong Kong's interests, and hence overall Patten was less trusted by the public than the colonial regime under his stewardship.

There are minimal socio-demographic differences between those who trusted the Hong Kong government and those who trusted the HKSAR government. Both were more trusted by older people and the less educated. There are however obvious differences between those who trusted Patten and those who trusted Tung. Generally speaking, in comparison with Patten's supporters, Tung's supporters were older, less educated and had a lower income. These people were more susceptible to the traditionalistic appeals of Tung. At the same time, they harboured greater expectations that Tung would respond to their material and normative aspirations.

DECLINING POLITICAL SUPPORT

In view of public attitudes toward the old and the new regimes, it is not surprising that Tung as the Chief Executive of post-colonial Hong Kong and the HKSAR government had an auspicious beginning. In a poll commissioned by the *Apple Daily* and undertaken by the Hong Kong Institute of Asia-Pacific Studies of the Chinese University of Hong Kong immediately after his electoral victory in December 1996, Tung received an average performance score of 71.3 (maximum score: 100; pass score: 50) from the respondents. This represents an all-time high. The score fell to 63.7 at the time of the handover of Hong Kong to China in July 1997. It however climbed to the post-handover high of 68 in September 1997. Since then it has fallen gradually and reached a low of 59.1 in January 1998, rebounding a little to 63.3 in February 1998 and 62.4 in March 1998.[11] Thereafter a situation of basically low popularity obtained,

---

[11] *Apple Daily*, 6 April 1998, p. A12.

being 60.1 in April 1998, 60.9 in June 1998, 56.5 in July 1998, 55.1 in September 1998, 60.6 in November 1998 and 58.6 in January 1999.[12] The trend of public support for the HKSAR government basically mirrors that for Tung.[13]

Declining public support for Tung and the HKSAR government shows that after some initial success, the governing strategy crafted by Tung has become increasingly ineffective in winning the hearts and minds of the people of Hong Kong. For one thing, the governing strategy of Tung represents in essence a project of depoliticization. By deliberately focusing on social, economic and standard of living issues, Tung seeks to draw people's attention away from political issues such as democratization and toward practical bread-and-butter concerns. By doing so Tung attempts to cultivate a base of political support different from that of his predecessor and shadow competitor — Chris Patten.

Just like any governing strategy, the long-term success of Tung's governing strategy hinges upon a set of favourable conditions. Foremost among these conditions is a prosperous economy that can provide the necessary resources to realize the policies originating from Tung's paternalistic goals. The early success of Tung's governing strategy is in no small measure due to public confidence in his determination and ability to keep his paternalistic promises. The congenial economic environment in the first four months of the HKSAR had boosted public confidence. Since October 1998, however, Hong Kong has been beset by the Asian financial turmoil, which has in turn brought about plunging asset values, a liquidity crunch, negative economic growth, a speculative raid on the Hong Kong dollar, increasing business bankruptcies, rising unemployment, wage and salary cuts, and falling government revenues. The sharp fall in property prices has hit the Hong Kong people — particularly the middle class — extremely hard. The property sector was the pivotal pillar of Hong Kong's economy for more than a decade before 1997. The 'bubble economy' created by the booming property market and the wealth it had produced were an important confidence-booster in a context of pervasive political anxieties. The

---

[12] Ibid., 27 July 1998, p. A2; 5 October 1998, p. A6; and 1 February 1999, p. A16.

[13] Home Affairs Department, *Report on a Telephone Opinion Poll in January 1999*, Hong Kong, Home Affairs Department, 1999, p. 25.

sudden shrinkage of wealth in society is totally unexpected and hence is psychologically and politically destabilizing. Under these circumstances, Tung was unable to improve the quality of life of the people or to promote economic growth in Hong Kong. To add to Tung's difficulties, a series of incidents of mismanagement had undermined public confidence in the competence of the civil service. Widespread social and economic grievances in Hong Kong have naturally been translated into political discontent with Tung and his government. In the end, the well-nigh exclusive reliance on performance as the means to cultivate political support by the HKSAR government has been rendered ineffective by deteriorating economic conditions.

In a situation of pervasive political discontent, the de-emphasis on politics in Tung's governing strategy aggravates the problem of declining political support for the HKSAR government. Tung's aversion to politics — mass politics in particular — has resulted in inattention to political coalition-building, creating channels and mechanisms to mobilize mass support and making use of the highly developed mass media to influence public opinion. As a result, concomitant with declining political support for the HKSAR government is a widening gap between the latter and the people. While the government and Tung are increasingly isolated from the people, political alienation and cynicism in society soar.

It is interesting however to note that Tung's popularity falls at a slower rate than the popularity of his government. This testifies to the fact that Tung's traditionalistic appeal and his 'fatherly' image still play a significant role in sustaining his own political status in the public mind. In addition, people have sympathy for his lack of political experience and the fact that he has not received dedicated and capable service from his officials. Still, in hard times, people desperately expect the new regime to behave as a *truly* paternalistic government. Apparently, declining public confidence in Tung as a political leader shows that he has largely failed to convince the people that his government is really a benevolent government for the people and not a pro-business government. Consequently, he has lost a substantial amount of moral authority over the people.

Under these circumstances, Beijing's seemingly unconditional support for Tung and its non-interference in local affairs are only marginally helpful as far as his popularity among the people is concerned. In effect, he can only bank on improvement in Hong Kong's economic conditions to reverse the political fortune of himself and his government. Unfortunately, his government has limited control over the economic fate of Hong Kong.

CONCLUSION

In contrast with other colonies, decolonization of Hong Kong did not lead to political independence. Instead, it became a special administrative region under Chinese sovereignty. The impossibility of achieving political independence explains the absence of independence movements in colonial Hong Kong and the associated rise of popular leaders in anti-colonial struggles. The limited amount of democratic reform in Hong Kong immediately before its return to China did not afford sufficient time to groom political leaders with substantial legitimacy among the people. China's mistrust of popular leaders has in effect excluded them from real power in post-1997 Hong Kong. The discord between China and Britain over Hong Kong before 1997 had prevented them from making joint efforts to groom political leaders for post-colonial Hong Kong. As a result, at the time of the handover, Hong Kong suffered from not only a dearth of political leaders but also some discontinuity in political leadership.

Tung Chee-hwa's ascent to the top post in post-1997 Hong Kong was due ironically to his lack of political ambition and experience and by the fact that he was politically nonpartisan. Nevertheless, as a political loner and newcomer given the power by China to preside over a highly politicized society where public mistrust of Beijing — Tung's political patron — and public suspicion of the intentions of the new regime obtain, the challenge Tung faces as a political leader is overwhelming. He has to use all the means at his disposal to craft a governing strategy which can bring him a decent degree of political legitimacy and allow him to govern with a certain level of effectiveness in adverse circumstances. As a successor regime to colonial rule, Tung's adminstration is quite unique.

Tung Chee-hwa more than once admitted that he was an admirer of the founder of independent Singapore, Lee Kuan Yew. As a believer in benign paternalistic rule based on solid performance, he is remarkably similar to Lee. According to Cho-oon Khong, Lee Kuan Yew 'addressed the urgent need to legitimate his rule through the promise of economic performance, which in turn required policies aimed at critical social problems if it was to achieve tangible results.'[14]

The PAP government sought to base its support, first, on the establishment of political order through an efficient and relatively, if not entirely, incorrupt government and, second, on its commitment to securing the welfare of its citizenry through a benevolent, if not entirely benign, paternalistic government.[15]

As typical Chinese politicians, both Tung and Lee are in fact practitioners of traditional Chinese statecraft with modern embellishments. As far as the problem of political support is concerned, the new regime in the HKSAR is in a much more difficult position than the regime led by the People's Action Party at the time of Singapore's independence. The Tung administration cannot draw upon a solid track record of achievements. It does not have a cohesive and popular political party to do its bidding. It cannot appeal to the people by means of modern ideologies. Tung himself as a political leader ironically abhors politics. Above all, Tung is not a popularly elected leader and is perceived to place China's interest above that of Hong Kong. Equally important, lingering public nostalgia for colonial governance is going to haunt the new regime for a long time to come.

In these unfavourable circumstances, there is probably no alternative to Tung's approach of creating a regime acceptable to the people of Hong Kong. With strong support from China, a vibrant economy, an efficient and dutiful civil service, and a large fiscal reserve, it is not impossible for the new regime to build a viable support base in society. My survey of public opinion on the eve of the establishment of the HKSAR shows that Tung's appeal to traditional values and promise of a competent government did help to place the new regime on a solid foundation. The political image projected by Tung had succeeded in raising public expectations of his government. Whilst it is true that people had more trust in the old regime than the new one, nevertheless Tung had won a surprise personal victory by surpassing his predecessor, Chris Patten, in popularity.

The effectiveness of the new regime's approach to seeking support

[14] Cho-oon Khong, 'Singapore: Political Legitimacy Through Managing Conformity', in M. Alagappa (ed.), *Political Legitimacy*, p. 112.

[15] Ibid., pp. 113–14.

is heavily dependent on the good performance of the economy and bountiful fiscal resources. The administration has to be efficient and capable of resolving Hong Kong's mounting social, economic, and livelihood problems. Also indispensable is a high degree of social and political stability. Only after these conditions are met can the new regime continue to draw political support from both the business community and the common people and to counter the challenge coming from the liberal elite. When the economy falters, the government lacks fiscal resources, and the administration is perceived to be incompetent, the whole strategy to win support can unravel with a consequent beginning of decline in political support.

[12]
Political Parties, Élite–Mass Gap and Political Instability in Hong Kong

LO SHIU HING

Political parties play a stabilizing role in any regime through articulation and aggregation of the interests of citizens. In the case of the Hong Kong Special Administrative Region (HKSAR), however, political parties that are popularly supported by citizens have minimal influence on the government. The HKSAR Government is increasingly dominated by clientelist parties and élites without strong grass-roots support. As the economic gap between the haves and have-nots becomes increasingly serious, and Hong Kong people are not satisfied with many livelihood issues such as housing, the unrepresentative and client-dominated polity will widen the gap between the government and ordinary citizens, with the potential danger of political instability.

Introduction

Political parties constitute either an "intermediary institution" between the government and ordinary citizens, or a link in the "élite–mass gap".[1] By aggregating and representing the interests of citizens, political parties not only narrow any communication gap between the élites and the masses but also play a crucial function of stabilizing a regime. As Alan Ball puts it succinctly:

> One of the most important functions of political parties is that of uniting, simplifying and stabilizing the political process. Political parties tend to provide the highest common denominator ... Parties bring together sectional interests, overcome geographical distances,

> and provide coherence to sometimes divisive government structures
> ... This bridging function of political parties is an important factor in
> political stability.[2]

Other political analysts point to the intermediary role of political
parties, which constitute a link between the state and civil society.[3]

Political parties are a relatively new phenomenon in Hong Kong,
which changed from a colony of Britain to a Special Administrative
Region of the People's Republic of China (PRC) on 1 July 1997. The
parties have emerged and mushroomed since the Tiananmen incident
of June 1989.[4] Although there are numerous political parties in the
Hong Kong Special Administrative Region (HKSAR), some of them are
dependent on the PRC for political influence upon the HKSAR Government, where a multiparty system without any dominant party exists.[5]
The HKSAR has a multiparty system because a number of small political parties, which aim at capturing political power, persist and compete among themselves. However, the HKSAR's multiparty system is
weak. None of the popularly supported parties can form a government
because of the partially directly-elected Legislative Council (LegCo),
and the determination of both the PRC and the HKSAR Government
to maintain an executive-dominant polity. Together with the senior
bureaucrats, the HKSAR's Chief Executive, Tung Chee-hwa, and his top
policy-making body, the Executive Council (ExCo), constitute the executive branch, which is politically dominant over any political party.

This article will also contend that political parties in the HKSAR
are increasingly divided into two types: popularly supported parties
which are politically powerless, and patron–client type of parties which
are backed by the PRC and the HKSAR Government to share some
political power.[6] While the popularly-supported political parties are
increasingly marginalized in the HKSAR's executive-dominant polity,
the masses regard the clientelist parties as unpopular and unrepresentative of public opinion. The inability of political parties to function
as an intermediary between the ruling élites and the masses contributes
to a widening élite–mass gap, precipitating a crisis of political turbulence in the HKSAR in the years to come.

Political Parties in the HKSAR:
Ideological Spectrum, Origins and Constraints

The ideological spectrum of political parties in the HKSAR is relatively
narrow (see Table 1). Most political parties are situated between the
liberal and conservative spectrum, with the exception of a political
group named April the Fifth Action Group (AFAG), which vows to
change the PRC's polity from authoritarianism to democratic socialism

Political Parties, Élite–Mass Gap and Political Instability in Hong Kong 69

TABLE 1
The Political Spectrum of Parties in the HKSAR

Radical	Liberal	Moderate	Conservative	Reactionary
AFAG	DP	ADPL	LP	–
	Frontier	DAB	HKPA	–
		CP		

Note: This political spectrum is derived from Leon P. Baradat, *Political Ideologies: Their Origins and Impact* (New Jersey: Prentice-Hall, 1994), p. 13.

and which often has confrontations with the police on the streets.[7] It is rumoured that the AFAG has been blacklisted by the Hong Kong police force, and that its activists are under the surveillance of the Security Branch.[8] Strictly speaking, the AFAG is a pressure group rather than a political party, for it aims at using street protests to oppose government policies instead of supporting candidates to participate in local elections.

Political parties which support civil liberties and advocate Western-style democracy tend to have a liberal ideology. A notable example is the Democratic Party (DP), whose leaders are Martin Lee Chu-ming and Szeto Wah and which was formed by merging two political pressure groups, namely, the Hong Kong Affairs Society and Meeting Point.[9] Another liberal political party is Frontier, led by Emily Lau Wai-hing, a former LegCo member and an outspoken critic of the PRC, and Lee Cheuk-yan, a leader of the independent Confederation of Trade Unions.[10] The third liberal-oriented party is the Citizens Party (CP), led by Christine Loh Kung-wai, a former LegCo member and a critic of the colonial administration. While the DP, Frontier and CP share liberal values, they cannot amalgamate because of different strategies. The DP leaders aim to achieve democratization by trying to negotiate with the PRC Government, but this approach is rejected by Frontier, which tends to be sceptical about the usefulness of any dialogue with PRC officials. However, the CP advocates dialogue with PRC authorities to a degree much greater than both the DP and Frontier.[11] These divergent strategies of the liberal political parties towards China are detrimental to their struggle for democratization, because internal competition in local elections undermines the overall solidarity of Hong Kong's democracy movement.

The moderate political parties include the Association for Democracy and People's Livelihood (ADPL) and the pro-Beijing Democratic Alliance for the Betterment of Hong Kong (DAB). Both the ADPL and

DAB emphasize dialogue and friendly relations with the PRC. The DAB has a very close relationship with PRC officials and some party members may actually be underground members of the Chinese Communist Party.[12] Unlike the liberal parties, the ADPL is moderate in the sense that it is willing to abandon the call for political reform in the HKSAR for the sake of maintaining harmonious relations with the PRC. For example, ADPL chairman Frederick Fung Kin-kee and vice-chairman Bruce Liu Sing-lee supported the PRC's decision to establish the provisional legislature, which was set up in January 1997 to replace the LegCo, which was elected in 1995 under Governor Christopher Patten's political reform blueprint. The position taken by Fung and Liu alienated some ADPL members, such as Eric Wong Chung-ki, who eventually withdrew from the party and joined the DP. In short, the ADPL adopts a very moderate stance on democratization, although its leaders occasionally fluctuate their political position from moderate to hardline.[13]

The DAB is also a moderate political party whose views have changed from anti-Patten to pro-HKSAR Government since 1 July 1997. During the Patten administration, the DAB acted as its critic and severely opposed democratization in the name of respecting the Sino-British Joint Declaration, the Basic Law and other earlier Sino-British agreements on Hong Kong. However, the DAB's stance has tilted towards supporting the HKSAR Government since the transfer of sovereignty. When the provisional legislature endorsed the HKSAR Government's decision to freeze indefinitely the collective bargaining bill, which had been passed by the former LegCo several days before the handover, the DAB members of the provisional legislature supported the decision. Since some DAB leaders are simultaneously leaders of the pro-Beijing Federation of Trade Unions (FTU), they are expected to fight for working-class interests. Unfortunately, both the DAB and FTU uncritically supported the decision of the HKSAR Government to abandon the collective bargaining bill. Even worse, the DAB tried to change the electoral system of LegCo elections in May 1997 in such a way as to benefit itself.[14] Obviously, the DAB is drifting towards conservatism at the expense of its populist image.

In terms of class background, the DAB is similar to other liberal-minded parties. While the CP is composed mainly of middle-class intellectuals and professionals, the DAB tends to project a lower middle-class image as some of the DAB members come from working-class unions. However, the other two liberal-oriented parties — the DP and Frontier — have lower middle-class backgrounds similar to the DAB. The DP members include some unionists, such as Lau Chin-shek of the Independent Trade Union, and Cheung Man-kwong of the

Political Parties, Élite–Mass Gap and Political Instability in Hong Kong   71

Professional Teachers' Association, while Frontier has Lee Cheuk-yan, also an independent unionist leader. Even the ideologically moderate ADPL projects a lower middle-class image, although its power base is located mainly in Shumshuipo district, unlike the other liberal parties and the DAB whose influence and appeal tend to be territory-wide. Because of the very similar class background of all the liberal and moderate parties, they often compete for support from the same source of voters in local elections.

Unlike the lower middle-class backgrounds of most liberal parties and the two moderate parties, the ideologically conservative parties in the HKSAR appeal to the members of the upper-middle class. The conservative political parties include the Liberal Party (LP) and the Hong Kong Progressive Alliance (HKPA). The LP is liberal in the narrow sense of supporting private property rights and societal free-dom, but it is not liberal in the broader sense of supporting political reform and equality of opportunity. The LP is actually conservative and tries to maintain the political as well as economic status quo. Funded by big business and founded by former pro-British élites, the LP can be seen as a party of big businessmen without grass-roots support.[15] The LP leader, Allen Lee Peng-fei, managed to be elected in the 1995 LegCo direct elections, but most LP members have traditionally per-formed poorly in local elections. The party's power base is located at the LegCo's functional constituencies, where the industrial and com-mercial groups have traditionally been captured by LP members. After the direct elections held for District Boards in September 1994, many members from various districts such as Sai Kung and Sha Tin were dissatisfied with the LP's leadership, and withdrew from the party. Ultimately, the LP is a clientelist organization politically dependent on the patronage of PRC officials. The PRC has appointed LP members to various committees, such as (1) the Selection Committee which elected both the Chief Executive in December 1996 and members of the provi-sional legislature in January 1997, and (2) the Selection Committee which selected the Hong Kong members to China's National People's Congress in December 1997. The LP members of the Selection Commit-tee that elected the Chief Executive in December 1996 fully supported Tung Chee-hwa, who in return appointed an executive committee member of the LP, Henry Tang, to the ExCo. Without the full support of its patrons — PRC officials and the Chief Executive — the LP cannot become politically influential in the HKSAR polity.

Apart from the LP, which can be viewed as a clientelist and business-dominant party, another client receiving top political posi-tions in the HKSAR from the patron is the HKPA, which in 1997 merged with a small pro-Beijing party, the Liberal Democratic Federation (LDF). Realizing that such a merger would expand its political influence, the LDF, led by former ExCo and LegCo member Maria Tam Wai-chu, made a strategic move to co-operate with the HKPA.[16] The LP, LDF and HKPA are all upper-class political parties. Nevertheless, the HKPA cannot amalgamate with the LP because of mutual competition and jealousy. For the HKPA leaders, the LP is composed of the former supporters of the British colonial administration and its patriotic sentiment towards the PRC is relatively weak. On the other hand, in the minds of LP leaders, the HKPA was formed by PRC officials from the New China News Agency and it competes with the LP for political support among the business élites.[17] As with liberal-oriented and middle-class parties like the DP, Frontier and CP, which cannot amalgamate because of strategic differences, the conservative-oriented and upper-middle class parties encounter a similar problem because of mutual jealousy and rivalry.

The inability of political parties, whether lower-middle or upper-middle class, to amalgamate perpetuates the political preponderance of the executive branch of the HKSAR Government. The executive branch is composed of the Chief Executive, the ExCo and senior bureaucrats. However, as long as the Chief Executive is not directly elected but merely chosen by an Election Committee whose composition tends to be controlled by a political patron — the PRC authorities — the Chief Executive is ultimately accountable to China rather than to any political party. Although the Chief Executive, like the Governor in Hong Kong during the final years of British rule, behaves in a way that is seen to be accountable to the public, his power is not really checked by the political parties. The Chief Executive consults the parties but has the discretion to ignore their views. Tung Chee-hwa meets with the DP leaders regularly to discuss Hong Kong affairs, but his action is not constrained by their opposition. Nor does Tung face any opposition from the conservative parties, like the LP, HKPA and DAB which fully support the HKSAR Government. To date, Tung has enjoyed a high degree of relative autonomy *vis-á-vis* any political party in the HKSAR.

At most, the ExCo takes into consideration the views of the HKPA, LP and DAB which together occupy a majority of the seats in the sixty-member provisional legislature. Because three ExCo members — Leung Chun-ying, Tam Yiu-chung of the DAB and Henry Tang of the LP — are also members of the provisional legislature, they listen to the views of other legislators and act as the executive-legislative intermediary. Unlike Governor Patten who abolished the overlapping membership between the ExCo and LegCo and relied on his own charisma as well as the lobbying efforts of senior bureaucrats to ensure that government bills were supported by the LegCo, Tung has reversed Patten's political approach. He has restored the overlapping membership of ExCo and LegCo in order to allow the former to tap the latter's views, but he refrains from attending LegCo meetings himself to answer questions from legislators, although the role of bureaucrats to lobby legislators for government bills has been retained.[18] Whenever the Patten administration put forward its bills in the former LegCo, its officials actively lobbied for support from different political parties, including the DP, LP, DAB and HKPA. The Tung government, however, tends to rely on the support of the LP, DAB and HKPA in the provisional legislature. If the DP returns some of its members to the LegCo after May 1998, the Tung administration will also likely lobby the DP legislators for support of government policies. Yet, the DP's role in the legislature will be curbed because the government can forge an alliance with the HKPA, LP and DAB to override any opposition from the DP. Therefore, the DP is unlikely to become a political force that can challenge the political coalition between the HKSAR Government and pro-establishment parties. In the event that the pro-establishment LP, DAB and HKPA cannot reach a compromise on certain issues, the DP may be able to exploit their division to lobby for support from independent LegCo members and to exert some degree of influence on the executive branch.

However, in the HKSAR's executive-dominant polity, the senior bureaucrats are still powerful actors who formulate government policies and they have the discretion to ignore the views of the political parties. Despite the fact that senior officials of the HKSAR Government are under constant attack by the mass media and elected politicians for their maladministration, they do not have to resign from their posts because of government scandals and blunders.[19] For the senior bureaucrats, the political parties can obstruct their powers to formulate and implement policies. Since the late 1980s, the colonial government had encouraged senior civil servants to treat citizens as clients. As a result of democratization and politicization, senior bureaucrats had taken into account the views of political parties during the policy-making process, ranging from the formulation of the annual budget to social welfare policy.[20] In the former LegCo and the current provisional legislature, senior bureaucrats have taken the views of the opposition seriously, and political acceptability of government policies has become a must since the early 1990s. Recent events have suggested, however, that some senior officials of the HKSAR Government seem to be turning a deaf ear to the views of elected representatives — a situation that, if it persists, may widen the élite–mass gap.[21]

In any case, senior bureaucrats are far more politically powerful than elected politicians because of three factors. First, as long as political parties cannot form a government in power even if they secure a majority in the legislature, senior bureaucrats will retain the capability to push forward government policies if they desire. Bureaucrats do not have to shoulder any political cost, such as resignation, when they choose to turn a deaf ear to the opposition of political parties. Secondly, because of the fragmented and divided nature of the political parties in the legislature, government bills can be passed easily by lobbying for support from pro-establishment parties, such as the LP, HKPA and DAB. As a result, the liberal-oriented parties remain politically isolated and have their influence curbed. Thirdly, Article 74 of the Basic Law — the HKSAR constitution — confers upon the LegCo's President and Chief Executive the power to disapprove private member's bills "relating to government policies".[22] This power was exercised by the President of the provisional legislature, Rita Fan Hsu Lai-tai, who ruled a bill, initiated by a member, as out of order.[23] Because the Basic Law gives considerable power to the LegCo President, any further exercise of this power by the President would cripple the power of any political party which seeks to use private member's bills as a means of changing government policies. In short, the multiparty system in the HKSAR is characterized by relatively weak and small parties, with minimal impact on the politically dominant and powerful executive.

Parties, Élite–Mass Gap and Political Instability

Given the fact that clientelist parties, which lack any power base at the grass-roots level, are enjoying far more political influence than the liberal-oriented and popularly supported parties, the HKSAR's political institutions are becoming increasingly unrepresentative. This politically unrepresentative phenomenon is testified by the ease with which the HKPA, DAB and LP capture seats in influential bodies, such as the ExCo, whose members are all appointed by the Chief Executive, the provisional legislature, and the Selection Committee, which chose thirty-six Hong Kong representatives to China's National People's Congress (NPC) (Table 2). Nevertheless, the DAB, LP and HKPA were far less popular than the DP in the LegCo's direct elections in 1995. The DP obtained 41.9 per cent of the votes in LegCo's direct elections; the DAB 15.4 per cent; the LP only 1.6 per cent; and the HKPA 2.8 per cent.[24] However, the critical stance of the DP jeopardizes its chances of becoming the beneficiary of political patronage. As mentioned before, the DP boycotted the election of members of the provisional legislature. On the other hand, the PRC regarded the DP as "subversive" and did not appoint its members to the Selection Committee which chose the Chief Executive and provisional legislators. Three DP members failed in their

Political Parties, Élite–Mass Gap and Political Instability in Hong Kong 75

TABLE 2
Parties and the HKSAR's Influential Political Bodies

Party	424-member Selection Committee for	60-member Provisional Legislature	15-member Executive Council
HKPA*	56 (7 elected)	9	1
DAB	43 (4 elected)	10	1
LP	12 (1 elected)	10	1
ADPL	4 (none was elected	4	0
DP	0	0	0

Notes: * The HKPA members here include members of the Liberal Democratic
Federation.
Sources: *Sing Tao Jih Pao*, 2 November 1997, p. A9; *Apple Daily*, 9 December 1997,
p. A22; *Eastweek*, 26 June 1997, p. 68; Chris Yeung, "Political Parties," in
Joseph Cheng, ed., *The Other Hong Kong Report 1997* (Hong Kong: The
Chinese University Press, 1997), p. 70; *Ming Pao*, 9+December 1997, p. A4;
and http://legco.gov.hk/.

attempt to obtain ten nomination signatures from the 424 members of
the Selection Committee that elected the thirty-six Hong Kong deputies
to China's NPC.[25] For the overwhelming majority of the members of the
Selection Committee, the DP adopted an "anti-PRC" stance that was
politically detrimental to them.[26] Thus, the most popularly supported
political party, the DP, remains at odds with the most powerful patron
in the HKSAR's political arena, namely PRC officials.

The lack of representation and the proliferation of clientelism in
the HKSAR polity will widen the élite–mass gap in the long run. After
the LegCo elections in May 1998, it is expected that a loose coalition
involving the DAB, LP and HKPA will check the influence of members
of liberal-oriented political parties such as the DP, Frontier and CP. The
LegCo that will be elected in May 1998 to serve a term of two years is
already structurally constrained by its composition: one-third of its
sixty members are directly elected from geographical constituencies,
half of them chosen by functional constituencies, and ten seats are
chosen from an Election Committee. A proportional representation
system is adopted for the twenty directly elected seats, which must
give the DAB more seats than what the party attained in the 1995
LegCo direct election.[27] In 1995, the LegCo's direct elections adopted a
single-member single-vote system which resulted in some popular DAB

candidates who obtained a considerable number of votes being defeated by DP candidates.[28] In Hong Kong where the legislature has merely one-third of the directly elected seats, using proportional representation to protect the interests of minorities — the business élites and pro-Beijing politicians who are already over-represented in the LegCo's functional constituency elections — is not really justified. Since the direct election method has been reshaped in such a way as to favour the representation of pro-Beijing and pro-HKSAR Government parties, the representative function of the popularly supported and liberal-oriented parties is undermined, thus directly or indirectly widening the gap between the ruling élites and the masses.

While the DAB must be able to prevent the liberal-oriented parties from capturing all of the twenty directly elected seats, the LegCo's functional constituency election perpetuates and legitimizes the political dominance of conservative-oriented parties — the LP and HKPA — in the legislature. Some functional constituency elections in the 1995 LegCo witnessed many business candidates being automatically elected.[29] Only a minority of functional constituencies give representation to grass-roots organizations, such as labour, teachers and social workers. The majority of the thirty functional constituency seats in the LegCo must be occupied by the pro-Beijing and pro-government élites, especially members of the LP and HKPA. Functional constituency elections tend to have a narrower franchise than direct elections and they are prone to being controlled by the business élites. It can be said that functional constituency elections are tailored for the LP and HKPA, which have become clientelist parties enjoying a relatively high degree of political influence out of proportion to their relatively low level of public support.

Another factor which will entrench the influence of conservative parties in the HKSAR's legislature is the ten seats to be elected from the Election Committee in 1998. The Election Committee is bound to be dominated by the business élites — a pattern seen in all previous committees responsible for selecting the Chief Executive and members of the provisional legislature, and for arranging the transitional affairs of Hong Kong (such as the Preliminary Working Committee set up in 1993). Even in the 1995 LegCo election, the Election Committee, which was composed of all directly-elected District Board members, had appointed six pro-Beijing legislators — an indication that patron–client relations and political compromise were commonplace in the electoral process.[30] In the 1998 LegCo election, the ten seats to be elected from the Election Committee will guarantee the political representation of pro-Beijing and pro-HKSAR Government élites. Since these élites within the HKSAR's political structure are mostly uncritical clients, the communication gap that exists between them and the masses is likely to widen.

If this is the case, the HKSAR polity will sooner or later encounter a crisis in political representation as conservative-minded élites who are not popular but who are the clients of a powerful patron, the PRC, can obtain a disproportionate degree of political influence. On the other hand, the liberal-minded élites, who have been popularly supported by citizens since the 1991 LegCo direct elections, are destined to be a minority opposition in the HKSAR legislature, and thus also the ExCo, which has been traditionally dominated by pro-government as well as conservative élites. Consequently, the functions of the political parties to aggregate the interests of citizens and to stabilize the polity are distorted and undermined. The conservative political parties in the HKSAR lack public support, but their dominance in the polity is artificial and destabilizing. For one thing, decisions made by the clientelist legislature from May 1998 to 2000 will not represent the wishes of ordinary citizens. An alarming signal came from the provisional legislature when it repealed the collective bargaining bill passed by the former LegCo several days before 1 July 1997. In the 1998–2000 LegCo, there is a real danger that the business-dominated legislature will again make unpopular decisions and pass legislation unfavourable to the working class.

What is more, at a time when the income gap between the rich and the poor is widening, and as the number of Hongkongers living below the poverty line is increasing,[31] inequality will become a political time-bomb for the HKSAR. In the event that more citizens feel that their interests cannot be represented in the existing political structure, their grievances could be translated into anti-government protests and even riots which would endanger the HKSAR's political stability. As one angry citizen, in a letter to a newspaper, wrote:

> Apparently, Hong Kong has been economically prosperous in recent years. Those people who benefit are not the hard-working ordinary people, but the officials and business people who collaborate with each other as well as the speculators. This situation easily builds up social dissatisfaction which has been accumulated and suppressed. If there are no civil liberties and if grievances cannot be voiced, social dissatisfaction will increase to a point where a riot will erupt in a way more serious than the anti-British riot in 1967. At that time, Hong Kong's economic prosperity will turn into smoke.[32]

So far, civil liberties in the HKSAR have generally been maintained and citizens can voice their grievances through various channels, particularly via the mass media.

However, if any political turbulence in the HKSAR does occur, it will likely take the form of protests and perhaps even riots initiated by ordinary citizens. Writing in 1986, Ian Scott argued that "political turbulence" in Hong Kong was partly caused by the emergence of political groups making demands on the colonial government, and partly attributable to the PRC's intervention in Hong Kong affairs.[33] Political turbulence in the HKSAR is unlikely to be influenced by the China factor, for the PRC has adopted a relatively non-interventionist policy towards the HKSAR, at least in the short run. For the central government in Beijing, proving the feasibility of the concept of "one country, two systems" is necessary in order to appeal to Taiwan for reunification, and to win the confidence of the international community. Judging from the rising social dissatisfaction of the Hong Kong people, political crises in the HKSAR will very likely stem from internal sources, namely, the unrepresentative nature of the political institutions and the widening communication gap between the government and ordinary citizens.

In retrospect, the colonial administration regarded the 1966–67 riots as an outcome of a widening communication gap between the government and ordinary citizens.[34] To put it in another way, the riots were the product of a huge élite–mass gap. The colonial government acted swiftly to narrow that gap by introducing administrative reforms, such as the establishment of the City District Offices which promoted and explained government policies to the public. In the early 1980s, District Boards were established to allow more élites to participate in politics, and to let the masses vote for representatives to these consultative bodies. Democratizing reforms in Hong Kong from the 1980s to the Patten administration actually narrowed the élite–mass gap further, forcing the colonial government to become more responsive, accountable and transparent than ever before.[35] The colonial rulers in Hong Kong after the 1966–67 riots learnt how to bridge the élite–mass gap through administrative and political reforms.

Nevertheless, the HKSAR Government has so far not fully appreciated the importance of bridging the élite–mass gap. Instead, it has been adopting the policy of patronage, granting an unprecedented degree of political influence to conservative-minded élites who lack mass support at the grass-roots level. Even worse, the current polity tends to minimize the influence of liberal-minded élites who are determined to fight for the interests of the populace. After the LegCo elections in May 1998, liberal élites are bound to be in a minority in the legislature, without effective checks on executive power. In the event that the pro-Beijing DAB, which can be considered a relatively populist party among all the clientelist parties, continues to adopt an uncritical stance

Political Parties, Élite–Mass Gap and Political Instability in Hong Kong 79

without a political conscience to speak for the interests of the increasingly alienated masses, the élite–mass gap will be widened further.

One factor sowing the seeds of political instability in the HKSAR is that, apart from popularly supported parties which cannot function effectively, the other "élite–mass linkages", such as elections and interest groups, are becoming increasingly ineffective.[36] Elections have traditionally served a legitimizing function for the Hong Kong government, channelling élite participation to such institutions as District Boards, Urban Council, Regional Council and LegCo. The Patten administration democratized the electoral system, abolishing appointed seats in District Boards while widening the franchise of the LegCo's functional constituency elections. Objectively speaking, Patten's reforms made elections more meaningful than ever before, for the élites and masses perceived elections as part and parcel of the process in which "Hong Kong people [are] ruling Hong Kong".

However, with the onset of the HKSAR Government, the rolling back of Patten's electoral reforms has produced a destabilizing effect that is underestimated by the Tung administration. In July 1997, appointed seats were reintroduced to District Boards, thus balancing the influence of liberal political parties, particularly the DP.[37] The re-injection of appointed clients to the HKSAR's political institutions has rendered elections meaningless, for the appointees are mostly pro-HKSAR government, or previously defeated candidates. Elections are increasingly meaningless also because of the structural constraints imposed on the liberal-oriented parties. No matter how well they perform in the LegCo elections, the over-complicated and manipulated electoral system makes it impossible for them to gain a majority of seats in the legislature, at least from now to 2007 when a review of political reform in the HKSAR will be undertaken in accordance with the Basic Law. If structural constraints on political parties are so great, elections become nothing more than window-dressing that creates an illusion of democratic development. In the event that more citizens perceive local elections as meaningless, the élite–mass gap will deteriorate further.

Objectively speaking, one factor that mitigates the gradually widening élite–mass gap is the civil liberties enjoyed by interest groups. Hong Kong has been traditionally a liberalized society where interest groups can articulate their interests, make their demands known, and criticize the government. This situation persists in the HKSAR. However, business interest groups have traditionally enjoyed far more influence than non-business groups. The HKSAR administration, as with the colonial regime, consults extensively with business groups through hundreds of advisory committees, producing a partnership between the government and business.[38] The crux of the problem is that this partnership is not accompanied by a corresponding relationship be-tween the HKSAR Government and non-business groups. Non-business groups, such as women, ethnic minorities and labour, have been and are under-represented in Hong Kong's political institutions. There is no functional constituency allocated to both women's groups and ethnic minorities, in spite of the fact that women are quite visible at the top levels of the bureaucracy, and ethnic minorities such as Indians have been contributing significantly to the territory's economic prosperity. Traditionally, Hong Kong's women's groups and ethnic minorities have been politically passive,[39] unlike some radical groups such as the April the Fifth Action Group which has vowed to challenge the authority and legitimacy of the ruling élites. Although some functional constituency seats in the LegCo are allocated for labour unions, working-class inter-ests are often under the threat of the business-dominated legislature. If non-business interest groups are artificially and permanently excluded from the HKSAR's power centre, and if they are dissatisfied with the government policies affecting them, they would probably resort to protests outside the increasingly unrepresentative polity.

Political Instability in the HKSAR

The lessons of the 1966–67 riots have to be learnt by the HKSAR Government. The riots not only exposed the élite–mass gap in Hong Kong under colonial rule, but they were also the result of a convergence of social and political problems. In the 1950s and the early 1960s, working-class interests were neglected, social welfare policy was underdeveloped, and the political structure was unrepresentative.[40] The danger for the present HKSAR is that working-class interests risk being neglected by the business-dominated polity, social welfare re-form is insufficient to appease the anger of the have-nots, and the political structure shows signs of bias towards the economically afflu-ent élites. It is urgent for the government to address the issues of social, economic and political inequality. Otherwise, economic inequality in the HKSAR may converge with social dissatisfaction and political alienation, producing an abrupt crisis which would be similar to the 1966–67 riots.

Recent events in the HKSAR have shown that the economic confi-dence of ordinary citizens is declining. The Southeast Asian and South Korean financial crises which began in the second half of 1997, and the drastic fluctuations in the Hong Kong stock market have shaken the confidence of many Hongkongers.[41] The HKSAR leadership's emphasis on the success of the "one country, two systems" approach has un-intentionally undermined its capacity to anticipate future problems, let alone any sudden crisis.[42] The declining economic confidence of Hongkongers and the self-indulgence of the HKSAR leadership in giving a gloss to the territory's social, economic and political circumstances will widen the gap between the masses and the élites in the long run.

The persistence of economic inequality will also sooner or later make social mobility more rigid, generating public sentiment that the masses cannot elevate themselves in society through hard work and perseverance. Many tycoons in the HKSAR were from humble origins, but they climbed up the social ladder swiftly in the 1960s and the 1970s. A case in point is property developer Li Ka-shing, who was originally a worker in a plastic factory.[43] However, as economic inequality worsened in the 1990s, social mobility from the lower classes to upper classes has become far more difficult. New immigrants who arrive in Hong Kong from the mainland are mostly from the lower classes. Livelihood issues, such as the high cost of housing, become a stumbling block to their attempts to move up the social ladder. To cope with the housing problem, the Tung Chee-hwa administration promises to build 85,000 flats a year from 1999 onwards.[44] He has also tried to achieve an ambitious ten-year housing plan by increasing land supply and encouraging tenants to purchase their flats in public housing estates at a relatively low price.[45]

By 2007, the results of Tung's ambitious housing programme will be seen. By that time, the masses and their elected representatives will assess critically the government's housing and social welfare policies. If the government is then able to demonstrate to the people that livelihood problems can be solved, a social crisis would be defused. If not, social dissatisfaction would probably converge, and liberal-oriented political parties would push for further democratization in the form of an entirely directly-elected legislature and a Chief Executive selected by universal suffrage.[46] On the other hand, it can be anticipated that conservative parties such as the LP and HKPA would try to prolong the existence of functional constituency elections in the LegCo and resist any move towards universal suffrage. Therefore, the year 2007 will potentially be an explosive watershed for the HKSAR's political development. Regardless of whether political disputes will converge with gradually increasing social dissatisfaction, a bone of contention between the liberal and conservative élites will undoubtedly be the reform of the HKSAR polity — an issue that was intentionally or unintentionally postponed by the drafters of the Basic Law in the late 1980s.

Prior to the start of Sino-British negotiations over Hong Kong's future in 1982, an official of the Hong Kong Government had commented that the colony's political stability rested on "a tripod of con-sents" between the Hong Kong people, Britain and the PRC.[47] This equation for political stability in Hong Kong has, however, changed with the termination of British rule since 1 July 1997. While the British factor is no longer significant, the Hong Kong people are politically divided into liberal and conservative groups; and the China factor remains uncertain and unpredictable. The division among the Hong Kong people sows seeds of political bickering, struggle, confrontation and instability. Given the reality that the conservative élites share some political power with the still powerful bureaucrats under the PRC's patronage, and that the liberal élites, popularly supported by the masses, are powerless and alienated, political conflict will be inevitable.

Conclusion

The HKSAR polity is increasingly showing the roots of a problem: the inherent bias in favour of clientelist parties that rely on political patronage and the prejudice against popularly supported parties that cannot function as a stabilizing force. Unable to articulate the interests of the masses effectively, the popularly supported parties, such as the Democratic Party, constitute a peripheral force which can exert pres-sure on the executive-dominant, clientelist-oriented regime. If these parties continue to be alienated in an unrepresentative polity, they could resort to more street protests to oppose government policies. The liberal-oriented opposition may also encounter an internal split be-tween the hardliners, who advocate confrontational tactics, and the softliners, who support continual negotiation with the government. On the other hand, the socially frustrated masses will increasingly find the liberal parties politically powerless. This may lead to the alienation of the masses, especially if the government is unable to deal effectively with livelihood issues.

The élite–mass gap, which had been narrowed as a result of ad-ministrative and political reforms initiated by the British from the late 1960s to the first half of the 1990s, now faces a real danger of being widened further. Preoccupied with promoting the success of the "one country, two systems" policy, the HKSAR leadership has ignored the politically destabilizing impact of the widening élite–mass gap. The government has recently taken the appropriate step to address the livelihood problems of the masses by putting forward the ten-year housing programme. Yet, whether this programme will be successful and whether the income gap between the rich and the poor narrows will ultimately affect the political debate in 2007, when liberal and conservative élites will argue over the pace and scope of the HKSAR's political reform.

Political Parties, Élite–Mass Gap and Political Instability in Hong Kong 83

NOTES

1. Robert D. Putnam, *The Comparative Study of Political Elites* (New Jersey: Prentice-Hall, 1976), pp. 154–64.
2. Alan Ball, *Modern Politics and Government* (London: Macmillan, 1993), p. 81. The representative function of political parties is also discussed in A. H. Birch, *Representation* (London: Macmillan, 1972), p. 97.
3. Theda Skocpol, "Bringing the State Back In: Strategies of Analysis in Current Research", in *Bringing the State Back In*, edited by Peter B. Evans, Dietrich Rueschemeyer and Theda Skocpol (Cambridge: Cambridge University Press, 1986), p. 23. Skocpol maintains that parties are "mediators between electorates and the conduct of state power" (p. 23). See also James Manor, "Parties and the Party System", in *India's Democracy*, edited by Atul Kohli (New Jersey: Princeton University Press, 1988), p. 63. Manor remarked that India's political parties "have provided the main links between state and society, state-society relations (p. 63)".
4. Louie Kin-shuen, "Politicians, Political Parties and the Legislative Council", in *The Other Hong Kong Report 1992*, edited by Joseph Y. S. Cheng and Paul C. K. Kwong (Hong Kong: The Chinese University Press, 1992), pp. 53–78; and Chris K. H. Yeung, "Political Parties", in *The Other Hong Kong Report 1997*, edited by Joseph Y. S. Cheng (Hong Kong: The Chinese University Press, 1997), pp. 49–70.
5. In the "Third World", there are no-party, one-party and multiparty systems. Within the multiparty systems, one political party may be dominant. Rod Hague, Martin Harrop and Shaun Breslin, *Comparative Government and Politics: An Introduction* (London: Macmillan, 1994), pp. 249–52. One example of the multiparty system with a dominant party is Japan where the Liberal Democratic Party once dominated the political arena. See Ronald J. Hrebenar, "The Changing Postwar Party System", in *The Japanese Party System: From One-Party Rule to Coalition Government*, edited by Ronald J. Hrebenar (Colorado: Westview, 1986), pp. 6–7. However, Hong Kong can be viewed as having a multiparty system without any dominant political party.
6. Patron–client relationship is unequal and "involves an exchange between a superior patron or patron group and an inferior client or client group". See Vicky Randall and Robin Theobald, *Political Change and Underdevelopment: A Critical Introduction to Third World Politics* (London: Macmillan, 1985), p. 52. It must be noted that patron–client politics also exist in some Southeast Asian states. For example, see Anek Laothamatas, "From Clientelism to Partnership: Business-Government Relations in Thailand", in *Business and Government in Industrializing Asia*, edited by Andrew Macintyre (New York: Cornell University Press, 1994), pp. 195–215.
7. Liberals tend to respect the concept of law, seek change in the political system and are confident that the application of reason will solve problems. Conservatives, however, tend to be more supportive of the status quo, are doubtful about whether change will bring about positive results, and are pessimistic about the impact of using reason. Leon P. Baradat, *Political Ideologies: Their Origins and Impact* (New Jersey: Prentice-Hall, 1994), pp. 19–25.
8. *Apple Daily*, 29 November 1997. One member of the April the Fifth Action Group, Leung Kwok-hung, is regarded as a radical by the Hong Kong police. He believes in "Utopian" socialism, advocating democracy and human rights in Hong Kong. See a special report on Leung in *Apple Daily*, 12 November 1997, p. A14.
9. Yu Wing-yat, "Organizational Adaptation of the Hong Kong Democratic Party", *Issues and Studies*, 33, no. 1 (January 1997): 96–98.
10. It must be noted that trade unions in Hong Kong are divided into left-wing, independent and right-wing. At present, leaders of the pro-China Federation of

Trade Unions (FTU) tend to be clients who receive the reward of political positions from the HKSAR Government. For example, the FTU Deputy Director Tam Yiu-chung is a member of the top policy-making ExCo. See *Apple Daily*, 18 November 1997, p. A14.

11. The CP's political platform includes advocating small government, environmental protection, educational reform, and "active cooperation" between Hong Kong and China. See *Hong Kong Economic Journal*, 5 May 1997, p. 5.

12. Shiu Sin-por, the Director of the China-funded One Country Two Systems Economic Institute, told a group of senior civil servants in September 1997 that the DAB was "under the management" of the Social Welfare Department of the New China News Agency in Hong Kong. Shiu's talk was on the PRC's organs in Hong Kong, at an advanced China course for senior Hong Kong civil servants at the Civil Service and Development Institute on 23 September 1997. Shiu has very close relations with PRC officials and his information on the role of the Chinese Communist Party in Hong Kong is reliable.

13. Fung and Liu, for example, were appointed by the PRC to a 424-member Selection Committee which would elect Hong Kong representatives to the PRC's National People's Congress. In a meeting of the Selection Committee which decided to approve the members of its presidium by a show of hands, Fung and Liu disagreed with the way in which the names of presidium members were endorsed and thus they voiced their opposition. Their action illustrated the dilemma of being co-opted by the PRC. On the one hand, they emphasize the necessity of communicating and co-operating with PRC officials. But, on the other, any unpopular action on the part of PRC authorities forces them to express opposition. For a discussion of the dilemmas of political élites who are co-opted by the PRC, see Wong Wai-kwok, "Can Co-optation Win Over the Hong Kong People? China's United Front Work in Hong Kong Since 1984", *Issues & Studies* 33, no. 5 (May 1997): 125–131. Wong said that some élites who were co-opted by China did not believe that they were responsible to the citizens (p. 126). But Fung and Liu appeared to act in such a way as to show that they spoke for the interests of the Hong Kong people.

14. The DAB attempted to include neighbourhood (*kaifong*) associations in the LegCo's social welfare functional constituency, which was traditionally occupied by social welfare groups. Alienated by the DAB's attempt to manipulate the electoral system, social workers advertised in the Chinese newspapers to denounce the DAB's action. See the advertisement in *Ming Pao*, 6 October 1997, p. A13.

15. Sonny S. H. Lo, "Legislative Cliques, Political Parties, Political Groupings and Electoral System", in *From Colony to SAR: Hong Kong's Challenges Ahead*, edited by Joseph Y. S. Cheng and Sonny S. H. Lo (Hong Kong: The Chinese University Press, 1995), pp. 53–55.

16. *Hong Kong Economic Journal*, 29 April 1997, p. 5.

17. Chris Yeung writes: "The close links between the HKPA and Xinhua (New China News Agency) have been an open secret". See his "Political Parties," in *The Other Hong Kong Report 1997*, edited by Cheng, p. 63.

18. Governor Patten separated the membership of the ExCo and the LegCo. For a review of executive–legislative relations in Hong Kong, see Norman Miners, "The Transformation of the Hong Kong Legislative Council 1970–1994: From Consensus to Confrontation", *Asian Journal of Public Administration* 16, no. 2 (December 1994): 224–48.

19. In November 1997, the failure of the Health Bureau to prevent dispensers in a government clinic in Cheung Sha Wan district from mixing up mouth wash with anti-fever medicine indicated the extent of public maladministration in the HKSAR.

Yet, the Health Department chief did not have to resign and she got the support of both Tung and the Secretary for Administration Anson Chan. In fact, colonial officials in Hong Kong did not resign from their positions despite numerous scandals. A notable example was the Legal Department in Hong Kong under British rule. See Ming K. Chan, "The Imperfect Legacy: Defects in the British Legal System in Colonial Hong Kong", *Journal of International Economic Law* 18, no. 1 (Spring 1997): 145–48.

20. For the political attitudes of senior bureaucrats, see Joseph Y. S. Cheng and Jane C. Y. Lee, "The Changing Political Attitudes of the Senior Bureaucrats in Hong Kong's Transition," *China Quarterly*, no. 147 (September 1996), pp. 912–37.

21. After 1 July 1997, there have been signs that some government officials have been less open than before. Elaine Chung, the Director of the Urban Services Department, argued with members and the chair of the Urban Council (UrbCo) over the architectural design of the central library. She claimed that the chair had approved her decision on the new architectural design, but the chair later denied this and some UrbCo members criticized her for not respecting the opinion of the elected representatives. See *Apple Daily*, 17 November 1997, p. A14.

22. *The Basic Law of the Hong Kong Special Administrative Region of the People's Republic of China* (hereafter cited as *The Basic Law*) (Hong Kong: Consultative Committee for the Basic Law, April 1990), p. 29.

23. *Express Daily*, 1 November 1997, p. A4.

24. Louie Kin-shuen and Shum Kwok-cheung, eds., *A Collection of Materials on Hong Kong Elections 1995* (in Chinese) (Hong Kong: The Chinese University Press, 1996), p. 117.

25. *South China Morning Post*, 27 November 1997, p. 4.

26. PRC officials have equated the DP with the Association in Support of the Patriotic and Democratic Movement in China, an organization formed by some Hong Kong liberal democrats like Szeto Wah and Martin Lee during the June Fourth Incident in the PRC. Although Lee is no longer the leader of the Association, some DP members are still affiliated with it. To PRC officials, the DP wants to "terminate the leadership of the Chinese Communist Party" in China. See *Hong Kong Economic Journal*, 13 November 1997, p. 7.

27. The proportional representation system "favours the representation of 'minorities' ... , [it] gives any well-organized pressure group — be it a union, a religion, an ethnic group, a profession, or an ideological faction — a chance to win seats". See Guy Lardeyret, "The Problem With PR", in *The Global Resurgence of Democracy*, edited by Larry Diamond and Marc F. Plattner (Baltimore: Johns Hopkins University Press, 1993), p. 161.

For a hypothetical situation of the DAB's performance where different proportional representation formulae were used in the 1995 LegCo's direct elections, see Lo Shiu-hing and Yu Wing-yat, "The Electoral System of Hong Kong's Legislative Council: Results under Different Proportional Representation Formulae", in *The 1995 Legislative Council Elections in Hong Kong*, edited by Kuan Hsin-chi, Lau Siu-kai, Louie Kin-shuen and Wong Ka-yin (Hong Kong: The Chinese University Press, 1996), pp. 97–134.

28. The DAB obtained 15.4 per cent of the total votes in the 1995 LegCo's direct elections, but it had only elected 2 members (10 per cent of the directly elected seats). On the other hand, the DP obtained 41.9 per cent of the total votes in the same direct elections but gained 12 directly elected seats (60 per cent of the directly elected seats). It is therefore understandable why the DAB wanted to change the direct election method from the single-member single-vote system to proportional

representation. For the DAB's performance in the 1995 LegCo elections, see Louie and Shum, eds., *A Collection of Materials on Hong Kong Elections 1995*, p. 117. For the 1991 LegCo elections, see Hung Ching-tin, "The Reasons for the Complete Defeat of Left-wing Participation in Elections", in *Media Strategy and Elections*, edited by To Yiu-ming and Nip Yee-man (in Chinese) (Hong Kong: Humanities, 1995), pp. 63–76.

29. This situation took place in the functional constituencies involving (1) commerce; (2) industry; (3) finance; (4) real estate and construction; (5) architectural, surveying and planning sector; (6) Urban Council; and (7) rural groups. See *Boundary and Election Commission: Report on the 1995 Legislative Council General Election* (Hong Kong: Government Printer, 15 December 1995), p. 136.

30. Ibid., p. 137. The DP had two members elected by the Election Commission while the DAB had two; the LP one; the ADPL one; the LDF one; the HKPA one. A pro-Taiwan One Two Three Alliance had one member elected; and finally a pro-Beijing politician from a group named Civic Force. See Louie and Shum, eds., *A Collection of Materials on Hong Kong Elections 1995*, pp. 117–18.

31. The number of people receiving Comprehensive Social Security Assistance in August 1995 was 119,000, and it increased to 179,000 in August 1997. See *South China Morning Post*, 13 November 1997, p. 30. Another report which uses a monthly income of HK$9,500 as the "median wage" estimates that there are 600,000 people living below this level. See *Apple Daily*, 24 May 1997, p. A8.

32. Letter to the editor, "Officials and Business Suppress and Eat into the Fruits of Prosperity", *Apple Daily*, 7 October 1997, p. F11.

33. Ian Scott, "Policy-making in a turbulent environment: The case of Hong Kong", *International Review of Administrative Sciences* 52 (1986): 447–69.

34. The riots prompted the colonial administration to reform social welfare in the late 1960s and the 1970s and to tackle the issue of police corruption later in the mid-1970s. See Frank Welsh, *A History of Hong Kong* (London: HarperCollins, 1997): 467–94.

35. Arguably, democratization has strengthened the legitimacy and governability of the Patten administration. For an opposite view that democratization in Hong Kong during the Patten era brought about the problem of "ungovernability", see Lau Siu-kai, "Decolonization a la Hong Kong: Britain's Search for Governability and Exit with Glory", *Journal of Commonwealth and Comparative Politics* 35, no. 2 (July 1997): 28–54.

36. For these linkages, see Putnam, *The Comparative Study of Political Elites*, pp. 154–60.

37. Different District Boards have different political complexions. Some are dominated by the DP; some by the DAB; and some without any party dominance. In the Central and Western District Board, for example, the former chairperson who was a DP member was not re-elected. He was replaced by a District Board member more supportive of the PRC's policy. In Kwun Tong District Board, the former chairperson was an independent supported by DP members, but she was ousted by a pro-Beijing member in July 1997. In short, even at the district level, political institutions seem to be influenced by patronage, undermining their representativeness and threatening the relations between the élites and the masses.

38. The ADPL Chairman, Fung Kin-kee, maintains that the provisional legislature has become "a rubberstamp of the business sector", for it had repealed the collective bargaining bill endorsed by the LegCo before 1 July 1997. See *Ming Pao*, 31 October 1997.

39. See Irene Tong, "Women", in *The Other Hong Kong Report 1994*, edited by Donald

Political Parties, Élite–Mass Gap and Political Instability in Hong Kong 87

H. McMillen and Man Si-wai (Hong Kong: The Chinese University Press, 1994), pp. 367–87. The Indians were only politically active over the issue of British nationality; see Rup Narayan Das, "A Nationality Issue: Ethnic Indians in Hong Kong", in The Other Hong Kong Report 1990, edited by Richard Y. C. Wong and Joseph Y. S. Cheng (Hong Kong: The Chinese University Press, 1990), pp. 147–58.

40. See Ian Scott, "The State and Civil Society in Hong Kong" (Paper presented at the conference on Political Development in Taiwan and Hong Kong, held at the University of Hong Kong, 8–9 February 1996, pp. 7–12.

41. For example, on the morning of 12 December, some citizens phoned the popular radio programme, "The Nineties", saying that they were really worried about Hong Kong's economic prosperity in the future. One citizen accused the HKSAR Government of "beautifying" the economic performance of the territory, but she said that the reality was the opposite.

42. Humans' mental capacities to grasp the complexity of policy problems are often limited, as Lindblom and Woodhouse warn us. See Charles E. Lindblom and Edward J. Woodhouse, The Policy-Making Process (New Jersey: Prentice-Hall, 1993), p. 5.

43. See Ha Ping, A Biography of Li Ka-shing (in Chinese) (Hong Kong: Ming Pao, 1996).

44. Building Hong Kong For a New Era: Address by the Chief Executive the Honourable Tung Chee-hwa at the Provisional Legislative Council Meeting on 8 October 1997 (Hong Kong: Government Printer, 1997), p. 20.

45. See Editorial, "Tenants Purchase Scheme", Ming Pao, 10 December 1997, p. D10.

46. Article 45 and Article 68 of the Basic Law state respectively that the HKSAR's Chief Executive and legislature will aim at the selection process of universal suffrage. See The Basic Law, p. 19 and p. 27.

47. Remarks made by Denis Bray, the former District Commissioner for the New Territories. See Norman Miners, The Government and Politics of Hong Kong (Hong Kong: Oxford University Press, 1982), p. viii.

LO SHIU HING is an Assistant Professor in the Department of Politics and Public Administration, University of Hong Kong.

[13]

CHANGES IN CONTINUITY:
GOVERNMENT AND POLITICS IN THE HONG KONG
SPECIAL ADMINISTRATIVE REGION

Steve Tsang
St. Antony's College
Oxford University

Despite the worldwide media coverage and lavish celebrations, the British handover of Hong Kong to the People's Republic of China (PRC) had long been intended to be an anticlimax. Both the Sino-British Joint Declaration (1984) and the Chinese Basic Law for the Hong Kong Special Administrative Region (1990) stipulate clearly that the existing system and way of life in this former British colony will be respected and maintained notwithstanding the transfer of sovereignty. The *raison d'être* for the two documents is to maintain continuity in Hong Kong. Indeed, in the days which followed the handover, the Government and the people of the Special Administrative Region (S.A.R.) tried to behave as if little had changed except for the flag and other symbols of British rule. However, in reality, important changes are taking place in the politics of Hong Kong, though there is a common wish among the local people to ignore them as best as they can. The focus of this article is to examine the forces behind such a transformation, identify the major pattern of changes, and assess how the forces for change and for continuity have been interacting in the early days of Hong Kong as a Chinese S.A.R.

44

"No Changes for Fifty Years" in Perspective

The Joint Declaration, the Basic Law, and the public rhetoric of the Chinese leaders in the last fifteen years appear to suggest the policy of keeping the status quo in Hong Kong for fifty years under the "one country, two systems" principle was devised to serve Hong Kong's interests. In reality this was only one of the objectives of this policy, for which the most important has always been that of furthering the interests of the P.R.C. and its ruling Communist Party.[1] What the P.R.C. policy really means is that the Hong Kong system and way of life are to be maintained after the handover in order to ensure the S.A.R. will continue to advance the wider interests of the P.R.C. as defined by the party.[2] It is unrealistic to expect the *status quo ante* to be maintained simply because it was so laid down in the Basic Law and the Joint Declaration. After all, to the P.R.C. leaders the Basic Law has no intrinsic value except to enhance their interests, and the rule of law is still an alien and essentially irrelevant concept. The same applies even more powerfully to the Joint Declaration, the primary function of which in Beijing's view was to secure the retrocession of Hong Kong, a *fait accompli* by July 1997. In other words, the emphasis put on continuity was never meant to prevent changes or to adhere to the Basic Law or the Joint Declaration for the sake of it. From the Chinese Communist Party's point of view, developments which it deems appropriate to the circumstances of the S.A.R. are entirely permissible, though it would prefer not to breach the two documents concerned.

The crux of the matter for Hong Kong and its people after the handover is, therefore, to reconcile their own needs with the Communist Party's preparedness to intervene, if necessary,

---

[1] For an analysis of China's policy toward Hong Kong, see Steve Tsang, "Maximum Flexibility, Rigid Framework: China's Policy Towards Hong Kong and Its Implications," *Journal of International Affairs* 49, no. 2 (Winter 1996): 414-33.

[2] *Renmin Ribao*, 1 July 1997 (speech by Jiang Zeming).

45

and its preference to uphold the Basic Law. Hong Kong must persuade the party leadership that it will abide by the Basic Law and, in so doing, respect the interests of Beijing. In terms of their protection for the way of life in Hong Kong, the Basic Law and the Joint Declaration are hardly stronger than glazed doors, but the shattering of them will have such significant symbolic importance that it is very much to the local interest to keep them in place.

Realistically, one must recognize that the transfer of sovereignty invariably involves altering the power relationship among Hong Kong, Beijing, and other power centers within the P.R.C. such as the Communist Party, the People's Liberation Army (P.L.A.), Guangzhou, and other regional authorities. Consequently, the continuity in Hong Kong's system and way of life can only be maintained by appropriate changes being made. A basic adjustment which is inherent in the situation is, for example, the need to redefine and work out the relations between the S.A.R. Government and the local branch of the Chinese communist Party headed by the Hong Kong and Macao Work Committee. It is impossible that the British Hong Kong Government's policy of formally ignoring and legally proscribing the existence of the Chinese Communist Party can be sustained after the transfer of sovereignty.[3] Furthermore, the derailing of the "through-train" as a result of the Sino-British dispute over political developments during the governorship of Chris Patten (1992-97) also requires a reorganization of some of the political institutions, particularly the local legislature. Above all, changes are unavoidable because the S.A.R. Government needs to work out a *modus operandi* with the P.R.C. leadership concerning its actual scope of autonomy.

---

[3] For the historical background to the Hong Kong Government's outlawing the Communist Party, see Steve Tsang, *Government and Politics: A Documentary History of Hong Kong* (Hong Kong: Hong Kong University Press, 1995), 283-4.

46

The Political Imperative

The P.R.C.'s assumption of sovereignty has introduced a political imperative to Hong Kong. The end of British rule means the Government in Hong Kong can no longer ignore the wishes of the metropolitan power in purely domestic matters. Prior to July 1, London by and large left the Hong Kong Government to manage all matters which did not involve defense or foreign policy. In contrast, Hong Kong's new sovereign power is much more actively concerned with developments within the territory. Since the handover, no policy of the S.A.R. Government can be successful unless it will prove tolerable to the Beijing leadership, in addition to it being acceptable to the local people. Such a political imperative has emerged in spite of the Joint Declaration and the Basic Law, by which the P.R.C. should simply have taken the place of Britain and behaved accordingly. This change is rooted in the fundamental differences between the natures of the political systems in Britain and the P.R.C. as well as in these nations' assessments of how Hong Kong fits into their respective national lives and politics.

To put the British policy in perspective, one must understand a basic irony of British rule in Hong Kong, which also applies elsewhere in the British Empire. It is that most British imperialists were at heart Little Englanders, who might have "basked in the glory of imperialism" while it lasted but "had no wish, and had made no serious attempt, to spread English civilization to their conquered lands," and thus "fought no war to prevent their colonies from detaching themselves from the empire" when the end loomed.[4] In other words, while the British were very proud of their contributions to the creation of modern Hong Kong, which after more than a century of imperial rule finally incorporated the British sense of rule of law and ideas of liberalism and democracy into its way of life

---

[4] Tsang, *Government and Politics*, 4.

47

voluntarily, London had next to no interest in the running of the colony. As a democracy, the elected Government in London did not want to be bothered about Hong Kong because the British electorate did not interest itself in it. Hong Kong was never a serious electoral issue. As long as there were no scandals or financial mismanagement—the last would have provoked intervention as a matter of course since a bankrupt colonial government would have to be bailed out by the British Exchequer—London left Hong Kong alone. The British Government never promised Hong Kong "a high degree of autonomy" but, in fact, let the colony enjoy as high a degree of autonomy as conceivable because it did not have an interventionist ethos in colonial affairs and Hong Kong was too peripheral to mainstream British life and politics. In terms of British politics, domestic Hong Kong policies seldom made it onto the agenda of the Cabinet unless and until a crisis was looming in the colony which might have implications for Britain or wider British interests.[5] The other occasions when Hong Kong would receive attention from top British policy makers were when its interests conflicted with those of the United Kingdom. The most notable examples were the negotiations over the sharing of the costs of the British garrison in the colony, and when London had to respond to the Lancashire textile lobby (when it still existed) over cheap textile imports from the colony. As a result, Hong Kong Government generally did not have to be concerned whether its policies would meet with approval in London, which it could by and large take for granted in the postwar era.

In sharp contrast, the political system in the P.R.C. is totally different from that in Britain and Hong Kong and occupies an incomparable position in the political calculations of the P.R.C. After the status of Hong Kong was placed on the

---

[5] For a detailed study of how the British Government in London supervised the making of a major policy, in this case, over constitutional reform, see Steve Tsang, *Democracy Shelved: Great Britain, China and Attempts at Constitutional Reform, 1945-1952* (Hong Kong: Oxford University Press, 1988).

48

political agenda between 1979 and 1981, the P.R.C. leadership increasingly elevated Hong Kong's significance in its political calculations.[6] Although Hong Kong was "neither a domestic nor a foreign issue" it remained so important that paramount leader" Deng Xiaoping took a personal interest" in it when he was alive.[7] Indeed, the communist regime long deemed it one of its missions in history to secure the retrocession of Hong Kong. Following the collapse of communism as a state ideology and as the source of legitimacy in the 1980s, particularly after the Tiananmen Incident of 1989, the Beijing leadership has intended to use the retrocession of Hong Kong to rally public support and to reaffirm its legitimacy by stressing that it is the only government in the last 150 years to have ended China's humiliation by the Western powers.[8] This consideration was largely behind the lavish, well orchestrated and yet carefully controlled celebrations for the handover, and was publicly confirmed by Chairman Jiang Zemin.[9] Furthermore, Hong Kong is the most important economic locomotive which pulls the mainland economy along and enables the party-state to deliver improvements in living conditions to its people—a crucial pillar for its rule after the collapse of communism as a source of legitimacy. Moreover, the P.R.C. leadership sees its policy over Hong Kong as a major inducement to the Republic of China Government to resolve the Taiwan issue. Hong Kong is, therefore, a matter of first-class importance for the communist regime which cannot afford to let the S.A.R. Government adopt any policy that may harm its interests. In addition, the

---

[6] For the placing of Hong Kong on the P.R.C.'s political agenda, see Steve Tsang, *Hong Kong: Appointment With China* (London: I.B. Tauris, 1997), 81-94, and Robert Cottrell, *The End of Hong Kong* (London: John Murray, 1993), 58-76.

[7] Michael Yahuda, *Hong Kong: China's Challenge* (London: Routledge, 1996), 15.

[8] *Renmin Ribao,* 1 July 1997 (leading article).

[9] *Renmin Ribao,* 2 July 1997 (speech by Jiang on 1 July). Jiang is rendered "Chairman," not "President," here because his position as head of state is Chairman of the P.R.C.

49

Leninist party-state in the P.R.C. has an interventionist ethos and believes in the exercise of authority as a confirmation that it has secured sovereignty over Hong Kong. Finally, the Chinese are not like the Little Englanders who pretended to be great British imperialists but at heart only cared about England. The Chinese sense of history gives rise to an unspoken urge to bring any lost territory deemed to be sacred Chinese land back to the fold, and to require its once stray people to learn to behave like "proper" Chinese again.[10] In light of the considerations above, it is impossible for the P.R.C. Government to let the S.A.R. Government enjoy as high a degree of autonomy as the Hong Kong Government had under the British, regardless of the Basic Law and the Joint Declaration. Consequently, the Government and politics in the S.A.R. must operate on the basis of a new political imperative, namely that the S.A.R's policies must not prove intolerable to Beijing.

A crucial question which follows is: how will this political imperative be met? The answer can be found in the P.R.C. policy of exercising maximum flexibility within the rigid framework of its sovereign authority not being compromised.[11] Although the "communists had no love for condescending Hong Kong capitalists," they have conceded the need for maximum flexibility because" they know they "needed those capitalists for their knowledge of business and technology" in order to enable Hong Kong to continue to play "the role of the leading economic locomotive" in their crucial reforms on the mainland.[12] This selfish concern is the most powerful inducement for Beijing to limit its interference into Hong Kong affairs. It does not mean no meddling, however. What it does mean is that Beijing is prepared to let the S.A.R. Government run domestic policies within Hong Kong and will keep a

---

[10] For the Chinese sense of history, see W.J.F. Jenner, *The Tyranny of History: The Roots of China's Crisis* (London: The Penguin Press, 1992).

[11] Elaboration of this concept can be found in Tsang, *Appointment With China*, 132-55.

[12] Tsang, *Appointment With China*, 134.

50

watchful eye on them in order to make sure its own wider interests are not impaired. This surveillance should, in accordance with the Basic Law, be based simply on holding the S.A.R. Government and its Chief Executive "accountable to the Central People's Government," and on the National People's Congress reserving the right to invalidate any law passed by the S.A.R. legislature deemed "not in conformity with the provisions" of the Basic Law.[13] This formal structure and procedure ignores a basic element in the P.R.C. political system and its control mechanism—the Communist Party. Whatever the Basic Law says, it is inconceivable that the party is prohibited from functioning as the Chinese leadership's ears and eyes in the S.A.R., a function which it had performed prior to the retrocession and one of its central tasks elsewhere in the P.R.C.

The Communist Party in Hong Kong

The exact roles of the Communist Party's Hong Kong and Macao Work Committee after the handover are not yet entirely clear. Its terms of reference are undoubtedly being revised as a result of the transfer of sovereignty, but it is possible that the Chinese leadership has not made a final decision on the matter. What is known is that the current Secretary of the Work committee, Zhou Nan, who also serves as Director of the local Xinhua News Agency, will be replaced in the latter capacity by a senior diplomat, Jiang Enzhu. The name of Zhou's successor as party secretary, if already chosen, is still a secret at the time of writing. Zhou holds ministerial rank in the bureaucracy and Central Committee rank in the Communist Party.[14] It is unlikely that his successor will be given a

---

[13] *The Basic Law of the Hong Kong Special Administrative Region of the People's Republic of China* (Beijing: National People's Congress of the People's Republic of China, 4 April 1990), articles 43 and 17, respectively.

[14] Tsang, "Maximum Flexibility, Rigid Framework," 415.

51

lower rank, even though a reorganization of the Xinhua branch is underway.[15] The most obvious aspect of this downsizing involves the transfer of Xinhua's quasi-diplomatic tasks to the office of the Foreign Ministry's Special Commissioner for Hong Kong. The commissioner is Ma Yuzhen, a former ambassador to Britain and roughly of assistant or vice-minister rank. Ma is responsible directly to the Foreign Ministry. It is not clear whether he has been or will be appointed Secretary of the Work Committee. It is possible that the new secretary will have a new cover such as serving as the Special Commissioner concurrently or retaining the title of Director of the Xinhua branch. Whatever public office the secretary may be given, it will not alter the fact that he will be the most important local representative of the Beijing leadership.

The position and role of the Work Committee Secretary are also related to that of the S.A.R. Chief Executive and the local Director of Xinhua. From Beijing's point of view, there are good reasons to adhere to the Basic Law provision that the Chief Executive should be made responsible for the running of the S.A.R. and be held accountable directly to the central leadership.[16] After Britain and China agreed to the retrocession, Hong Kong was seen by many in the P.R.C. as a lucrative prize in which they would like to share come the actual transfer of sovereignty. Building up a direct chain of command between the top leadership and the Chief Executive will, therefore, minimize if not preempt other interested groups within the P.R.C. power structure from pressuring Hong Kong for their selfish, departmental, or localized gains. This was undoubtedly one of the main reasons why the Chief Executive was given a bureaucratic rank higher than that which usually is granted to a provincial governor. Unlike a provincial governor who enjoys the status of a minister, the S.A.R. Chief Executive's rank has deliberately been left undefined but clearly at

---

15 *Wen Wei Bao* (Hong Kong), 12 November 1996.
16 *Wen Wei Bao*, 20 December 1996.

52

a grade above the ministerial level, allowing him to hold the rank of either a State Councilor or Vice-Premier, and thus less susceptible to pressure from heads of provincial governments. The central leadership takes this matter so seriously that Jiang Zemin repeatedly stressed in his public speeches at the time of the handover that "no central department or locality may or will be allowed to interfere in the affairs" of the S.A.R.[17] Hong Kong is simply far too important to the top Chinese leadership to risk it being messed up by lower-level (though still senior) cadres.

Although Tung Chee-hwa was for all practical purposes hand-picked by the Chinese leadership to serve as the Chief Executive, he is not a party member and will always remain a little suspect.[18] To the top communist leaders, their heavy reliance on a nonparty member who is a major capitalist cannot be more than a necessary evil. If the top leadership cannot trust senior communist cadres over Hong Kong, it is doubtful that it will have faith in Tung without reservations. The publicly projected close relations between Jiang and Tung should not be taken at face value. Hence, it is questionable whether the central leadership can resist for long the temptation to have the party secretary keep an eye on Tung. The apparent delay in naming a successor to Zhou Nan as party secretary reflects the central leadership's desire to reserve its decision on the revised terms of reference for the secretary as it assesses how satisfactory Tung performs as Chief Executive. This is in line with the policy of exercising maximum flexibility within a rigid framework.

While the new party secretary may retain the title of Director of Xinhua, one should not assume the two offices will always be combined. However, the Xinhua office will, in any event, retain much of its previous nonclandestine responsibili-

---

[17] *South China Morning Post,* 2 July 1997 (Jiang's speech at S.A.R. establishment ceremony).

[18] For an account of how Beijing picked Tung, see Yang Zhongmei, *Hongding Fuhao* (Taipei: Sunbright Publishing Co, 1997), 125-40.

ties, less those transferred to the Special Commissioner's office.[19] It will undoubtedly also continue to provide cover for some of the P.R.C.'s clandestine activities, particularly in intelligence. The main thrust of its political (as distinct for its news agency or clandestine) tasks will remain united front work in the S.A.R. Given the semi-official nature of Xinhua as an agency of the party-state, it will probably also be turned into a major point of contact with the Taiwanese. Such a role will involve heavy additional united front work focusing on the Taiwanese. It may also require the use of some of Hong Kong's facilities, particularly by regulating access to them to the Taiwanese, as chips for the semi-official bargaining across the Taiwan Strait. In the event that the Secretary of the Work Committee is not appointed concurrently as the Director of the Xinhua branch, the secretary will unquestionably still be given the responsibility to supervise the director's contacts with Taipei, and to serve as the liaison between Xinhua and the Chief Executive in order to make sure what the negotiators need from the S.A.R. Government as inducements to the Taiwanese will indeed be available. The appointment of Jiang Enzhu, Ma Yuzhen's successor as ambassador to Britain until earlier this year, to head the Xinhua branch confirms Beijing's intention to retain it for much more important functions than that of a news agency.[20]

Apart from being two of the P.R.C.'s best diplomats, both Jiang and Ma are well regarded party men and are of the same bureaucratic rank. Either one of them may be promoted to become Zhou Nan's successor as Secretary of the Work Committee. Although it is possible for a third person to be selected instead, such a prospect is slim. The choice of the Chinese

---

[19] For an analysis of the Work Committee's work, see Tsang, *Appointment With China,* 138-44; for an insider's account, see Xu Jiatun, *Xu Jiatun Xianggang Huiyilu* (Taipei: Lianjing chubenshe, 1993).

[20] *The Times* (London), 26 July 1997. The *Times* correspondents are not certain whether the appointment meant a demotion of Jiang or the retention of important functions for Xinhua.

54

leadership remains unclear at the time of writing, but it is an important issue. The party rank of the new secretary will reflect the thinking of the central leadership and the degree of autonomy which the Chief Executive will be allowed to enjoy in reality. In terms of its importance, Hong Kong does not rank below Shanghai or Tianjin, whose party secretaries are of Politburo standing and clearly senior to their respective mayors. If the party secretary for Hong Kong should also be given a Politburo seat, the implication must be that the central leadership would like to keep a relatively close watch over the S.A.R. Government and allow the more usual method of control over China's regions to prevail. The appointment of someone of the Central Committee rank will signify a willingness to let the Chief Executive have a freer hand in the management of domestic affairs within the S.A.R. If no new senior cadres are sent to the S.A.R. in the next few months and either Ma or Jiang should be elected to membership of the Central Committee in the Fifteenth Party Congress, the promotion will probably indicate who will have been made the local party secretary. The choice of either Ma or Jiang with Central Committee rank will suggest a compromise between the usual practice within the P.R.C. and the special requirements of the S.A.R. The Chief Executive's unusually high bureaucratic rank will compensate for the S.A.R. party secretary not being given a place in the Politburo, but it will also create an anomaly. The lower-ranked party secretary will still monitor and report to the top leadership the performance of the higher-ranked Chief Executive, but he will be discouraged if not deterred from interfering with the work of the Chief Executive. Such an arrangement will imply that the top leadership is implementing, earnestly in its own way, the policy of "one country, two systems."

In spite of the transfer of sovereignty, it is very likely that the central leadership will continue to keep the existence of the Work Committee an open secret, rather than bring it into the open. This is partly because the Communist Party gener-

55

ally keeps the work of its leadership organs confidential. A more immediate consideration is that removing the veil of the Work Committee, like the posting of a Politburo member to the S.A.R. may cause a stir and provoke unwanted speculation as to whether it will undermine the "high degree of autonomy" that has been promised. A formal announcement of the appointment of the party secretary is therefore unlikely.

Yearnings for Democracy and the Political Reality

The fact that the attitude of the Chinese Communist leadership has become the political imperative since July does not negate the importance of the indigenous political forces which have emerged. After Britain entered into negotiations with the P.R.C. over the Joint Declaration in 1982, the issue of democratization was put back on the political agenda after a hiatus of some thirty years. It took most of the rest of the decade to gather sufficient momentum to become a major focus in local politics.[21] The turning point came in 1989. The Tiananmen Incident had such a powerful impact on the Hong Kong people that it caused "a big change of opinion" and produced "a widespread desire for a substantial proportion of the legislature to be directly elected."[22] The first-ever direct elections to the legislature in September 1991 showed the new attitude of the local people, in spite of the fact that the number of seats available for contest amounted to only 30 percent of the total. By giving an overwhelming victory to the prodemocracy candidates who were "perceived to be the group which has stood up strongly for Hong Kong interests and in opposition to the Chinese government," the local people demonstrated

---

[21] For a monograph study of democratization in the 1980s, see Shiu-hing Lo, *The Politics of Democratization in Hong Kong* (Basingstoke: Macmillan, 1997).

[22] Percy Cradock, *Experiences of China* (London: John Murray, 1994), 231; for an analysis of the impact of Tiananmen on Hong Kong, see Tsang, *Hong Kong: Appointment With China*, 156-80.

56

clearly their political inclinations.[23] This trend continued. Confirmation came in the 1995 Legislative Council elections. Contrary to the expectation that the imminence of the handover would deter voters from supporting the prodemocracy parties, the voters again gave them a landslide victory. The elections "demonstrated beyond reasonable doubt that the idea of democracy had taken root in Hong Kong."[24]

Although it is an exaggeration to say that the genie of democracy had come out of the bottle since Hong Kong was never turned into a democracy under the British, the head of the genie was allowed to pop out briefly before being pushed back into the bottle. During the governorship of Chris Patten, the idea of democracy was nurtured in Hong Kong. In spite of the fact that directly elected legislators never numbered more than a third of the total, the manner by which Governor Patten treated them in public and in the legislative chamber gave them and the rest of the community the feeling that they had a taste of democracy. In Patten, a first-class British politician, the legislators and the people of Hong Kong saw how a political leader used to a democratic environment behaved on a daily basis.[25] The Legislative Council transformed itself from "a tame talk shop into something akin to a real parliament."[26] The reversal of Patten's limited democratic reform proposals in July did not remove the marks which his tenure had left on the local people. The democratic activists, headed by Martin Lee, finally mustered enough public support to mount a credi-

---

[23] Ian Scott, "An Overview of the Hong Kong Legislative Council Elections of 1991," in R. Kwok, J. Leung and I. Scott, eds, *Votes Without Power: The Hong Kong Legislative Council Elections of 1991* (Hong Kong: Hong Kong University Press, 1992), 20.

[24] Steve Tsang, "A Famous Victory Leaves Hong Kong with an Hangover," *Parliamentary Brief* 4, no. 1 (October 1995); 77.

[25] For Patten's own views of his governorship, see Johnathan Dimbleby, *The last Governor: Chris Patten and the Handover of Hong Kong* (London: Little, Brown & Co., 1997).

[26] *South China Morning Post,* 29 June 1997 (leading article).

57

ble campaign to press for greater democracy notwithstanding the transfer of sovereignty. While the democratic movement cannot and should not be deemed a political imperative in Hong Kong, it is nevertheless a major political force in the S.A.R. which should not be ignored.

Public yearnings for democracy has encouraged Martin Lee and his colleagues in the Democratic Party to take a course which may challenge the new political imperative. The first issue which arose following the handover was the legality of the Provisional Legislative Council (P.L.C.), which was appointed by the P.R.C. Government in December 1996. Although the Chinese appointed thirty-three of the sixty serving legislative councilors to the P.L.C., including four members of the prodemocratic Association for Democracy and People's Livelihood, they excluded everyone from the Democratic Party, the largest elected party in the last British Legislative Council.[27] The legality of the P.L.C. is not beyond debate because the Basic Law not only makes no provision for such a body but also specifically lays down the rules for forming the first S.A.R. legislature.[28] This was a legacy of the Sino-British cooperation, based on the principle of "convergence," when the Basic Law was promulgated in 1990. The shared objective then was to ensure that the political institutions in Hong Kong would be kept essentially unchanged at the transfer of power in an arrangement known locally as the "through train." Immediately after the handover, Martin Lee and the Democratic Party tried to challenge the legality of the P.L.C. but had their case "thrown out of court."[29] Lee's move is resented by Beijing. From the latter's point of view, the legality of the P.L.C. is not in question because a decision of a committee on the National People's Congress had authorized and laid down the principles for the appointment of such a council in March

---

[27] *Xianggang Shang Bao* (Hong Kong), 22 December 1986.

[28] *The Basic Law*, 27, 59-60, 65-7.

[29] *South China Morning Post*, 6 July 1997.

58

1996.[30] The P.R.C.'s position was put publicly by Vice-Premier and Foreign Minister Qian Qichen. He stressed that this new body was created as a result of "the through train having been derailed," that the P.L.C. and the last British Legislative Council were based on "two different sources of legal legitimacy *(fatong)*" and that the legitimacy for the P.L.C. came from the relevant decision of the National People's Congress.[31] Although the Democratic Party agreed not to challenge the legality of the P.L.C. again after its initial attempt failed, Lee encouraged anyone who might come to be affected negatively by legislation passed by the P.L.C. to challenge the constitutionality of such ordinances.[32] Indeed, a second challenge was made by a group of lawyers but the legality of the P.L.C. was upheld by the Appeal Court of the S.A.R.[33] The issue of the legality of the P.L.C. cannot at this stage be considered a closed book in the politics of the S.A.R. Other challenges may be launched based on, for example, the controversy of whether the P.L.C. has the authority to bar the offsprings of Hong Kong fathers born to P.R.C. mothers on the Chinese mainland from having the automatic right of abode in the S.A.R.

Insofar as the question of democratization is concerned, the P.R.C.'s position was reaffirmed by Jiang Zemin during the handover ceremonies. He pledged the P.R.C's support for a "gradually improved democratic system *suited to Hong Kong's reality*" and to develop such a system "in accordance with the Basic Law."[34] A key difference between the P.R.C. leadership and Hong Kong's advocates for democracy lies in what constitutes a "democratic system suited to Hong Kong's reality." The latter's assessment is in line with that made by

---

[30] Yiguo Liangzhi Chongyao Wenxien Xuanbian (Beijing: Zhongyang wenxian chubenshe, 1997), 281-2.

[31] Ibid., 306.

[32] *Ming Bao* (Hong Kong), 6 July 1997.

[33] *The Independent*, 30 July 1997.

[34] *South China Morning Post*, 2 July 1997 (Jiang's speech at the establishment of the S.A.R.). Emphasis added.

59

most Western observers and scholars, which is that either a parliamentary or a presidential democracy will suffice. Neither system will, however, be deemed to suit Hong Kong's reality as far as the P.R.C. leadership is concerned. The reality, as seen from Beijing, is that democratic developments in the S.A.R. must not undermine either China's sovereign authority over the region or the Communist Party's dominant position within the P.R.C. as a whole.

This requirement defines the perimeter for democratic politics in the S.A.R. It is based on the policy originally laid down by Deng Xiaoping, under which the basic qualification in selecting the local people to administer Hong Kong is that they must "love the motherland and love Hong Kong." In his elaboration, Deng explained that "in choosing them, one must of course include local left-wingers but their number should be kept as small as possible, one should also include a few right-wingers, but most ought to be chosen from those who take the middle ground."[35] In other words, the kind of democracy "suited to Hong Kong's reality" should preferably be based on the principle of democratic centralism which can usually deliver the desired electoral results. Failing that, a kind of guided democracy is also acceptable. In a normal Western style liberal democracy, it is simply impossible to meet Deng's first requirement—to guarantee the election of a legislature with the right kind of balance needed to meet the wider interests of the country, as defined by the party leadership.

The concern was, therefore, raised by keen and perceptive observers even before the handover that the P.R.C. appointed P.L.C. would "change . . . Hong Kong's electoral laws and legislative structure."[36] This worry quickly turned into a reality after retrocession. In the second week of July, the P.R.C. decided to supplant Hong Kong's first-pass-the-post electoral

---

[35] *Deng Xiaoping Wenxuan,* vol. 3 (Beijing: Renmin chubenshe, 1993), 74 (my translation).

[36] Jamie Allen, *Seeing Red: China's Uncompromising Takeover of Hong Kong* (Singapore: Butterworth-Heinemann Asia, 1997), 126.

60

system which had consistently given the prodemocratic parties resounding successes since direct elections were introduced to Legislative Council elections in 1991. Its replacement will be a "unique proportional representation system" supplemented by a ban on all foreign passport-holders from running in geographic constituency elections. The changes are expected to "reduce the presence of pro-democracy candidates" in the legislature to be elected in 1998.[37]

The prodemocratic parties of Hong Kong have been asked to face a political irony. It is that, to increase democratic representation in the S.A.R., they will have to proceed within the perimeter defined by Beijing. The alternative is to face the same fate which befell the reform package put forward by Governor Patten in October 1992. Presented without their blessings, the P.R.C. authorities systematically dismantled all of the more democratic institutions which Patten established and reduced the extent of democracy in Hong Kong to the status before Patten became Governor.[38] The new political imperative has imposed a severe limit to the scope for genuine democratization in Hong Kong.

Institutional Responses

Hong Kong's civil service has been a force wholly dedicated to keeping the local political system as little affected by the transfer of power as possible. Built on the tradition of its modern British counterpart, the Hong Kong civil service has always been expected to detach itself from party politics. Although its insularity to politics has been undermined since the 1980s when the Hong Kong Government responded to the slowly but steadily rising demand for accountability to the general public by requiring its top officials to take on quasiminis-

---

37 *The Independent* (London), 9 July 1997.

38 For a dispassionate analysis of the Chinese handling of the Patten reforms, see Tsang, *Hong Kong: Appointment With China*, 181-208.

61

terial duties, the ethos of the civil service has remained essentially the same.[39] It is rooted in the old benevolent and paternal authoritarianism before the 1980s. Its essence is, in the words of a retired senior official, to "get on with our 'work' as defined and determined by ourselves" in accordance with what senior officials believe to be the best interests of Hong Kong.[40] This tremendous self-confidence of the civil service is based on the fact that the government which it constituted had "by then, after almost a century and a half . . . finally reached the standard of 'as good a government as possible in the traditional expectations of the Chinese.'"[41] This self-confidence and its political neutrality mean the local civil service is programmed to continue to run the Administration with minimal regard being paid to the changes in sovereignty. This is a tremendous force for continuity.

The effectiveness and value of the Hong Kong civil service have also been enhanced by two factors. To begin with, the forthright stand which Governor Patten took in defense of this policies against vehement attacks from the P.R.C. in the past five years restored the credibility of the Hong Kong Government and, with it, maintained the morale of its civil service. Unlike the preceding five years when Sir David Wilson was Governor (1987-92), the Government under Patten was no longer regularly portrayed as a lameduck. The reversal of Patten's reforms by the P.R.C. did not destroy the credibility of the Hong Kong Government. The second factor is Tung Chee-hwa's success in securing the blessings of the P.R.C. to allow all top civil servants under Patten to remain in office, excepting Attorney General Jeremy Mathews who, being a non-Chinese person, cannot remain as a head of department under the

---

[39] Tsang, *Government and Politics,* 160.

[40] James Hayes, *Friends and Teachers: Hong Kong and Its People 1953-87* (Hong Kong: Hong Kong University Press, 1996), 298.

[41] Steve Tsang, "Government and Politics in Hong Kong: A Colonial Paradox," in J.M. Brown and R. Foot, eds., *Hong Kong's Transitions, 1842-1997* (Basingstoke: Macmillan, 1997), 78.

62

Basic Law. This has allowed the civil service to function basically as before and prevent morale from collapsing.

The maintenance of the integrity, efficiency, and effectiveness of the civil service, particularly of its police force, means
the basis for stability and good order has been kept intact.
This is a matter of critical importance since it is the basic factor
which will dissuade the P.R.C. from interference. As Deng
Xiaoping put it, "if turmoil should occur, the central government would intervene."[42] He added that, if Hong Kong should
ever be used as a subversive base against the P.R.C. and the
S.A.R. Government should fail to deal with it satisfactorily,
"the [PLA] garrison would be deployed."[43] This policy has
since been enshrined in the P.R.C.'s S.A.R. Garrison Act of
December 1996.[44] The ability of the police and the civil ser-
viced generally to respond effectively and efficiently to prevent or contain disturbances in the S.A.R. is, therefore, a vital
factor for the survival of Hong Kong's "system and way of
life." Fortunately for Hong Kong, one of the most valuable
legacies of British rule is the creation of a police force which is
"probably . . . the world's most effective anti-riot force, one
which did not resort to excessive use of violence."[45] Since the
handover, the S.A.R. Government and its police are proving
that they are keeping the efficiency and tact which characterized their predecessors in handling protests and demonstrations in the last three decades.

The significance of this matter cannot be exaggerated
because of a subtle but crucial change to the relations between
the head of government and the general officer in command of
the forces in Hong Kong. Under the British, the Governor
was also the commander-in-chief of the armed forces, even

---

[42] Deng, *Wenxuan, vol. 3,* 74 *(my translation).*

[43] Ibid., 221.

[44] *Yiguo Liangzhi Congyao Wenxian Xuanbian,* 297.

[45] Tsang, *Government and Politics,* 174. For a general introduction to the
police, see Kevin Sinclair, *Asia's Finest: An Illustrated Account of the Royal
Hong Kong Police* (Hong Kong: Unicorn Books, 1993).

63

though operational matters rested with the Commander of British Forces. When troops had to be called out, they were deployed in support of civil power and by authority of the Governor exercised through the British Commander. Except in times of war, the troops were, as a general rule, required to play a subordinate and subsidiary role in support of the police, and the local chains of command for both led ultimately to the Governor who remained in overall control. Under P.R.C. sovereignty, the arrangements are quite different. To begin with, the P.L.A. garrison in the S.A.R. is directly responsible to the Central Military Commission (C.M.C.) in Beijing, and the Chief Executive is not concurrently or even nominally commander-in-chief of the garrison. Under the Garrison Act, if the S.A.R. should ask for assistance from the garrison, it will have to send its request to the central Government. It would be up to the C.M.C. to "send troops to perform the tasks of giving assistance to maintaining social order and performing disaster-relief" as appropriate.[46] The fact that the Chief Executive is completely by-passed in the chain of command means the civil authority in the S.A.R. has absolutely no say over the deployment of the garrison.

Another fundamental change which has occurred concerns the nature of the armed forces in Hong Kong and their positions in the national life and politics of the metropolitan powers. Under the British, the armed forces belong to the state despite the royal trappings and are not expected to be an instrument for power struggle in British politics. They have had no political roles either in Britain or in Hong Kong. In the P.R.C., the situation is very different. The P.L.A. is, in fact, the armed forces of the Chinese Communist Party. It has traditionally and frequently been used as an instrument for policy and for power struggle nationally and within the party. One must expect the Central Military Commission to deploy the

---

[46] Article 14 of Garrison Act. Translation by BBC, *Summary of World Broadcast*, FE/281G/8, 9 January 1997.

64

S.A.R. garrison for party or wider national interests if it so wishes. Indeed, the Garrison Act provides that if "the National People's Congress Standing Committee decides to declare . . . a state of emergency for the HKSAR . . . the Hong Kong garrison troops shall perform their duties in accordance with the . . . central people's government's decisions."[47] Furthermore, unlike the constitution of the civil government, no organizational concession of real significance has been made to the special circumstances of the S.A.R. in the command structure of the P.L.A. garrison. Its commander, Major-General Liu Zhenwu, has to operate with the presence of his alterego—a political commisar of equal military rank. There can be no question of the party's determination and ability to retain absolute control over the garrison within the S.A.R.

Since there is nothing which the S.A.R. Government can do about the command structure or the nature of the P.L.A., it is vitally important that it should eliminate the need for its police to call on the support of the garrison. Hence, the S.A.R. must ensure that its civil service, in general, and its police, in particular, will keep stability and good order in order to minimize the chance of involvement in local affairs by the garrison. All indications from the S.A.R. so far suggest the civil service and its police are ready to meet this challenge.

The basic factor which has enabled the civil service to meet the internal security needs so far also applies to its resistance to the spread of corruption from the north. It is the ethos of the civil service which has gained wide acceptance not only within the service itself but also in the community at large. After the immensely successful public campaigns to fight organized corruption and to enhance the Government's responsiveness to public opinions in the 1970s, the great self-esteem which the Hong Kong Civil Service (particularly its Administrative Service) had for itself finally gained public rec-

---

[47] Article 6 of Garrison Act. BBC translation.

65

ognition and acceptance.[48] This *esprit de corps* is also sustained by the strong sense of pride which the people of Hong Kong feel in their Government and society when they compare them with those on the mainland. This is a vital and the strongest force which helps the civil service to resist the temptation of corruption. However, the infiltration of corrupt practices has increased following the integration of the local and the mainland economies in the 1980s.[49] Whether Hong Kong society and its civil service will be able to hold the line against corruption in the long term remains an open question.

Conclusions

The transfer of power in Hong Kong has caused much greater changes than most people expected when the Sino-British Joint Declaration stipulating the protection of its system and way of life for fifty years was signed in 1984. Many of the changes could have been foreseen if one had looked completely dispassionately and rationally at the political situation in Hong Kong, the real thrust of the P.R.C. policy, and the reality inherent in Britain's conceding sovereignty to a communist party-state. However, Hong Kong is an emotional subject for most people concerned with its future—scholars, journalist, and policy-makers alike. Since the future of Hong Kong was raised as a subject for negotiations between Britain and the P.R.C. in 1979, there always has been a strong element of wishful thinking involved, as everyone wants to see Hong Kong continue to prosper. The P.R.C. policy of "keeping the

---

[48] Tsang, "Government and Politics in Hong Kong: A Colonial Paradox," 74-7.

[49] *Report by the Commissioner of the Independent Commission Against Corruption 1993* (Hong Kong: government Printer, 1994), 9. For the significance of the anticorruption campaigns in the 1970s, see Tsang, *Government and Politics*, 186-92. The most recent single-volume work on the subject is T.W. Lo, *Corruption and Politics in Hong Kong and China* (Buckingham: Open University Press, 1993).

66

status quo for fifty years" under the "one country, two systems' idea was often misunderstood, usually either taken at face value or dismissed as a propaganda ploy.

The reality has always been more complicated, cruel and, in an important sense, more promising. Changes would have to happen after the handover despite all the agreements and commitments to ensure continuity. The fundamentally different nature of the metropolitan powers, their contrasting assessments of Hong Kong's value and place in their respective national lives and politics, and the sheer impossibility for a dynamic society to retain its character and yet resist changes for half a century should have raised doubts as to the viability of the P.R.C. policy as it was proclaimed. Once one understands the true nature of the P.R.C. policy—in the shape of exercising maximum flexibility within a rigid framework of furthering its sovereignty and national interests as defined by the Communist Party—it follows that certain changes will be unavoidable though they will be limited by the need—again, motivated by the party's self-interest—to maintain continuity in Hong Kong. The direction and extent of the changes have been guided primarily by two factors. First and foremost is what will enhance the P.R.C.'s interests most. The other factor is what will be needed to keep the people of Hong Kong sufficiently satisfied that they as a society will continue to function as "the goose which lays golden eggs" for the P.R.C. Ultimately, it is the attitude of the P.R.C. leadership which decides the changes permitted in Hong Kong—the political imperative of the S.A.R. This is a cruel fact from which there is not escape.

The limits of Britain in protecting the status quo enshrined in the Joint Declaration become obvious if one puts the British policy since 1979 into perspective. As soon as negotiations for Hong Kong's future were raised, the British were engaged in a rearguard operation. Its basic objective was to minimize damage to Hong Kong and to Britain. There was no real prospect for lost grounds to be recovered, since the Joint Declaration

67

and its negotiation process demonstrated clearly the lack of will and resources for Britain to insist on a settlement over Hong Kong which it and the people of Hong Kong really wanted.[50] A subtle but fundamental realignment of power among Britain, the P.R.C., and Hong Kong occurred once the Joint Declaration was signed.[51] As the declining power in this triangular relationship, Britain tried hard to persuade the P.R.C. to adhere to the Joint Declaration but it had no resources to require the P.R.C. to back down when the latter departed from the agreement, as happened, for example, over the appointment of the P.L.C. In other words, the Joint Declaration and, for that matter, the Basic Law as well, merely provides a guide to what government and politics should be like in the S.A.R. The continued existence of the Sino-British Joint Liaison Group until 2000 and Britain's moral responsibility to the people of Hong Kong under the Joint Declaration should create no illusion concerning the role which Britain can play with regard to the government and politics of the S.A.R. Britain will monitor the situation in Hong Kong and will offer its views to Beijing but it will have little influence over political events within the S.A.R.

The real source of strength in Hong Kong's transformation into a Chinese S.A.R. is the resilience of its people. What enabled Hong Kong to respond adroitly to adversity and to prosper in the last fifty years were the resourcefulness, entrepreneurship, and pragmatism of its people, including particularly its business community. These are human qualities which flourish under pressure. The tremendous flexibility which allowed Hong Kong to prosper without Government subsidy or the Government playing a leading role in running

---

[50] Margaret Thatcher, *The Downing Street Years* (London: Harper Collins, 1993), 489-90.

[51] For a detailed analysis of this process, see Steve Tsang, "Re-alignment of Power: The Politics of Transition and Reform in Hong Kong," in P.K. Li, (ed.), *Political Order and Power Transition in Hong Kong* (Hong Kong: Chinese University of Hong Kong Press, 1997).

68

the economy in the past will also enable it to adapt to the changes implied in the transition. The resilience of Hong Kong's economy will help it remain economically vital to the P.R.C.—the basic inducement to Beijing for making an exception of Hong Kong. The resilience of its people means that they are realistic enough to come to terms with whatever restrictions may be imposed on democratic development. Given the strength of the local people's desire for democracy, such an eventuality will provoke some public protests and agitations. However, the overall reaction of the local people is more likely to be one of pragmatism. In much the same way as they adjusted to the prospect of retrocession, their realism is likely to steer them away from open confrontation with Beijing and toward tolerating a degree of political interference.[52]

In terms of politics within the S.A.R., the prodemocracy activists will remain a potent force. They will continue to put pressure on the S.A.R. and P.R.C. Governments. As long as they do not go beyond the P.R.C. leadership's bottom line, they will be tolerated. Their activism, if properly harnessed, will help to minimize erosion of the S.A.R.'s autonomy and the scope of democracy, as politics is a dynamic process of give-and-take. However, if they should push beyond the limits laid down by Beijing, they will provoke a strong reaction which will set their cause back, as happened to Patten's 1992 reform package. On their own, the prodemocracy politicians are prone to raise the stakes in confronting Beijing. In reality, since they cannot sustain their movement if they should divorce themselves from the general public, they will probably exercise a degree of restraint in accordance with the pragmatism prevailing in Hong Kong society.

In short, in spite of the commitment of "no changes within fifty years" under the terms of the Joint Declaration and the Basic Law, changes in government and politics are unavoidable and have already been happening in the S.A.R. The gen-

---

[52] Tsang, *Appointment With China*, 216.

69

eral acceptance of this commitment means that changes, so far, are within the framework of maintaining continuity.

Part IV
Economy and Society

[14]

THE POLITICAL ECONOMY
OF HONG KONG'S INDUSTRIAL UPGRADING:
A Lost Opportunity

ALEX HANG-KEUNG CHOI*

Nicholas Owen's widely read article, "Economic Policy in Hong Kong", is renowned for its critique on the functioning of the self-regulating market in the colony (Owen 1971b). A less noted aspect, however, is the discussion on the relation between industrial upgrading and export-oriented industrialization (EOI), and the inconsistency in Owen's treatment of Hong Kong. Typical of his generation of writers, Owen was unreservedly optimistic about EOI. His optimism was founded on the argument that this form of industrialization provided "the opportunity of 'manufacturing up', that is moving into the better quality end of a product range through experience, better design and improved quality control" (p. 153). But in a later section of the same article, he contradicted his own expectation by providing a list of figures showing that industrial deepening had not taken place in Hong Kong. He concluded that,

> we would *expect* industry to become more capital-intensive as the economy develops ... the surprising result ... is that there is no evidence of any capital deepening in the seven-year period 1960-7. (pp. 189-190, emphasis in the original)

Although Owen did not explain why there was no capital deepening, it should be pointed out that his expectation of a more mature industrial structure for Hong Kong was largely justifiable and widely shared.[1] By the time of his writing, Hong Kong had gone through two full decades of rapid industrialization, and signals for the need for industrial upgrading, such as labor shortage, protectionism and foreign competition, had been flashing for almost ten years since the end of the 1950s. But why was the industrial structure of Hong Kong so stubbornly entrenched in the labor-intensive stage? Why had all the signals for upgrading been ignored? Before tackling this puzzle, it is necessary to review two theories on industrial development

---

* The author is attached to the Department of Political Studies, Queen's University, Canada. This article has benefitted from discussion with various participants in the International Workshop on Hong Kong: Polity, Society, and Economy Under Colonial Rule, organized by the Documentation and Research Center for Contemporary China, Sinological Institute, Leiden University, the Netherlands, 22-24 August 1996. He would especially like to thank Tak-Wing Ngo, Stephen Chiu, Hui Po-keung and Wu Yongping. He is also grateful to an anonymous reviewer who raised many interesting issues. Finally, he wants to acknowledge his debt to Vivian Hung who has rendered unfailing help over the years.

[1] According to one authority, industrial deepening is "one of the oldest topics [...] in the history of its industrial development". See Ho 1992, p. 161.

158

— free marketism and interventionism — which have to a great extent shaped many people's conception of industrial restructuring.

Market and State

The colonial government of Hong Kong followed a policy of free marketism.[2] According to this theory, industrial restructuring will automatically take place in a free market once the labor resources have been exhausted (Balassa 1981, p. 22). Since the late 1980s and 1990s, however, the idea that industrial upgrading can be left to market forces has lost much appeal and credibility, at least in the case of Hong Kong. It was possible for the neoclassicists to maintain that Hong Kong had an "appropriate" capital goods sector in the late 1970s and early 1980s, because industrial deepening had not yet been aggressively pursued by the other Newly Industrializing Countries (NICs).[3] However, this argument cannot stand up to the realities of the 1990s. It is widely accepted that Hong Kong's industrial structure has lagged far behind its heavily interventionist neighbors and competitors, such as Taiwan, Singapore and South Korea.[4] Even some of the most faithful neoclassicists such as Edward Chen have acknowledged that the failure of Hong Kong to upgrade its industries can be attributed to "the apparent under-intervention" of the government (Chen 1989).

The introduction of the topic of state intervention has opened up an entirely new line of inquiry into industrial restructuring, as it shifts the focus of the debate from how a free market works, to why the state lets the market work in a particular way. In the case of Hong Kong, the question, then, is: what has prevented the colonial state from interfering in the market? Or, from another angle: why was the Hong Kong state so impervious to the social pressure for intervention? Recently, Stephen Chiu has made an important contribution to answering these questions.

Chiu concentrates on two interrelated variables: state-capital alliance and state capacity. The early existence of a powerful and close-knit commercial and financial class, and their success in dominating the two most important political institutions, the Legislative and Executive Councils, have limited the colonial state's capacity to raise tax revenue and constrained its interest to the pursuance of an interventionist "selective industrial policy". The state, as a result, "aims at providing general support to economic development rather than aiding the growth of industrial sectors in particular" (Chiu 1994, p. 42).

---

2   The terms "free market", "free port", "free trade" and "*laissez-faire*", are used interchangeably in this article to denote an economic structure with a high degree of openness and low degree of government intervention. The position of the Hong Kong government on this policy is discussed in Miners 1997, pp. 43-44.

3   See, for example, the debate between Martin Fransman (1982) vs John S. Henley and Mee-Kau Nyaw (1985).

4   Wade 1990; Ho 1993; Yeh and Ng 1994; Choi 1994b; Kim 1993.

159

Chiu's theoretical framework is innovative in highlighting the importance of social alliance and state capacity in Hong Kong's non-interventionism. However, it also has certain shortcomings. Based on the statist-institutionalist approach, he conceptualizes the state as a neutral, efficiency-enhancing organization clearly demarcated from society (Chiu 1994, p. 42). This theory not only denies the state as a set of social relations mediated by class, race and gender, but has also led Chiu to misinterpret some of the historical developments in Hong Kong.

For example, Chiu maintains that the financial and commercial bourgeoisie derived their political power from their ability to capture strategic institutions that bridged state and society. Since the beginning of this century, according to Chiu, the "elite settlement" process has developed "a dense institutional network" that has, more or less permanently, tied the state and the bourgeoisie together (Chiu 1994, p. 85). This explanation, however, cannot account for the post-war development. Since the early 1960s, the industrial bourgeoisie has been incorporated into state institutions, and by 1968, they more or less dominated the Legislative and Executive Councils in terms of numbers.[5] But this domination has failed to translate into an interventionist industrial policy. As clearly shown by the economic policies of the 1960s and 1970s, the colonial state chose to associate with the traditional and declining financial and commercial bourgeoisie, rather than with the emerging one based on manufacturing industries, irrespective of the former's level of representation in these political institutions. Indeed, throughout the entire span of colonial history, the state never switched allies and never carried out any policies harmful to the fundamental interests of the British trading and banking elite. This situation demonstrates that the critical factor lies, not in the control of institutions, but in the very nature of the colonial state. The issue of the control over political institutions, as pointed out by so many critics, arises largely from a penchant for studying visible elements of the state at the expense of structural constraints and social forces working behind these institutions.[6]

Another interesting point raised by Chiu which is useful to demonstrate the class and race nature of the colonial state is its alleged limited financial capacity.[7] It is indisputable that, immediately after the Second World War, the resources of the

---

[5]   In King's study on the composition of the Chinese unofficials in the Legislative and Executive Councils, he pointed out that in 1968-69, the industrial elite ("the new rich", as he calls them) took the same number of seats as the trading and banking elite ("the old rich"). Two years later, the industrial elite gained the upper hand with 53.7% of the seats, while the old families had been reduced to 30.8% (King 1975).

[6]   Robison 1991; Hewison 1989, Hawes and Hong 1993; Cammack 1990, 1992.

[7]   Although the race factor is not one of the main topics of this article, race and class are closely intertwined in Hong Kong. British colonialism was profoundly a racial system in which the British held political power over the Chinese. The economic sectors were racially segregated: export manufacturing was almost exclusively an occupation of the Chinese, while British capital was predominantly concentrated in banking, trading and public utilities. This means that industrial policy inevitable had a racial dimension.

160

government were limited, but with the rapid expansion of the economy since the early 1950s, tax revenues saw a tremendous expansion, even without any increase in tax rate. Government surplus accumulated year after year, despite an ambitious public works program. With a large reserve under government control, it is questionable to claim that the government lacked the financial resources to upgrade industry to a greater extent. Chiu argues that Hong Kong's colonial bureaucrats were inclined towards "'hoarding' official resources for rainy days" because of the economy's susceptibility to the ups and downs of the world economy. This echoes the official view.[8] Even if one accepts this argument, one would argue that such a policy is shortsighted, since a more proactive way to safeguard against economic downturn is to invest in upgrading the industrial structure so that industries can maintain competitiveness even in unfavorable world economic conditions.

However, in order to understand the "hoarding" phenomenon, one has to venture beyond the official line, and ask questions such as, hoarding for whom? and, if the reserves are spent, spending for whom? In regard to the former question, a strong case can be made for the argument that the hoarding took place in the interest of Britain, which, unlike Hong Kong, has had many "rainy days" since the end of the Second World War. Hong Kong performed its proper role as a colony in supporting the economy of the metropole (see below for further discussion). As for the latter question, if the reserves would have been spent in the local economy, they would most probably have benefitted the industrial bourgeoisie. Chiu, however, disapproves of such a policy, not only because it would have diminished Hong Kong's preparation against rainy days, but because it would have constituted a kind of "special assistance" to "only small segments of the business community" (Chiu 1994, p. 63). But is this really true? One has to take into consideration the economic background of Hong Kong in the 1960s and 1970s. Given the facts that industry was the mainstay of the economy, and that its rapid growth since the 1950s has been widely viewed as a miracle which saved Hong Kong from economic disaster after the collapse of the entrepôt trade, it seems strange to hold that supporting industry would have meant yielding to special interests. It can in fact be shown that the strict adherence to free port, free trade and non-intervention was not aimed to provide "general support" for private business, but to cater to the needs of the financial and commercial bourgeoisie. In other words, "general support" was actually a policy serving the special interest of the financial and commercial bourgeoisie.

The Resistance Against Industrial Upgrading

By the end of the 1950s, when Hong Kong was celebrating its industrial success, signs of industrial crises already began to surface. Britain imposed restrictions on Hong Kong's exports under the 1959 Lancashire Pact, and since then, protectionism became a serious menace to export-oriented industries. Britain's attempt to enter the European Community in 1961 created another serious threat to Hong Kong's indus-

---

[8]   Hinds 1991, p. 38; *Far Eastern Economic Review* (hereinafter *FEER*), 12 November 1964.

161

tries. In order to become a member of the Community, Britain had to terminate the Imperial Preference system under which Hong Kong products were given tariff concessions when entering the British market.[9]

The problem was made worse by additional factors. Perhaps a victim of their own success, the land and labor costs of Hong Kong industries had rapidly increased during the 1960s.[10] Moreover, since the mid-1960s, Taiwan, South Korea and Singapore had one by one entered the export manufacturing race. With a much lower cost structure, these new players in the export game presented Hong Kong with formidable competition.[11] Despite the fact that Hong Kong had had an early start in EOI, its industrial structure still lingered in the labor-intensive stage, and doubt was cast on its ability to meet the challenge posed by its competitors. Symptomatic of its weakness was that Hong Kong industrialists preferred to engage in so-called "cut-throat competition", i.e. the scramble to produce low-priced goods at the expense of quality and durability.[12] Since short-run cost-cutting was the primary concern, there was little incentive to invest in design, mechanization and labor training to raise productivity.

This problem seemed to be endemic to the industrial structure as a whole. Even the most advanced and capitalized textile sector only maintained its share of the British market, not by high productivity and efficiency, but by relying on the Imperial Preference tariff concession, and, most importantly, "by using its spindles and looms twice as intensively as in Europe" (*FEER*, 27 December 1962).

In spite of all these problems, the Hong Kong government was determined not to pursue industrial upgrading. This was clearly indicated by Governor Robert Black in his remark that Hong Kong could never hope to catch up with the developed countries, "because our only competitive advantage is clearly lower labor costs", and, he continued, "I do not see how we can compete on the basis of capital inten-

---

[9] Since the late 1950s, numerous reports in the *FEER* were devoted to the issue of protectionism, a typical one by Kayser Fung appearing on 18 October 1962, pp. 130-131. For a detailed analysis of the trend of protectionism, see Mok 1979.

[10] For example, a *FEER* report on 3 January 1963 claimed that "Hong Kong is no longer a cheap labor economy". In the next issue (10 January 1963), *FEER* further reported that high land costs had not only deterred foreign investment, but also forced local industrialists to consider relocation overseas.

[11] The first *FEER* report that mentioned Taiwan as a competitor appeared on 13 May 1965. On 14 October 1965, a full *FEER* analysis reported that Taiwan's export processing zone offered free port facilities, cheap labor and low land cost. All these together meant that "Taiwan might present one of the biggest single threats to its [Hong Kong] economy in the future". It was further reported that Taiwan made an intense effort to woo investment from "overseas Chinese in Hong Kong". Several months later (14 July 1966), Singapore and South Korea were added to the list of competitors in a mid-year review of Hong Kong industries.

[12] This kind of vicious competition led Nicholas Owen (1971a) to argue that too much competition could retard the development of large and efficient production units, and "the small firm usually escapes only through its extinction" (p. 144).

162

sive industries" (*FEER*, 5 March 1964). Governor Black's lack of vision closely paralleled that of his predecessor, Alexander Grantham, who declared in 1949, a time when fundamental socio-economic changes were already well under way, that he was proud to be the "Governor of a Colony of shopkeepers", and found the prospect of Hong Kong's industrialization "obscure".[13]

While these remarks could be used to show that the colonial administrators were conservative, uninnovative and complacent, one should not lose sight of the fact that in adhering to the entrepôt principle and refusing to redistribute resources for industrial upgrading, the government acted in the interest of the commercial and financial bourgeoisie. It will be argued below that these interests were one of the key factors in explaining the colonial state's indifference to industrial restructuring. It will also be argued that these interests were mediated by a whole set of structural and non-structural factors including the crisis of colonialism; the ambivalent relation between the trading and industrial elites; the internal division of the industrial class; and the weakness of the labor movement.

By demonstrating the relevance of these factors in the making of post-war economic policy, it is hoped that the pitfall of what Irfan Habib (1985) has called "studying a colonial economy without perceiving colonialism" can be avoided. Looking at Hong Kong's economic policy through the prism of colonialism, we can see that the colonial state's agenda and its resource distribution policy were strongly influenced by the interests of the metropole and the resident British commercial and financial interests, which were not necessarily the same as those of the majority of the Hong Kong people. By showing that these interests were intimately tied up with the *laissez-faire* doctrine, this study challenges the hegemonic colonial discourse, appropriately named "The Barren-Rock-Turned-Capitalist-Paradise Legend" by Tak-Wing Ngo in his contribution to this volume, which has painted a deceptively gentle and benevolent picture of colonialism.

The Colonial Structure

In the literature on Hong Kong, there is no shortage of studies on the "colonial state". What is interesting, however, is that many of these studies actually try to deny that the colonial state was really "colonial", by ignoring its real interests and constituency and the claim that the colonial state shared a "general interest" with the

---

[13]   Not more than five years later, when the industrial transformation was proceding rapidly, Grantham regretted that "we were all at one time too ready to think of Hong Kong only as an entrepot". See Grantham 1965; Choi 1994a.

163

local inhabitants.[14] Due to space limitations, only two examples will be presented here.

Alvin Rabushka has argued that the colonial state realized very early that "the fate of the native inhabitants must not be entrusted to a comparatively few British residents". Since the Chinese population was an absolute majority, doing so would jeopardize political stability. Consequently, the colonial state resisted the desire of the British residents to control local affairs and maintained a high degree of autonomy. The Chinese, at the same time, "received substantial benefit from British rule" (Rabushka 1973, pp. 45, 49).

Lau Siu-Kai, via a slightly different approach, arrived at the same conclusion. On the one hand, he argued that the colonial state was a "bureaucratic polity", and hence insulated from the British trading interests. On the other hand, he pointed out that the value of Hong Kong in the eyes of Britain did not lie in territorial gain or natural resources, since it offered neither, but in its potential to serve as a trading post. Thus, Lau states, the role of the colonial government was "to secure a stable environment wherein *all* groups can live together peacefully" (Lau 1982, p. 33, emphasis added). An important characteristic of the colonial power in Hong Kong was said to be its function of "guardian of common interests" (*ibid.*).

One of the pillars supporting this benevolent image of colonialism is the free trade policy, which is portrayed as the engine generating economic prosperity for the entire population. In colonial discourse, free trade has been credited with numerous positive contributions, such as laying the foundation for Hong Kong's industrialization and rapid economic growth. However, upon closer examination, it turns out that the free trade policy was neither the driving force behind Hong Kong's industrial transformation in the 1950s, nor amenable to the upgrading of the industrial sector in the 1960s. The real reason why the state steadfastly adhered to this policy was that it economically benefitted the British commercial and financial elite.

The orthodox claim has been that free market was the foundation of Hong Kong's industrial transformation in the 1950s.[15] This claim has come under attack, and the most forceful challenges come, not from those who argue that a free market scarcely existed[16], but from those who point to its irrelevance. For example, Ian Scott has pointed out that if free market was really a sufficient condition for industrial transformation, then Hong Kong would have industrialized long before the

---

[14] Note that this position is more pro-colonialist than modernization theory. Modernization theory recognizes the exploitative aspect of colonialism, but claims that exploitation is needed to change traditional society and bring modernity. See Geertz 1963 on Dutch colonial rule in Java, and Brewer 1980 for a general treatment on colonialism and imperialism. Many studies deny that British colonial rule in Hong Kong brought exploitation.

[15] There are numerous studies which highlight the important role of free market in Hong Kong's industrial success. For two typical accounts, see Riedel 1974 and Sung 1985.

[16] This is argued in Lui 1985; Nthenda 1979; Henderson 1993; Schiffer 1991.

164

Second World War, since the free trade regime had been in existence ever since Hong Kong became a colony in the 1840s (Scott 1989, p. 71; *FEER*, 30 June 1966).

While labor-intensive, export-oriented industries may survive, and even flourish, under a free trade regime, industrial upgrading is another matter altogether. As demonstrated by Taiwan, South Korea and Singapore, the state has to play an active role in channeling resources to targeted industrial sectors (Cheng 1990; Woo 1989; Rodan 1989). Although in itself, it does not guarantee success, it is the first step towards that goal. In Hong Kong, the colonial state has been averse to the re-distribution of resources to the Chinese industrial elite since this was incompatible with its class nature. To analyze this constraint on industrial upgrading, three factors need further elaboration: the insecurity of Hong Kong's colonial structure; the crisis of British imperialism; and the dominance of the financial and commercial elite in the colonial state.

The Insecurity of Hong Kong Colonialism

Political insecurity was the hallmark of Hong Kong's postwar colonialism. Discussion on the political future of Hong Kong is not new: in at least two critical periods after the Second World War the security of the colonial system was in doubt, first after the Communist victory in China in 1949, and then around the time of the Vietnam War and the Cultural Revolution in the second half of the 1960s. These incidents showed that Hong Kong's political status did not depend solely on the expiration of the New Territories lease — although no one doubted that, if Hong Kong could last for so long under British rule, 1997 would be a day of reckoning between China and Britain — but also on the rise and fall of the great powers and Cold War politics. The good old days of British hegemony were over, and China, going through Communist transformation, emerged as a world power reluctant to tolerate imperialistic intrusion. In this situation of Britain's decline and China's ascendancy, Hong Kong's continued existence as a British Colony came to rely on the goodwill of China, and China alone.

To a certain extent, the sense of insecurity in the Colony was ameliorated by the fact that China derived considerable economic benefits from the *status quo*, because the Colony became a major source of exchange earnings.[17] Pro-China sources in Hong Kong conveyed messages in the 1960s indicating that China had no intention of altering the Colony's status in the near future.[18] Despite these reassurances, the Hong Kong people, living in the shadow of the Cold War, knew that at any time, strategic needs could become more important than economic considerations. Especially since the Sino-Soviet dispute broke into the open, the

---

[17] It was said that as much as 50% of China's foreign exchange was contributed by Hong Kong. See *FEER*, 20 June 1963. Rabushka (1973, p. 32) suggested a smaller figure of 40%.

[18] For instance, K.C. Wong, Chairman of the pro-Beijing Hong Kong Chinese General Chamber of Commerce assured Hong Kong people in 1964 that Beijing wanted the colony's political status "[to] be maintained for many years" (*FEER*, 27 August 1964).

165

Soviet Union constantly accused China of collaborating with capitalism and colonialism in Hong Kong.[19] Insecurity was also aggravated by the Vietnam War, at least in its early stage. It generated a great deal of anxiety in Hong Kong because the war might explode into a major conflict between the US and China, potentially affecting the status of the Colony (*FEER*, 10 February 1966).

In this situation, Hong Kong's capitalist class was caught in a huge dilemma. They were totally powerless to protect the private property system which was fundamental to their survival. While an editorial in the *Far Eastern Economic Review* claimed that they regarded the situation calmly as "an accepted fact of local life" (*FEER*, 6 February 1964), their uncertainty was naturally reflected in their investment decisions. Since the long-term future was insecure, industrialists did not invest in projects with long maturation dates, and bankers did not extend loans for such purposes. Joe England and John Rear commented that this "fast buck" mentality (i.e., quick profit, short vision) dominated the Hong Kong entrepreneur, who "is committed to making money, not to making goods for sale" (England and Rear 1975, p. 34). Many attempts have been made to explain this phenomenon from the perspective of the psychology of the Chinese entrepreneur. However, this psychology is very much conditioned by the insecurity of the colonial and capitalist system. The colonial state never tried to mitigate and compensate for political uncertainty by providing some kind of profit guarantee for long-term industrial investments.

What is more, the colonial state itself, as the investor of collective resources, was affected by the same logic engendered by this political environment. Investing social resources in industrial restructuring projects carried great risks, since by the time such projects would mature, Hong Kong might no longer be under British control. At the same time, the colonial government had to take care of the immediate interests of Britain and the resident financial and banking class, which were not in agreement with the idea of an industrial Hong Kong.

The Crisis of British Imperialism

The decline of Britain's industrial economy and the concomitant rise of Hong Kong as a world export champion created, in the words of Susan Strange, a "bizarre" situation whereby foreign exchanges earned by tiny Hong Kong had become a major pillar supporting the ailing sterling empire (Strange 1971, p. 112). This meant that Hong Kong's surplus was not made available, wholly or in part, for re-investment in Hong Kong. Britain was apparently not interested in using the surplus to upgrade Hong Kong's industrial structure since it needed the funds for its own use, and, given Hong Kong's uncertain future, such investments did not seem politically sound.

The debate on Hong Kong's surplus underlies the bigger issue of the level of London's control over Hong Kong's affairs. Contrary to what is commonly claimed,

---

[19]   See the discussion on the Soviet factor in the Kowloon Wall City crisis in *FEER*, 24 January 1963.

166

the Hong Kong government was not autonomous *vis-à-vis* Britain. Usually, the claim is supported by two pieces of evidence: the first is that Britain never opposed any legislation in recent Hong Kong history, and the second is that financial autonomy was returned to Hong Kong in 1958 after almost a decade of Treasury control (*Hong Kong Annual Report* (hereinafter *HKAR*) 1959, p. 51). However, these events are deceptive because only visible institutional controls were involved. That Britain chose not to make use of them does not mean that the colonial state was really autonomous — it may simply mean that Britain did not need to employ these formal mechanisms to secure compliance. Indeed, the sheer power wielded by London in the appointment and dismissal of the Governor ensured that the latter had to follow London's instructions carefully, if he did not want to jeopardize his career.[20]

The political subordination of the Colony to the metropole is one of the most important factors in accounting for the phenomenon of the colonial state's surplus accumulation, and its unwillingness to direct resources toward industrial upgrading. Economically, Hong Kong's subordination to the metropole was manifested by the fact that Hong Kong's sterling balance was kept in London. Initially, the sterling balance accumulated as a result of the debts which Britain incurred *vis-à-vis* the colonies and Commonwealth members during the Second World War. Given Britain's very weak balance of payments after the War, repaying these balances would have further threatened the value of the sterling, because balance holders were expected to convert them quickly into US dollars to buy imports. The management of these balances evolved into an elaborate set of monetary and exchange controls binding colonies and independent Commonwealth countries into a Sterling Area. Sterling balances could be accumulated by various means. In the case of Hong Kong, the overwhelming majority were built up in three ways. The first was the legal requirement of full exchange backing for each Hong Kong dollar issued, and such backing, as stipulated by law, had to be in sterling. Secondly, the Hong Kong government was obliged to bank its budget surplus in Britain, in sterling. And finally, the reserves of banks were also customarily placed in Britain.[21]

Recent debates on the post-war British empire have revealed that various post-war British governments, both Labor and Conservative, were preoccupied with maintaining the sterling's international position. The relation between the sterling and decolonization is too complex an issue to be dealt with here.[22] The size of Hong Kong's reserves and the role they played in the overall system still await more detailed studies and analyses. Nevertheless, it is possible to divide the impact of Britain's sterling policy on Hong Kong's industrialization into three phases.

---

[20] For a revealing account of London's tight control over the governor, see Brian St. Clair's paper in *FEER*, 9 April 1964.

[21] For detailed discussion of these mechanisms, see Lin 1970; Jao 1974, Crick 1965; Miners 1975, p. 8; *FEER*, 10 October 1966.

[22] For fascinating discussions, see Cain and Hopkins 1993; Schenk 1994; Hinds 1986 and 1991; Krozewski 1993 and 1996; Darwin 1988; Porter and Stockwell 1988P; Strange 1971.

167

After the Second World War, Britain entered a stage of decline. Its weakness was manifested in a serious shortage of US dollars, and the consequent inconvertibility of the sterling. Having already proved their value during the War, the colonies once again provided the solution for Britain's woes. Ernest Bevin, the Foreign Secretary of the Labor government, declared that he

> was not prepared to sacrifice the British Empire because if the Empire fell ... it would mean that the standard of living of our constituents would fall.[23]

The new mission which was defined at that moment was to turn the colonies into a dollar earning powerhouse by intensifying local development and expanding exports. After the sterling suffered another disaster in the 1947 convertibility crisis, a full-fledged program aimed at exploiting the export potentials of the colonies, embodied in the Colonial and Development Act, was launched. The 1949 sterling devaluation was staged with an eye to increasing the competitiveness of colonial exports (Krozewski 1993, p. 250). Policies were also devised to conserve dollars. Apart from strict import controls, industrialization of the colonies was encouraged for the first time in imperial history because domestic production could replace imports. This attempt to remold the colonial empire is known as its "Second Occupation" (Darwin 1988, p. 159; Krozewski 1993, p. 248).

The colonial officials in Hong Kong were of course affected by Britain's crisis. Short of exportable commodities, they encouraged labor-intensive industrialization, which was in essence the export of the abundant refugee labor power. The new economic policy encouraged the colonial officials to take a new look at Chinese manufacturing enterprises. Neglected, belittled and condemned as vain endeavors a few years previously, as Ngo shows in his article in this volume, these industries were now actively revived and promoted through policies such as the preferential allocation of raw materials, concessionary landleases, overseas export-promoting missions, etc. (Choi 1993). In light of post-war contingencies, colonial industrialization was no longer viewed as causing potential competition for British industries, but as a valuable asset that could not only save Hong Kong from economic collapse, but also make a precious dollar contribution to the British empire.

The Korean War was a blessing for the Sterling Area which saw export revenues soaring as a result of rising commodity prices. By the mid-1950s, thanks to the contributions from the colonies, the sterling crisis had temporarily receded. The significance of their contribution could be gauged from the changing composition of the sterling balances. Previously, major holders were ex-colonies and independent Sterling Area members such as India, Ceylon, Pakistan, Australia and South Africa. Their share had now been reduced, and that of colonies such as the Gold Coast (Ghana), Nigeria, Malaya and Hong Kong was increased (Schenk 1994, p. 26).

---

[23] Worsley 1960, pp. 110-111. See also Curtis 1995, p. 14; Porter 1984, p. 313; Cain and Hopkins 1993, p. 277.

168

With most of the balances directly under the control of British officials in the colonies, the prospect of a crisis precipitated by a run on the balances was greatly reduced. Britain's sterling crisis entered its second phase after the mid-1950s. New developments, within and outside of the empire, convinced Britain that empire and sterling were incompatible. A fateful decision was made to opt for the latter, a foregone conclusion to many who were aware of "the City's" domination in British politics.[24] In 1958, the sterling was made convertible. Between the late 1950s and early 1960s, the colonial empire was hastily dismantled in a period widely known for its "Winds of Change".

Britain had begun to wonder about the economic value of maintaining the empire since the mid-1950s. In terms of trade, the purchasing power of the Sterling Area shrank after the collapse of commodity prices in the post-Korean War era, while trade between West European countries expanded dynamically. At the same time, the costs of maintaining the empire became unbearable, and severely sapped Britain's limited reserves. To gain the loyalty of independent members of the Sterling Area, Britain encouraged capital investment in these countries, worsening its own external account. The cost of policing and maintaining the formal empire rose steeply in the face of challenges from nationalist movements, stirred up by the intensification of exploitation disguised as colonial development. The pivotal event that changed Britain's course was the 1956 Suez Crisis. This incident, which ignominiously terminated after the refusal of the US to extend aid, ended any lingering hope of Britain's regaining its former glory, and convinced it to accept a "special relationship" with the US. In effect, Britain became a subordinate partner of the US in Europe. In a sequence of quickly unfolding events, it accepted US assistance in making the sterling convertible. In the colonies, it exchanged political independence for the commitment of colonial leaders to manage their sterling balances "responsibly". And finally, Britain launched its first bid for European Community membership in 1961.[25]

The question pertinent to our study is why Britain did not give up Hong Kong, especially when it had given up colonies with greater economic value. There are three plausible reasons. In the first place, the strategic value of Hong Kong as an outpost against, and an intelligence-gathering center on, Communist China would have made the US reluctant to allow Britain to abandon the colony (Tang 1994). Secondly, since there was no credible threat to British rule within Hong Kong, defense expenditure was not likely to be heavy. And finally, this expenditure had to be weighed against the potential gains. Hong Kong's blooming industrial economy set it apart from other colonies which predominantly relied on primary commodity production. In the final phase of the sterling crisis which started in the early 1960s, the costs of retaining Hong Kong were far outweighed by the profits.

---

[24]   Cain and Hopkins 1993, p. 276; Anderson 1992; Ingham 1984.

[25]   Maynard 1971; Krozewski 1993, p. 254; Porter and Stockwell 1988, p. 27.

169

Unfortunately, however, the dismantling of the empire did not cure the ailing sterling. Before long, Britain was hit by another series of crises: "[e]verything went wrong between 1959 and 1965", according to John Darwin (1988, p. 241; see also Cain and Hopkins 1993, p. 292). Analysts have attributed the crisis in this phase to Britain's weak industrial capacity, which was hurt by an overvalued sterling. London's obsession with defending the sterling's status as an international reserve currency made it resist devaluation, thus keeping the value of sterling artificially high. All this put heavy pressure on Britain's reserves. Successive governments subsequently deflated the economy so as to reduce imports, and raised interest rates to attract larger inflows of foreign capital. After implementing these rescue measures, the value of the pound was stabilized, but not without harming industry. Triggered off by another period of low reserves shortly afterwards, the same cycle repeated itself, giving rise to a pattern called the "stop-go" economy. Convertibility aggravated this situation because, since capital could move in and out more freely, the pound was more exposed to speculation pressures.[26]

It was during this period of currency crisis, decolonization and the uncertain allegiance of former colonies to the sterling, that Hong Kong's contribution acquired increasing significance. The political control over Hong Kong by British officials was the best guarantee that the mechanism for transferring Hong Kong's reserves to London would operate without hindrance. Indeed, the mechanism functioned so efficiently that, by 1967, an astonishing £350 million, which represented about one-third of Britain's total reserves, had been deposited in London.[27] The sheer size of these reserves gave Hong Kong a prominent place in the calculations of political decision-makers. Susan Strange commented that if "the fat surplus" had not been of such a major concern to successive British governments, "their policy towards Hong Kong might not have been petrified for so long into almost total immobility" (Strange 1971, p. 112).

From this perspective, an altogether different interpretation of the Hong Kong government's lack of finances is in order. While it is true that the government practiced a tight fiscal policy, it is naive to explain this policy purely in terms of the ideological commitment of the Financial Secretary to fiscal conservatism. In reality, tremendous benefits were reaped by Britain in keeping the Hong Kong government lean, if not mean. The question of reserves also helps explain the low priority given by the colonial state to allocating resources for industrial upgrading.

Critical observers have long pointed out that the placing of Hong Kong's reserves in London represented a direct exploitation of the Hong Kong people. Joe England put forward this idea most bluntly. According to him, "Rarely has ex-

---

[26] For the "stop-go" economy, see Leys 1989; Glynn and Booth 1996; Cains and Hopkins 1993, p. 282.

[27] *FEER*, 30 November 1967 and 4 April 1968. See also Jao 1974, p. 142, Table 6.3; and Schenk 1994, p. 52, Appendix to Chapter 2.

170

ploitation of a colony by the metropolitan power been so direct."[28] This, however, is in need of some qualification, since formally, the reserves were still nominally owned by the Hong Kong government, which earned interest on investments in securities and other financial instruments. On the other hand, Britain did derive benefits from the Colony's reserves, and exercised considerable control over them. In the first place, Hong Kong was not allowed to remove these funds from Britain even if the sterling was unstable. In the second place, the mere fact that Hong Kong was required to put its reserves in London rather than leave them in Hong Kong meant that its resources were "on loan" to Britain and could not be used for its own purposes (Lin 1970). To put it in more stark terms, Britain used Hong Kong's colonial status to force a developing territory to extend a loan to a developed country. In a milder tone, the *Far Eastern Economic Review* criticized the reserve system as an imposition on Hong Kong of the "unpleasant task" of supporting the sterling, a task which "is a potentially damaging drain on funds which Hong Kong could make fruitful use of" (*FEER*, 2 May 1968, 6 June 1968).

The Class Interests of the Colonial State

Because Hong Kong was designed as a trading post, an elite of British traders and bankers long dominated the Colony. By the end of the 19th century, however, their economic position had been gradually overshadowed by the Chinese trading elite, who had become the top taxpayers supporting the colonial government.[29] Their ethnic and political connections enabled them to play a dominant role in the colonial state. This remained the case even after the main form of accumulation had switched from commercial and entrepôt activities to export-oriented manufacturing. The commercial and financial sector reoriented its business to serve the rapidly expanding industrial economy. However, the political power of the Chinese business elite lagged far behind the increase in its economic power. This can only be explained by the particularities of the colonial state, which protected the position of the British banking and trading elite.

The extent of the latter's power was illustrated in the 1967 devaluation episode.[30] The Hong Kong dollar initially followed the sterling in a full 14.3% devaluation. After four days, the Hong Kong government revaluated the dollar by 10%, resulting in an effective devaluation of 5.7%. Most incredibly, the colonial state compensated in full the loss incurred by the commercial banks on their sterling

---

[28]   England 1976, p. 10; Djao 1976, p. 93; Benton 1983, p. 9; Hong Kong Research Project 1974, p. 32.

[29]   An excellent analysis of the class relationship in prewar Hong Kong is to be found in Chan 1991.

[30]   This episode also indicates that the interests of Britain and resident British capital were not always in harmony. After the devaluation, Hong Kong-based British banks strongly opposed London's demand that all Hong Kong reserves be deposited in sterling.

171

reserves with money from the Exchange Fund. This compensation, together with the loss of the Exchange Fund itself, cost the Hong Kong people a total of HK$450 million, or nearly HK$120 per person.[31]

With the colonial officials readily defending the interests of the financial and commercial sector, the industrial transformation had to be carried out in an economic regime originally designed for a commercial economy. Rather than replacing the trading system, the industrial sector was superimposed onto it. Although initially, the free port structure was beneficial to the export-oriented industries, it eventually became an impediment to the latter's further development. Through their privileged access to the colonial state, the commercial elite stonewalled almost every request from the industrial elite that might have affected the free trade system. The calls for the modification of the free port policy, the setting up of protective tariffs, and the formation of an "industrial commission" to direct industrial and trade development all either fell on deaf ears or were dismissed out of hand.[32]

A rare glimpse into the conflicts between the industrialists and the banking elite, and the position of the colonial government, can be obtained from a study of the debate on the setting up of an industrial bank. The prospect of inexpensive long-term industrial loans was the primary driving force sustaining their campaign for more than one and a half decades, from the mid-1950s up to the early 1970s.[33] The industrialists argued that the bank was needed, not just for their own interests, as a kind of investment in a brighter and more prosperous future for society as a whole. A sympathizer stated:

> An industrial bank is not merely an organization for the dissemination of financial assistance; it is also a centre for economic study and thought, and a source of guidance to the government ... (*FEER*, 12 March 1965).

This vision collided with the bankers' preference for *laissez-faire*, who regarded the industrialists' demand, in the words of R.G.L. Oliplant, the deputy chief of the Hongkong and Shanghai Banking Corporation, as being a

> mere thinly disguised plea for a bank which will give loans to industry against inadequate security. ... It is not the duty of a bank to accept the equity risk in a new business, which is the responsibility of the proprietor and shareholders ... (*FEER*, 25 March 1965).

---

[31] See HKAR 1968, pp. 46-47; Strange 1971, p. 114; Jao 1974, pp. 36-37, Miners 1975, p. 9.

[32] For an example of these requests, see *FEER* 6 December 1962; Miners 1975, pp. 227-231.

[33] In 1973, a small hire purchase loan scheme for machinery was launched by the government for small and medium-size enterprises. For an overview of the debate on the industrial bank issue, see Mok 1981.

172

What is interesting about these remarks is not so much Oliplant's steadfast adherence to the common banking principle of never borrowing short and lending long, but the implied assumption that if a commercial bank would not take such risks, the Hong Kong state should also not do so.

The ability of the banking community to impose its will on the state was clearly indicated in the composition of the 1959 Industrial Bank Committee, whose Chairman was the deputy Financial Secretary, and which included five "unofficials".[34] None of them had any industrial background, but three came from the banking sector (Chiu 1994, p. 82). The dominance of the committee by members from the banking sector almost guaranteed a negative recommendation, because they "resented the notion of setting up a public or semi-public financial institution that would compete with them" (Chiu 1994, p. 83).

In conclusion, the uncertain colonial future, the crisis of British imperialism, and the dominance of the financial and commercial class presented a formidable obstacle against industrial upgrading. Nevertheless, these structural constraints were not deterministic. If the industrial bourgeoisie had possessed the power and the resolve to change the structure, this could have significantly affected the outcome. But the industrial elite did not push for change. On the one hand, they believed they benefitted from the low cost structure maintained by the colonial state, and on the other hand, they were weakened by their own internal division and disorganization. Throughout the 1950s and 1960s, the leading stratum of the industrial elite was a domesticated class, as will be shown in the next two sections.

The Ambivalent Relationship
Between the Industrial and Commercial Classes

Although the Hong Kong state was loath to alter the free port structure, it did care for the well-being of the industrial economy. The Hong Kong state was very willing, and also largely successful, in suppressing the rise of production costs and assisting the diversification of Hong Kong's overseas markets, as long as the fundamental interests of the financial and commercial bourgeoisie and of Britain were left untouched.

The Hong Kong government's exceptionally active approach in trade promotion was not only an attempt to expand industrial production, but also a way of deflecting the anger of the Hong Kong people over British protectionism, and to cater directly to the interests of the commercial sector which had been marketing a large portion of its industrial products overseas.[35] The large investments in infrastructure

---

[34]   These were unofficial members of the Executive and Legislative Councils, who, prior to the mid-1980s, were appointed by the governor.

[35]   The market development activities were eventually centralized in 1966 under the Hong Kong Trade Development Council (*FEER*, 20 October 1966). It is also interesting to point out that, while the Chinese Manufacturers' Association argued that the main problem facing the industrial sector in the 1960s was industrial upgrading, the General Chamber of Commerce

173

projects, including the building of reservoirs, the airport, the ocean terminal, and new towns consumed a large part of government revenue, but with the growing tax revenue and repressed welfare costs, infrastructural expenditure was found to be a good investment within the means of a small budget.

The most important means of maintaining the low cost structure was the colonial state's apathy, if not outright hostility, toward any call to improve the livelihood of the working people. Demands for higher wages, better working conditions and improvement of social welfare were all branded as a conspiracy aimed at turning the Colony into a "welfare state", which was deemed unrealistic and unreasonable for a developing society. On this issue, the two main factions of the capitalist class maintained a high degree of consensus, which reinforced the colonial state's resistance to calls for fairer redistribution.

Another pillar supporting Hong Kong's low cost structure was China. Up to the early 1970s, China supplied food and consumer goods to Hong Kong at stable prices below the world market level. Combined with the rise of price levels for Hong Kong's industrial exports due to the general inflationary tendency of the world market in the 1960s, China thus enabled Hong Kong manufacturers to make increasing profits: the people in China directly subsidized Hong Kong's industries.[36]

The dedication of the colonial state to maintaining the low cost economy is perhaps not too difficult to understand, because keeping costs low is essential for the flourishing of labor-intensive, export-oriented industries. What is more significant, however, is that the continued success of these industries was not only in the interest of the Chinese industrialists, but also in the interest of the British financial and commercial elite. After the trade embargo, the latter reoriented themselves to serve the rapidly growing industries, either as distributor of industrial materials and machinery, exporters of finished products, financiers of loans and credits, or agents of shipping lines.[37]

The close relationship between the two sectors can be easily gauged from the changing nature of the business of banks and trading firms, for example the largest bank in the Colony, the Hongkong and Shanghai Banking Corporation. Between 1966 and 1970, loans extended directly to the manufacturing sector accounted for over 30% of its portfolio. The manufacturing sector "remained the largest sector in terms of credit allocated" (Jao 1983, p. 538). The banks also made huge profits in the import/export business generated by industry (Jao 1974, p. 94), and the big

---

stated that it was the search and development of new markets. Its demand to the Hong Kong Government was the setting up of more overseas trade representative offices (*FEER*, 4 April 1963).

[36]   The importance of inexpensive and stable supplies from China for the Hong Kong economy has been raised by many observers. See England and Rear 1975, p. 41; England 1976, p. 24; Benton 1983, p. 9; *FEER*, 9 November 1094, 10 August 1967, and 14 September 1967.

[37]   My position on the relations between the industrial and the commercial and banking sectors is different from Chiu's. Chiu argues that although co-existing, they were separated because there were few "institutional linkages" between them. See Chiu 1994, p. 65.

174

trend in the banking sector during the industrial boom was to establish branches in industrial districts such as Tsuen Wan and Kwun Tong (*HKAR* 1960, p. 66).

Some big *hongs*[38], not satisfied with merely serving the manufacturing industries, branched out and set up factories of their own. In 1964, Jardines joined hands with the South China Textile Mill to form a new company, the Textile Alliance, which innovatively integrated all the separate production processes from spinning to garment-making in one big operation (*FEER*, 20 March 1964). Another big *hong*, Wheelock Merden, was also reported to have an extensive involvement in the textile and toy industries, although its main business remained shipping (*FEER*, 9 January 1964).

The high integration between the trading and the industrial sectors is further confirmed by Victor Sit's 1979 study on small industries, which found that as many as 44.8% of small firms obtained orders solely, and another 12% partly, from trading firms. Sit (1982) also suggests that these percentages must have been higher in the 1960s because firms tended to develop their own marketing network once they were more established.

Indeed, the business of marketing Hong Kong's industrial products became so important for traders that when textile quotas were imposed on Hong Kong in the early 1960s, the traders engaged in a fierce battle with the manufacturers for control over these quotas, which was essentially a fight over the access to export markets. They successfully persuaded the government to reject the Chinese Manufacturers' Association's (CMA) proposal to divide the quota in two equal halves, one for traders and one for manufacturers, and to accept instead a system based on past performance. The latter system worked against the manufacturers because a large number of them had relied on exporters to market their outputs. Under the scheme proposed by the traders, they were disqualified for quota entitlement.[39]

The manufacturing industries became Hong Kong's most important and dynamic sector, and their ups and downs had a direct impact on the entire economy. But why did the sector fail to convert its economic might into political power? This article argues that, among other factors, the industrial elite was a class domesticated by the colonial state.

The Domesticated Industrial Bourgeoisie

Already in 1977, Davies noted that "the voices that are heard in government circles come from [big] firms which [employ more than 200 workers] or which are non-manufacturing concerns occupied in banking, accounting, stockbroking or import/export business" (Davies 1977, p. 63).

---

[38] The large expatriate-owned trading firms.

[39] *FEER*, 17 January 1963, 7 February 1963. For the operation of the quota system, see Morkre 1979.

175

The issue of the discrepancy between the political and economic powers of the industrial bourgeoisie was also raised by Jeffrey Henderson (1991). He noted that "in spite of the significance of manufacturing industry to the colony's economy, the corridors of power continue to be the almost exclusive preserve of representatives of banking and commercial interests" (p. 172). He further speculated that this lack of political power was the underlying reason why the colonial state was unwilling to support economic restructuring, but no investigation had yet been undertaken in this important phenomenon (p. 174).

To account for the political weakness of the industrial bourgeoisie, structural factors such as those discussed in the previous section are very crucial, although they should not be overstated. As the political development in the colony has shown, the industrial bourgeoisie did not mobilize themselves resolutely to fight for their own place in the sun. Instead, they were depoliticized and demobilized by internal division, susceptibility to manipulation, and its sense of insecurity toward the capitalist system. The more organized and better endowed upper stratum was co-opted, absorbed and tamed by the colonial state. The small industrialists, deprived of leaders, were cheaply bought off by the colonial state which met their basic need for suppressed labor costs. As a result, the colonial state was able to domesticate the bourgeoisie, which became a class more committed to the *status quo* than to the active creation of the preconditions for its own development.

In order to understand the political impotence of the industrial bourgeoisie, it is important to be aware of its internal divisions and the adept way in which the colonial state manipulated these divisions to its own advantage. Contemporary observers noted the "disunity" between the larger, better capitalized industrialists and the small manufacturers who ran labor-intensive businesses.[40] Some tended to see such internal divisions primarily in terms of a sub-ethnic conflict between the large immigrant Shanghainese spinners, and the native Cantonese small manufacturers. Although this sub-ethnic division may be real (see below), it should not be exaggerated because the persistent rivalry was probably not due to ethnicity *per se*, but rather to the substantial differences in interests and capital needs. That ethnicity cannot in itself be a sufficient factor to account for intra-elite rivalry is shown in a study by Wong Siu-lun, who noted that the Shanghainese industrialists were on friendlier terms with the expatriates than with the Cantonese, and preferred to join the former's trade association, the General Chamber of Commerce, rather than the association of their Cantonese compatriots, the CMA, which represented the small industrialists (Wong 1988).

A review of the statements made by the CMA published in the *Far Eastern Economic Review* in the 1960s shows that the CMA consistently advocated a more interventionist, less export-oriented, industrial policy, which was in sharp disagreement with that of the colonial government. Apart from making persistent demands for cheap land and subsidized loans, it urged the state to expand and protect the domestic market by ending the free port system. Finding the colonial government

---

40   *FEER*, 6 June 1963, 10 December 1964, and 6 May 1965.

176

entirely unresponsive to its cause, the CMA launched a "Buy Hong Kong Campaign", first through their Annual Exhibitions, and then through "Hong Kong Weeks". When Britain strengthened its protectionist measures in the mid-1960s, the CMA annoyed the colonial government by demanding retaliation.[41] At the same time, they voiced the theory that the free port system was the main obstacle to diversification, a policy which the Hong Kong government had actively promoted as a countermeasure to protectionism. Hsin Sutu was perhaps the most vocal and articulate defender-cum-theorist of the CMA position. He argued,

> In a world of protective tariffs and quotas, Hong Kong, being a free port, will be left devoid of bargaining power in trade negotiation. Hong Kong's industries are left unprotected against foreign dumping of goods in domestic market. In the absence of tariff duties, vigorous competition by established foreign enterprises usually prevent new local industries from breaking into the fields, in spite of the fact that such industries are ones to which Hong Kong is best adapted. There is some truth in the claim that the free port status results in Hong Kong specializing too narrowly. Hong Kong needs a certain degree of protection in order to diversify its industries and thereby add to its industrial stability. (Hsin 1963; see also Hsin *et al.* 1977)

The colonial state dealt with this opposition in a classical divide-and-rule manner. It set up and subsidized the Hong Kong Federation of Industries (HKFI) as a counterweight to the CMA. Through a legal restriction confining voting rights to firms employing more than 100 workers, the government, in effect, invited large Shanghainese firms to control the HKFI. These firms did not share the same difficulties as their smaller counterparts: for example, they had better access to capital, and their land problem was less acute because they had either purchased land or secured concessionary 25-year leases from the government when they relocated to Hong Kong in the late 1940s and early 1950s (Wong 1988, p. 89). Most importantly, as an immigrant community without a massive local political base, the Shanghainese industrialists chose to collaborate with the colonial state rather than stand on the side of their native counterparts. They had a relationship with the colonial government not too unlike the one that existed between the overseas Chinese and the colonial state in pre-independent Southeast Asia.[42] Viewed from this perspective, it is no surprise that the HKFI behaved, in the words of the *Far Eastern Economic Review* (6 June 1963) like a "lap-dog" of the government.

With the biggest and most organized stratum of the industrial bourgeoisie standing firmly on its side, the Hong Kong government could safely disregard the

---

[41]   *FEER* 17 Jan 1963, 12 December 1963, 16 December 1965, 17 February 1966, 19 October 1967 and 14 December 1967.

[42]   On this point, see Mackie 1976; Budiman 1988, McVey 1992.

various demands of the small manufacturers without much fear of political repercussions. Its position was further strengthened by the favorable conditions in world
trade prevailing in the 1950s and 1960s. With plentiful opportunities to make short-
term profits, it was even more difficult for small manufacturers to organize themselves towards distant goals with uncertain rewards.[43]

The political timidity of the industrial bourgeoisie was also enhanced by the
China factor. The classical method employed by the national bourgeoisie to coerce
the colonizer to make concessions to their demands is through a movement for
self-rule or outright independence. However, in the peculiar context of Hong Kong,
such a strategy was deemed unfeasible because the colonial state was regarded as a
necessary evil because it acted as a buffer against a Communist takeover. Although
the colonial state would probably not have been able to resist China if it had
demanded the return of the Colony, it was widely believed that without British rule,
Hong Kong would certainly have been reclaimed by China.[44]

The industrial bourgeoisie also regarded the demand for greater democracy as
a way to increase their power as too risky because they believed that, if elections
were held, the strong Communist unions in Hong Kong would win.[45] Given these
limitations, real or imagined, the CMA favored an exit strategy by sponsoring relocation to low cost production areas in neighboring countries such as Taiwan and
Singapore.[46] In a period of brisk growth, even this exit strategy posed little threat
to the colonial state, which viewed the threat of exit as little more than a political
nuisance.

In conclusion, internal division, external manipulation, and the fear of a change
in the capitalist *status quo* discouraged any political actions that could have brought
about a change in the colonial state's industrial policy. Unwilling to back their
demands with credible threats, the industrial bourgeoisie became complacent with
the generally good business environment and largely ignored the needs for industrial
upgrading.

The Externally Oriented Working Class Movement

A serious attempt to shift the industrial structure away from low cost,
labor-intensive, export-oriented industrialization was undertaken, not by the
industrial bourgeoisie, but by the working class, in the form of a mass strike in

---

[43]   On this point, see *FEER* 21 November 1963 and 14 June 1965.

[44]   A message that was repeatedly impressed on the Hong Kong people was China's objection
to countries promoting a two-Chinas foreign policy with regard to Taiwan. China would not
tolerate Hong Kong becoming independent and effectively a "third" China.

[45]   The conventional wisdom why Hong Kong had no democracy is that truly democratic reform
would invite the pro-Communist labor unions to power. For some insightful comments
refuting this view, see Rear 1971; and Benton 1983, p. 29.

[46]   See the reports of such attempts in *FEER* 21 May 1963, 20 March 1964.

1967. By undermining the political stability of the Colony, the working class delivered a clear message that the *status quo* was not to be tolerated forever. This message, had it been taken seriously, might well have generated some momentum to redesign the industrial system in a more capital- and technology-intensive way. Unfortunately, after the end of the Cultural Revolution, both the pro-Beijing Federation of Trade Unions (FTU), the largest labor organization, and the leaders of the strike quickly reverted to the conciliatory welfarism of the previous period. This gave the colonial government an excuse to interpret the 1967 incident as a purely externally instigated event, delaying major changes as if the socio-economic structure of the Colony needed a facelift rather than surgery. It was this strategy of withdrawal, reconciliation, and quiet collaboration by the leftist-organized labor sector that deprived Hong Kong society of an important impetus for change.

Hong Kong achieved rapid industrialization in the two decades immediately after the Second World War. High growth rate together with almost full employment earned the Colony a reputation of "growth with equality", which was used to justify the colonial system (Chau 1979; Chow and Papanek 1981). Lau even stated that, "it is rare to find a colonial government in history giving out so much for the governed" (Lau 1982, p. 44). However, despite the rapid growth, the mass majority of the workers did not feel that they enjoyed the fruits of economic progress. Even some members of the colonial government admitted that the living conditions of the working class were bad, and warned that this could breed social unrest. The Labor Commissioner, R.M. Hetherington, reported in 1961 that the working conditions of a large proportion of workers were "the least tolerable". They had to work 12 hours a day, seven days a week. And even so, "all but the simplest pleasures are beyond the reach of many families" (Hetherington 1963, pp. 31, 36).

The open economic system coupled with the *laissez-faire* social policy created much economic insecurity among the working class. There was little protection against unemployment caused by fluctuations of the world market. Depression, or a slight change in consumer taste in a distant market, could cause a slump in Hong Kong, and trigger off mass lay-offs. In the words of England and Rear, "many families live on a knife edge and sickness or injury, bereavement, or old age, push them to the margins of extreme poverty ..." (England and Rear 1975, p. 66).

While the working class harbored a strong sense of grievance, there were few avenues for them in the political structure of the colonial system to make their voices heard. Democracy was not possible for Hong Kong, according to the colonial government, because China would not allow it. Instead, a consultation system was set up, but the government consulted only those Chinese leaders who were employers and capitalists who opposed political reform because a democratic system would inevitably weaken their power in the colonial system. All the reforms proposed by the unofficials in the 1960s aimed, in one way or the other, at

179

enhancing their own power within the government rather than opening the polity for mass participation.[47]

Although the Cultural Revolution in China did play a role, the 1967 incident may be seen as working class resistance against the state-sponsored pattern of economic development.[48] Scott (1989) made the following penetrating remark: "the cause of the riots lay in economic and social conditions which were, in turn, a product of the colonial regime's political and class structure" (p. 89).

When political stability was at stake, the ruling elite were prepared to negotiate changes. They also fully realized that the production pattern that relied on cheap labor would no longer be sustainable if real concessions were granted. In view of working class activism, a perceptive *Far Eastern Economic Review* reporter stated that the industrial future for Hong Kong was industrial upgrading, because "better products fetch better prices", and industries could then afford better wages (*FEER*, 1 June 1967).

With the unrest in the labor sector, the CMA again sought the opportunity to drive home their usual demands (*FEER*, 14 December 1967). At that moment, the state gave in. In 1969, it reinvestigated the possibility of setting up an industrial bank, and eventually, in 1972, created a loan scheme to help small industries purchase machinery.[49] A package of labor reforms was also formally put on the government agenda, some of which had already been raised and debated for more than ten years without any result.[50] Henry Lethbridge reluctantly acknowledged that the Communist unions had been the "midwife of social progress" (*FEER*, 1 February 1968).

If the organized workers had continued to assert themselves forcefully, they might well have given rise to some permanent changes in the economic and political structure. However, their pressure on the government slackened from early 1968 onwards. Many factors contributed to the unraveling of the movement. It is beyond question that the end of the Cultural Revolution in China had a tremendous impact. When political normality returned, the Chinese leadership once again appreciated Hong Kong's important contribution to its exchange earnings, and the desirability of the export-oriented industries in the Colony.

There has been much discussion on the weakness of the labor sector in Hong Kong. One theory explains the phenomenon by referring to the division of the labor sector into two rival factions — the pro-Taiwan Trade Union Council (TUC) and the pro-Beijing Federation of Trade Unions (FTU). The problem with this theory

---

[47]   For the unofficials' opposition to the first post-war democratic reform program, the Young Plan, see Tsang 1988. For the versions of reform proposed by the unofficials in the 1960s, see *FEER*, 31 October 1963, 18 March 1965.

[48]   For discussions of the 1966 and 1967 incidents, see Scott 1989; Rear 1971; and Heaton 1970.

[49]   See Mok 1981; *FEER*, 12 April 1969, 21 August 1969.

[50]   See *FEER*, 22 February 1968; Scott 1989, p. 125.

180

is that it entirely discounts the possibility of active unionism generated by competition. In order to understand the weakness of the unions, one has to look at the nature of each of these organizations, rather than at the relation between them. As Djao has pointed out, the TUC espoused labor-capital harmony and its main function was to carry out Taiwan's anti-Communist strategy in Hong Kong (Djao 1976). As for the FTU, its primary function, except briefly in 1967, was to stabilize Hong Kong and support China's modernization. Since it was controlled by China, the FTU was criticized for not being a real representative of the Hong Kong workers (Zeng 1976, p. 63). The conciliatory attitude on the part of FTU, the largest labor union in the Colony, goes much further towards explaining the timidity of the labor sector than the theory of factional conflict within the labor sector.[51]

Realizing that the FTU was once again under firm Chinese control, the Hong Kong government was reasonably certain that the most important threat to its rule had gone, and shrewdly decided to attribute the 1967 incident entirely to external instigation without dealing with its local roots (*HKAR* 1967, pp. 1-19). With China safely back on the side of the colonial *status quo*, the economic and political structure of Hong Kong could be perpetuated without much adjustment. The post-1967 social reforms, including those in the labor sector, were more or less cosmetic measures which placed neither undue strain on the low cost structure, nor opened channels for mass participation in the political system.

Conclusion

Hong Kong's failure to restructure its industrial system in the 1960s cannot be attributed simply to the institutional domination of a financial and commercial elite over a neutral state. While it cannot be denied that the influence of this class was strong, and that they continued to resist the pressure to divert resources toward industrial upgrading, their influence was based, not on the number of seats they could obtain in the Legislative and Executive Councils, but on the class and race nature of the colonial state. At the same time, we must not treat the colonial state too deterministically, nor should we regard it as unchangeable. Its power prevailed because the industrial bourgeoisie and the organized working classes were not willing to challenge its political foundation. The industrial bourgeoisie did not do so because they had to rely on the protection from the colonial state against Communist China. The organized workers did not do so because they were primarily Beijing-oriented, and Beijing had a vested interest in the EOI structure of the Colony.

Because industrial restructuring was not pursued in the 1960s, Hong Kong lost the best, if not its only, opportunity for industrial deepening. From the 1970s onward, strategic considerations led China to a gradual reestablishment of economic and political links with the capitalist world. Foreign goods and capital reentered the

---

[51]   For interesting observations on FTU's divided loyalty, see Cooper 1978; England and Rear 1975, p. 4.

181

Chinese market on a large scale. The reincorporation of China into the Western order led to the rapid resuscitation of Hong Kong's once eclipsed entrepôt trade, and the fast development of its financial market to serve China's insatiable need for capital. The cheap labor and land in China provided Hong Kong's labor-intensive industries with a second lease of life, and these industries have virtually turned Southern China into Hong Kong's industrial backyard. Under these conditions, it was understandable that industrial upgrading did not figure prominently either on the state's agenda, or in the mind of the industrialists.

The failure to seize the opportunity for industrial upgrading in the 1960s carries profound implications for Hong Kong's industrial development and the well-being of the industrial workers. Without a capital- and technology-intensive sector, the relocations of industries to the north of Hong Kong means de-industrialization rather than regional specialization. This hollowing process has adversely affected the livelihood of many workers who, since the early 1980s, have been increasingly hit by massive lay-offs and wage stagnation. For the working people, the hope for prosperity under colonial industrialism remained as illusive in the 1980s and 1990s as it was in the 1960s.

REFERENCES

Anderson, Perry, 1992. *English Questions*. London: Verso.

Balassa, Bela, 1981. *The Newly Industrializing Countries in the World Economy*. New York: Pergamon Press.

Benton, Gregor, 1983. *The Hong Kong Crisis*. London: Pluto Press.

Brewer, Anthony, 1980. *Marxist Theories of Imperialism: A Critical Survey*. London: Routledge.

Budiman, Arief, 1988. "The Emergence of the Bureaucratic Capitalist State in Indonesia". In Lim Teck Chee (Ed.), *Reflections on Development in Southeast Asia*, Singapore: Institute of Southeast Asian Studies.

Cain, P.J., and A.G. Hopkins, 1993. *British Imperialism: Crisis and Reconstruction 1914-1990*. London: Longman.

Cammack, Paul, 1989. "Review Article: Bringing the State Back In?", *British Journal of Political Science*, Vol. 19, pp. 261-290.

——— 1990. "Statism, New Institutionalism and Marxism", *Socialist Register 1990*, pp. 147-170.

182

——— 1992. "The New Institutionalism: Predatory Rule, Institutional Persistence and Macro-Social Change", *Economy and Society*, Vol. 21, No. 4, pp. 397-429.

Chan, Wai Kwan, 1991. *The Making of Hong Kong Society: Three Studies of Class Formation in Early Hong Kong*. Oxford: Oxford University Press.

Chau, Laurence L.C., 1979. "Economic Growth and Reduction of Poverty in Hong Kong", *The Philippine Economic Journal*, Vol. 18, No. 4, pp. 570-615.

Chen, Edward K.Y., 1989. "Export-led Economic Development in Chinese Societies: The Existence of Transferability of the NIC Model". In Y.C. Yao, Victor Mok and Lok-Sang Ho (Eds.), *Economic Development in Chinese Societies: Models and Experience*. Hong Kong: Hong Kong University Press.

Cheng, Tun-jen, 1990. "Political Regime and Development Strategies: South Korea and Taiwan". In Gary Gereffi and Donald L. Wyman (Eds.), *Manufacturing Miracles: Paths of Industrialization in Latin America and East Asia*. Princeton: Princeton University Press.

Chiu, Stephen, 1994. *The Politics of Laissez-faire: Hong Kong's Strategy of Industrialization in Historical Perspective*. Hong Kong: Hong Kong Institute of Asia Pacific Studies, Occasional Paper No. 40.

Choi, Alex H., 1993. "The Industrial Transformation of Hong Kong 1945-58: Market or State?" M.A. Thesis, Queen's University, Canada.

——— 1994a. "Beyond Market and State: A Study of Hong Kong's Industrial Transformation", *Studies in Political Economy*, No. 45, pp. 28-64.

——— 1994b. "The Deindustrialization of Hong Kong and the Demise of a NIC". Unpublished paper presented to the Tenth Annual Conference, Canadian Association for the Studies of International Development, Calgary, Canada, 12-14 June.

——— 1996. "Statism and Asian Political Economy — Is There a New Paradigm?" Unpublished manuscript.

Chow, Steven C., and Gustav F. Papanek, 1981. "Laissez-Faire, Growth and Equity: Hong Kong", *The Economic Journal*, Vol. 91 (June), pp. 466-485.

Cooper, Eugene, 1978. "The Politicization of Chinese Craft Organization in Post World War II Hong Kong", *Journal of the Hong Kong Branch of the Royal Asiatic Society*, Vol. 18, pp. 83-100.

Crick, W.F. (Ed.), 1965. *Commonwealth Banking Systems*. Oxford: Clarendon.

183

Curtis, Mark, 1995. *The Ambiguities of Power: British Foreign Policy Since 1945*. London: Zed Books.

Darwin, John, 1988. *Britain and Decolonization: The Retreat from Empire in the Post-War World*. London: Macmillan.

Davies, S.N.G., 1977. "One Brand of Politics Rekindled", *Hong Kong Law Journal*, Vol. 7, No. 1.

Djao, Angela Wei, 1976. "Social Control in a Colonial Society: A Case Study of Working Class Consciousness in Hong Kong". Ph.D. Thesis, University of Toronto.

England, Joe, and John Rear, 1975. *Chinese Labor Under British Rule: A Critical Study of Labor Relations and Law in Hong Kong*. Hong Kong: Oxford University Press.

England, Joe, 1976. *Hong Kong: Britain's Responsibility*. Britain: Fabian Society.

Far Eastern Economic Review (FEER), various issues.

Fransman, Martin, 1982. "Learning and the Capital Goods Sector under Free Trade: The Case of Hong Kong", *World Development*, Vol. 10, No. 11, pp. 991-1014.

Geertz, Clifford, 1963. *Agricultural Involution*. Berkeley: University of California Press.

Glynn, Sean, and Alan Booth, 1996. *Modern Britain: An Economic and Social History*. London: Routledge.

Grantham, Alexander, 1965. *Via Ports: From Hong Kong to Hong Kong*. Hong Kong: Hong Kong University Press.

Habib, Irfan, 1985. "Studying a Colonial Economy — Without Perceiving Colonialism", *Modern Asian Studies*, Vol. 19, No. 3, pp. 335-358.

Hawes, Gary, and Hong Liu, 1993. "Explaining the Dynamics of the Southeast Asian Political Economy: State, Society and the Search for Economic Growth", *World Politics*, Vol. 45, pp. 629-660.

Heaton, William, 1970. "Marxist Revolutionary Strategy and Modern Colonialism: the Cultural Revolution in Hong Kong", *Asian Survey*, Vol. 10, No. 9.

184

Henderson, J., 1991. "Urbanization in the Hong Kong-South China Region: An Introduction to Dynamics and Dilemmas", *International Journal of Urban and Regional Research*, Vol. 15, No. 2, p. 172.

————— 1993. "Against the Economic Orthodoxy: On the Making of the East Asian Miracle", *Economy and Society*, Vol. 22, No. 2, pp. 200-217.

Henley, John S., and Mee-Kau Nyaw, 1985. "A Reappraisal of the Capital Goods Sector in Hong Kong: The Case for Free Trade", *World Development*, Vol. 13, No. 6, pp. 727-748.

Hetherington, R.M., 1963. *Industrial Labour In Hong Kong*. Hong Kong Economic Papers, No. 2.

Hewison, Kevin, 1989. *Power and Politics in Thailand*. Manila: Journal of Contemporary Asia Press.

Hinds, Allister H., 1986. "Sterling and Imperial Policy, 1945-1951", *Journal of Imperial and Commonwealth History*, Vol. 15, No. 1, pp. 150-176.

————— 1991. "Imperial Policy and Colonial Sterling Balances, 1943-56", *Journal of Imperial and Commonwealth History*, Vol. 19, No. 1.

Ho, C.K., 1993. "Industrial Restructuring and the Dynamics of City-State Adjustment", *Environment and Planning A*, Vol. 25, No. 1, pp. 27-46.

Ho, Yin-ping, 1992. *Trade, Industrial Restructuring and Development in Hong Kong*. London: Macmillan.

Hong Kong Research Project, 1974. *Hong Kong: A Case to Answer*. Nottingham: Spokesman Books.

Hsin, Sutu, 1963. "Whither Hong Kong's Industry?" *United College Journal*, Vol. 2.

Hsin, Sutu; Chien-min Chang; and Kin-Yu Cheng, 1977. "A Summary of Industries in Hong Kong with Special Reference to Their Structure", *Journal of the Chinese University of Hong Kong*, Vol. 4, No. 1, pp. 185-205.

Ingham, Geoffrey, 1984. *Capitalism Divided: The City and Industry in British Social Development*. London: Macmillan.

Jao, Y.C., 1974. *Banking and Currency in Hong Kong: A Study of Postwar Financial Development*. London: Macmillan.

185

_____ 1983. "Financing Hong Kong's Early Postwar Industrialization: The Role of the Hongkong and Shanghai Banking Corporation". In Frank H.H. King (Ed.), *Eastern Banking: Essays in the History of the Hongkong and Shanghai Banking Corporation*. London: Athlone Press.

Kim, W.B., 1993. "Industrial Restructuring and Regional Adjustment in Asian NIEs", *Environment and Planning A*, Vol. 25, No. 1, pp. 27-46.

King, Ambrose Yeo-chi, 1975. "Administrative Absorption of Politics in Hong Kong: Emphasis on the Grass Roots Level", *Asian Survey*, Vol. 15, No. 5, pp. 412-439.

Krozewski, Gerold, 1996. "Finance and Empire: The Dilemma Facing Great Britain in the 1950s", *The International History Review*, Vol. 18, No. 2, pp. 48-69.

Krozewski, Gerold, 1993. "Sterling, the 'Minor' Territories, and the End of Formal Empire, 1939-1958", *Economic History Review*, Vol. XLVI, No. 2, pp. 239-265.

Leys, Colin, 1989. *Politics in Britain: From Labourism to Thatcherism*. Revised Edition, London: Verso.

Lin, T.B., 1970. "A Theoretical Assessment of the Currency System of Hong Kong", *The New Asia College Academic Annual*, Vol. 12, pp. 179-194 (in Chinese).

Lui, Tai Lok, 1985. "Industrialization of Hong Kong: All Laissez-Faire?" *Asian Exchange*, Vol. 3, No. 2/3, pp. 33-48

Mackie, J.A.C. (Ed.), 1976. *The Chinese in Indonesia: Five Essays*. Melbourne: Nelson.

Maynard, Geoffrey, 1971. "Sterling and International Monetary Reform". In Paul Streeten and Hugh Corbet (Eds.), *Commonwealth Policy in Global Context*. Toronto: University of Toronto Press.

McVey, Ruth (Ed.), 1992. *Southeast Asian Capitalists*. Ithaca: Southeast Asia Program, Cornell University.

Miners, Norman J., 1975. *The Government and Politics of Hong Kong*. Hong Kong: Oxford University Press.

Mitchell, Timothy, 1991. "The Limits of the State: Beyond Statist Approaches and Their Critics", *American Political Science Review*, Vol. 85, No. 1, pp. 77-96.

186

Mok, Victor, 1979. "Trade Barriers and Export Promotion: The Hong Kong Example". In Tzong-biau Lin, Rance P.L. Lee and Udo-Ernst Simonis (Eds.), *Hong Kong: Economic, Social and Political Studies in Development*. Armonk: M.E. Sharpe.

—— 1981. "Small Factories in Kwun Tong: Problems of Strategies for Development". In Ambrose Y.C. King and Rance P.L. Lee (Eds.), *Social Life and Development in Hong Kong*. Hong Kong: Chinese University Press.

Morkre, Morris E., 1979. "Rent-Seeking and Hong Kong Textile Quota System", *Developing Economies*, Vol. 17, No. 1, pp. 110-118.

Nthenda, Louis, 1979. "Recent Trends in Government and Industry Relationship in Hong Kong". In Tzong-biau Lin, Rance P.L. Lee and Ernst Simonis (Eds.), *Hong Kong: Economic, Social and Political Studies in Development*. Armonk: M.E. Sharpe.

Owen, Nicholas C., 1971a. "Competition and Structural Change in Unconcentrated Industries", *Journal of Industrial Economics*, Vol. 19, No. 2, pp. 133-147.

—— 1971b. "Economic Policy in Hong Kong". In Keith Hopkins (Ed.), *Hong Kong: The Industrial Colony*. Hong Kong: Oxford University Press.

Porter, A.N., and A.J. Stockwell, 1988. *British Imperial Policy and Decolonization, 1938-1964*. Volume II: 1951-1964. London: Macmillan.

Porter, Bernard, 1984. *The Lion's Share: A Short History of British Imperialism, 1850-1983*. London: Longman.

Rabushka, Alvin, 1973. *The Changing Face of Hong Kong: New Departures in Public Policy*. Washington: American Enterprise Institute for Public Policy Research.

Rear, John, 1971. "One Brand of Politics". In Keith Hopkins (Ed.), *Hong Kong: the Industrial Colony*. Hong Kong: Oxford University Press.

Riedel, James, 1974. *The Industrialization of Hong Kong*. Tübingen, Germany: Institut für Weltwirtschaft an der Universität Kiel.

Robison, Richard, 1991. "Culture, Politics and Economy in the Political History of the New Order". In Richard Robison, *Power and Economy in Suharto's Indonesia*. Manila: Journal of Contemporary Asia Press.

187

Rodan, Gary, 1989. *The Political Economy of Singapore's Industrialization*. London: Macmillan.

Schenk, Catherine R., 1994. *Britain and the Sterling Area: From Devaluation to Convertibility in the 1950s*. London: Routledge.

Schiffer, Jonathan R., 1991. "State Policy and Economic Growth: A Note on the Hong Kong Model", *International Journal of Urban and Regional Research*, Vol. 15, No. 2, pp. 180-196.

Scott, Ian, 1989. *Political Change and the Crisis of Legitimacy in Hong Kong*. Hong Kong: Oxford University Press.

Sit, Victor F.S., 1982. "Dynamism in Small Industries: The Case of Hong Kong", *Asian Survey*, Vol. 22, No. 4, pp. 399-409.

Strange, Susan, 1971. *Sterling and British Policy: A Political Study of an International Currency in Decline*. London: Oxford University Press.

Sung, Yun-Wing, 1985. "Economic Growth and Structural Change in the Small Open Economy of Hong Kong". In Vittorio Corbo, Ann O. Krueger and Fernando Ossa (Eds.), *Export-Oriented Development Strategies: The Success of Five Newly Industrializing Countries*. Boulder: Westview.

Tang, James T. H., 1994. "From Empire Defence to Imperial Retreat: Britain's Postwar China Policy and the Decolonization of Hong Kong", *Modern Asian Studies*, Vol. 28, No. 2, pp. 317-337.

Tsang, Steven Y.S., 1988. *Democracy Shelved: Great Britain, China and Attempts at Constitutional Reform in Hong Kong, 1945-1952*. Hong Kong: Oxford University Press.

Wade, Robert, 1990. *Governing the Market: Economic Theory and the Role of Government in East Asian Industrialization*. Princeton: Princeton University Press.

Wong, Siu-lun, 1988. *Emigrant Entrepreneurs: Shanghai Industrialists in Hong Kong*. Hong Kong: Oxford University Press.

Woo, Jung-en, 1991. *Race to the Swift. State and Finance in Korean Industrialization*. Irvington: Columbia University Press.

Worsley, Peter, 1960. "Imperial Retreat". In E.P. Thompson (Ed.), *Out of Apathy*. London: New Left Books, pp. 110-111.

188

Yeh, A.G-O., and M.K. Ng, 1994. "The Changing Role of the State in High Tech Industrial Development: The Experience of Hong Kong", *Environment and Planning C: Government and Policy*, Vol. 12, pp. 449-472.

Zeng Shuji, 1976. *Xianggang yu Zhongguo zhi jian* (Between Hong Kong and China). Hong Kong: Yi Shan Books.

[15]

The Organization of Industrial Relations in Hong Kong: Economic, Political and Sociological Perspectives*

Stephen W.K. Chiu, David A. Levin

Abstract

In this paper we focus on the question of why the organization of industrial relations in post-World War II Hong Kong has remained much less formalized and centralized than in other industrial societies. We assess the merits of three analytical perspectives — transaction cost economics, political conflict and neo-institutional sociology — in accounting for these characteristics. We argue that the economic and political perspectives, despite their limitations, contribute to the understanding of the Hong Kong case but that it is the institutional environment, analyzed from a neo-institutional sociological perspective, which has constituted the overarching framework within which political and economic variables operate. We conclude by suggesting that the contrast between the continuity in the organization of industrial relations in postwar Hong Kong and the reorganization of industrial relations in post-colonial Asian societies shows that political arguments may apply best during periods of major political transformation when significant shifts occur in the power structure.

Descriptors: industrial relations, governance structures, transaction costs, political conflict, neo-institutional sociology, Hong Kong

Introduction

The growing literature on organizational fields and sectors has demonstrated that they can be structured in diverse ways. One major difference is in their governance structures, which can range from the 'more spontaneously equilibrating operation of markets to various types of self-enforcing mechanisms such as alliances or network forms, to externally enforced hierarchies and regulative structures' (Scott 1995: 105). A key issue in the analysis of organizational fields is to identify those historical and contextual conditions under which distinctive governance structures emerge and stabilize.

The theoretical literature bearing on governance structures in organizational fields derives mainly from the analysis of systems for regulating economic transactions at the level of the economy, or for particular industries, or among a set of business firms (Hollingsworth et al. 1994). With a few recent exceptions (Johnson 1994; Seidman 1994), less attention has been given in this literature to another aspect of the organizational structuring of economic life, namely the emergence and stabilization of systems governing industrial relations. With its focus on the diversity of rules and rule-making arrangements governing internal and external labour market transactions, industrial relations offers a rich site for testing the adequacy of different theoretical approaches to the question of why governance structures emerge and take particular forms.

The comparative study of industrial relations on the other hand has contributed substantially to the identification of cross-national diversity in the systems regulating industrial relations (Adams 1995; Bean 1985; Clegg 1976; Poole 1986). Ingham (1974) for example stressed the substantive differences between the highly centralized and regulated systems governing industrial relations in Scandinavian countries and the relatively more decentralized, fragmented and lower degree of formal normative regulation characteristic of the British industrial relations system, a theme which has been further developed in subsequent comparative studies of industrial relations in the European social democracies (Ferner and Hyman 1992; Fulcher 1988; Korpi and Shalev 1979; Swenson 1991). This approach has been less successful, however, in developing theoretical explanations of the origins and persistence of national differences in industrial relations and of national cases where collective bargaining is rare or non-existent (Bean 1985: 8; Fulcher 1991: 18–19; Poole 1986: 30).

We maintain that Hong Kong represents a case of the organization of industrial relations without collective bargaining. From a comparative perspective this is surprising because Hong Kong's postwar system of industrial relations emerged within the broader context of an industrializing competitive market economy and a British colonial political system that granted freedom to workers to organize unions. Given these two conditions, Hong Kong's industrial relations might have been expected to evolve towards a system characterized by a growing volume of procedural and substantive rules governing labour market transactions as an outcome of institutionalized bargaining among organizations representing employees and employers at enterprise, industry and/or national levels (Kerr et al. 1973: 258–277). Since this scenario did not materialize, it poses a problem for analysts of Hong Kong's industrial relations. Why has the organization of industrial relations throughout the post-World War II period remained far less formalized and centralized than in other industrial societies at a comparable stage of economic development?.

In contrast to the more idiographic approach found in industrial relations accounts of the Hong Kong case (England and Rear 1981; Turner et al. 1980), we start from the flourishing theoretical writings on the origins of economic institutions. These writings are diverse, encompassing contributions from economics, political science and sociology and with different strands within each disciplinary approach. We contend, however, that these writings share in common the analytical problem of explaining why rules governing economic transactions emerge, take particular forms and then persist or undergo change but that the examples we use from these disciplinary approaches are distinctive in what each considers to be the key causal variables shaping governance structures. Briefly, the economic perspective views governance structures as the product of choices by economic actors between more and less efficient arrangements in the context of competitive market structures and processes. The political perspective stresses the shifting balance of power among class actors and the state as the key determinant of governance structures. The sociological perspective highlights the significance of cognitive and normative variables in framing the ways in which governance structures come to be designed.

The first section, below, sketches the institutional characteristics of Hong Kong's industrial relations over the past 25 years during which the economy has developed rapidly. Our purpose is to establish empirically that it has continued to remain organized in a decentralized manner and with a low degree of formal regulation of labour market transactions. The second section compares the economic, political and sociological perspectives on the emergence and stabilization of economic governance structures and the third section applies these perspectives to explain the organization of Hong Kong's industrial relations. Our concluding section argues that these three perspectives need not be mutually exclusive. Each contributes to our understanding of the organization of Hong Kong's industrial relations, but at a distinct level of analysis.

Hong Kong's Industrial Relations System

Over two decades ago, an observer of Hong Kong's industrial relations remarked that Hong Kong had become a 'highly industrialized society with an underdeveloped system of industrial relations' (England 1971: 256). This assessment was based on the low density of trade-union membership (less than 20 percent) as well as on movement fragmentation including the prominent division between the pro-Chinese Hong Kong Federation of Trade Unions (FTU) and the pro-Taiwanese Hong Kong and Kowloon Trades Union Council (TUC) whose historic origins were tied to Chinese politics, the near absence of collective bargaining and the minimal role played by the British colonial government in regulating the labour market and employment relations. As a result, unilateral workplace regulation by employers (or occasionally jointly by employers and the government) prevailed.

Despite substantial changes in Hong Kong's economy and polity since the 1970s, the organization of industrial relations has shown considerable continuity in three respects. First, the structure of the trade-union movement has continued to be characterized by a relatively low membership density and fragmentation. The union density rate (declared membership of employees' unions divided by the number of salaried employees and wage earners) was about 16 percent during most of the 1980s. It has risen since the late 1980s due to substantial growth in civil service unionization, but has levelled off in the 1990s at about 21 percent. The fragmentation of the trade-union movement has been even more pronounced in the 1990s than it was in the early 1980s. The number of registered employee trade unions rose from 276 in 1971 to 491 at the end of 1993 (Registrar of Trade Unions 1972, 1994). Moreover, three new peak union federations have been formed since the early 1980s.

Second, collective bargaining continues to be rare as a method for regulating labour market transactions. A recent study (World Bank 1995: 83) attributes the flexibility in Hong Kong's labour market to its decentralized enterprise bargaining, but this is misleading because the 'bargaining' that occurs at the enterprise level is typically in the form of unilateral determination of conditions of employment by the employer or occasional negotiations between individual workers and management. There is little collective representation by trade unions and such negotiations that do occur are rarely governed by formal procedural rules. A substantial proportion of union resources are instead devoted to providing welfare services, some in the form of running businesses such as canteens, restaurants, shopping cooperatives, housing schemes and recreation centres as selective incentives to entice workers to join (England 1989: 138–146).

Third, although the colonial state expanded its regulation of substantive conditions of employment over the last 25 years of its existence (England 1989: 233–234), it abstained from intervening into core elements of the employment relationship. For example, although the Trade Boards Ordinance of 1940 empowered the state to set minimum wages, it never invoked this law in practice, insisting that wage determination should be a matter to be decided between employer and employee. The authority of the employers to hire and fire was also unchallenged by legal statutes and remained a managerial prerogative until the final days of colonial rule when an unfair dismissal ordinance was passed but was then suspended after the July 1997 handover and subsequently repealed. The state also adhered to the 'voluntarist' framework by not introducing statutes compelling employers to recognize trade unions as bargaining partners. Even if collective agreements between employers and unions were concluded, they did not carry any legal significance, but were, in effect, 'gentlemen's agreements'. In short, the legal framework has not been conducive to more formalized or centralized patterns of industrial relations.

Some of the large private-sector business organizations in the relatively sheltered sectors of the economy including utilities, transport and communication have created joint consultation committees but these do not accord unions a formal representational or participatory role (Levin and Chiu 1997). The large firms in these sectors do not have formal bargaining relationships with unions with one notable exception discussed later.

Chinese (mostly Cantonese) employers who predominate in the vast number of medium and small-sized enterprises in manufacturing and the service sector rely on more informal labour relations practices (Levin and Ng 1995: 122–125). As in the large private-sector firms, the individual employees in these enterprises can negotiate with employers over wages and other terms of employment, but the terms of employment are normally decided unilaterally by the employers (England 1989: 145). When the employer unilaterally changes the terms of employment, the only options usually open to workers are to accept the new terms or quit.

The organization of industrial relations thus continues to remain highly decentralized in the private sector in that decisions with regard to pay are made primarily at the enterprise level and are determined by individual agreements between employers and employees. Because there are rarely formal rule-making arrangements that guide decision making on pay and other conditions of employment or for resolving individual grievances, the organization of industrial relations also remains characterized by a low degree of formalization.

The principal exception to this picture is the Hong Kong government which, with about 180,000 employees in 1994 (6.5 percent of the labour force), is Hong Kong's largest employer. The organization of industrial relations in this case involves formal joint consultation bodies at both the central and departmental levels of government and, unlike the private sector, civil service trade unions are accorded a representational and participatory role on these bodies which enables them to exert some influence on changes in their conditions of employment. Why it became exceptional is taken up later.

The Emergence and Stabilization of Governance Structures: Theoretical Perspectives

Within economics, it is the new institutional economics that is centrally concerned with the 'rule and governance systems that develop to regulate or manage economic exchanges' (Scott 1995: 25). Analyses of this burgeoning literature emphasize the theoretical and methodological diversity among its various strands such as the work on property rights, public choice processes, transaction cost economics, evolutionary economics and game theory (Eggertsson 1990; Rutherford 1994; Vromen 1995). We focus here on one of these strands, the transaction cost approach associated with the work of Williamson, since one of his influential contributions concerns the diversity of organizational arrangements for regulating labour market exchanges (Williamson 1975: 57–81; Williamson 1985: 240–272; Dow 1997: 57–63).

A basic assumption of the transaction cost approach is that governance structures regulating economic exchange emerge to reduce uncertainty by establishing a stable structure of transactions. Governance structures persist when they contribute to the reduction of transaction costs in economic exchange. A given set of institutional arrangements always confers some benefits to the society or the agents concerned, and institutional persistence and innovation, whether due to conscious design or the workings of the 'invisible hand' of the market, are attributed to such benefits (Rutherford 1994: 81–128). These benefits reflect the mix of economic incentives and disincentives which in turn derive from competitive market structures and processes.

The transaction cost approach implies that whatever rules exist for regulating labour market transactions, they will be a product of the net economic benefits conferred on the relevant actors. If industrial relations actors have not sought to create more formal rules for regulating their relationships under conditions in which they are free to do so, then it is presumably because, within a given economic and organizational environment, the costs of changing the rules relative to expected benefits would not make the actors better off. In his analysis of the organization of labour markets, for example, Williamson (1985: 245) argues that only under conditions of human asset specificity (i.e. where skills are acquired in a learning-by-doing fashion and are imperfectly transferrable across employers) and non-separable work relations (i.e. where individual productivity is difficult to measure) will employers and workers have an incentive to devise formal governance arrangements to sustain their relationship.

The political approach includes a broad class of arguments centring on the distribution of political power and the influence of political conflicts on the contours of governance structures. We focus specifically on the 'political conflict' perspective, which stresses the distribution of power between capital and labour, as well as their interactions and relationship to the state. One line of analysis in the relevant literature emphasizes the interaction and balance of power between capital and labour and their respective influence on public policy (Ingham 1974; Korpi and Shalev 1979; Stephens and Stephens 1982; Swenson 1991) but recent research also highlights the autonomous role of the state and political power in shaping the governance of industrial relations: managers and trade unions are not necessarily the most important architects of institutions (Skocpol 1980; Zeitlin 1985). In a similar vein, Knight's (1992) 'social conflict' explanation highlights the emergence of institutions as a result of 'distributional' conflicts among rational actors who are motivated by their material interests. One proposition derived from the political approach is that the organization of industrial relations will be determined by the political interests of the state and the relative influence of class actors in shaping public policies. A second proposition is that the organization of industrial relations will change only with significant shifts in the political structure or major threats to the maintenance of the existing power structure.

The sociological perspective we outline here draws on neo-institutional sociology in organizational analysis (DiMaggio and Powell 1991; Biggart 1991; Scott 1995) as well as on recent contributions on the embeddedness of national economic institutions. Neo-institutional sociology views actors as 'subjectively constituting their world views by combining and recombining the symbolic and cognitive elements available to them in their culture but in ways that are also constrained by the availability of these elements' (Campbell 1994: 10). Regarded by a leading proponent as a 'cultural model' of organizations (Dobbin 1994a), neo-institutional sociology argues that the availability of culturally meaningful institutional models in the institutional environment significantly influences how actors view and interpret the world by limiting the range of 'reasonable' possibilities. Actors tend to follow these pre-existing models in their construction of practices to tackle new problems. Instead of strategic behaviours, the neo-institutional perspective emphasizes the cognitive and normative effects of the institutional environment on social behaviour (Fligstein 1990). We consider this perspective to be particularly relevant to the colonial situation since colonizers and colonized can be expected to bring different and conflicting cognitive and normative frames to bear on matters of economic governance. Neo-institutional sociology views the forging of governance structures as a process shaped by the institutional environment through the various mechanisms of isomorphism — coercive, mimetic, and normative — by which dominant institutional models are spread across an organizational field (DiMaggio and Powell 1983). The state (through public policies) and professional communities are often regarded as important agents of this institutional diffusion (Baron et al. 1986; Meyer 1983). Public policies can also shape the goals and interests of firms and trade unions. This suggests that conflict and competition between interest groups and strategic behaviours can be viewed as an intermediate variable between the institutional environment and organizational outcomes (Dobbin 1992). Dobbin argues that public policies and organizational forms are also subject to the influence of the institutional environment (Dobbin 1994a, 1994b).

Recent sociological approaches to explaining cross-national variations in economic governance structures also focus on questions of how broader institutional environments influence the historical development, characteristics and effects of these structures (Hollingsworth and Boyer 1997; Whitley 1992a, 1992b). In theorizing about the institutional environment, Whitley (1992a: 19–36) highlights the effects of both 'proximate' institutions such as state structures and policy frameworks which embody certain moral principles as well as 'background institutions' of family and educational systems that socialize actors into distinctive normative and behaviourial orientations.

A schematic comparison of the differences among the economic, political and sociological perspectives outlined above is shown in Table 1.

Applying the Three Perspectives

The Economic Perspective

The new institutional economics suggests that market structure and competition, through its effects on economic incentives, will shape the contours of the system governing industrial relations. If this structure changes, it will eventually lead to the emergence of new forms of governance as actors seek more efficient organizational arrangements to govern their transactions.

This perspective appears at first glance to be relevant in explaining the emergence of the postwar organization of Hong Kong's industrial relations. Up to the Second World War and until the early 1950s, entrepot trade was

Table 1 Schematic Comparison of Differences Among Explanations of Governance Structures	Issues/ Approaches	Economic	Political	Neo-institutional sociology
	Focus	Economic transaction	Political conflict	Social construction of reality, including actors, interests and contexts
	Problem institutions are designed to solve	Reduce transaction costs	Maintain social and political stability; resolve distributional conflicts	Establishing and applying cognitive frames for structuring social life
	Driving force of change	Competitive pressures	Struggle for power	Institutional isomorphism
	Process of change	Conscious designs (visible hand) and/or natural selection (invisible hand)	Intentional policy innovation	Application of available institutional models to new problems and organizational fields
	Outcome of change	More efficient forms of governance structures	Structure instrumental to maintaining stability and distributing benefits favourable to powerholders	Diffusion and maintenance of shared meanings; legitimization and institutionalization of actions and organizational forms

the mainstay of Hong Kong's economy. This basis for capital accumulation was relatively more conducive to the development of a militant and strong union movement as indicated by the peaks of union activities in the 1920s and the late 1940s. The structural features of the major industries, which mainly served the local community and the trading sector, proved to be congenial to the growth of unions: large firm size, a relatively stable employment relationship, and a skilled and predominantly male labour force. Within this economic structure and propelled by economic deprivations caused by severe price inflation following both world wars, workers discovered that forming unions to pursue higher wages and striking or threatening to strike could be effective strategy for shaping the terms of the employment relationship (England and Rear 1981: 124–125, 185–188). Following the spread of industrial disputes in the years immediately following the Second World War, more institutionalized forms of bargaining appeared to be emerging between major (often British) employers and trade unions in the more sheltered sector of the economy such as the public utility companies (England and Rear 1981: 189–190; Turner et al. 1980: 89–90). From the early 1950s, however, the decline of entrepot trade, the influx of capital and labour from China, and the opening up of western markets triggered off a process of rapid export-led industrialization. This process had major implications for the union movement and subsequent patterns of labour relations. First, previous union strongholds in industries such as ship-building and repairing declined. Second, the emergence of export-oriented light industries created a 'preponderance of industrial jobs in the light-industry export sectors characterized by low-skill, minimal advancement opportunities, and job insecurity, along with a workforce dominated by young women who do not anticipate long-term industrial employment', thus setting in motion a process of structural demobilization that 'militate[s] against effective class politics or organization' (Deyo et al. 1987: 51; Deyo 1987, 1989).

The expanding manufacturing sector facilitated inter-firm mobility among semi-skilled workers due to the spatial density of firms, the relative ease of transferring skills across firms and industries and the use of piece-rate pay systems (Levin 1979). The path to economic advancement for these semi-skilled workers was mainly through the external labour market. Despite substantial economic inequality and exploitation, a high rate of economic growth combined with a tightening labour market generated a steady improvement in workers' standard of living by providing jobs and real wage growth. The rapid development of an export-dependent, labour-intensive manufacturing sector along with a gender stratified labour force undermined unionization and the development of formal collective bargaining. Employers in the labour-intensive export manufacturing sector also had no incentive to recognize trade unions or bargain collectively with workers. An individualized mode of bargaining with employees allowed them to tap into an active and competitive external labour market and adjust their labour force according to market conditions. Large firms preferred to develop a quasi-internal labour market and, in a small number of cases, rudimentary forms of joint consultation with employees. This postwar shift to a deregulated system governing industrial relations can thus be viewed as a product of the postwar economic environment which reduced the economic incentives for creating more formalized and centralized mechanisms for regulating labour market transactions because of the low level of 'human asset specificity' (Williamson 1985: 256).

This economic explanation is incomplete, however, in three respects. First, while it successfully predicts the low level of centralization and formalization in the competitive export manufacturing sectors, it is unable to explain why collective bargaining remained equally underdeveloped in the sheltered sectors where employment stability is higher, workers are more skilled and where even trade unions have been stronger (England and Rear 1981: 189–190). If skill-specificity was low in the multitude of small and medium-sized export factories in Hong Kong, the same cannot be said for the major private employers in the relatively more sheltered sectors such as public utilities and transport. The early postwar momentum towards collective wage bargaining in the utilities, for example, slacked in the early 1950s as negotiations over terms of employment became informal and sporadic. This shift predated the explosive development of small-scale export manufacturing in the late 1950s.

Second, the new institutional economics focuses on the structure of transactions mainly at the firm level in order to deduce the most efficient governance structure. This focus largely brackets the effect of public policy on firm-level governance structures as well as those in the organizational field. The state's role in Hong Kong appears to fit economists' prescription for a minimal regulatory framework for employment relations, hence its reputation for *laissez-faire* policies (Chiu 1996). Nevertheless, this begs an explanation of why public policy has been non-interventionist in Hong Kong while public regulation of employment has been more widespread in other industrial or newly industrializing economies (World Bank 1995: 70–71). For example, Hong Kong has continued to lack minimum wage legislation or legal stipulation of rights to collective bargaining.

Third, a prominent feature of the Hong Kong trade-union movement is its political orientation. The economic perspective takes this as a given, and fashions an explanation on this basis (Y. Wong 1988). This, however, begs the question of why Hong Kong's trade unions have been primarily interested in realizing political objectives rather than seeking economic benefits for their members by establishing formal rule-making arrangements with employers. One reason why collective bargaining never spread in Hong Kong is clearly the unions' lack of interest in it. The fact that even trade unions operating in favourable circumstances — highly unionized, sheltered industries — seldom attempted to push for formal collective bargaining suggests that factors other than utilitarian economic calculations are at work. An examination of the political and institutional context shaping actors' objectives and strategies is thus necessary to explain what is taken for granted under proximate economic explanations.

The Political Perspective

The political perspective directs us to consider firstly the nature of British colonial rule in Hong Kong. To maintain the stability of the colonial system, the colonial state refrained historically from intruding into the indigenous society by adopting non-interventionist policies in the social and economic spheres, thus creating a 'minimally integrated socio-political system' (Lau 1982: 17–21). The state did not, however, govern in isolation from all social classes. From the early twentieth century onwards, a process of 'elite settlement' set in when the local business class began to develop a close partnership with the colonial state (Chiu 1996). When manufacturing was about to spurt in the 1950s, the state's coalition with the dominant financial–commercial bourgeoisie was supported by a dense institutional network, inducing a coalescence of interests between the two parties. One consequence of this alliance was the state's support for capital accumulation in general. The peripheral position of the industrial bourgeoisie in the state–capital alliance allowed the colonial state to resist their demands for a more active and selective industrial policy to aid industrial growth.

It is within this broader political context that the British voluntarist tradition towards the governance of industrial relations became entrenched in Hong Kong. The dominance in postwar colonial politics of the capitalist class, comprised of leaders of the major trading houses, public utilities, banks and a small number of large-scale Shanghainese and Cantonese industrialists with privileged access to the state, meant that the state had little incentive to challenge employers' managerial prerogatives at the workplace. This class also supported the state's non-interventionist stance towards the employment relationship because it gave them considerable freedom to set contractual terms with individual employees. The maintenance of the authoritarian colonial political system until the mid-1980s marginalized the labour movement's influence in the political arena apart from its occasional and largely ineffective disruption tactics. A political explanation of the origin of Hong Kong's industrial relations system therefore highlights the conflictual interests between labour and capital, but their contrasting positions in the political structure. The 'collusion' between state and capital engendered a voluntarist institutional framework for industrial relations that effectively gave employers power over the labour process and employment relations by leaving the employment contract to bargaining between individual employers and workers.

Prior to the onset of political reforms in the 1980s, only the overt threats to social stability in 1966–67 induced the colonial state to deviate from its rhetorical and policy commitment to voluntarism. These disturbances posed a major crisis of legitimacy for the colonial regime (Scott 1989: 81–126). The colonial government responded by improving the economic and social conditions of the working class in order to build a new basis of legitimacy for its rule.

The merit of this political perspective is that it explains how political relations shaped the state's regulatory framework for employment matters by linking it to the interests of the dominant actors — the colonial state elite and the business elite — in the polity. In this respect, Hong Kong business elites were acting in a politically rational manner (cf. Adams 1994: 50). Given the political exclusion of labour and trade-union ineffectiveness in challenging this arrangement, the dominant actors were capable of projecting their interests upon the industrial relations system.

The political perspective is not without its shortcomings, however, one of which is that it predicts the adoption by the union movement of a political strategy aimed at the state and its policies. There are two reasons for this. First, the postwar economic changes created a situation that would logically favour a polity-oriented strategy by the labour movement. As the size of the working class grew, especially that of the semi-skilled group of workers, political action would perhaps have been the most effective route for the labour movement to advance the welfare of the working class (Sturmthal 1967: 59). Political action would also have been a rational strategy to compensate for the unions' weakness at the workplace level. Second, the failure of the colonial state elites to grant political citizenship rights to the working class or to broaden secondary industrial citizenship rights should have encouraged the union movement to pursue political and economic goals through a political strategy of action (Lipset 1983).

In one respect, the movement's strategy did become political, although not as the term is normally defined in industrial relations (Levin and Chiu 1994). The behaviour of both the FTU and the TUC had strong political overtones in the 1950s and 1960s due to the prominence of their allegiances to the two Chinese political parties, their participation in the propaganda war between the Taiwanese and Mainland Chinese governments, and their involvement in Left–Right clashes in Hong Kong, particularly during the 1950s (Leung 1986a, 1986b). Such behaviour was political primarily on a symbolic level, however, because it did not entail taking action against the colonial authorities. Except for participation in mass disturbances in 1967, influenced by the Cultural Revolution then underway in China, and despite its radical rhetoric, the FTU rarely challenged the existing power structure in Hong Kong, demanded an end to colonial rule, or pressured the state for improvements in labour legislation and redistribution of resources to the working class. For most of the 1950s and 1960s, the politics of the TUC and the FTU were of an 'extrovert' nature, focusing on competition between the two Chinese governments rather than using their resources to advance workers' interests through domestic political action. This pecularity of the Hong Kong trade-union movement is a critical lacunae in political explanations.

The second shortcoming of this political perspective lies in its explanation for the *laissez-faire* policies of the colonial state towards industrial relations. It follows an interest-based logic in explaining these policies as being consistent with employers' interest in maintaining sole control over their workforce and the interest of state elites in minimizing intervention into the economy and the civil society. Yet it can be argued that more state intervention in industrial relations would also have been in their interests. Adams (1994: 51) argues that state suppression of unions and collective bargaining is 'to be expected where the following factors are operative: (a) a low level of economic development, (b) a totalitarian government in power, (c) a radically market-oriented government in power, (d) a government in power which seriously fears rebellion, and (e) a country in which the business elite has a great deal of influence on government policy'. Since all these factors were present in Hong Kong in the late 1940s and 1950s and since other newly industrializing Asian countries such as Malaysia, Singapore and South Korea had adopted restrictive policies towards trade unions and industrial relations when embarking on their industrial takeoff, one might have expected the colonial state to exert much tighter control over industrial relations and trade-union activities than it did (Deyo 1989; Sharma 1985; Siddique 1989). Another reason for expecting government suppression of the union movement was the threat to public order posed by the union movement's involvement in Chinese political struggles. The political schism between the pro-Chinese Communist Party and pro-Kuomintang wings of the union movement could easily get out of hand and endanger social stability, as happened on several occasions in the late 1940s and 1950s (Leung and Chiu 1991; Tsang 1988: 81–91, 124–137).

In the late 1940s, the colonial government was clearly concerned about the potential for communist-inspired politically motivated strikes. It was for this reason that the Hong Kong Governor sought permission from the

Colonial Office to revive the Illegal Strikes and Lockouts Ordinance of 1927, which had been repealed by the Trade Unions and Trade Disputes Ordinance of 1948 (Colonial Office 37/3729). Yet direct suppression of the union movement was not seriously considered. An early postwar memo from a Hong Kong Labour Officer to the Colonial Office, while expressing concern about the influence of Chinese politics on union policies and practices, considered it to be the duty of the Labour Office (subsequently renamed the Labour Department) to 'give all the advice and assistance [to the unions] that lies in its power but to refrain, as far as possible, from imposing any direct measures of control' (Colonial Office 537/2188: 16).

The framework created by the Hong Kong government in the late 1940s for regulating trade unions was notable for its liberal character. It was premised largely on helping unions to rationalize their internal operations (e.g. finances and elections) rather than imposing tight controls over their activities and political orientation. The sporadic repression of union political activities did not amount to a wholesale suppression of them *as unions,* but was more in the form of patchwork remedial measures directed at (from the state's point of view) their excesses. For example, while the Registrar of Trade Unions had the power to de-register unions under the trade unions ordinance, the Registrar did so only in the case of unions with chaotic administration and financial scandals. No politically oriented unions on the left and the right were ever de-registered in the postwar period. The Labour Department chose not to 'pursue an active policy to combat political infiltration of the labour unions by the two Chinese parties. It adopted instead a policy which is best described as "positive non-intervention"' (Tsang 1988: 87–88). The Trade Unions and Trade Disputes Ordinance facilitated the formation of new unions, thus contributing to union movement fragmentation (Turner et al. 1980: 107). The policy of allowing multiple unions to form in the same trade, industry and enterprise appears at first glance to be a rational strategy for reducing the potential for a united opposition from organized labour to colonial rule. On the other hand, it can also be argued that this policy lacked political rationality because, as noted above, allowing ideologically divided union groupings to operate created the basis for intense inter-union struggles which posed major challenges to the maintenance of social stability and public order especially in 1956 and 1967 (England and Rear 1981: 17–21; Leung 1996: 143–146). This anomaly can be explained by focusing on the institutional origins of the regulatory framework for industrial relations.

Neo-Institutional Sociology

Institutional Isomorphism and the Regulatory Framework

Making sense of British colonial policies *vis-à-vis* the field of industrial relations entails starting with the ways in which British political culture framed conceptions of economic life. By the 19th century, British political culture was organized around the principle of the sovereignty of individuals. The principles guiding economic order had come to resemble this political culture by according economic sovereignty to masses of individual economic actors in the belief that this would promote general economic welfare (Dobbin 1994b: 164). British conceptions of the relations between employers and their workers in the marketplace closely paralleled this dominant economic principle. The precepts of economic individualism, with the individual as the best judge of his or her own interests rather than any other outside authority (i.e. the state), underpinned the system of industrial relations as it did other spheres of economic life (Fox 1985: 6). Collective organization (i.e. trade unions and employer associations) and collective bargaining came to be viewed as an offshoot of the bargaining between sovereign individuals in the market. British individualism also dovetailed with the common-law tradition in protecting the freedom of association. As long as a group of consenting individuals form an organization and that organization represents the interests of their constituency, it should be protected by law.

The Hong Kong colonial government's approach to labour relations was strongly influenced by these British conceptions of voluntarism in the making of employment contracts (England 1989: 159–160). The common law tradition is a major source of labour laws in Hong Kong so that English legal precedents on contracts of employment have been generally applicable to Hong Kong. The legal framework governing trade unions was similarly influenced by British conceptions (Roberts 1964: xvi).

The Trade Unions and Trade Disputes Ordinance of 1948, based on the British Trade Disputes Act of 1906 and a prewar model trade unions ordinance circulated among British colonial territories (England and Rear 1981: 136), reflected a number of principles of the voluntarist ideology of labour relations: (1) relative freedom for employees and employers to form trade unions and to bargain collectively, but also freedom for employers not to recognize and bargain with unions of employees; (2) exemption of trade unions, provided they are registered, from the law of criminal conspiracy for actions taken in contemplation or furtherance of a trade dispute; and (3), lack of enforceability in the courts of collective agreements reached between unions and employers. Also adopted was the British reliance on voluntary conciliation to resolve industrial disputes (Brown 1960: 177–190; England and Rear 1981: 319; Fox 1985: 131–134).

The establishment of the Labour Department in 1947 was influenced by a combination of prewar Colonial Office dispatches on the expansion of labour departments in dependent territories and recommendations made by special labour advisers from Britain (Colonial Office 129/615). Its organization closely mirrored that of other British territories; so did that of the postwar reactivated Labour Advisory Board. British influence is also evident in the ways the colonial government constructed labour relations for its own employees. It was an accepted principle of British colonial administration that since 'Colonial Governments themselves are often the largest employers of labour in their territories, a special responsibility is placed on them and they are frequently obliged to initiate and lead wage movements'

(Central Office of Information 1956: 12). The Colonial Office encouraged the opening of direct, formal channels of communication with civil servants where their trade unions were relatively mature (Roberts 1964: 373). In the late 1950s, the Hong Kong government made a formal commitment to consult with two major staff associations of civil servants over changes in conditions of service, although it delayed establishing a formal consultative body until 1968. The Hong Kong government's practice of linking civil-service to private-sector pay trends and the use of commissions of inquiry to resolve occasional major controversies between the government and civil-service unions were also standard practices in the British Commonwealth (Roberts 1964: 374).

The 'voluntarist' framework of governance for private-sector industrial relations as well as the development of more formalized labour relations practices in the civil service thus resulted from the diffusion of British conceptions and institutional models to Hong Kong. They reflect the process of coercive isomorphism resulting from 'both formal and informal pressures exerted on organizations by other organizations upon which they are dependent and by cultural expectations in the society within which organizations function' (DiMaggio and Powell 1983: 150). The colonial link between Hong Kong and Britain led to the direct and indirect imposition of standard operating procedures and legitimated rules and structures based on British ideals and conceptions of the role of the state in the economy.

There were limits to this coercive isomorphism, however. For example, advice from the Colonial Office in 1950 that Hong Kong should establish formal joint negotiating machinery and set statutory minimum wages was not accepted by the Hong Kong Governor, who argued that neither measure was workable, given the educational gap between employers and workers, the Chinese workers' preference for unofficial and personal methods of resolving disputes, the rivalries among unions and their leaders, the fluctuating conditions of the labour market and the unsettled political situation at the time (Colonial Office 537/6032). This suggests that the principles of the voluntarist approach had become sufficiently institutionalized and valued by local colonial officials to propel them to defend the existing system from proposals that would compromise these principles. The Colonial Office accepted this rationalization for existing voluntarist policies and practices, since it did not pursue the matter further. This episode illustrates that once a governance framework is institutionalized, it becomes 'meaningful' to the actors concerned, and any subsequent changes will have to 'make sense' in terms of this legitimated and rationalized framework.

The 'voluntarist' framework for governing industrial relations did not entail the complete absence of legal regulation of employment contracts. Legal restrictions applying to the employment of women and young persons were first enacted in the 1920s and 1930s on the instructions of the Secretary of State for the Colonies. In the postwar period, legislative 'qualifications' of freedom of contract were expanded and eventually crystallized in a unified labour code in the form of the Employment Ordinance of 1968, with industrial safety and health regulated in separate legislation. While the early legislative interventions dealt mainly with more 'vulnerable' categories of workers — children and women — their scope was gradually expanded to provide a statutory floor of standards of employment covering working hours, holidays and health and safety (England 1989: 166–215). The political approach to the origins of systems for the governance of industrial relations would explain the expansion of regulation by the need for state elites to shore up social and political stability in the face of potential and actual mass discontent and unrest among the working class. This argument is not entirely convincing, however, because some labour legislation preceded the mass labour unrest of 1967 and many of the major industrial conflicts before 1967 were not directed at either the colonial state, its labour legislation or substantive terms of employment.

This co-existence of voluntarism and restrictive/protective labour legislation is also linked to the diffusion of British traditions of public policy to the colonies. Paternalistic government action, designed to protect the more vulnerable against the consequences of an unregulated market economy, was a strong, although subterranean, current running from the early Victorian factory acts to Disraeli's social reforms (Roberts 1979). The diffusion of British institutional models to Hong Kong did not, however, guarantee the reproduction of the whole edifice of British employment and industrial relations (England and Rear 1981: 204–205). Our main point, rather, is that much of the formal institutionalized regulation of employment can be traced to conscious and unconscious modelling upon the British conceptions of free labour markets, voluntarist collective bargaining and remedial protection of workers.

Orientations of Employers and Workers
The diffusion of British conceptions of voluntarism helps to account for the decentralized system of industrial relations in Hong Kong. Since the law does not positively prescribe for the recognition of unions, collective bargaining or enterprise joint consultation, Hong Kong employers are free to negotiate with *individual* workers at their respective workplaces. Insofar as Hong Kong employers see 'labor as a commodity to be hired at the cheapest price and laid off or dismissed as the market dictates' (Nishida 1992: 189), this mirrors the British invention of labour as a fictitious commodity since the repudiation of paternalism in the pre-industrial era (Fox 1985: 48–54; Hobsbawm 1968: 344–345; Polanyi 1957).

This 'loose' regulatory framework also enabled the emergence of diverse employment practices within the labour market. Given Hong Kong's open economy and the 'permissiveness' of the regulatory framework, employers of varied national origins became agents of diffusion of somewhat different institutional models of employment practices with respect, for example, to payment and training systems (England and Rear 1981: 69–100: Nishida 1992; Tomita 1985). Most employers, however, combined features of authoritarianism and benevolent paternalism in labour management and resisted sharing decision making about employment conditions with unions (England and Rear 1975: 273–274; Redding 1990: 127–135; S. Wong 1988:

137). The one exceptional case in the private sector of a collective agreement that covers substantive and procedural matters involves Cable and Wireless Limited, a public enterprise incorporated in the United Kingdom. In this case, it was the top management, then comprised of British expatriates who were familiar with formal labour relations from their home country experience, who not only encouraged local staff to form a union in 1970 but also initiated the proposal for having a collective agreement with the union (Leung 1983: 125–126). This exception illustrates that the critical factor was not the market that determined the outcome, but rather a foreign management's institutional conceptions of how relations should be structured.

The underdeveloped legal and institutional framework could not have been sustained without the acquiescence of the expanding industrial working class. A key issue in this respect concerns the nature of worker orientations to the employment relationship. One line of argument is that 'neo-Confucian' values, claimed to be reproduced through the family and educational institutions in the newly industrialized East Asian economies, predisposed workers, both male and female, to comply with managerial authority and avoid confrontation (Redding 1990: 47–48; Tai 1989: 19–20; Vogel 1991: 98–100). Another cultural explanation, one tailored more specifically to Hong Kong, stresses the impact of the 'refugee mentality' on workers' industrial attitudes and behaviour. A substantial portion of the rapidly growing post-World War II population was comprised of refugees fleeing from the political turmoil and changes in China (Podmore 1971: 25–26). As first generation migrants, their primary interest was in maximizing their short-term income (cf. Piore 1979: 108–114). Encountering a *laissez faire* economy without a state-sponsored safety net, they also had to depend to a large extent on the familial group for economic survival.

Salaff (1981: 8) describes the Hong Kong Chinese family system as a centripetal type which 'gathers in its forces by demanding the primary loyalty of its members and mobilizing their labor power, political and psychological allegiances on behalf of kinsmen'. A consequence was an ethos of 'utilitarianistic familism' that gave precedence to familial interests over societal or other group interests and produced a relatively low level of social involvement with co-workers and neighbours (Lau 1982: 72; Mitchell 1972a: 434–435, 1972b: 319–320).

The primacy of family loyalties, short time-horizons and social detachment from co-workers militated against social investment in organizational efforts aimed at changing the rules of the labour market. A low level of 'social capital', regarded as a key factor underpinning political apathy, can also account for the low level of unionization and collective action among workers (Mitchell 1998). This type of orientation prompted workers to seek alternative opportunities for improving their economic status in the dynamic postwar economy. If employers did not meet their substantive expectations, they could exercise the exit option individually by changing jobs. The effectiveness of this option was enhanced as establishments proliferated in the growing sectors of manufacturing such as garments, electronics and plastics. High labour mobility in manufacturing thus became an alternative to collective organization and negotiation for improving their economic status (England and Rear 1975: 44).

The colonial system also dampened the development of working-class interest in a more formal representational and participatory role for labour. The absence of a democratic ideology or democratic political institutions meant there was no model on which to base the demand for an extension of democratic practices to industrial relations. An apolitical orientation was also fostered through the educational system. In line with practices in Britain and in its colonies, the postwar Hong Kong colonial educational system was geared to maintaining an apolitical image of the school curriculum. Concerned about the growing influence of pro-communist educational efforts in the colony, the colonial government established a Special Bureau within the Education Department in 1949 to counter those efforts. Other forms of control were also imposed on textbooks, syllabuses and the registration of schools to prevent communist infiltration and the politicization of education (Sweeting 1993: 53, 197–201). Civic education, created as a school subject in 1953, emphasized proper conduct and the responsibilities of a good citizen rather than civic rights, participation and democracy (Leung 1997; Morris and Chan 1997: 249). Morever, opportunities for students to participate in school governance were minimal (Tse 1997: 34). As a result, Hong Kong's postwar younger generation was socialized through schooling into a culture of political alienation (Leung 1997). This was hardly conducive to the development of conceptions of industrial democracy.

The impact of the Chinese family, the refugee mentality and the colonial educational system in limiting the potential for collective challenges to managerial authority should not be exaggerated, however. Strong anti-colonial and anti-foreign sentiments were present among a sizable minority during the postwar years. These sentiments sustained political unionism and, on occasion, could be activated to challenge employers' and even the state's authority. Nevertheless, the combination of the legal and regulatory institutions of voluntarism and a cultural aversion to collective action, derived from the influence of background institutions of family and the education system, fostered the dominance of individualist and instrumental labour market strategies (Chiu 1992). The political character of the main union movement during most of the postwar period further reinforced the use of these strategies. This type of unionism was not regarded as useful by most workers, who preferred either to avoid involvement with politically oriented associations or else were strongly alienated from the pro-communist unions in particular (Lau 1982: 12). It is necessary, therefore, to turn to this institutional construction of political unionism.

Union Orientations

The colonial state's permissive and non-interventionist approach to labour relations has both reinforced, and been reinforced by, the orientation and character of the Hong Kong trade-union movement. During the early post-World War Two period 'a pattern of British-style collective bargaining seemed to lie ahead' as major unions presented the larger firms with concerted wage demands (England 1989: 145). It is the subsequent retreat of trade unions from collective bargaining that needs to be explained.

The main reason why collective bargaining never became a major objective of Hong Kong trade unions can be traced to the institutional environment in which they first emerged. Hong Kong trade unions shared with their Chinese counterparts the same indigenous sources of institutional models: triad societies, craft guilds and political parties (Deng 1949; Zhongguo Laogong Yundongshi Xubian Biancuan Weiyuanhui 1984). The influence of these traditional organizational forms and practices on modern labour organizations in China and Hong Kong had two repercussions. First, it contributed to the fragmentation of the labour movement. Second, guild origins imbued Hong Kong's trade unions with a primary emphasis on mutual aid functions (Chan 1975: 180, 265; England and Rear 1975: 85).

In addition to their welfare orientation, the unions' involvement as principal agents in the political rivalry between the Kuomintang and Chinese Communist Party also deflected their resources and commitments away from developing a strong workplace presence or from seeking formal relationships with employers (Levin and Ng 1993: 42–43). This cannot be explained in terms of market processes and economic incentives and disincentives. While the economic approach predicts that weak unions will turn to politics in order to reshape industrial relations governance structures, the type of political activities in which Hong Kong's trade unions engaged did not historically conform at all to this outcome.

The peculiar nature of the political orientation of Hong Kong's trade-union movement can be traced to the institutional environment during its formative period in the first two decades of this century (Chesneaux 1968). While the link between the guild origin of Hong Kong trade unions can be described as mimetic, the impact of Chinese political parties on unions can be viewed as a variant of coercive isomorphism. The political environment when unions emerged in China in the early decades of the twentieth century was replete with revolutionary and nationalistic movements. Within this environment, the dominant goals and objectives of nascent secondary associations became infused with nationalism. Whatever their other pursuits, the promotion of China's national interest became the universal principle legitimizing their existence. Trade unions were no exception. Nationalism soon came to be identified with support for one of the two dominant political parties in China. Influenced by both the guild model and the Chinese political parties, the basic structure and orientation of Hong Kong's labour movement took shape by the 1920s:

'. . . a multiplicity of associations in the crafts and traditional industries of the Colony each heavily influenced by guild traditions of exclusiveness combined with mutual aid for members; and in addition a group of large unions organized on an industrial basis — notably amongst seamen, dockers, tramway workers and printers — motivated by political ideals of nationalism and anti-imperialism.' (England and Rear 1975: 80)

As the civil war in China intensified after the Second World War, this basic structure of trade unionism was more or less reproduced intact in Hong Kong.

This politicized model of organizational development has impeded the union movement's effectiveness in advancing workers' employment interests (cf. Sabel 1981: 236–237). The political character of the unions also gave both the government and the employers grounds for resisting formal bargaining relations with the unions. The other 'pillar' of voluntarism in Hong Kong has thus been the indifference of the FTU and the TUC towards formal collective bargaining and the absence of sustained and coordinated pressures by the trade-union movement on the colonial state for legislative changes and political reforms beneficial to labour. Only by taking into consideration the influence of the institutional environment during the formative period of the Hong Kong union movement can we make sense of its peculiar political orientation.

Conclusions

We have sought to demonstrate and assess the relevance of economic, political and sociological approaches to the origins of economic governance structures for explaining why the organization of Hong Kong's industrial relations did not develop the degree of formalization or centralization found in other industrialized market economies. The economic perspective would explain this outcome in terms of the smooth operation of the largely unregulated competitive labour market and workers' ability to capitalize on the market to maximize their individual welfare in the context of an expanding economy. This perspective takes for granted, however, what needs to be explained, namely why the labour market operates in such a competitive manner. It also fails to explain why collective bargaining did not develop, even in circumstances that elsewhere are considered favourable.

The political explanation would account for the emergence of an unregulated labour market by focusing on how, within an authoritarian political system that excluded representation of labour, the interests of the colonial state and the dominant financial–commercial business elite meshed to support the British voluntarist tradition as a policy and strategy for the governance of Hong Kong's industrial relations. However, this perspective is unable to account for the trade-union movement's failure to pursue working class interests by attempting to influence public policies through extra-political channels. A political approach would also predict that the state (as well as employers) would exercise much tighter control over industrial relations, including suppression of trade-union activities, than it actually did.

In our view, neo-institutional sociology is best able to solve these puzzles generated by the application of the economic and political perspectives. The diffusion of British conceptions and institutional models for the governance of industrial relations explains the 'positive non-interventionism' or 'permissiveness' characteristic of the Hong Kong colonial government's approach to industrial relations. This approach was premised on the belief that workers and employers, as autonomous individuals, should be left to solve their problems individually or collectively as they preferred, and normally without interference by the state, while at the same time the state had a duty to protect the more vulnerable employees from threats to their health and safety at the workplace and from excessive exploitation. The permissiveness of this governance framework created space for the emergence of somewhat diverse employment practices at the enterprise level.

The absence of state regulation of wages and protection against unfair dismissal, as well as the lack of unemployment benefits, helps to explain why workers became highly sensitive to labour market incentives and developed individual strategies for taking advantage of market opportunities. Furthermore, within the voluntarist framework and buttressed by the influence of the family and educational systems, employees adopted an individualist and instrumental orientation to work and the labour market, while employers have sought to maximize their control of the labour process and maintain flexibility. The sociological neo-institutional perspective can also explain why the Hong Kong union movement did not act in ways that might be expected from the rational choice approach to the study of trade unionism (Crouch 1982). The effects of the institutional environment at the time that modern unions emerged left an enduring imprint on their objectives and strategies that diverted them from seeking more formalized and centralized rule-making arrangements. This fits with the argument that an organization's character is initially shaped by the historical conditions at the time of its founding, and that, once formed, this character is likely to continue to exert a profound influence on subsequent behaviour (Stinchcombe 1969: 168–169).

The institutional environment thus constituted the overarching framework within which the distinctive nature of Hong Kong's industrial relations system developed. Our analysis testifies to the impact of the colonial political economy on Hong Kong's labour relations, but also that politics and economics have operated within the broader, earlier established institutional framework. Sometimes, they were inconsistent with the institutionalized rules and occasionally that led to modifications of the framework governing industrial relations. In other instances, political power and economic interests reinforced the existing institutional framework. For example, the power structure and political imperatives under British colonialism were consistent with a largely arms-length approach to economic and social affairs. *Laissez-faire* was a rational strategy of rule given Hong Kong's governing coalition and power structure. However, the 'crisis of legitimacy' facing colonial rule in the late 1960s led to a departure from the voluntarist tradition as the colonial state expanded legal protection and improved labour standards and benefits for workers. Hong Kong's pattern of industrial development also generated an industrial and enterprise structure that proved inimical to unionization and collective bargaining. While the *practice* of collective bargaining has not been institutionalized, Hong Kong's decentralized and competitive labour market is entirely consistent with the *formal structure* of employment relations which embodied the 'institutional myth' of freedom of contract and the view of labour as a commodity.

In this respect, the three perspectives on the explanation of governance structures are not mutually exclusive, but contribute to our understanding of the Hong Kong case at a distinct level of analysis. The sociological neo-institutional perspective highlights the social constitution of actors and their interests and the dominant policy framework. The political perspective then pitches its explanation at the intermediate level by showing how political struggles link the different actors and the broader policy framework to specific institutional outcomes. The economic perspective examines actions and calculations at the proximate level, i.e., how certain forms of transactions become 'rational' and 'efficient' given the political and institutional contexts.

Due to the postwar continuity of British colonial rule over Hong Kong, coupled with the union movement's acceptance of the industrial relations governance framework, Hong Kong's industrial relations were not restructured in ways comparable to that of other post-colonial states in the region, such as those in Malaysia, Singapore and South Korea, which outlawed left-wing unions, imposed stricter controls over union activities and finances, or sought to politically demobilize labour by restructuring the union movement (Arudsothy and Littler 1993: 113–114; Ayadurai 1993: 63–65; Choi 1989: 28–29; Deery and Mitchell 1993: 9–12; Deyo 1989: 106–130; Kim 1993: 137). The divergence between Hong Kong and Singapore in the organization of their industrial relations is especially striking given the similarities in terms of the legacy of British colonial rule and the Chinese ethnicity of the majority of their populations. The People's Action Party (PAP), on taking power in Singapore in 1959, amended the existing trade-union law to prevent the development of splinter unions and multi-unionism, supported a pro-government union federation while refusing to register an oppositional federation, passed new legislation in 1960 and again in the late 1960s that restricted unions' freedom to bargain collectively, and, in 1982, amended the trade-union law once again by redefining unions as having objectives of consensus and productivity improvement instead of the 'confrontational objectives of the original ordinance' (Leggett 1993: 223–225). The union movement was thus gradually depoliticized through state action, and unions became 'absorbed into promoting national goals' (Thurley 1983: 224).

The contrast between the re-organization of industrial relations in post-colonial Singapore with the continuity of the framework governing industrial relations in Hong Kong suggests that political arguments may apply best during a period of major political transformation, when there are significant shifts in the power structure. Under such conditions, the pre-existing institutional assumptions may be called into question and the intensification of conflicts among political actors may then allow new models of industrial relations to gain ascendancy (cf. DiMaggio 1988).

This raises the question of what will happen to the governance of industrial relations following the end of British colonial rule and Hong Kong's return to Chinese sovereignty on 1 July 1997. The severing of Hong Kong's colonial link to Britain may result in a new process of institutional isomorphism. China's modelling of industrial relations has been more paternalistic than the British approach and relies less on the institutional myth of voluntarism and freedom of contract. The Chinese government also has a much lower level of tolerance for an autonomous and militant union movement. A possible scenario is that the organization of industrial relations will gradually become more centrally and formally regulated from the top down after 1997 in ways that will bring the governance of Hong Kong's industrial relations closer to that of other post-colonial societies in the region.

Note

* We wish to thank Ronald Calori, Stewart Clegg, Frank Dobbin, Stephen Frenkel and the two anonymous reviewers for their comments on earlier versions of this paper. We gratefully acknowledge support from the Hong Kong Research Grants Council for an earmarked research grant for the project: 'Hong Kong Trade Unions Bracing for the Future: Environmental Changes and Organizational Dynamics' (HKU 391/96H).

References

Adams, Roy
1994 'State regulation of unions and collective bargaining: An international assessment of determinants and consequences' in *The future of industrial relations: Global change and challenges.* J. Niland, R. Lansbury and C. Veveris (eds.), 41–62. Thousand Oaks, CA: Sage.

Adams, Roy
1995 *Industrial relations under liberal democracy: North America in comparative perspective.* Columbia, SC: University of South Carolina Press.

Arudsothy, Ponniah, and Craig R. Littler
1993 'State regulation and union fragmentation in Malaysia' in *Organized labor in the Asia–Pacific region.* S. Frenkel (ed.), 107–132. Ithaca, NY: ILR Press.

Ayadurai, Dunston
1993 'Malaysia' in *Labour law and industrial relations in Asia.* S. Deery and R. Mitchell (eds.), 61–95. Melbourne: Longman Cheshire.

Baron, James, Frank Dobbin, and P. Devereaux Jennings
1986 'War and peace: The evolution of modern personnel administration in U.S. industry'. *American Journal of Sociology* 92/2: 350–383.

Bean, Ron
1985 *Comparative industrial relations.* London: Croom Helm.

Biggart, Nicole
1991 'Explaining Asian economic organization: Towards a Weberian institutional perspective'. *Theory and Society* 20/2: 199–232.

Brown, Henry Phelps
1960 *The growth of British industrial relations: A study from the standpoint of 1906–1914.* London: Macmillan.

Campbell, John L.
1994 'Recent trends in institutional analysis: Bringing culture back into political economy'. Mimeograph.

Central Office of Information, Reference Division
1956 *Labour in the United Kingdom dependencies.* London: Central Office of Information.

Chan, Ming K.
1975 *Labor and empire: the Chinese labor movement in the Canton delta, 1895–1927.* Ann Arbor MI: University Microfilms International.

Chesneaux, Jean
1968 *The Chinese labor movement, 1919–1927.* Stanford, CA: Stanford University Press.

Chiu, Stephen Wing Kai
1992 *The reign of the market: Economy and industrial conflicts in Hong Kong.* Occasional Paper No. 16. Hong Kong: Institute of Asia–Pacific Studies, Chinese University of Hong Kong.

Chiu, Stephen Wing Kai
1996 'Unravelling the Hong Kong exceptionalism: The politics of laissez-faire in the industrial takeoff'. *Political Power and Social Theory* 10: 229–258.

Choi, Jang Jip
1989 *Labor and the authoritarian state: Labor unions in South Korean manufacturing industries, 1961–1980.* Seoul: Korea University Press.

Clegg, Hugh
1976 *Trade unionism under collective bargaining.* Oxford: Blackwell.

Colonial Office, Original Correspondence
1948 Series 37/3729, 'Measures to meet communist activity in Hong Kong'.

Colonial Office, Original Correspondence
1947 Series 537/2188, 'Proposed new constitution'.

Colonial Office, Original Correspondence
1946 Series 129/615, 'Labour Department'.

Colonial Office, Original Correspondence
1950 Series 537/6032, 'Labour: Strikes'.

Crouch, Colin
1982 *Trade unions: The logic of collective action.* London: Fontana.

Deery, Stephen, and Richard Mitchell
1993 'Introduction' in *Labour law and industrial relations in Asia.* S. Deery and R. Mitchell (eds.), 1–19. Melbourne: Longman Cheshire.

Deng, Zhongxia
1949 *Zhongguio zhigong yundong jianshi* (A short history of the Chinese labour movement). Beijing: Renmin Chubanshe.

Deyo, Fred, *editor*
1987 *The political economy of the new Asian industrialism.* Ithaca, NY: Cornell University Press.

Deyo, Fred
1989 *Beneath the miracle: Labor subordination in the new Asian industrialism.* Berkeley, CA: University of California Press.

Deyo, Fred, Stephen Haggard, and Hagen Koo
1987 'Labor in the political economy of East Asian industrialization'. *Bulletin of Concerned Asian Scholars* 19/2: 42–53.

DiMaggio, Paul
1988 'Interest and agency in institutional theory' in *Institutional patterns and organizations.* L. Zucker (ed.). 3–21. Cambridge, MA: Ballinger.

DiMaggio, Paul, and Walter Powell
1983 'The iron cage revisited: Institutional isomorphism and collective rationality in organizational fields'. *American Sociological Review* 48/2: 147–160.

DiMaggio, Paul and Walter Powell
1991 'Introduction' in *The new institutionalism in organizational analysis.* P. DiMaggio and W. Powell (eds.). 1–38. Chicago, IL: University of Chicago Press.

Dobbin, Frank
1992 'The origins of private social insurance: Public policy and fringe benefits in America, 1920–1950'. *American Journal of Sociology* 97/5: 1416–1450.

Dobbin, Frank
1994a 'Cultural models of organization: The social construction of rational organizing principles' in *Sociology of culture: Emerging theoretical perspectives.* D. Crane (ed.). 117–141. Oxford: Blackwell.

Dobbin, Frank
1994b *Forging industrial policy: The United States, Britain, and France in the railway age.* Cambridge: Cambridge University Press.

Dow, Gregory K.
1997 'The new institutional economics and employment regulation' in *Government regulation of the employment relationship.* B. Kaufman (ed.), 57–90. Madison, WI: Industrial Relations Research Association.

Eggertsson, Thrainn
1990 *Economic behavior and institutions.* Cambridge: Cambridge University Press.

England, Joe
1971 'Industrial relations in Hong Kong' in *Hong Kong: The industrial colony.* K. Hopkins (ed.), 207–259. Hong Kong: Oxford University Press.

England, Joe
1989 *Industrial relations and law in Hong Kong,* 2d. Ed. Hong Kong: Oxford University Press.

England, Joe, and John Rear
1975 *Chinese labour under British rule.* Hong Kong: Oxford University Press.

England, Joe, and John Rear
1981 *Industrial relations and law in Hong Kong.* Hong Kong: Oxford University Press.

Ferner, Anthony, and Richard Hyman
1992 'Industrial relations in the new Europe: Seventeen types of ambiguity' in *Industrial relations in the new Europe.* A. Ferner and R. Hyman (eds.), xvi–xxxviii. London: Blackwell Business.

Fligstein, Neil
1990 *The transformation of corporate control.* Cambridge, MA: Harvard University Press.

Fox, Alan
1985 *History and heritage: The social origins of the British industrial relations system.* London: Allen and Unwin.

Fulcher, James
1988 'On the explanation of industrial relations diversity: Labour movements, employers and the state in Britain and Sweden'. *British Journal of Industrial Relations* 26/2: 246–274.

Fulcher, James
1991 *Labour movements, employers and the state: Conflict and co-operation in Britain and Sweden.* Oxford: Clarendon Press.

Hobsbawm, Eric
1968 *Labouring men: Studies in the history of labour.* London: Weidenfeld and Nicolson.

Hollingsworth, J. Rogers, Philippe C. Schmitter, and Wolfgang Streeck, *editors*
1994 *Governing capitalist economies: Performance and control of economic sectors.* New York: Oxford University Press.

Hollingsworth, J. Rogers, and Robert Boyer, *editors*
1997 *Contemporary capitalism: The embeddedness of institutions.* Cambridge: Cambridge University Press.

Ingham, Geoffrey K.
1974 *Strikes and industrial conflict: Britain and Scandinavia.* London: Macmillan.

Johnson, Paul
1994 *Success while others fail: Social movement unionism and the public workplace.* Ithaca. NY: ILR Press.

Kerr, Clark, John T. Dunlop, Frederick Harbison, and Charles A. Myers
1973 *Industrialism and industrial man.* Harmondsworth: Penguin.

Kim, Hwang-Joe
1993 'The Korean union movement in transition' in *Organized labor in the Asia–Pacific region.* S. Frenkel (ed.), 113–161. Ithaca, NY: ILR Press.

Knight, Jack
1992 *Institutions and social conflict.* Cambridge: Cambridge University Press.

Korpi, Walter, and Michael Shalev
1979 'Strikes, industrial relations and class conflict in capitalist society'. *British Journal of Sociology* 30/2: 164–187.

Lau, Siu-kai
1982 *Society and politics in Hong Kong.* Hong Kong: Chinese University Press.

Leggett, Chris
1993 'Corporatist trade unionism in Singapore' in *Organized labor in the Asia–Pacific region.* S. Frenkel (ed.), 223–246. Ithaca, NY: ILR Press.

Leung Po-lung
1986a 'A brief history of the Federation of Trade Unions' in *Dimensions of the Chinese and Hong Kong labor movement.* M. Chan, Leung Po-lam, Leung Po-lung, Chiu Wing Kai and Luk Fung Ngo (eds.), 127–131. Hong Kong: Hong Kong Christian Industrial Committee [in Chinese].

Leung Po-lung
1986b 'A brief history of the Trades Union Council' in *Dimensions of the Chinese and Hong Kong labor movement.* M. Chan, Leung Po-lam, Leung Po-lung, Chiu Wing Kai and Luk Fung Ngo (eds.), 132–137. Hong Kong: Hong Kong Christian Industrial Committee [in Chinese].

Leung, Benjamin Kai Ping
1996 *Perspectives on Hong Kong society.* Hong Kong: Oxford University Press.

Leung, Benjamin Kai Ping, and Stephen Chiu Wing Kai
1991 *A social history of industrial strikes and the labour movement in Hong Kong, 1946–1989.* Occasional Paper 3. Hong Kong: Social Sciences Research Centre, University of Hong Kong.

Leung, Sai-wing
1997 *The making of an alienated generation: The political socialization of secondary school students in transitional Hong Kong.* Aldershot: Ashgate.

Leung, Shiu Hung
1983 'Industrial relations in Cable and Wireless: A unionist's view' in *Contemporary issues in Hong Kong labour relations.* S.H. Ng and D.A. Levin (eds.), 123–133. Hong Kong: Centre of Asian Studies, University of Hong Kong.

Levin, David.
1979 'Industrial conflict in Hong Kong: recent trends' in *Social tensions and industrial relations arising in the industrialization process of Asian countries.* The Japan Institute of Labour (ed.), 145–169. Tokyo: The Japan Institute of Labour.

Levin, David, and Stephen Chiu
1994 'Decolonization without independence: political change and trade unionism in Hong Kong' in *The future of industrial relations: Global change and challenges.* J. Niland, R. Lansbury and C. Verevis (eds.), 320–348. Thousands Oaks, CA: Sage.

Levin, David A. and Stephen Chiu
1997 'Empowering labour? The origins and practice of joint consultation in Hong Kong' in *Innovation and employee participation through works councils: International case studies.* R. Markey (ed.), 280–306. Aldershot: Avebury.

Levin, David, and Ng Sek Hong
1993 'Hong Kong' in *Labour law and industrial relations in Asia: Eight country studies.* S. Deery and R. Mitchell (eds.), 20–60. Melbourne: Longman Cheshire.

Levin, David, and Ng Sek Hong
1995 'From an industrial to a post-industrial economy: Challenges for human resource management in Hong Kong' in *Employment relations in the growing Asian economies.* A. Verma, T. Kochan and R. Lansbury (eds.), 119–157. London: Routledge.

Lipset, Seymour Martin
1983 'Radicalism or reformism: The sources of working-class politics'. *American Political Science Review* 77/1: 1–18.

Meyer, John
1983 'Conclusion: Institutionalization and the rationality of formal organizational structure' in *Organizational environments: ritual and rationality*. J. Meyer and R. Scott (eds.), 261–282. Beverly Hills, CA: Sage.

Mitchell, Robert Edward
1972a *Family life in urban Hong Kong*, Vol. II. Taipei: Orient Cultural Service.

Mitchell, Robert Edward
1972b *Levels of emotional strain in Southeast Asian cities*, Vol. II. Taipei: Orient Cultural Service.

Mitchell, Robert Edward
1998 *Velvet colonialism's legacy to Hong Kong: 1967 and 1997*. Occasional Paper No. 76. Hong Kong: Hong Kong Institute of Asia–Pacific Studies, Chinese University of Hong Kong.

Morris, Paul, and K.K. Chan
1997 'The Hong Kong school curriculum and the political transition: Politicisation, contextualisation and symbolic action'. *Comparative Education* 33/2: 247–264.

Nishida, Judith
1992 'Technology transfer and East Asian business recipes: The adoption of Japanese cotton spinning techniques in Shanghai and Hong Kong' in *Reworking the world*. J. Marceau (ed.), 181–204. Berlin: Walter de Gruyter.

Piore, Michael J.
1979 *Birds of passage: Migrant labor and industrial societies*. Cambridge: Cambridge University Press.

Podmore, David
1971 'The population of Hong Kong' in *Hong Kong: The industrial colony*. K. Hopkins (ed.), 21–54. Hong Kong: Oxford University Press.

Polanyi, Karl
1957 *The great transformation: The polit-*
[1944] *ical and economic origins of our time*. Boston: Beacon Press.

Poole, Michael
1986 *Industrial relations: Origins and patterns of national diversity*. London: Routledge and Kegan Paul.

Redding, S. Gordon
1990 *The spirit of Chinese capitalism*. Berlin: Walter de Gruyter.

Registrar of Trade Unions
1972 *Annual departmental report 1971–72*. Hong Kong: Government Printer.

Registrar of Trade Unions
1994 *Annual departmental report 1993*. Hong Kong: Government Printer.

Roberts, Ben
1964 *Labour in the tropical territories of the Commonwealth*. London: Bell.

Roberts, David
1979 *Paternalism in early Victorian England*. London: Croom Helm.

Rutherford, Malcolm
1994 *Institutions in economics*. Cambridge: Cambridge University Press.

Sabel, Charles
1981 'The internal politics of trade unions' in *Organizing interests in Western Europe*. S. Berger (ed.), 209–244. Cambridge: Cambridge University Press.

Salaff, Janet.
1981 *Working daughters of Hong Kong: Filial piety or power in the family?* Cambridge: Cambridge University Press.

Scott, Ian
1989 *Political change and the crisis of legitimacy in Hong Kong*. Hong Kong: Oxford University Press.

Scott, W. Richard
1995 *Institutions and organizations*. Thousand Oaks, CA: Sage.

Skocpol, Theda
1980 'Political response to capitalist crisis: Neo-Marxist theories of the state and the case of the New Deal'. *Politics and Society* 10/2: 155–201.

320   Stephen W.K. Chiu, David A. Levin

Seidman, Gay W.
1994 *Manufacturing militance: Workers'
     movements in Brazil and South
     Africa, 1970–1985.* Berkeley, CA:
     University of California Press.

Sharma, Basu
1985 *Aspects of industrial relations in
     ASEAN.* Singapore: Institute of
     Southeast Asian Studies.

Siddique, S.A.
1989 'Industrial relations in a third world
     setting: A possible model'. *The
     Journal of Industrial Relations* 31/3:
     385–401.

Stephens, Evelyne H., and John D.
Stephens
1982 'The labor movement, political
     power, and workers' participation in
     Western Europe'. *Political Power
     and Social Theory* 3: 215–249.

Stinchcombe. Arthur L.
1969 'Social structure and the invention
     of organizational forms' in
     Industrial man. T. Burns (ed.),
     153–195. Harmondsworth: Penguin.

Sturmthal. Adolf
1967 'Industrialization and the labor
     movement: A set of research
     hypotheses' in *Labor relations in the
     Asian countries.* Japan Institute of
     Labour (ed.), 51–62. Tokyo: The
     Japan Institute of Labour.

Sweeting. Anthony
1993 *A phoenix transformed: the recon-
     struction of education in post-war
     Hong Kong.* Hong Kong: Oxford
     University Press.

Swenson. Peter
1991 'Bringing capital back in, or social
     democracy reconsidered: Employer
     power. cross-class alliances, and
     centralization of industrial relations
     in Denmark and Sweden'. *World
     Politics* 43/4: 513–544.

Tai. Hung-chao
1989 'The oriental alternative?' in
     *Confucianism and Economic
     Development.* H.C. Tai (ed.), 2–27.
     Washington, DC: The Washington
     Institute Press.

Thurley, Keith
1983 'Concluding   remarks'   in
     *Contemporary issues in Hong Kong
     labour relations.* S.H. Ng and D.A.
     Levin (eds.), 220–231. Hong Kong:
     Centre of Asian Studies, University
     of Hong Kong.

Tomita, Teruhiko
1985 'Japanese management in Hong
     Kong'. *Southeast Asian Studies*
     22/4: 391–405.

Tsang, Steve
1988 *Democracy shelved: Great Britain.
     China, and attempts at constitu-
     tional reform in Hong Kong.
     1945–1952.* Hong Kong: Oxford
     University Press.

Tse, Thomas K.C.
1997 *The poverty of political education in
     Hong Kong secondary schools.*
     Occasional Paper No. 69. Hong
     Kong: Hong Kong Institute of
     Asia–Pacific Studies, the Chinese
     University of Hong Kong.

Turner, Herbert, Patricia Fosh, Margaret
Gardner, Keith Hart, Richard Morris, Ng
Sek Hong, Michael Quinlan, and Dianne
Yerbury
1980 *The last colony: But whose? A study
     of the labour movement, labour mar-
     ket and labour relations in Hong
     Kong.* Cambridge: Cambridge
     University Press.

Vogel, Ezra
1991 *The four little dragons: The spread
     of industrialization in East Asia.*
     Cambridge,   MA: Harvard
     University Press.

Vromen, Jack J.
1995 *Economic evolution: An enquiry into
     the foundations of new institutional
     economics.* London: Routledge.

Whitley, Richard, editor
1992a *European business systems: Firms
     and markets in their national con-
     texts.* London: Sage.

Whitley, Richard
1992b *Business systems in East Asia:
     Firms, markets and societies.*
     London: Sage.

Williamson, Oliver
1975 *Markets and hierarchies: Analysis and antitrust implications.* New York: Free Press.

Williamson, Oliver
1985 *The economic institutions of capitalism: Firms, markets, relational contracting.* New York: Free Press.

Wong, Siu-Lun
1988 'The applicability of Asian family values to other sociocultural settings' in *In search of an East Asian development model.* P.L. Berger and H.H.M. Hsiao (eds.), 134–152. Brunswick, NJ: Transaction Books.

Wong, Yue-chim
1988 'The economics of organized labour, with some reference to Hong Kong' in *Labour movement in a changing society: the experience of Hong Kong.* Y. Jao, D. Levin, S. Ng and E. Sinn (eds.), 90–112. Hong Kong: Centre of Asian Studies, University of Hong Kong.

World Bank
1995 *World development report 1995: Workers in an integrating world.* Oxford: Oxford University Press.

Zeitlin, Jonathan
1985 Shop floor bargaining and the state: a contradictory relationship' in *Shopfloor bargaining and the state.* J. Zeitlin and S. Tolliday (eds.), 1–45. Cambridge: Cambridge University Press.

Zhongguo Laogong Yundongshi Xubian Biancuan Weiyuanhui
1984 *Zhongguo Laogung Yundongshi* (A history of the Chinese labour movement), Vol. 1. Taipei: Zhongguo Wenhua Daxue Laogong Yanjiusuo Lishihui.

[16]

Whither Hong Kong's Unions: Autonomous Trade Unionism or Classic Dualism?

Ed Snape and Andy W. Chan

Abstract

This paper presents an evaluation of the ideologies and behaviour of the main union groupings in Hong Kong. The CTU's 'adversarial' line and 'inclusive' approach has threatened to break with the traditional industrial pacifism of Hong Kong's unions, while the FTU's recognition of its 'dual functioning' role has been balanced by the need to maintain credibility as an effective representative of workers' interests. The TUC has been passive in recent years, and the independent unions tend to be rather inward-looking. Unions remain weak, particularly at the workplace, and it is uncertain to what extent union autonomy can be sustained.

1. Introduction

As of 1 July 1997, the Basic Law provides Hong Kong with quasi-autonomous status as a 'Special Administrative Region' (SAR) of the People's Republic of China (PRC) and guarantees that 'the previous capitalist system and way of life shall remain unchanged for 50 years' (Basic Law, article 5). However, in spite of such guarantees, which include the legal system and labour policies (art. 147), China's resumption of sovereignty after 150 years of British rule is likely to be the key factor shaping the future of Hong Kong's unions.

Beijing has an interest in the industrial and political stability of Hong Kong, which pro-Beijing groups within the territory, including the pro-communist Hong Kong Federation of Trade Unions (FTU), are careful not to undermine. Thus, the FTU might be expected to adopt a conciliatory approach to employers and government, and to draw on the example of the official mainland unions, whose traditional responsibilities have included worker education, discipline and the co-ordination of productivity campaigns (Ng and Ip 1994). In contrast, the FTU's main rival federation, the Hong Kong Confederation of Trade Unions (CTU), espouses a more

Ed Snape and Andy Chan are in the Department of Management at the Hong Kong Polytechnic University.

© Blackwell Publishers Ltd/London School of Economics 1997. Published by Blackwell Publishers Ltd, 108 Cowley Road, Oxford, OX4 1JF, and 350 Main Street, Malden, MA 02148, USA.

40   *British Journal of Industrial Relations*

radical approach, emphasizing the importance of union autonomy and championing the pro-democracy cause.

This paper presents an evaluation of the current state and likely future development of Hong Kong unions. After examining the structure of the union movement, we review the classic adversarial and dual models, followed by a detailed evaluation of the ideologies and behaviour of the main union groupings. Although Hong Kong's industrial relations system resembles neither, the CTU and FTU espouse strategies which draw on the adversarial and dual models respectively. Both strategies face key challenges, largely because of Hong Kong's unique political position. A better educated and more demanding work-force and an emergent democratic political culture have fuelled demands for representative, even adversarial, trade unionism, to which even the FTU has responded. However, unions remain weak, particularly at the workplace, and it is uncertain to what extent union autonomy can be sustained.

The study is based on interviews, a postal survey of union officers, and a content analysis of Labour Department files on labour disputes. Interviews were conducted with officials of the main union federations and selected unions,[1] and with government Labour Officers responsible for conciliating in labour disputes. The research was conducted during 1994–6.

2. The structure of the union movement in Hong Kong

After a decade of decline or stability, union density has recovered, rising to 21 per cent by 1992, with membership at an historic high (Table 1). However, these aggregate figures conceal significant changes in the distribution of membership and structural change poses a serious challenge to many unions. The 1990s have seen the loss of a third of a million manufacturing jobs, largely because of relocation to the People's Republic of China (Suen 1994: 155). While this has been offset by the growth in services employment, and there have been gains for unions in the public sector, the main employment growth has been in private sector services (Chau 1993: 134), which remain largely non-union.

The union movement in Hong Kong comprises three main federations. The largest is the FTU, followed by the CTU and the historically pro-nationalist Hong Kong and Kowloon Trades Union Council (TUC); there are also many 'independent' unions (either unaffiliated or affiliated to other federations), which are usually smaller and sector-specific. After a decline · during the late 1970s and early 1980s, FTU membership has increased, reflecting a more active approach to membership servicing and recruitment and the approach of 1997 (see Table 2). TUC membership declined following the 1984 Sino–British Joint Declaration, although it has subsequently recovered. Historically, the TUC was the FTU's main rival, but its place has been taken by the CTU, established in 1990 as a federation of hitherto independent unions. The independent category has seen the most rapid growth, especially in the public sector, owing to employment growth,

© Blackwell Publishers Ltd/London School of Economics 1997.

TABLE 1
Union Membership in Hong Kong, 1973–1994

Year end	No. of unions	Membership	Growth (% p.a.)	Density (%)
1973	283	295,735	17.48	21
1974	292	317,041	7.20	22
1975	302	361,458	14.01	24
1976	311	388,077	7.36	25
1977	313	404,325	4.19	24
1978	327	399,995	−1.07	23
1979	340	399,392	−0.15	21
1980	357	384,282	−3.78	19
1981	366	345,156	−10.18	16
1982	378	351,525	1.85	16
1983	382	352,306	0.22	16
1984	384	357,764	1.55	16
1985	391	367,560	2.74	16
1986	403	367,345	−0.06	16
1987	415	381,685	3.90	16
1988	430	416,136	9.03	17
1989	439	437,939	5.24	18
1990	452	468,746	7.03	19
1991	469	486,961	3.89	20
1992	481	525,538	7.92	21
1993	491	543,800	3.47	21
1994	506	562,285	3.40	21

Sources: *Report of the Commissioner for Labour;*
Registrar of Trade Unions *Annual Reports*, various
years.

TABLE 2
Affiliated Unions and Membership of the Main Union Federations, 1979–1994

Year	FTU Unions	FTU M-ship	TUC Unions	TUC M-ship	CTU Unions	CTU M-ship	Independents Unions	Independents M-ship
1979	66	196,543	74	36,941			200	165,908
1980	66	182,601	71	36,723			220	164,958
1981	69	169,647	70	35,927			227	139,582
1982	71	171,073	70	35,521			237	144,931
1983	71	167,933	70	34,564			241	149,809
1984	73	166,461	71	35,535			240	155,768
1985	70	167,832	70	35,116			251	164,612
1986	72	169,802	68	31,447			263	166,066
1987	78	168,550	69	28,827			268	184,308
1988	81	173,956	71	17,835			278	224,345
1989	81	173,820	70	18,176			288	245,943
1990	82	175,746	70	30,693	21	74,038	279	188,269
1991	84	181,498	69	30,648	24	70,524	292	204,291
1992	87	192,019	69	30,769	27	75,212	298	227,538
1993	89	199,862	66	30,714	28	78,498	308	234,726
1994	91	205,916	65	31,143	28	79,537	322	245,689

Sources: *Report of the Commissioner for Labour*, and Registrar of Trade Unions
Annual Report, various years.

© Blackwell Publishers Ltd/London School of Economics 1997.

42   *British Journal of Industrial Relations*

rising expectations and concerns about job security, and to the fact that public-sector employers have been less hostile to trade unionism than those in the private sector (Leung 1992; England 1989).

Table 3 details the membership composition of the union groupings. The CTU's strength is mainly in social services (teachers and social workers) and in transport and communications, with a smaller presence in manufacturing and construction. The largest affiliated union is the Professional Teacher's Union, with 56,000 members in schools, colleges and universities.[2] The FTU is stronger in manufacturing and construction, and has significant membership in most sectors, although it is relatively weak in community, social and public services. The TUC has particular strengths in the restaurant trade, in transport and in some sectors of manufacturing. The remaining independent unions are found mainly in the community, social and public services, with significant independent union membership also in transport and communications, trade and construction, and to a lesser extent in manufacturing.

TABLE 3
Distribution of Union Members by Sector and by Affiliation, 1994

Sector (ISIC)	FTU	TUC	CTU	Independent	Sub-total of members by sector
Agriculture, Hunting, Forestry & Fishing	963 (78.0%) [0.5%]	0	0	272 (22.0%) [0.1%]	1,235 (100%) [0.2%]
Manufacturing	56,775 (84.6%) [27.6%]	4,548 (6.8%) [14.6%]	2,460 (3.7%) [3.1%]	3,317 (4.9%) [1.4%]	67,100 (100%) [11.9%]
Electricity, Gas & Water	4,432 (83.4%) [2.2%]	104 (2.0%) [0.3%]	0	781 (14.7%) [0.3%]	5,317 (100%) [0.9%]
Construction	11,331 (61.6%) [5.5%]	213 (1.2%) [0.7%]	10 (0.1%) [0.01%]	6,843 (37.7%) [2.8%]	18,397 (100%) [3.3%]
Wholesale & Retail Trade, Restaurants & Hotels	24,479 (48.9%) [11.9%]	15,314 (30.6%) [49.2%]	2,260 (4.5%) [2.8%]	8,006 (16.0%) [5.5%]	50,059 (100%) [8.9%]
Transport, Storage & Communication	51,304 (58.3%) [24.9%]	6,979 (7.9%) [22.2%]	10,386 (11.8%) [13.1%]	19,326 (22.0%) [7.9%]	87,995 (100%) [15.6%]
Financing, Insurance, Real Estate & Business Services	29,069 (96.7%) [14.1%]	215 (0.7%) [0.7%]	0	976 (3.2%) [0.4%]	30,260 (100%) [5.4%]
Community, Social & Personal Services	27,563 (9.1%) [13.4%]	3,770 (1.2%) [12.1%]	64,421 (21.3%) [81.0%]	206,168 (68.3%) [83.9%]	301,922 (100%) [53.7%]
Total number of members	205,916 (36.6%) [100%]	31,143 (5.5%) [100%]	79,537 (14.1%) [100%]	245,689 (43.7%) [100%]	562,285 (100%) [100%]

Key: ( ): row percentage
     [ ]: column percentage
Source: Registrar of Trade Unions *Annual Report*, 1994.

© Blackwell Publishers Ltd/London School of Economics 1997.

Unions have been weak in Hong Kong. 'Neo-Confucian' values emphasize obedience to management authority and limit the development of unions (Levin and Chiu 1994: 155), while individualism and 'utilitarian familism' provide an alternative strategy for social advancement (Wong and Liu 1994: 79). Hong Kong's industrial structure is also unfavourable to unionization, with a relative absence of heavy industry and a preponderance of small-scale workplaces. Apart from in the public sector and some large corporations, unions remain largely ineffective at workplace level (Levin and Chiu 1993), with fewer than 5 per cent of the work-force covered by collective bargaining (Yeung 1988: 57; Tang *et al.* 1955: 22). In the civil service, management–union relations take the form of joint consultation, but even this is lacking in private-sector workplaces (Lee 1994). Instead, Hong Kong unions have traditionally concentrated more on direct services to members (Levin and Chiu 1993: 202– 3). During the 1980s, unions also expanded their political activity, acting as lobbyists on labour and social issues and taking a role in elections. In this, they have responded to an increasingly demanding work-force and electorate, as improved living standards have led to higher expectations, while democratization has created opportunities for participation in the political process and contributed to demands for greater social justice.

Traditionally, membership of FTU and TUC unions was as much an expression of political loyalty to the Chinese Communist Party (CCP) or the Kuomintang (KMT) as an attempt to secure representation at work, and such unions were largely dominated by party political interests. Since the 1970s, there has been a growth of unions affiliated to neither the FTU nor the TUC, much (but not all) of it in the public services. This is symptomatic of the greater articulation of middle-class interests, which is now providing strong support for the pro-democracy cause (Yahuda 1996: 51–2). The leaders of the non-FTU/TUC unions tend to be younger and the membership is generally better educated.[3] Such unions have focused more on the employment-related interests of their members and generally have avoided partisan CCP or KMT loyalties. Chai (1993) sees this as providing a new 'independent model' in contrast to the 'political model' trade unionism of the FTU and TUC. However, she recognizes that the non-FTU/TUC unions 'cover a wide spectrum of orientations and styles ranging from traditional inward-looking orientation and inactivity to fairly aggressive activism emanating from the Western assumption of conflict-based unionism' (Chai 1993: 130). The latter tendency is represented by the CTU, and the relationship between the CTU and the independent unions is an issue to which we return below.

3. The theory and practice of trade unionism

Given the transfer of sovereignty, Hong Kong has been characterized as standing 'between two societies' (Turner *et al.* 1991), and Hong Kong's

© Blackwell Publishers Ltd/London School of Economics 1997.

44   *British Journal of Industrial Relations*

unions are perhaps similarly torn. One way to visualize this is to compare the traditional Western capitalist model of unions with the classic dualism of Communist unions (Table 4). The former sees unions in an adversarial relationship with employers, as an autonomous representative of labour interests within a pluralist industrial relations framework.

However, in spite of its capitalist economy, Hong Kong lacks the pluralist tradition and workplace unionism of those Western economies on whose experience the adversarial model is based. Frenkel (1993), for example, suggests that, while Hong Kong shares the features of institutional autonomy and a market economy with Australia and New Zealand, the state has traditionally pursued a very different strategy in Hong Kong, excluding unions from major influence, in comparison with the 'bargained corporatism' of these other two countries. Add to this the strong employer resistance to unionization, and we may question the feasibility of adversarial unionism based on collective bargaining in Hong Kong.

The dual model was typical of communist-controlled states. On the one hand, unions play a *representative* function, safeguarding labour's rights and interests (although, given the assumption that basic workplace conflicts of interest cannot exist, this role may be limited to dealing with disagreements over short-term priorities and errors by local management); on the other hand, unions play a *production* function, involving the maintenance of labour discipline, the mobilization of workers in productivity drives, and labour education on production, management and ideological matters (Pravda and Ruble 1986a: 3). However, while the FTU may espouse elements of the dual model, it is less able than PRC unions to perform the production function, since it lacks the level of institutionalization found in the PRC and needs to compete with other unions to attract members.

TABLE 4
Capitalist and Communist Unions Compared

Communist unions (Classic dualism)	Capitalist unions (Classic adversarial model)
1. Unitary view of economic interests: it is axiomatic that no 'industrial conflicts' exist.	1. Pluralist view of economic interests.
2. The production function is paramount.	2. Representative function is paramount.
3. Protection of members' rights is secondary.	3. Production function is either (a) not acknowledged, or (b) secondary.
4. Subordination to the party.	4. Autonomous organizations.
5. No collective bargaining. Union practices exclude the use of adversarial means.	5. Adversarial collective bargaining is the typical process of industrial relations.

Source: Littler and Palmer (1986: 265).

© Blackwell Publishers Ltd/London School of Economics 1997.

Neither the adversarial nor the dual model is descriptive of the industrial relations situation in Hong Kong. However, the existing ideological differences between unions, and particularly between the CTU and FTU, appear to draw on the adversarial/dual models. Below, we analyse the ideology and conduct of the four union groupings, particularly in terms of the emphasis given to the production and representative functions, and to union autonomy. We refer mainly to the federations rather than to individual unions, because most Hong Kong unions are small and the federations play a key role, providing services directly to individual members and office space and staff resources to some individual unions, and each has distinctive political and ideological characteristics.

4. The main union groupings

Hong Kong Confederation of Trade Unions (CTU)

The CTU was established in August 1990 and has been closely associated with the Christian Industrial Committee (CIC). The CIC itself was established in 1968 by the Hong Kong Christian Council, initially focusing on labour education and on advising workers on grievances, but developing a wider role as a critic of government policies and in fostering the development of independent unions. As a CTU official explained,

> It was one of the objectives of CIC to finally . . . group all these independent unions together to form an independent federation. So in 1988 we called a two-day conference, bringing together 21 independent unions, and discussed for the first time seriously . . . whether they wanted an independent union centre or not. . . . After June 4th [the Tiananmen Square incident], of course, the process was hastened. Everybody felt the need to have it really strongly . . .

The CTU leadership aims to build autonomous unions and to encourage Hong Kong workers to see union membership as being about the protection of their interests. However, in spite of distancing themselves from the communist–nationalist divide, the CTU does not advocate an apolitical approach. Its leaders actively lobby the government on labour and social issues, and its members have been involved in Legislative Council (LegCo) and District Board elections, with Lau Chin-shek, CTU chairman, and Lee Cheuk-yan, CTU chief executive, winning election to LegCo. Both are active members of the Hong Kong Alliance for the Promotion of the Patriotic and Democratic Movement in China and are critical of PRC labour policies. Indeed, it appears that the controversial political stance of CTU leaders has deterred some unions from affiliating, and there are similar concerns within the CTU, which leaders fear may increase after 1997.

   The CTU leadership has a three-fold strategy for the future. First, it aims to build grass-roots support by effectively representing workers' interests to government and to employers:

© Blackwell Publishers Ltd/London School of Economics 1997.

46   *British Journal of Industrial Relations*

> We still have to remain as *the* organization that is most able to deal with labour disputes. If we can maintain that image, if we can maintain that trust, we can still gain the respect of workers. (CTU official)

CTU officials see their organization as playing a proactive role in labour disputes and as adopting an inclusive approach to unionization. As one of them put it,

> when workers have a problem, you can't just say 'Go to hell, you're not my affiliate'. You want them to grow also, and you want them to be, one day, your affiliate, but that may take a long time. (CTU official)

CTU officials claim that the FTU is hampered in its ability to represent workers effectively in disputes by a desire to maintain stability, and that the CTU is often cast in the role of troubleshooter:

> If they [workers] feel that they are confident with the FTU, we don't go in; we pull out. But the problem is very often that the FTU is not doing anything, and then the workers come to us, and they want us to get involved. . . . We have to go there even if we are being labelled as 'militants', 'troublemakers' or anything. (CTU official)

The CTU operates a telephone 'helpline' for workers; it received around 5,000 calls in 1994, over 1,000 of which required follow-up action by CTU officials.

The CTU claims to have organized ten new unions between 1990 and 1994, seven of which have affiliated to the CTU (*Turning Point*, June 1994: 3). Recent organizing efforts include security guards, retailing, hotels and transport. The usual pattern is for workers to approach the CTU with a dispute and for the CTU to try to form an enterprise or industrial union. The key is to build a core group of committed lay activists, which is often difficult in the face of employers' victimization of unionists. The CTU's approach does not necessarily lead to employer recognition, even where the FTU and TUC are recognized, and some employers deal with employee representatives but refuse to meet CTU officials.

The second element in the strategy is to develop the political consciousness of union leaders and members through an education programme, emphasizing the importance of the CTU's political autonomy. As one official put it,

> We want to confront them with politics, because we don't think we can avoid that in the future. And so we hope that through the seminars we can have more of a consensus in our political positioning and also a consensus on the independence, true independence, of the CTU.

The third element is to contribute to the development of political democracy in Hong Kong, as a means of preserving trade union and other freedoms. This is expressed through CTU support for election candidates and in its association with the pro-democracy lobby. The CTU appears to have had some success here, benefiting from the popularity of the pro-

© Blackwell Publishers Ltd/London School of Economics 1997.

democracy cause. In the 1995 LegCo elections, CTU-endorsed candidates did rather better than those of the FTU, particularly in the new occupationally based functional constituencies, which effectively had a mass work-force electorate (Table 5).

TABLE 5
Performance of Union-Endorsed Candidates in the 1995 Legislative Council Elections

	All constituencies			*New occupational functional constituencies only*		
	Candidates	Successful (%)	% of vote won	Candidates	Successful (%)	% of vote won
FTU	9	3 (33)	35	6	1 (17)	27
CTU	7	5 (71)	45	6	4 (66)	42
FLU	1	1 (100)	34	0	—	—
TUC	1	0 (0)	17	0	—	—

Source: Authors' analysis of election results; union endorsements based on press and campaign materials.

CTU leaders see their organization as a forceful representative of labour interests and are critical of the services-based approach of many Hong Kong unions. Such a strategy, combined with very low subscription rates (HK$5 per month is common), has contributed to the growth of civil service unions in particular, and the FTU is noted for its range of services. CTU leaders argue that this encourages a short-sighted instrumentalism among members, who shop around for discount deals and services with little understanding of the union representative function, while low subscriptions contribute to organizational weakness. Some of the newer CTU affiliates have adopted higher subscription rates (up to $30 per month), but there is resistance to a general increase in subscriptions and the CTU cannot ignore recruitment rivalry. While the CTU lags the FTU in the provision of direct benefits, the largest CTU affiliate, the Professional Teachers' Union, has a well developed range with a low membership subscription, and the CTU itself is responding to pressure from affiliates and developing a range of discount purchase deals.

The Hong Kong Federation of Trade Unions (FTU)

The Hong Kong FTU was established in April 1947. After a period of industrial militancy, the unions had suffered setbacks by the early 1950s (England 1989: 112–13), and the FTU subsequently placed less emphasis on industrial disputes, focusing on the provision of welfare services (Ng 1984: 20; England 1989: 113–14). The labour riots and civil unrest of 1967 have been attributed in part to FTU influence (Ng and Cheng 1993: 217; Leung 1994: 187), but from the mid-1970s the FTU again adopted a more moderate

48   *British Journal of Industrial Relations*

stance, in line with Beijing's conciliatory open door policy (Leung and Chiu 1991: 54–5; Leung 1994). Writing in 1980, H. A. Turner and his colleagues concluded that:

> Although not recognised by employers in the Western sense of formalised joint relations with managements, it [the FTU] has apparently established an accept-ance by them in the trades where its membership is concentrated which is sufficient . . . to enable it to claim some credit for wage increases and grievance settlement. In effect, it has attained a certain equilibrium position which . . . FTU leaders appear reluctant to disturb by vigorous recruiting campaigns or the adoption of major social-reformist policies. (Turner *et al.* 1980: 149)

In their subsequent study, conducted in the late 1980s, they suggest that 'time had reinforced its situation as we described that in our previous study' (Turner *et al.* 1991: 83).

At its 24th General Meeting in 1986, the FTU set out its policy for the transitional period as 'participating in social affairs, striving for reasonable rights and benefits for workers, expanding our patriotic solidarity and promoting social stability and economic prosperity in Hong Kong' (Hong Kong FTU 1988: 2). Official FTU statements on disputes have emphasized the need for a 'practical and realistic attitude' and the 'co-operation of all parties concerned' (Hong Kong FTU 1988: 2).

The FTU has become increasingly active in politics in response to the political reforms since the mid-1980s. It has endorsed election candidates, associating itself with the pro-Beijing Democratic Alliance for the Betterment of Hong Kong. Although the FTU has co-operated with other labour groups in campaigning on labour and social policy, such co-operation has been intermittent and the political divisions remain. The FTU has been more comfortable campaigning on such issues as pensions and labour importation; but on more politically sensitive areas, such as the pace of democratic reform and the future political system of the SAR, it has tended to follow a pro-Beijing line, in contrast to some other labour groups (Levin and Chiu 1993: 213).

Leaders of the CTU have been critical of the FTU, accusing it of concentrating on discount and welfare services, and dodging the thornier issue of industrial grievances and disputes (*FTU Press*, June 1994: 2). The allegation is that the FTU places greater emphasis on maintaining stability than on advancing the interests of workers (Xu Jia-tun 1993; *Turning Point*, January 1994: 1–3). The FTU's response is that it offers a comprehensive range of direct services for members, while also handling labour disputes and protecting workers' interests *vis à vis* employers and the government. The FTU reported in 1994 that, along with its affiliates, it handled 16,000 enquiries in the previous two years, and followed up 662 labour claims and disputes, involving a total of 24,200 people and HK$300 million in settlements (*FTU Press*, June 1994: 2). Since 1990 the FTU has developed a network of four Labour Service Centres, dealing with labour matters such as claims, disputes and industrial accidents, and nine District Services Offices,

© Blackwell Publishers Ltd/London School of Economics 1997.

providing assistance on housing, health and labour matters. Furthermore, since the mid-1980s the FTU has become more pro-active on recruitment and organizing, including attempts to move beyond its blue-collar origins to recruit service-sector and white-collar workers.

The FTU officers we interviewed were sensitive about accusations that the pro-Beijing stance compromises its ability to represent members effectively. Officers in the Labour Service Centres, for example, argued that they had discretion in their daily work to deal with claims and disputes as they saw fit, and that, while the pro-Beijing stance might colour the FTU's approach to certain political issues, they must be able to provide tangible gains for members in order to recruit and retain members in competition with the CTU.

In fact, our interviews with FTU officials revealed a range of opinions on the organization's priorities. Some emphasized the 'production' function. For example, one official spoke of the need to be 'fair' and 'reasonable' in dealing with labour disputes and to

> balance the power between the two sides, if the workers are too demanding, too unreasonable or too strong or militant, whilst the employer [is] being gentle. (FTU official)

Others emphasized the 'representative' function:

> Unions must serve workers and safeguard justice. A union should not emphasise harmony in [the] workplace, or stability, peace. Otherwise the union must die. (FTU official)

Such sentiments received top-level official approval in FTU Chairman Cheng Yiu-tong's declaration on the FTU's role after 1997. Speaking after the 1995 Annual General Meeting, he said:

> to prevent the SAR Government from formulating wrong policies ... unions should monitor the government ... be brave to voice out workers' demands to have an even distribution in society. The real role of unions is as the workers' representative, not the government's! (*FTU Press*, June 1995: 7)

Clearly, there is evidence of the dual union role in the views of FTU officials, but this includes a recognition of the representative function, and developments in the FTU's infrastructure and services have moved the organization on since Turner *et al.*'s assessment.

Hong Kong and Kowloon Trades Union Council (TUC)

Membership of a TUC affiliate has been seen as an expression of loyalty to the Nationalists, and the TUC and most of its affiliates display the Nationalist flag in their offices. Over the years, the TUC's influence has come mainly through its participation in the channels of communication with the government, such as the Labour Advisory Board. In recent years, the TUC has adopted a low profile. The association with the KMT may

© Blackwell Publishers Ltd/London School of Economics 1997.

50   *British Journal of Industrial Relations*

contribute to this, with TUC leaders avoiding a high profile for fear of incurring the wrath of Beijing. The emphasis has been on maintaining the existing membership, and there has been relatively little recruitment activity. The TUC was established in 1948, and most affiliates date from the 1950s or earlier and are concentrated in the declining areas of manufacturing and in traditional restaurants. The leadership comes mainly from the older age group, and there have been difficulties recruiting younger workers.

The KMT legacy is at once a source of both strength and weakness: strength in that nationalist loyalties bind affiliates to the TUC; weakness in that it places the TUC in a vulnerable position after 1997 and may impair new organizing efforts. The TUC plans to continue after 1997, and was redeveloping its Mong Kok premises with the promise of improved facilities and income on completion in 1998. However, unless the next few years see a major change in TUC policies, including a shift to a more outward-looking approach and perhaps even a de-emphasizing of the Nationalist tradition, it is difficult to see it playing a major role in industrial relations or politics after the transfer of sovereignty.

The Independent Unions

The independent unions are the fastest growing, particularly in the public sector, and their combined membership exceeds that of the FTU. The largest — indeed, Hong Kong's largest union overall — is the Hong Kong Chinese Civil Servants' Association (CCSA), which has grown from 12,000 members in 1981 to over 108,000 by the end of 1994. However, civil service unions remain fragmented, with many small unions recruiting only in specific grades or departments and having a few hundred members.[4] Dual membership is common, as individuals may join a departmental grade union as well as the CCSA, often shopping around for union-provided benefits and discount bargains.

The ability to deal with members' occupation-specific problems, the political neutrality of the union and the range of direct services (e.g. retail discounts) are cited by leaders as important reasons for the growth of such unions. The first point underlines the attraction of organization-specific unions for many Hong Kong workers, while the second suggests that many are reluctant to be associated with controversial political views. Thus, while independent unions have occasionally been involved in industrial action,[5] they tend to be rather inward-looking (Chai 1993; Leung 1992: 390), often representing relatively secure and privileged groups, and in general they are less visible than the FTU and CTU.

5. Union ideology and conduct

Our discussion so far has described the alleged 'militance' and adversarial approach of the CTU, the 'moderation' and dual functioning role of the

© Blackwell Publishers Ltd/London School of Economics 1997.

FTU, the low-profile of the TUC and the generally inward-looking approach of most independent unions. We now subject these stereotypes to greater scrutiny through an analysis, first, of the beliefs of union leaders on the proper role for a union and, second, of union conduct in labour disputes.

Union Leaders' Views on Union Roles

Questionnaires were mailed to the senior officials (e.g. chairman, president) of all unions registered with the Registrar of Trade Unions as at the beginning of 1996, a total of 522 unions.[6] Completed questionnaires were received from 141 unions, a 27 per cent response rate. Of these, 23 were

TABLE 6
Factor Structure of Union Functions and Responsibilities

Statement	*Autonomy*	*Production function*	*Representative function*	
			Strong	*Weak*
Autonomy				
Unions should not be under the influence of employers	**0.88551**	−0.03062	0.12865	0.06057
Unions should not be under the influence of government	**0.88544**	0.03929	0.11632	0.14425
Unions should not be under the influence of a political party	**0.83748**	0.06508	−0.03512	0.02758
Production function				
Unions have a responsibility to resolve disputes with as little conflict as possible	0.05305	**0.82659**	0.02184	0.01261
Unions have a responsibility to encourage workers to improve productivity	0.18064	**0.72901**	−0.13296	0.07535
Unions have a responsibility to educate workers on their contribution to the stability and economic development of Hong Kong	−0.05333	**0.70485**	0.16404	−0.06875
A union's role is to be a neutral mediator between workers and the employer	−0.09359	**0.63674**	−0.16485	0.48116
Representative function				
Strong form				
Unions have a responsibility to pursue the interests of workers whatever the consequences	−0.06632	−0.07503	**0.72138**	0.17036
Unions' primary responsibility is to workers rather than to the general public, the government or employers	0.12238	0.08948	**0.66967**	0.10905
The interests of workers and their employers are basically in opposition	0.08989	−0.02976	**0.65813**	−0.27147
Weak form				
Unions have a responsibility to pursue improved terms and conditions of employment for workers	0.11002	−0.03684	0.17370	**0.86394**
Unions have a responsibility to protect workers against exploitation by employers	0.45126	0.19161	−0.09356	**0.54435**

Source: Authors' union survey, 1996 (*N*=141).

© Blackwell Publishers Ltd/London School of Economics 1997.

52   *British Journal of Industrial Relations*

affiliated to the CTU (including 4 FLU affiliates), 23 to the FTU and 12 to the TUC; 83 were independent.

Respondents responded to 12 statements on the role of unions according to a five-point scale (strongly agree to strongly disagree, with a neutral mid-point) (Table 6). Factor analysis (principal-components analysis, varimax rotation) of the 12 statements yielded a four-factor solution, accounting for 64 per cent of the variance. The factors are interpretable as follows:

1. *Autonomy*: the belief that unions should be independent of external influence
2. *Production function*: the belief that unions should act as a neutral mediator and educator of workers to minimize conflict and improve productivity
3. *Representative function — strong form*: the recognition of a basic conflict of interest between employers and employees and the belief that unions should pursue workers' interests regardless of the consequences
4. *Representative function — weak form*: the belief in the need to represent the interests of workers, based on more moderate statements than those loading heavily on factor 3.

We compared the union groupings on the four factor scores (Table 7). In terms of 'Autonomy', there were no significant differences on the analysis of variance, although a t-test between CTU and FTU unions alone suggested that officials of CTU affiliates agree with the need for autonomy more strongly than do officials of FTU affiliates ($P < 0.05$). TUC officials identify with the production function more strongly than do officials of CTU and FTU unions. This is perhaps surprising, but may reflect an historic right-wing attitude to management and an anti-leftist view of industrial conflict. While there were no significant differences in the weak-form representative function, TUC union officials identified more with the strong form than did officials of FTU and independent

TABLE 7
Differences in Attitudes on the Functions and Responsibilities of Unions by Affiliation

	Mean factor scores			
	Autonomy	*Production function*	*Representative function*	
			Strong	*Weak*
(1) Independent	−0.0027	0.0521	−0.2089	−0.0650
(2) CTU	0.1975	−0.4184	0.4869	0.1744
(3) FTU	−0.3888	−0.1951	−0.1620	−0.0216
(4) TUC	0.3491	0.8634	0.8291	0.1549
Group comparisons				
$P \leq 0.05$	No	(4) > (2)	(4) > (1)	No
Modified LSD	significant	(4) > (3)	(4) > (3)	significant
	differences		(2) > (1)	differences
F value	1.8379	4.7871***	6.3811***	1.4775

* $p \leq 0.05$; ** $p \leq 0.01$; *** $p \leq 0.001$

© Blackwell Publishers Ltd/London School of Economics 1997.

unions, and CTU officials did so more than independent unions. There was no significant difference between CTU and FTU officials on the analysis of variance, although again a pairwise *t*-test suggested that officials of CTU affiliates identified with the strong-form representative function more strongly than did officials of FTU affiliates ($P < 0.05$).[7]

Thirty questionnaires also contained open-ended comments (28 in Chinese, 2 in English). The Chinese responses were translated into English by one of the researchers and checked for interpretation by an assistant. All were then content-analysed.

CTU, TUC and independent unions referred to the possibility of a suppression of unions after the transfer of sovereignty in 1997. For example:

> Will the Chinese government observe the 'One Country, Two Systems, HK People Rule HK' after 1997? If not, giving more opinions means you will more easily be accused of making mistakes. There won't be any valuable opinions.

There were no such references by FTU affiliates.

CTU unions were particularly prone to mention union weakness as a key problem, attributing this to a lack of resources and activists and the attitudes of government and employers. This reflects the CTU's aspiration to be an active representative of labour interests, combined with their slim resource base and vulnerability after 1997.

There was general support for the representative function of unions, along with some recognition on the part of FTU and independent unions of the need for unions to be 'reasonable' in their demands and to contribute to the stability and prosperity of Hong Kong. Thus:

> Union representatives should communicate, consult and mediate with employers or employers' associations on labour matters and problems in order to avoid conflicts. (FTU affiliate)

> Unions should educate workers to increase productivity . . . (Independent)

> Unions should play a part in maintaining HK's stability and prosperity. (Independent)

In addition, some independent unions complained about unions being overly political in recent years.

Overall, the attitudes of union officials are to some extent in line with our expectations. Leaders of individual CTU unions share their federation leadership's endorsement of representative unionism, along with their concerns about the ability to sustain this. The evidence on FTU officials is less clear-cut: they identify with the representative function less strongly than do TUC officials, but, contrary to our expectation, there is no evidence that they are more strongly committed to the production function than are officials of other unions. Independent union officials appear rather lukewarm in their attitudes, consistent with an inward-looking approach which avoids a generalized identification with labour movement issues. This group identifies less strongly with the representative function (strong-form) than

© Blackwell Publishers Ltd/London School of Economics 1997.

54   *British Journal of Industrial Relations*

do CTU officials, which may explain why the CTU has failed to attract major new affiliates from among the independents in recent years. The main surprise is the TUC's high rating on both the production and representative functions, which belies its reputation as a compliant, moderate organization. Perhaps what we are seeing here is a statement of ideals as opposed to actual practice.

Union Involvement in Labour Disputes

So much for attitudes. What of differences in union behaviour? We can approach this by looking at unions' involvement in labour disputes.

While working days lost due to strikes in Hong Kong are among the lowest in the world, conflicts of interest do emerge, and the government provides, through the Labour Department, a voluntary conciliation service to help employees and employers resolve disputes and claims.[8] Conciliation cases are concentrated in manufacturing, construction, the wholesale and retail trades, restaurants and hotels. These sectors are relatively dispute-prone because of the high level of redundancies and layoffs, and they lack the institutionalized consultative mechanisms of the public sector, for example. The majority of disputes do not involve a union, although unions tend to deal with larger cases (see Table 8). The FTU and the CTU have a much higher level of involvement than the TUC and independent unions, while the TUC's minimal involvement reflects the organization's low profile in recent years.[9]

In the five-year period 1990–4, the CTU and its affiliates were involved in 90 disputes, compared with 118 for the FTU. Given that the CTU is smaller,

TABLE 8
Trade Union Involvement in Labour Disputes, 1990–1994

All disputes	With union involvement	Of which:[a]			
		FTU	CTU[b]	TUC	Independents
Number of disputes					
797	206	118	90	2	25
(100%)	(26%)	(15%)	(11%)	(0%)	(3%)
Number of workers involved					
47,513	20,079	11,714	8,029	40	3,706
(100%)	(42%)	(25%)	(17%)	(0%)	(8%)

[a] Includes 9 cases with both FTU and CTU involvement; 3 with both FTU and Independents; 11 with both CTU and Independents; 2 with FTU, CTU and Independents; 1 with FTU, CTU and TUC.
[b] CTU includes the Federation of Hong Kong and Kowloon Labour Unions (FLU).

Source: authors' analysis of unpublished Labour Department files.

© Blackwell Publishers Ltd/London School of Economics 1997.

it has a higher ratio of disputes per 1,000 members (0.22 for the CTU; 0.12 for the FTU in 1994) and of workers involved per 1,000 members (22.76 and 13.48 respectively in 1994). There were no significant differences between the two organizations in the cause or industry pattern of disputes.[10] This is surprising, since the FTU has more members in the dispute-prone sectors, and suggests that the CTU was seeking out disputes and/or was being sought out by non-members. This outward-looking approach represents a commitment to inclusive unionism on the part of CTU officials which goes back to the days of the CIC's involvement in organizing, but is controversial within the CTU itself:

> In the CTU some of our members did express [the view] that a lot of the CTU energy is spent in solving problems which are not our problems. . . . That is why we . . . hold a forum regularly for our members to understand the problems faced by workers in other sectors so that they can support . . . the CTU spending resources on helping other workers . . . (CTU official)

FTU disputes have a lower incidence of industrial action, and particularly strikes, than CTU disputes, while non-union disputes show an even lower level (Table 9). Disputes in which both the FTU and CTU become involved have a relatively high level of industrial action, perhaps reflecting a tendency for both to become involved in the more intractable cases. Surprisingly, 50 per cent of disputes involving unions other than the FTU and CTU resulted

TABLE 9
The Conduct of Labour Disputes by Union Status, 1990–1994

	(1) Non-union	(2) FTU	(3) CTU	(4) Joint FTU/CTU	(5) Other unions
% of disputes involving industrial action	8	21	37	50	50
% of disputes involving strikes	3	5	15	8	10
N = 797 (All disputes)	591	106	78	12	10
Mean settlement per worker involved*	10.4597 (1) < (2)	25.6593 (2) > (1), (3)	13.5948 (3) < (2)	24.9804 —	7.000 —
N = 249 (Disputes resolved by conciliation with a specified cash settlement)	160	46	37	4	2

Notes:
(1) Cases without union involvement.
(2) All FTU cases without CTU involvement.
(3) All CTU cases without FTU involvement.
(4) Cases with both FTU and CTU involvement.
(5) Cases without either FTU or CTU involvement.

* Analysis of variance: $F = 5.6324$, $P = 0.0002$. Null hypothesis of equal means across groups is rejected. Modified LSD multiple range tests ($P \leq 0.1$) shown under each group mean. $X < Y$ shows that mean of group X is significantly smaller than that of Y.

Source: As Table 8.

© Blackwell Publishers Ltd/London School of Economics 1997.

56 *British Journal of Industrial Relations*

in industrial action, a higher proportion than the CTU, although there was only one strike.[11] This must be seen in the context of a low overall level of disputes for such unions, and may reflect more institutionalized employment relationships, particularly in the public sector, with only the highly contentious issues resulting in a dispute.

However, the CTU's relatively militant approach to disputes does not necessarily translate into more favourable outcomes for workers. Looking at disputes fully settled by conciliation for a specified sum, the FTU outperforms the CTU ($p<0.1$) and non-union ($p<0.05$) cases in terms of the settlement per worker (Table 9). We must be careful not to draw too much from one type of settlement,[12] but it is hardly consistent with the view of the FTU as a relatively ineffective representative of labour's interests.

Furthermore, we encountered some evidence that the FTU was now showing greater militance. Labour officers and FTU officials referred to several recent disputes involving industrial action or a particularly tough line being taken by an FTU official. As we noted earlier, this may be driven by competition with the CTU. FTU officials themselves agreed that recruitment competition was important, and a CTU official suggested that the FTU was copying the CTU's approach with an eye to the 1995 LegCo elections. Another suggestion was that there is a 'generational' effect, pushing the FTU towards a more energetic pursuit of the representative function. A younger FTU official we interviewed suggested that many of the leaders of FTU affiliates, often men in their forties or older, adopted a strongly pro-China stance, something he feared might undermine workers' confidence in the FTU.

6. Summary

Hong Kong unions have generally been weak at the workplace, and from the 1950s played only a limited role in employee representation and industrial conflict. This was compounded by the FTU's industrial pacifism, especially once China opened its doors to the West. However, since the 1980s, largely in response to greater democratization and public awareness of social issues, unions have developed their political role, participating in elections and campaigning on social and labour issues. The Nationalist–Communist split has been eclipsed by a new rivalry between the FTU and the CTU. This time it is a rivalry which makes an issue of union strategy and the conduct of disputes. The concerns are now those of Hong Kong itself, rather than of the Nationalist–Communist struggle.

The CTU–FTU rivalry appears to reflect the classic adversarial–dualism distinction, but Hong Kong lacks the socio-political conditions necessary for either — on the one hand an industrial relations system operating under pluralist assumptions, on the other a one-party communist state with institutionalized unions playing a transmission-belt role. The result is that, while the CTU and FTU may espouse strategies that owe much at the

© Blackwell Publishers Ltd/London School of Economics 1997.

ideological level to the adversarial and dual models, in practice there are real problems implementing such strategies.

The CTU is a militant representative of workers in disputes and a champion of labour and social welfare interests in the political arena. The basis of the CTU/CIC project, particularly as applied to blue-collar workers, is that such groups have grievances, but lack the resources and proletarian community support to mount sustained collective action, expressing their grievances instead through individual exit, or at most in sporadic outbursts of collective protest (Deyo 1989). Since the 1970s, middle-class activists such as those of the CIC/CTU have provided support for workers' action (Chai 1993: 140). The aim of CTU leaders is thus to raise workers' consciousness of the meaning and the importance of trade union 'autonomy', and to focus grass-roots pressure for union representation.

The CTU strategy depends on the development of class consciousness as a basis for collective action, something that has been lacking in Hong Kong. However, the apoliticized nature of Hong Kong society has been changing, and it is possible that interest in political democracy will spill over into the industrial sphere. Levin and Chiu (1994: 157) argue that this is possible only if the unions exert greater pressure for workplace participation than they have done historically. On this reading, the CTU represents a break with the traditional industrial pacifism of the older unions and may yet succeed in extending union representation.

However, the CTU's inability to win recognition from employers, along with the uncertainties surrounding 1997, mean that there is doubt as to its ability to develop and sustain its brand of adversarial trade unionism. The CTU's approach depends on the continuance of political pluralism. The leadership recognizes this and aims not only to win workers' support in the industrial relations arena, but also to link this with popular political causes and the maintenance of democracy, trusting in popularity and mass support as a safeguard against repression. Holding the federation together may itself prove to be a major challenge, particularly if there is growing concern among existing and potential affiliates not to be too closely associated with the controversial political line of the CTU leadership.

In contrast, the FTU has a reputation as a moderate organization with a relatively co-operative approach to employers and pro-Beijing politics. However, we cannot write off the FTU as an effective representative of worker interests. First, in spite of some recognition of the production function in our interviews and in FTU policy pronouncements, there was no evidence in our survey that FTU officials were significantly more committed to the production function than were officials of other unions. Second, while the FTU adheres to the pro-Beijing line on key political questions, this has left room for social reformism on labour and social issues, and this seems set to continue beyond 1997. Third, the FTU plays a significant role in labour disputes. While it is often regarded as less militant than the CTU, and its disputes are less likely to involve industrial action, we found no evidence that the outcomes for workers are less favourable. Fourth, the FTU is

© Blackwell Publishers Ltd/London School of Economics 1997.

making itself more accessible to workers and has been adopting a more active and perhaps more militant approach to labour representation.

The TUC has had a lower profile than either the CTU or the FTU. The organization has held together, reflecting the traditional loyalties, but an ageing organization and the challenge of structural economic change make it unlikely that the TUC will play a major role in future, short of a dramatic shift in the organization's strategy. One possibility might be a closer relationship, perhaps even a merger, with the CTU. Both CTU and the TUC are affiliated to the International Confederation of Free Trade Unions, and as such may be expected to share the ideal of autonomous trade unionism. However, links between them have been limited and there are barriers to a closer relationship owing to differences in strategy and ideology, and also in terms of leadership compatibility.[13] The CTU has failed to attract major new affiliations even from independent unions in recent years, partly because of its controversial political and industrial relations image, which the highly cautious leadership of the TUC is unlikely to find appealing.

The independent unions, although the largest group, are more inward-looking than the CTU and FTU in terms of industrial conflict, recruitment and political activity. They can be assertive representatives of their members' interests when the occasion requires it, but their identification with the broader union 'movement' is generally limited. There is also some evidence that they have an aversion to political controversy, making a major extension of CTU affiliation unlikely.

7. Future prospects

Hong Kong is not alone in East Asia in experiencing developments in autonomous union action in recent years. South Korea and Taiwan, for example, both saw new and relatively militant unions challenging the status quo in the 1980s, linked to democratization in Korea and the end of martial law in Taiwan (Frenkel 1993a; Wilkinson 1994; Kuruvilla and Venkataratnam 1996). Such developments have been associated with social change, rising expectations and a better educated and more highly skilled work-force, as the economy emerges from the initial period of low-cost, export-oriented industrialization. The progress of autonomous unions is unlikely to be even, and government and employers may attempt to exert their influence, through suppression, incorporation or union substitution strategies. Nevertheless, to the extent that the demand for autonomous trade unionism is linked to the level of economic and social development, it may be difficult to reverse.

Given the return to Chinese sovereignty, what insights does the experience of the PRC offer for the future of Hong Kong unions? The production function was paramount for PRC unions (Wilson 1986; Ng and Ip 1994), but market reforms have called this into question (Ng and Ip 1994; Zhu 1995). With the devolution of authority to enterprise management, it is difficult to

© Blackwell Publishers Ltd/London School of Economics 1997.

argue that management and workers have no clash of interests, and the All-China Federation of Trade Unions (ACFTU) faces the threat of autonomous grass-roots action should it fail to convince an increasingly demanding work-force that it is competent to perform the representative function (Chan 1993; Zhu 1995). There are signs that the ACFTU is responding (Ng and Ip 1994: 14–19), and some ACFTU cadres have called for unions to be relieved of their social welfare functions, to adopt a more adversarial representational role, and for membership to be made voluntary (Chan 1993: 55).

However, unions have been slow to negotiate collective labour contracts with foreign-invested enterprises (*China Staff*, April 1995: 10), and union cadres are still conscious of the need to perform an 'ambassadorial' role with foreign investors (*South China Morning Post*, 22 February 1994: 1). Furthermore, the response to the Tiananmen Square protests and the suppression of the independent unions (Chan 1993: 57), show that the Communist Party will not tolerate radical labour action. Thus, unions in the PRC are in a state of flux, with market reforms pushing them towards a more vigorous pursuit of the representative function, but with continuing political pressures to play a co-operative role in economic development plans. Hong Kong unions may face a similar dilemma.

It has been suggested that there will be a weakening of trade union autonomy in Hong Kong, as Beijing favours the pro-China unions, along with a convergence towards the PRC model of socialist 'state corporatism', involving state control of unions and the suppression of political opposition to minimize dissent and maintain a compliant labour force (Frenkel 1993). Certainly a more assertive state is likely, but for over a decade Hong Kong has seen a growing political consciousness, and while Beijing has rejected the 1995 LegCo, elected under Patten's democratic reforms, gradual democratization is still envisaged after 1997. The Basic Law provides for freedom of association, and while Beijing has old loyalties to the FTU, it has also been building links with other unions, including those of the CTU and TUC. The CTU's present strategy is problematic; however, to the extent that a degree of political pluralism is maintained, autonomous trade unionism is likely to find space, albeit more limited than in recent years. The SAR government might seek a corporatist arrangement in the more highly organized public services, but here the FTU is, relatively speaking, at its weakest, and the more assertive and highly skilled work-force might require a bargained approach. Whether a future SAR government could develop a corporatist labour policy along Singaporean lines is questionable, given the existing degree of pluralism in political attitudes.

The role of unions must be seen in the context of the political future of Hong Kong. The 'one country, two systems' concept in principle offers a high degree of autonomy to the SAR government, but the extent to which this can be realized in practice is uncertain. Two contrasting views exist on how Hong Kong's future autonomy might best be safeguarded (Wong and Levin 1995). The liberal view is that democratization and the development

© Blackwell Publishers Ltd/London School of Economics 1997.

60   *British Journal of Industrial Relations*

of a stronger political culture will deter Beijing from excessive intervention. The strategy of the CTU leadership draws on this. On this view, the survival of autonomous unions in Hong Kong will be an indicator of the survival of democracy and political freedom.

In contrast, the business lobby and pro-China groups have emphasized the need to reassure Beijing of the economic benefits of a capitalist Hong Kong, and democratization, including autonomous trade unionism, may be seen as a threat. First, such developments, along with an attendant increase in dissent and social protest, may prove unacceptable to the Communist regime, which appears more comfortable with economic than with political reform. Second, there is a risk of increasing demand for state welfare and other services and for greater regulation of business, thus undermining the attractiveness of Hong Kong as a place to do business.

The business lobby generally sees no role for unions, but, to the extent that unions are accorded a role by pro-Beijing groups, it is likely to emphasize social responsibility and the production function. However, the difference between playing such a role in Hong Kong and playing it under communism is that it is more difficult to appeal to a community of interests between capital and labour in a capitalist enterprise. This is becoming clear under the PRC's own economic reforms, and Ng (1994) argues that the Hong Kong FTU's experience in dealing with a capitalist economy may provide lessons for its All-China counterpart. The issue facing the Hong Kong FTU is the extent to which it can perform the representative function at the expense of a more co-operative line. In recent years, anxious to maintain its membership and to secure support for its candidates in elections, the FTU has tried to convince workers and voters that it is an effective representative of their interests. Whether such representative unionism will be sustained, especially if political freedoms and the CTU's activities are restricted, remains to be seen.

Acknowledgements

The authors gratefully acknowledge the financial support of the Hong Kong Polytechnic University (project code 340/183) and wish to thank the labour organizations, the Labour Department and the officials interviewed for their help. Thanks are also due to Ms Lee Wai-yi and Mr Jacky Cheung, who provided excellent research assistance.

Final version accepted 11 October 1996.

Notes

1. These were: the CTU, the FTU, the TUC, the Federation of Hong Kong and Kowloon Labour Unions (FLU), the Hong Kong Christian Industrial Com-

© Blackwell Publishers Ltd/London School of Economics 1997.

mittee (CIC), the Hong Kong Professional Teacher's Union and the Hong Kong Chinese Civil Servants' Association.

2. The CTU claims a total membership of 110,000, the Registrar's figures being an understatement because they exclude the FLU and the Federation of Hong Kong Transport Workers' Organizations, both of which are registered under the Societies Ordinance.

3. Our survey of union leaders (Section 5) found the mean number of years of union membership to be: TUC, 25 years; FTU, 18; CTU, 11; and Independents, 10.

4. There were 119 unions in the 'Public administration and defence' classification at the end of 1994.

5. For example the Cathay Pacific strike by the in-house Flight Attendants' Union in 1993.

6. The questionnaire was written in English, based on statements from the interviews and the theoretical literature, and then back translated into Chinese.

7. We also investigated this in a multivariate context. Four regression equations were estimated with the factor scores as dependent variables. Independent variables were: CTU, FTU and TUC dummies (independent union was the base case); industry dummies for transport and the 'community, social and public services'; total membership of the union; and the length of membership and union activity of the respondent (each measured in years). While none of these equations was particularly successful (the best R^2 adj was 0.10 in the case of factor 3; the worst was factor 1, where the overall F was not significant), it is notable that the union federation dummies were most often significant and were consistent with the results of the analysis of variance. Some industry dummies were significant: officials from public services unions were less likely to identify with the strong-form representative function, and transport unionists were more likely than others to agree with the weak form, as were officials in larger unions. The official's length of union membership and union activity were not significant. This suggests that the ideology of a federation may have more influence on officials' attitudes than does the length of the individual's union experience, and that such an influence may be independent of industry influence.

8. 'Disputes' are collective grievances involving more than 20 employees, while 'claims' are individual or small-group grievances, usually arising from a breach in the contract of employment or an alleged failure to adhere to the requirements of the Employment Ordinance. In our analysis, we focus on disputes.

9. This is even more significant given that the TUC's areas of membership concentration in manufacturing and restaurants are particularly prone to disputes.

10. This finding is based on chi-square tests comparing the distributions of FTU and CTU cases. Cases involving both the FTU and CTU were excluded so as to avoid understating any differences, as were all cases where the cause or industry was not clearly reported. Given the small number of cases in some industries, grouping of data was performed for this chi-square test: ISIC major divisions 1, 2, 3, and 4 were grouped as 'production', division 5 as 'construction', 6 as 'trade', and 7, 8 and 9 as 'other services'.

11. We group TUC and independent cases together because of the small numbers.

12. CTU officials allege that the FTU has turned away cases that do not have a good legal basis, in contrast to the CTU's declared willingness to help workers even in

© Blackwell Publishers Ltd/London School of Economics 1997.

62   *British Journal of Industrial Relations*

the absence of a clearly 'winnable' claim. To the extent that this is true, the above figures understate the CTU's relative achievements.

13. The TUC leadership has been dominated by an older generation of activists, many of them coming into the organization before the 1970s. The CTU, by contrast, tends to have younger and often better educated leaders. The two groups are likely to have very different ideas on organizational priorities.

References

Chai, B. Karin (1993). 'The politicization of unions in Hong Kong'. In D. H. McMillen and M. E. DeGolyer (eds.), *One Culture, Many Systems: Politics in the Reunification of China.* Hong Kong: Chinese University Press, pp. 121–57.

Chan, Anita (1993). 'Revolution or corporatism? Workers and trade unions in post-Mao China'. *Australian Journal of Chinese Affairs*, no. 29: 31–61.

Chau Leung-chuen (1993). 'Labour and employment'. In Choi Po-king and Ho Lok-sang (eds.), *The Other Hong Kong Report 1993.* Hong Kong: Chinese University Press, pp. 127–46.

Deyo, Frederic C. (1989). *Beneath the Miracle: Labor Subordination in the New Asian Industrialism.* Berkeley: University of California Press.

England, Joe (1989). *Industrial Relations and Law in Hong Kong*, 2nd edn. Oxford: Oxford University Press.

Frenkel, Stephen (1993a). 'Variations in patterns of trade unionism: a synthesis'. In Frenkel (1993b: 309–46).

—— (ed.) (1993b). *Organized Labor in the Asia–Pacific Region: A Comparative Study of Trade Unionism in Nine Countries.* New York: ILR Press.

Hong Kong (1991). *The Basic Law of the Hong Kong Special Administrative Region of the People's Republic of China.* Hong Kong: Joint Publishing (HK) Ltd.

Hong Kong FTU (1988). *Carry Forward the Fine Traditions, Open Up a New Prospect.* The Executive Committee's Work Report delivered by Mr Cheng Yiu-tong at the 26th Session of the Federation's General Meeting, 24 April.

Kuruvilla, Sarosh and Venkataratnam, C. S. (1996). 'Economic development and industrial relations: the case of South and Southeast Asia'. *Industrial Relations Journal*, 27(1): 9–23.

Lee, Chun-keung (1994). 'Lao Zi Xie Shang Huo Dong' ('The practice of joint consultation in Hong Kong'). *Labour Relations Newsletter (Hong Kong Labour Department)*, no. 35, December, p. 2.

Leung, Benjamin K. P. (1994). 'Social inequality and insurgency in Hong Kong'. In Leung and Wong (1994: 177–96).

—— and Chiu, Stephen (1991). *A Social History of Industrial Strikes and the Labour Movement in Hong Kong, 1946–1989.* Occasional Paper no. 3, Social Sciences Research Centre/Department of Sociology, University of Hong Kong.

—— and Wong, Teresa Y. C. (eds.) (1994). *Twenty-five Years of Social and Economic Development in Hong Kong.* Hong Kong: Centre of Asian Studies, University of Hong Kong.

Leung Hon-hoi (1992). 'The growth of white-collar unionism in Hong Kong'. In E. K. Y. Chen, R. Lansbury, Ng Sek-hong and S. Stewart (eds.), *Labour –Management Relations in the Asia–Pacific Region.* Hong Kong: Centre of Asian Studies, University of Hong Kong, pp. 384–93.

© Blackwell Publishers Ltd/London School of Economics 1997.

Levin, David A. and Chiu, Stephen (1993). 'Dependent capitalism, a colonial state, and marginal unions: the case of Hong Kong'. In Frenkel (1993b: 187–222).

—— —— (1994). 'Hong Kong's other democracy: industrial relations and industrial democracy in Hong Kong'. In Leung and Wong (1994: 133–76).

Littler, Craig R. and Palmer, Gill (1986). 'Communist and capitalist trade unionism: comparisons and contrasts'. In Pravda and Ruble (1986b: 253–71).

Ng Sek-hong (1984). 'Where are trade unions destined to be?' *The Hongkong Manager*, May: 19–22.

—— (1994). 'Industrial relations in joint ventures in China'. In S. Stewart and N. Campbell (eds.), *Advances in Chinese Industrial Studies*, Vol. 4. Greenwich, Conn.: JAI Press, pp. 13–28.

—— and Cheng Soo-may (1993). 'Transition to more cooperative and consensual patterns of labour-management relations: Singapore and Hong Kong compared'. *Asia Pacific Journal of Management*, 10: 213–27.

—— and Ip, Olivia K. M. (1994). 'The public domain and labour organizations'. In M. Brosseau and Lo Chi-kin (eds.), *China Review 1994*. Hong Kong: Chinese University Press, pp. 14.1–14.33.

Pravda, Alex and Ruble, Blair A. (1986a). 'Communist trade unions: varieties of dualism'. In Pravda and Ruble (1986b: 1–21).

—— —— (eds.) (1986b). *Trade Unions in Communist States*. Boston: Allen & Unwin.

Suen Wing (1994). 'Labour and employment', in D. H. McMillen and Man Si-wai (eds.), *The Other Hong Kong Report 1994*. Hong Kong: Chinese University Press, pp. 149–64.

Tang, Sara F. Y., Lai, Edmond W. K. and Kirkbride, Paul S. (1995). *Human Resource Management Practices in Hong Kong: Survey Report*. Hong Kong: Hong Kong Institute of Human Resource Management/Ashridge Management College.

Turner, H. A. *et al.* (1980). *The Last Colony: But Whose?* Cambridge: Cambridge University Press.

—— Fosh, Patricia, and Ng Sek-hong (1991). *Between Two Societies: Hong Kong Labour in Transition*. Hong Kong: Centre of Asian Studies, University of Hong Kong.

Wilkinson, Barry (1994). 'The Korea labour "problem"'. *British Journal of Industrial Relations*, 32: 339–58.

Wilson, Jean (1986). 'The People's Republic of China', in Pravda and Ruble (1986b: 219–51).

Wong, Thomas and Levin, David A. (1995). 'The social structure', in Ng Sek-hong and David G. Lethbridge (eds.), *The Business Environment in Hong Kong*, 3rd edn. Hong Kong: Oxford University Press, pp. 44–63.

—— and Lui Tai-lok (1994). 'Morality and class inequality'. In Leung and Wong (1994: 76–93).

Xu Jia-tun (1993). *Xu Jia-tun Xianggang Hui Yi Lu* (Xu Jia-tun's Memoirs of Hong Kong). Hong Kong: United Daily News Limited.

Yahuda, Michael (1996). *Hong Kong: China's Challenge*. London: Routledge.

Yeung Chi-kin (1988). 'Joint consultation, collective bargaining and trade union recognition: status and prospect'. In Y. C. Jao, D. A. Levin, Ng Sek-hong and E. Sinn (eds.), *Labour Movement in a Changing Society: The Experience of Hong Kong*. Hong Kong: Centre of Asian Studies, University of Hong Kong, pp. 54–66.

Zhu Ying (1995). 'Major changes under way in China's industrial relations', *International Labour Review*, 134(1): 37–49.

© Blackwell Publishers Ltd/London School of Economics 1997.

[17]

AFTER THE HONG KONG MIRACLE

Women Workers under Industrial
Restructuring

─────────── Stephen W. K. Chiu and Ching Kwan Lee

Hong Kong's transformation from an entrepôt into an industrial economy after World War II has often been cited as an outstanding example of the "East Asian miracle." It has received much attention from other parts of the world, both inside and outside the academy. The World Bank's magnum opus on the East Asian economies—China, Japan, the four "little dragons," and the high-growth Southeast Asian economies—published in 1993 provides a definitive statement of the hows and whys of the East Asian miracle. The irony is that just when the report was hot off the press, the Hong Kong "miracle" seemed to be running out of steam. The emergence of new low cost competitors in the Asian-Pacific region and the erosion of Hong Kong's own cost advantage caused growth in manufacturing industries to slacken in the 1980s. By the late 1980s, a trend of absolute decline set in and a wave of outward investment commenced, leading to deindustrialization and structural transformation in Hong Kong's domestic economy. Although this economic restructuring is receiving some scholarly attention, its impact on labor has not been widely recognized. How is Hong Kong's allegedly flexible labor market coping with the backbreaking pace of restructuring?

This article evaluates the effect of Hong Kong's economic restructuring on the lives of manufacturing workers, especially women. Women workers have always been the backbone of local industries; they account for the majority of rank-and-file production workers. Hong Kong's working mothers

─────────── Stephen W. K. Chiu and Ching Kwan Lee are Professors in the Department of Sociology, the Chinese University of Hong Kong. The study was funded by the Hong Kong Federation of Women. The authors are grateful to Chiu Man-yiu, Fung Wai-hing, Hung Ho-fung, Lau Yin-ngo, Wong Yan-yin, and especially Vivien Leung Hiu-tung for research assistance.

© 1997 by The Regents of the University of California

and daughters have improved their own families' standards of living and contributed significantly to Hong Kong's remarkable record of economic growth. How has the reversal of fortune in manufacturing affected them? Our study contributes to an emergent literature on industrial restructuring in the newly industrializing economies (NIEs) by highlighting the effect of gender through both quantitative and qualitative data and unraveling the "hidden injuries" of restructuring that have reshaped the ethos of Hong Kong's working women.[1] Our central argument is that familial responsibilities, gendered age discrimination, and Hong Kong's specific forms of restructuring have combined to undermine living and working conditions among women workers.

De-industrialization in Hong Kong

By the 1980s, Hong Kong's days of record-setting economic growth were over. Though the average annual growth rate of real gross domestic product (GDP) for the decade was 7.5%, the 5.22% average for 1990 to 1994 indicates that the economy has slowed. More important than this macroeconomic trend, however, is the economy's tremendous structural transformation. In the 1980s, the manufacturing industries' share of the GDP declined relative to other sectors, especially the tertiary sector. More recently, financial and business services have shown very impressive growth in their share of the GDP, as has the commercial sector.

Employment trends mirror these changes in the GDP's sectoral distribution. The number of workers employed in manufacturing was almost halved from 1987 to 1994, sliding from 918,600 to 558,300, due to the relocation of manufacturing production to countries with lower wage costs. The commercial sector became the largest employer, increasing from 627,900 to 849,000 workers over the same period, and there was growth in financial and business services sector employment as well. In the decade ending in 1994, manufacturing employment slumped by 39.2%, financial and business services shot up 132%, transport and communication rose 64%, and commercial and community services saw equally impressive growth.

---

1. Most recent studies focus on structural transformation rather than human consequences. For example, see Stephen W. K. Chiu, K. C. Ho, and Tai-lok Lui, *The City-States in the Global Economy* (Boulder, Colo.: Westview, 1997); Gordon Clark and W. B. Kim, *Asian NIEs and Global Economy* (Washington, D.C.: Johns Hopkins University Press, 1997); and Hugh Patrick and Larry Meissner, eds., *Pacific Basin Industries in Distress* (New York: Columbia University Press, 1991). For studies that pay special attention to employment implications of restructuring, see John Bauer, "Industrial Restructuring in the NIEs: Prospects and Challenges," *Asian Survey* 32:11 (November 1992), pp. 1012–1025; and Committee for Asian Women, *Silk and Steel: Asian Women Workers Confront Challenges of Industrial Restructuring* (Hong Kong: Committee for Asian Women, 1995).

754   ASIAN SURVEY, VOL. XXXVII, NO. 8, AUGUST 1997

Deindustrialization has been central to the labor market's structural transformation. Parallel to the decline in employment, the number of manufacturing establishments dropped from a high of 50,606 in 1988 to 33,863 by the end of 1994.[2] At least one-third of all manufacturing establishments either closed, moved elsewhere, or went broke in 1988, causing tremendous disruption to workers' lives. While the relocation of Hong Kong capital has been fairly easy due to China's geographical proximity and the adoption there of an open door policy in the late 1970s, it has not been so for labor. Although the specter of mass unemployment has not yet hit Hong Kong, the real employment problem is more complex than the "frictionless" picture painted by the government and many economists.[3]

Gender Differences in the Changing Labor Market

It is easy to conclude from official unemployment and underemployment statistics that the impact of this rapid industrial restructuring has been minimal and does not differ across gender. The overall unemployment rate stood at only 3.5% in the fourth quarter of 1995, and what's more it appeared to be less serious for women than men, with the figure for women standing at 3% versus 3.7% for men.[4] Nevertheless, this rosy picture is deceptive as unemployment figures only report those who say they are "actively seeking" work.[5] As we shall see below, women can be out of work for a variety of reasons that are often misinterpreted as "not actively seeking," which results in their not being officially classified as unemployed.

The General Household Survey also does not disclose detailed breakdowns by gender in the unemployment statistics. As a first approximation of the gender differences in unemployment, we shall first examine those official

---

2. See Industry Department, *Hong Kong's Manufacturing Industries* (Hong Kong: Government Printer, 1994); and Census and Statistics Department, *Quarterly Report on General Household Survey: October to December 1994* (Hong Kong: Government Printer, 1994).

3. See Wing Suen, "Sectoral Shifts: Impact on Hong Kong Workers," *Journal of International Trade and Economic Development* 4:2 (July 1995), pp. 135–52, for a more positive assessment of the impact of the restructuring process on workers.

4. Census and Statistics Department, *Quarterly Report on General Household Survey: October to December 1995* (Hong Kong: Government Printer, 1996). Female unemployment rate in Hong Kong has been consistently lower than, or at most equal to, that of male workers. In 1986, for example, the male and female unemployment rates were 3% and 2.5%, respectively. See Census and Statistics Department, *Annual Digest of Statistics 1990* (Hong Kong: Government Printer, 1991).

5. In recent years, the Census and Statistics Department claims to have included those who are without a job but not seeking work because they "believe that work is not available to them." It is unclear, however, how they operationalize this and how much of the unemployment figure it publishes can be accounted for by this reason.

STEPHEN W. K. CHIU AND CHING KWAN LEE   755

statistics that do offer gender information. According to the 1991 census, 31% of the entire female labor force was employed in manufacturing, and these women accounted for approximately 41% of the total manufacturing workforce.[6] Hence manufacturing has traditionally been a major source of employment for women. Over the past few years, however, restructuring has reduced significantly the number of women employed in this sector. According to government employment statistics, the number of female employees in all manufacturing industries dropped from 430,376 to 164,248, or by more than 61% between the end of 1987 and the end of 1995. In contrast, the number of male employees dropped by a substantial but smaller 52%—from 437,571 to 211,518 (see Table 1).[7] Consequently, the share of women workers in manufacturing dropped from 49.6% in 1987 to 43.7% in 1995.

In three of the five industries that employ the most people in manufacturing, the percentage decrease in the number of female employees far exceeded that of male employees. The rate of decrease was greater for men only in textiles, where restructuring started much earlier.[8] Even these figures might underestimate the magnitude of shrinkage in female manufacturing employment because they reflect the number of "persons engaged" in "manufacturing establishments," which includes clerical and other nonproduction workers whose numbers have increased with the transformation of many assembly plants to regional operations headquarters or trading offices. These workers are often female, and their increase is very likely to have concealed the percentage decrease of female employment in manufacturing.

Naturally, employment trends affect wages. In 1987, the real average daily wage in manufacturing at the craftsperson and operatives level including fringe benefits was HK$198.3 (HK$7.8 = US$1) for women and HK$267.5 for men. By 1995, the average wage for the men had increased to HK$288.9 while that of the women had declined to HK$184.2 (see Table 2). Correspondingly, the female-to-male wage ratio slumped from 77.3% in 1987 to 63.8% in 1995. This pattern appeared to repeat throughout the major manufacturing industries; for example, male workers in electronics manufacturing saw a hefty 16% increase in real wages over the 1987–95 period versus a 4.3% drop for female workers. Restructuring has clearly resulted in both a

6. Census and Statistics Department, *Hong Kong 1991 Population Census: Main Tables* (Hong Kong: Government Printing Department, 1992).

7. Census and Statistics Department, *Employment and Vacancies Statistics (Industrial Sector)*, various years. Third-quarter figures are reported because the department gives detailed breakdowns by gender only for third-quarter figures.

8. Stephen Chiu and David Levin, "The World Economy, State, and Sectors in Industrial Change: Labor Relations in Hong Kong's Textile and Garment-Making Industries," in Stephen Frenkel and Jeffrey Harrod, eds., *Industrialization and Labor Relations* (Ithaca, N.Y.: ILR Press, 1995), pp. 143–75.

756 ASIAN SURVEY, VOL. XXXVII, NO. 8, AUGUST 1997

TABLE 1 *Changes in Employment in Selected Manufacturing Industries by Gender, 1987–1995*

Industry		1987	1995	1987–1995 (%)	Female/Male Difference[1] (%)
Apparel, except footwear	total	258,221	80,222	−68.9	
	male	79,162	24,477	−69.1	
	female	179,059	55,745	−68.9	0.2
Textiles	total	119,081	58,787	−50.6	
	male	67,013	30,877	−53.9	
	female	52,068	27,912	−46.4	7.5
Plastic products	total	77,963	14,511	−81.4	
	male	41,322	9,078	−78.0	
	female	36,641	5,433	−85.2	− 7.1
Electrical and electronic machinery	total	125,841	34,823	−72.3	
	male	45,753	15,037	−67.1	
	female	80,088	19,786	−75.3	− 8.2
Fabricated metal products, except machinery and equipment	total	60,800	24,435	−59.8	
	male	40,551	17,299	−57.3	
	female	20,249	7,136	−64.8	− 7.4
All manufacturing industries[2]	total	867,947	375,766	−56.7	
	male	437,571	211,518	−51.7	
	female	430,376	164,248	−61.8	−10.2

SOURCE: Census and Statistics Department, *Report of Employment, Vacancies and Payroll Statistics* 1988, 1996.
[1]A negative percentage indicates a larger decline in female employment than male employment.
[2]The total is for all manufacturing industries including the above five.

substantial decline in real wages for women and a widening of the income gap between genders in the manufacturing industries.

We can infer from official statistics that the impact of restructuring is markedly different for men than women. However, official statistics rarely provide information beyond the aggregate level and cannot satisfactorily answer questions regarding the impact of industrial restructuring on women workers. With the large-scale contraction in industrial employment, who remain and who are finding new jobs in other sectors? Is unemployment a serious problem among former women manufacturing workers? What effects do changes in employment status have on their standards of living and families? How are they coping with the changes and how do they perceive their own situations?

STEPHEN W. K. CHIU AND CHING KWAN LEE　757

TABLE 2 *Changes in Real Average Daily Wages by Gender in Selected Manufacturing Industries, 1987–1995 (*constant 1992 Hong Kong dollars*)*

		1987	1995	Change 1987–1995 %
Garments	male	256.6	225.9	−11.9
	female	214.7	184.2	−14.2
	total	222.0	189.7	−14.5
	female/male %	83.7	81.5	
Cotton spinning and weaving	male	249.3	242.5	−2.7
	female	212.9	203.1	−4.6
	total	231.1	225.9	−2.2
	female/male %	85.4	83.8	
Knitting	male	249.3	230.6	−7.5
	female	225.6	170.8	−24.3
	total	231.1	182.6	−21.0
	female/male %	90.5	74.1	
Electronics	male	229.3	265.3	15.7
	female	189.2	181.1	−4.3
	total	198.3	197.6	−0.4
	female/male %	82.5	68.2	
Metal products	male	258.4	282.6	9.4
	female	163.8	164.5	0.5
	total	216.5	260.6	20.3
	female/male %	63.4	58.2	
Manufacturing	male	267.5	288.9	8.0
	female	198.3	184.2	−7.1
	total	220.2	220.4	0.1
	female/male %	74.1	63.8	

SOURCE: Census and Statistics Department, *Report on Survey of Wages, Salaries, and Employee Benefits* 1988, 1996.
Note: All wage statistics are September figures. The Hong Kong dollar has been pegged to the U.S. dollar at the rate of HK$7.8 = US$1.

Since Hong Kong's official statistics do not furnish breakdowns sufficient for a more detailed examination of the conditions of women manufacturing workers, we conducted a telephone survey in January 1995 and interviewed a random sample of 1,004 production workers employed in the manufacturing sector since 1989. While our target group was women workers, for comparative purposes we generated a random sample that included roughly the same

758   ASIAN SURVEY, VOL. XXXVII, NO. 8, AUGUST 1997

TABLE 3 *Changes in Employment Status of Manufacturing Workers by Gender, 1989–1995*

Current Employment Status	Male (%)	Female (%)
Remain in manufacturing	55.7	27.9
Switch to other sectors	25.9	14.5
Exit from full-time employment	5.6	30.9
Unable to find full-time job	12.8	26.7
Total	100	100

SOURCE: Survey sample.

Note: This table reports the current employment status for all male and female respondents who worked full-time in manufacturing industries five years before the interviews took place with or without job change.

proportion of men and women (499 male, 505 female).[9] We also conducted in-depth interviews with 40 women respondents working in different employment situations. The data from these interviews allow us to understand what economic restructuring means to those most affected by it.

Our survey questionnaire examined changes in self-reported employment status types during the period 1989–95, including *remain* employed full-time in the manufacturing sector; *switch* to full-time job in nonmanufacturing sector; *exit* from factory with no intention of seeking full-time employment; and *unable* to find a full-time job. A summary of the paths of employment change among our respondents during the five years leading up to the survey is presented in Table 3.

Striking differences emerge when we compare the paths of employment change across genders. Industrial restructuring has pushed out (exited or unemployed) far fewer men who had been working full-time in manufacturing five years before (18.4%) than women (57.6%). The most common reasons for exiting among men were retirement (30.8%) and health problems (19.2%) in contrast to child care for women. A significant difference can also be seen among respondents unable to find a full-time job (52%), with the figure for women (26.7%) being more than double that of men (12.8%). Some 61% of

---

9. The telephone poll was conducted by the Social Sciences Research Centre at the University of Hong Kong based on a sampling frame of the entire residential telephone directory of Hong Kong. The success rate was around 60% out of the eligible respondents (those who worked in manufacturing five years ago) contacted. We have in fact conducted another wave of telephone interviews with another 141 women workers, but since including them in statistical analysis would present sampling problems, we have decided to exclude them from the following analyses.

the women in this category had been seeking work for six months or more, versus only 39.3% of the men. Male workers were more likely to remain employed full-time in manufacturing industries, and a majority (60.4%) also reported an improvement in their standard of living, while only 43% of the women said the same. Moreover, a higher proportion of men were able to find new jobs in nonmanufacturing sectors, and of the workers who switched, only 38.9% of the men reported difficulty in finding their new job as opposed to about half of the women.

Fading of the Hong Kong Dream

Our survey and interviews also revealed other aspects of women workers' collective predicaments: employment fluidity, a declining living standard, severe demoralization, and a trend of involuntary retreat to domesticity. In the following sections, we will present the women's own articulation of these "hidden injuries."[10] Overall, we have found that the decline of the economic sector that provided these women a stable place in the working class has pushed them both down the class ladder and backward to domesticity as their gender destiny. These conditions are fostering a collective sentiment among a substantial segment of Hong Kong's working class that we describe as the fading of the "Hong Kong dream." Widely espoused by the Hong Kong people and recognized by social scientists as one of the central components of the ethos of the Hong Kong Chinese, the Hong Kong dream is a "success ethic" that places a premium on opportunity and portrays economic advancement as a matter of individual effort, hard work, and perseverance. Cultural wisdom has it that these eventually pay off as both improvements in standards of living and optimism for the future generation. But among women workers and their families caught up in the economic restructuring, the discrepancy between life situation and social ideology has become so wide that the latter can no longer be sustained.

Employment Fluidity

The interview data suggest that we need to rethink conventional conceptual tools before we can properly understand the women's difficulties; it may be misleading to focus on their current employment status as an indicator of their employment situation. The status recorded in government statistics and survey responses is only a snapshot of the instant at which the data were collected. The fluid and unstable labor market can create employment situations for women that defy easy description by conventional categories like "employed" or "unemployed." Economic restructuring has cut loose the as-

---

10. This term is borrowed from Richard Sennett and Jonathan Cobb, *The Hidden Injuries of Class* (New York: Knopf, 1972).

sumed stability, or the sense of relative permanence, inherent in the notion of "employment status." Many women reported that they went from "employed" to "unemployed" and then back to "part-time employment" in the five months between their first and second interviews with us. Therefore, it may make more sense for government officials, labor unions, women's advocates, and academics alike to consider "stability and quality of employment" rather than content themselves with a simplistic notion of employment status. The following profiles show that women shared very similar attitudes toward employment irrespective of their employment status when we met them in 1995: all wanted to work. Nevertheless, all were subjected to forces of circumstance (emanating from the labor market and their families) that pushed them to change jobs (both within and between sectors) and mode of (un)employment frequently over the past five years.

Women remaining in full-time manufacturing jobs.   Among those workers who remained in manufacturing (27.9% of the entire sample), the majority (63%) were in the textiles and apparel industries, with a substantial minority (17.3%) in basic metal and machinery industries (mainly electronics). Close to half (52%) of the workers had moved from one factory to another at least once in the past five years and another 37% changed firms more than once, reflecting the instability of the manufacturing sector. Factory relocation (to China or elsewhere) caused 33.3% of our respondents to change factories, followed by plant closure (20.3%) and slack work (15.9%). The fluid employment situation of those who remain in manufacturing is illustrated by the case of Yuk-fu (35), a packing worker in a printing factory (all interviewee names are fictitious). She immigrated from mainland China in 1980 and is now the mother of three school-age children. Over the past 15 years, she has worked in factories making watches, toys, and stationery, and also did a stint as a salesperson before shifting to this printing factory. When asked whether her present job was temporary or permanent she replied:

> I don't think it can last. Sooner or later, my department will disappear. Our boss has told the four of us in the packing section 'either I don't give you a pay raise or I close this department.' All because he has opened a factory in China. . . . We replied, 'If you like, you can dismiss us (and compensate us with severance payment). . . .' Therefore, he gave us a $5 per day pay raise and fired a worker who has worked for him for nine years. . . . because it seems a new legislation is to come out and he's afraid that we would meet the 10-year qualifying condition for more compensation payment.

In short, even those workers we classify as "remaining in full-time manufacturing" articulate a strong sense of insecurity. They have seen the relocation mega-trend and heard about the difficulty of those over 30 years of age have switching to service industry work. Many also stay on in anticipation of the

STEPHEN W. K. CHIU AND CHING KWAN LEE   761

end of the factory work era, hoping to collect the statutory lump sum severance payment when the factory eventually closes down.

Women having switched to service jobs.   The sense that employment is insecure and fluid is equally widespread among women who have switched to service-type jobs. The collective mood of this group is one of uncertainty and vulnerability; they often attribute landing their present service job to "luck" or "accident," usually after having made a strenuous job search effort. Their stories about the fluid nature of the service job market warn against any complacency in interpreting official statistics for "switched employees." In fact, only 72 (14.5%) of our respondents who had been working in manufacturing five years before managed to find nonmanufacturing jobs. Within this small group, 25% found work in the commercial (wholesale, retail, hotel, and restaurants) sector; 24% were in the community, social, and personal services sector (mainly the public sector); and 18% went into the finance, business, and commercial services sector. The most frequently mentioned reason given by those who have gone to another sector is relocation of factories to the mainland and elsewhere (30%); underemployment ("not enough work") came next with 24%, and 10% got out of manufacturing due to factory closure.

Lai-hing is an example of the women in this category. In her mid-40s with a son (18) and a daughter (13), she had been working as a full-time shop assistant in a convenience store for less than a month when we interviewed her. She had worked in sweater manufacturing for some 30 years before quitting factory work in 1994 when her factory closed down. She first took up temporary work in 1991, when the factory began suspending production on occasion owing to a drop in orders received; the extra jobs she held, which each lasted for only a few months, were working as a fruit shop assistant and a part-time position in a convenience store. In 1992, she worked at McDonald's for two days and applied for supermarket cashier positions. At the time of the interview, she was working the night shift and earning $5,000 per month. She explained that she was usually able to find a job only when there was an urgent demand for labor; otherwise, employers considered her too old and chose to wait for other applicants. When asked if she would keep her present job, she replied: "I believe there will be changes in the near future. I have heard that the rental lease expires in October. When that happens, the store must move elsewhere. . . . I wish they would stay in this neighborhood. . . . For us, location is a big constraint." Experiences such as this show that even when service jobs are available, they may not offer stable employment.

Reluctant exits from full-time employment. The situation these women face well illustrates the extent to which economic restructuring has disrupted the conventional meaning of employment status categories. Many of these women had held a series of service sector jobs before reluctantly exiting from full-time employment due to a combination of push and pull factors: low pay, balancing family time with work time, and physical exhaustion. Their work histories from the five years under discussion suggest that having switched to the service sector does not preclude the possibility of imminent exit from full-time employment. Their frequent job changes show that the line between being employed, unemployed, or reluctantly exiting is indeed volatile and thus has limited conceptual value. That many are engaged in some sort of informal work such as baby-sitting or direct sales when they are not formally employed only complicates the picture. Of the women surveyed, 30.9% who worked in manufacturing five years before had exited from full-time employment; most (65.3%) left the manufacturing sector between 1992 and 1994. The main reasons for this exit (with no intention of seeking a new job) were child care (51.1%), retirement (12.3%), health conditions (7.7%), and to care for family members (6.5%). This offers a strong indication that familial responsibilities fall mainly on women, compelling them to leave gainful employment and, in combination with the unfavorable labor market under industrial restructuring, pushing them back into full time domesticity.

It is common for women who seem to have successfully switched to the service sector to have changed jobs frequently over several years, only to quit full-time work again due to other constraints. Siu-wah's work history offers an illustration. Aged 30, with two children in kindergarten, Siu-wah worked in the clothing industry from 1982 to 1989 and then as a waitress for four different restaurants in just four years between 1991 and 1994. Although the pay for waitressing was good, averaging $8,000 per month, she quit when her daughter started refusing to be baby-sat by others, "crying every morning on leaving home." The long work hours at the restaurant, ending at 11 P.M., made it difficult for her to help with her daughter's school work. At the time of the interview, she was baby-sitting a six-year-old for $2,500 a month. As she explained her employment plans, her familial priority was clear: the "primary one is a transition period for her. I'll see how she manages her academic work and then decide whether to work or not. Even if I work, it may not be full-time."

Yet, another factor—age discrimination—frustrated the intentions of another woman's return to full-time employment. Siu-guan (45) had resumed garment making in 1986 when her two children were "old enough to look after themselves." When the owner of the factory where she worked closed it in 1992 to relocate it and emigrate, she found a job as an assistant in a cake shop. Yet, in 1994, she was laid off because of her age: "[Afterwards] I tried to look for other jobs. I even went to the Labor Department to register a job search. There was one with a suitable work schedule, 11 A.M. to 3 P.M. It turned out that they only wanted those below 40. That's why I was not accepted. I am so much older that I have lost interest in finding a job." Such examples illustrate just two of the reasons why some women cannot find full-time employment.[11] The choice of these two not to pursue full-time employment should be read as a response to unfavorable circumstances emanating from the labor market and their families that have left them with no other option. This point also makes the dividing line between this group—the reluctant exits—and the next—an elusive one.

Unemployed women. This last group comprises those who, at the time of our interviews, wanted to find full-time employment but could not do so. These women feel the greatest insecurity about the future and are those most victimized by age discrimination. A total of 26.7% of the women surveyed reported they could not find full-time jobs. Among them, 62 (46%) were employed part-time, yielding a total of 73 unemployed women workers, or 14.5% of all respondents. This figure is much higher than the 2.7% unemployment rate reported in the General Household Survey carried out by the government for the first quarter of 1995, with the caveat that the two sets of figures are not directly comparable.[12] Our figure, however, is consistent with the findings of surveys using different methods and with different sample sizes, including these conducted by the Federation of Trade Unions (FTU) on its members, that typically suggest an overall unemployment rate in the 10%–15% range.[13] Almost a quarter of our respondents (23.1%) had lost the

---

11. While the conflict between women's familial responsibilities and their labor force participation has been a long-standing phenomenon, it was relatively easier for women factory workers to juggle between family and work in the past because of the relative stability in working hours in manufacturing and the proximity of most factory premises to public housing estates. In the past, for example, women workers sometimes went grocery shopping during lunch breaks. Furthermore, the relative stability of manufacturing earnings sometimes allowed women workers to pay for baby-sitting (often by neighbors and relatives), which relieved them of some familial responsibilities. Now with a much reduced and unstable income, it is difficult for women to afford baby-sitting or other childcare services.

12. Our questions mainly tap the perception of the women workers themselves, while the official surveys attempt to measure employment status by a combination of objective and subjective indicators. Census and Statistics Department, *Quarterly Report on General Household Survey: January to March 1996* (Hong Kong: Government Printer, 1996).

13. For example, the Lai Chi Kok Organization of Workers' Concern for Society surveyed 297 interviewees and reported a 11.4% jobless rate in August 1995. The Hong Kong Association for Democracy and People's Livelihood also showed that 13% of its respondents were unemployed. See *Hong Kong Staff*, "Hong Kong Worried About Jobs" (September 1995), pp. 3–5.

last jobs they had held in 1992, 19.2% in 1993, and 20% in 1994; and another 30.8% claimed to have been unable to find full-time work since 1991.

Owing to the declining condition of the labor market, only 54.1% of the unemployed women were still seeking full-time jobs: a large number appear to have been too discouraged and frustrated by labor market conditions to continue their searches. Considerable difficulties face those who were searching; the majority (67%) had been actively seeking work for six-months or more, including 51.4% who have been doing so for more than a year. A typical example is Shun-ming (45), a divorced single mother living with her adopted son (14), who came from China in 1985. She worked in the garment and the basic metal industries until 1990 when she injured her ankle at work. The metal factory was relocated to the mainland and she found a part-time job selling honey:

> If it was not relocated, I thought I was to work at that factory until I died. . . . That part-time job only lasted for two months. They did not need anyone after the honey promotion period. . . . I also tried working as an amah in a construction site, making coffee, cleaning and cooking for the workers. When I went to the site, I realized that I had to walk up to [a] great height and I was very scared. . . . Every time I asked about other jobs, they would ask for my age and then tell me to wait for their calls. It means they don't want me.

She since gave up her search and was living on the $4,364 she received monthly in public assistance and as a disability allowance.

While we have presented only a fraction of the case histories we gathered for illustrative purposes, we can see that women in all employment status categories express a strong sense of insecurity about their financial and employment prospects. The overall employment situation for women workers is characterized by fluidity and instability to the extent that those with jobs are as insecure as those without. The results of our study also imply that the mass of unemployed women workers is in aggregate terms much larger than what is reflected in government statistics, given the increasingly blurred boundaries between the categories of employed, unemployed, and underemployed.

Downward Mobility: Declining Standard of Living

Besides fluidity of employment, Hong Kong's women workers suffer from severe downward mobility. Secure livelihoods and a steady rise in living standards over the decades for Hong Kong's working class have helped legitimize a social order founded on laissez-faire competition and staggering income inequalities. When economic restructuring pushes workers out of their secured niche in the labor market, it also removes a whole way of life that has been treasured for its security. Downward mobility has become widespread

STEPHEN W. K. CHIU AND CHING KWAN LEE 765

and painful, as many women workers have found that their decades of work experience and multifaceted production skills have become unmarketable. Wages and overall income plummeted for those with jobs, while others became welfare dependent. Moreover, the families of women who lost their jobs suffer as well. Working class families have to cut back on groceries, leisure, and social life, and many have found it necessary to borrow from relatives to pay for family medical and educational expenses. Since women are usually the ones responsible for dividing the shrinking household finances, they are the family member most painfully aware of such a downturn of fortunes. Among those women who did manage to hang onto jobs in manufacturing, about 30% said they did not have enough work, which suggests underemployment has indeed been quite serious in this sector. Close to half of all women workers who remained in manufacturing also reported a deterioration in their standards of living, a phenomenon consistent with the aforementioned decline of real wages among female, rank-and-file manufacturing employees.

Our interviews abound with the expressions of women suffering from declining standards of living. Nostalgic for the days when she worked as a skilled sewing machine operator in the garment industry, Sim-jing's (39) interviews echoed with the loss of pride she had in her occupation and the income it once brought. She explained that she agreed to be interviewed not for the recompense but for the chance to put down on record what "her" industry was like:

> The golden years were 1976 to '78. . . . Even in 1989, I was earning more than $300 a day. . . . This was one-third more than my husband's clerical job. . . . In 1991, I moved here [to Shatin] and found this sales job. Working from 1 P.M. to 8 P.M., I only got $3,800 a month! You can imagine how unhappy I was. But that's the reality.

Another woman, Kwan-ying (45), found herself unemployed after having worked for an electronics factory since 1968, the only job she'd ever held. She was very proud of her "jack-of-all-trades" experience in microchip production and was equally angry at being abandoned by society because employers came to consider her "underqualified" but "overexperienced." Her co-workers' job search experiences convinced her that she was unlikely to break out of this predicament:

> I was earning $10,000 a month [in 1990]. If, say, we worked an extra of four hours a day, we could have up to $15,000 a month. . . . Alas, now, the more experienced we are, the more difficult it becomes for us to find work. Employers find our experience too expensive, and experience is worth nothing when you don't have qualifications. . . . Some time ago, I was looking for a $5,000 a month job (half of what I used to earn!), and I found nothing.

Children's educational expenses impose difficulty for working class parents like Shun-hing (45), who was unemployed and felt guilty at being unable to buy a computer for her son: "He wants very much to buy a computer. I want to buy him one, too. At the time when I got my severance payment, it cost about $10,000. But then I wasn't sure how long my welfare assistance would last. . . . I thought if I bought the computer, it would use up two months' living expenses. How could we live? So, I didn't buy any computer."

Not only do educational expenses become a financial burden, but normal social life and leisure also have to be consciously cut due to reduced income. As a result, women workers increasingly find themselves socially isolated. For instance, Lai-yee (39) had been unemployed for four years and reported that she had frequently declined to meet with her former co-workers; maintaining friendship had become too costly for her. Other unemployed workers also reported the need to cut back on entertainment and social activities. Declining standards of living and uncertainty about the future mean cutbacks on essentials as well. Yuk-ying (34) reported that she cut food purchases and had to resort to borrowing when emergency needs arose. Her family, like many others in our study, simply did not have enough surplus to save: "I cut back on groceries, buying lower quality stuff. And I do not eat as much. I skip breakfast and lunch. . . . At one point, I have lost 30 pounds. I used to weigh 150 pounds. . . . Once my eldest daughter was sick and hospitalized, and I had to borrow from my mother."

Reductions in real income are also accompanied by an intensification of work for those still employed. Women who have taken up service work have discovered that their new workplaces permit less autonomy than the factory shop floor, especially that given those who had done piecework. The burden of service work has also sometimes had a serious impact on women workers' personal lives. Because of her overnight work schedule and her domestic responsibilities, one woman working in a 7-Eleven had divided her sleeping time into three sessions: "I get off from work at 7 A.M. I take care of my daughter for a while, have breakfast, take a shower and then sleep for about three hours. Then I take my daughter to school and sleep again in the afternoon. Around dinner time, I wake up again. . . . When I am tired, I sleep. When I am not tired, I wake up to do housework." Examples such as these make it clear that women workers as a whole suffer from a general decline in the quality of their working, family, and social lives. Their children are also implicated in this downward mobility trend, which adds even more psychological burdens to women who have always put their children's welfare first.

STEPHEN W. K. CHIU AND CHING KWAN LEE 767

Demoralization and Moral Critique

Deterioration of material life is but one aspect of the collective experience of women workers on their gradual descent of the class ladder. A less visible injury caused by economic restructuring may be demoralization and its corollary of serious affronts to the dignity and moral self-worth of these women. In the course of our interviews, we heard their expressions of anger and frustration, directed first at the unjust socioeconomic circumstances and then at themselves for being useless and helpless in the prime of their lives. A moral critique of Hong Kong's social order can be found behind this collective demoralization; it creeps in between the lines of the women's stories and is not without ambiguities and contradictions.

To understand the ethos of the working class being uprooted by the events of the past decade, one must recognize that work has always been more than a means of survival. Like workers elsewhere, our women interviewees have made moral and emotional investments in their jobs and workplaces and these have become integral parts of their identities. That is why, when their way of life is threatened, much more than a deterioration in standards of living is at stake. Demoralization is the day-by-day sapping of confidence and a prolonged weakening of will, the result of several years of job search that ends in vain. Lai-yee's situation after her four year search for a stable job was representative. Having worked as a skilled machine operator in a textile mill for 23 years and gained the respect of her "sisters" there, Lai-yee was forthright in her pride about her work and the bleak prospects of finding work again.

> I started at the age of 11 [the legal age for work was 14] and after four years I already became an assistant to the chief spinner. The chief spinner operated a big machine that controlled all other spinners. It's lots of responsibility. Later, I even became a chief spinner. . . . Now, after so many attempts, I really lost all motivation to find work. I have been to many interviews, they either say I am too old or they have already hired someone else. I went for a job selling jeans, and the guy said it to my face, 'I only want those between eighteen and twenty-two. Those younger ones, not your age.' Alas, how frank he was!

The experience of age discrimination was mentioned almost universally with anger by all interviewees. Many had, by the time of the interview, frozen out by a labor market that places a premium on younger age groups. Having been a skilled worker making leather apparel for a decade, Sin-fong was very frustrated by the unreasonable qualifications demanded by many employers. "Open the pages of a newspaper and you find that you are overaged and underqualified. There was an opening for a kitchen helper, preparing meals. I thought I had to settle for that since I could not find other things. I had never expected that even for this kind of job, people wanted you to know English too. . . . If they give you a job, the pay will be too low. There was a part-time cleaning job I asked about. Monday to Friday, 5 P.M. to 9 P.M., $1,500 per month. Only a bit more than $10 per hour and you have to work more than 20 days a month!"

Another target of the anger of women like Siu-guan was the unreasonably low pay rates of service jobs. "I went to an interview for an office assistant [job]. Only $170 per day, without any welfare or subsidies. It's not even a monthly paid, stable job. I don't care about long working hours but the pay must be reasonable. . . . I think it's age discrimination and that is very frustrating. I am so angry about this that I think I'd rather spend less than taking up the job."

Some interviewees told us they had nightmares from the stress about problems in the workplace, worrying that they would lose their jobs in the precarious labor market. Others pondered about the day when they would have to depend on government public assistance to survive, though most were of the attitude that they would not want to depend on the government unless absolutely necessary. One woman described applying for welfare as an option of last resort better only than dying. In the meantime, most found themselves dependent on their husbands, like Chui-king:

> In the past, I thought I was still young and there was no need to worry about unemployment. As long as I had the energy, I could get work anywhere. Now, my mentality is that since I cannot find work, I can only rely on my husband and my children. I depend on my husband for allowances and I have to be frugal. I go to the tea house less often.

This sense of helplessness is aggravated by the women's perception of the inefficacy of government assistance. About a quarter of the respondents had no idea of how the government could assist them, and only a small percentage (7.2%) had participated in government-sponsored retraining programs. Not surprisingly, only 6.0% of the respondents regarded retraining programs as helpful. One respondent admitted frankly that she participated in one such program merely to get the retraining allowance.[14] She did not believe that the training was useful for job hunting, a view corroborated by other sources. A small scale study conducted by the Federation of Trade Unions, for example, showed that fewer than half of the retrainees actually found jobs after completion of their training course.[15]

Involuntary Retreat to Domesticity
Finally, some women workers in our study have been deprived of stable employment and forced to withdraw from the formal labor market that once

---

14. A full-time retrainee receives a monthly allowance of HK$4,000, or about US$516.

15. *Apple Daily*, 4 March 1996.

STEPHEN W. K. CHIU AND CHING KWAN LEE   769

provided them with jobs, independent incomes, friends, social ties, and a sense of self-worth. Their involuntary retreat to full-time housewifery does not provide an easy way out. Like Betty Friedan's famous formulation of "the problem without a name" among American housewives in the 1950s and 1960s, we found our subjects trying hard to articulate a problem that Hong Kong society does not recognize. Kwan-ying, for example, always interjected "do you understand?" or "you won't understand" in her responses. She talked at length about the pains of being deprived of a job after so many years of work; staying home was compared to "becoming a disabled" person:

> In 1990, I started looking for jobs but no one wanted me. You know, I had a working life for more than twenty years and suddenly there was no more work to do. It's like becoming an idiot because there's too much dead time. . . . It feels like you're to explode. Even if I don't get paid, I'd like people to give me something to work on, to make use of my time. Do you understand? I have always worked. . . . Now, the whole thing is getting so depressing. It's not about whether I need a job for survival. I just cannot stand becoming disabled. Do you understand?

In contrast to the conventional wisdom that staying home is an easy solution to female unemployment, many women found full-time homemaking to be oppressive and detrimental to their mental and physical well-being. Workers like Fung-tai preferred the hectic schedule of her garment worker days to being left isolated at home. One woman told us how she became sick more often now that she was unemployed and at home all the time in comments echoing Fung-tai's. "I am now at home all the time and I feel so uncomfortable. There is pain here and there. When I was working, I was like an 'iron woman.' Staying home all day, getting bored and stuck, makes me feel sick. . . . My husband always says I have become very dumb and unresponsive. In the past, I could give him advice on his work. Now, strangely, I am no longer capable of doing that."

The retreat to domesticity incurs not only physical and psychological costs; it also throws women back into dependence, mostly on their husbands. The gender balance of power within the family, though never very even, is shifted further toward the husband, and the women feel deep indignation at such dependence. When Lai-tao became unemployed and the family income was reduced by $5,000 a month, she had to ask her husband and sometimes even her son for a larger allowance: "It's very insulting, face-losing. It makes you look like a useless person, if you don't work. If you work, at least, you can buy yourself a dress." Furthermore, in many cases the husband of a woman who loses her job becomes less willing to share in childcare or perform household chores. Women have thus found themselves confined again to a traditional feminine role from which they had at least a modicum of liberation when they held full-time jobs. The disturbing trend we have seen as economic restructuring unfolds is that working class women are being forced back into the home with domesticity their only option. In terms of the quest for equal opportunities for women, this is definitely a step backward.

Conclusion

Industrial restructuring in Hong Kong has dealt a heavier blow to the female segment of the working class than its male counterpart. Compared to their male, former co-workers, female workers have suffered from a higher unemployment rate, spent more time when searching for new jobs, and had to accept declining real wages. The collective predicament of women workers has four major aspects: fluidity and instability of employment, declining standards of living, severe demoralization, and involuntary retreat to domesticity. In this article, we have documented women's articulation of these hidden injuries, "hidden" because their predicament cannot be captured adequately by the categories used in official statistics, and because Hong Kong's society misconceives the home to be a panacea for the problem of women's unemployment.

This case study of the relationship between women and industrial development in Hong Kong suggests that the relative liberation of women from domestic dependence through formal employment can easily be reversed when the path of industrialization changes course. Today we see that, even though Hong Kong's economy remains robust and its populace enjoys one of the highest per capita income levels in Asia, women workers are once again being "marginalized" and viewed as "disposable" factors of production. From the perspective of women workers, the Hong Kong miracle has become a mirage and the Hong Kong dream has faded. The consequences of this collective demoralization among a sizable segment of Hong Kong's working class are yet to be explored.

[18]

The fraying of the socio-economic fabric of Hong Kong

Lau Siu-kai

Abstract Since the 1980s, Hong Kong has undergone momentous socio-economic changes, which in turn have greatly affected public attitudes toward society and the economy. Interpersonal trust and the sense of community have weakened. Hong Kong as a society is increasingly seen as unfair in the sense that it is not perceived as a land of opportunities for the hardworking. The capitalist rules of the game are increasingly considered by the people to be unacceptable. Public demands for more governmental intervention in the economy, particularly in the area of income redistribution, are increasingly raised. Nascent feelings of class antagonism are palpable as economic inequalities are getting worse. As social conflicts of various kinds proliferate, public anxieties about Hong Kong's fraying socio-economic fabric have come to the fore. People expect the government and the legal institutions to strengthen social order. At the same time, however, public trust of all social, economic and social authorities is declining. Accordingly, as social discontent and anxieties accumulate, the socio-economic system of Hong Kong will face serious challenge in the years ahead.

Keywords Hong Kong; social change; social conflict; trust; alienation.

When China regains its sovereignty over Hong Kong on 1 July 1997, the Basic Law, the mini-constitution of Hong Kong which stipulates the territory's political structure and core public policies will be put into effect. The primary objective of the Basic Law is to implement the principle of 'one country, two systems' in Hong Kong, the crucial component being the preservation of Hong Kong's economic and social *status quo*. Article 5 of the Basic Law explicitly states that '[t]he socialist system and

Professor Lau Siu-kai is professor and chairman, department of sociology, and director, the Hong Kong Institute of Asia-Pacific Studies, both at the Chinese University of Hong Kong.

Address: Hong Kong Institute of Asia-Pacific Studies, The Chinese University of Hong Kong, Shatin, NT, Hong Kong

© Routledge 1997

0951-2748

policies shall not be practiced in the Hong Kong Special Administrative Region, and the previous capitalist system and way of life shall remain unchanged for 50 years.' [1] The enshrinement of the socio-economic system as of 1984, when the Joint Declaration on the Question of Hong Kong, which sealed the political future of the territory, was signed by China and Britain, is thrashed out in minute detail in the Basic Law. The preponderant *raison d'être* of the Basic Law as a political contract between China and the people of Hong Kong is to allay the latter's anxieties and fears, the most glaring of these being that the *status quo* will be drastically changed to the detriment of their interests.

The nature of Hong Kong's socio-economic system as conceived by the Basic Law drafters consists of a free-wheeling capitalist system and a fair society, at the centre of which is a limited government practising *laissez-faire* and social non-interventionism. As late as the early 1980s there was by and large a consensus on the essentials of such a system, though social development had already produced some changes and further changes were in the offing.[2] Since the signing of the Joint Declaration, however, Hong Kong has continued to experience breathtaking changes which have left their imprints on people's social and economic attitudes. To anticipate the following discussion, the direction of attitudinal changes is the appearance of more public doubts about the essentials of Hong Kong's socio-economic system and the concomitant erosion of interpersonal trust and social cohesion. The fraying of Hong Kong's socio-economic fabric adds difficulty to the already difficult process of political transition. The long-term implication is even more far-reaching and troublesome, for it has to do with the adequacy of the Basic Law as a legal framework to foster social integration in and facilitate the further development of Hong Kong. It has also to do with the legitimacy of the mini-constitution in the territory.

This paper aims to document the changes in social and economic attitudes among the Hong Kong people. The data used come from a number of questionnaire surveys conducted by the author and others in the last decade,[3] particularly that for 1994.[4]

Social change since the early 1980s

In order to make sense of the changes in people's social and economic attitudes, it is necessary to describe briefly the pattern of social change that has engulfed Hong Kong since the early 1980s.[5] These changes have to a certain extent transformed the economic system and the structure of life chance opportunities in Hong Kong, and in the process also altered interpersonal relationships and organizational patterns in society. The end result of these changes is weakened public identification with the dominant values of Hong Kong in the postwar period.

Among these changes, of predominant importance is the slowdown in economic growth, which has produced wide-ranging effects on society.

Hong Kong used to pride itself on the postwar economic miracle based on export-driven industrialization. Nevertheless, since the early 1980s, the economic growth rate has gone down. Economic performance in recent years has further faltered, despite the uplifting effects of economic takeoff in China. Until a few years ago, Hong Kong had recorded impressive growth in its real gross domestic product (GDP). The average annual growth rate of GDP between 1961 and 1973 was 9.5 per cent, moderating to 8.9 per cent between 1974 and 1983. In 1984–88, GDP grew by 8.1 per cent a year. Since then, the growth rate has plummeted. GDP grew by 2.6 per cent in 1989, 3.4 per cent in 1990, 5.1 per cent in 1991, 6 per cent in 1992, 5.9 per cent in 1993 and 5.5 per cent in 1994.[6]

Unfortunately, concomitant with sluggish economic growth is inflation, 'which has much to do with the imported inflation produced by the pegging of the Hong Kong dollar with the US dollar, the brain drain and the shortage of labour. Most of the time since the early 1980s, Hong Kong has been plagued by a high inflation rate (high by comparison with the past). The percentage rise in the Composite Consumer Price Index in 1982, 1987, 1988, 1989, 1990, 1991, 1992, 1993 and 1994 is 10.9, 5.7, 7.8, 10.3, 10.2, 11.6, 9.6, 8.8 and 8.8.[7] The skyrocketing of the cost of housing represents the most conspicuous and damaging dimension of inflation. The prohibitive cost of housing has barred a lot of people from home ownership, with the younger generation particularly hard hit. Inflation has fuelled speculative activities of various sorts, generating wealth redistributive effects of a kind not favourable to public respect for those time-honoured moral values such as honesty, diligence, modesty, prudence and frugality. Jittery efforts by everybody to get the fast buck and to beat inflation at each other's expense have created much stress and strain in society.

Hong Kong has also witnessed de-industrialization since the early 1980s. There is an irreversible trend for local industries to move into China to capitalize upon the cheap labour there. This led to a huge 50 per cent reduction in the manufacturing workforce of the territory in the period 1984–94.[8] De-industrialization has created increasing unemployment or partial employment among manufacturing workers, whose skills cannot easily be transferred to the expanding service sector, which ironically suffers from a labour deficit, part of which is filled by labour import. An underclass consisting of the old and the poorly educated has emerged, which will prove to be enduring. Even within the service sector, most of the jobs available are low-paying ones, whereas only a minority of people with professional, managerial and entrepreneurial skills are able to benefit disproportionately from expanding opportunities in the financial, investment and commercial fields.

Economic changes in Hong Kong have tremendously widened the gap between the rich and the poor. Industrialization in the early postwar period went hand in hand with rising economic equality. However, the continuous decrease in income inequality in the first three decades of the postwar period was interrupted in the mid-1970s; since then income inequality has worsened. The Gini Index is .49 in 1960; it falls to .43 in 1971. Between 1971 and 1976 it remains at .43, gradually rising to .45 in 1981. In 1991 it rises to .48, and it appears that income inequality will continue to grow in the future.[9] In all likelihood, the Gini Index must have substantially underestimated the actual magnitude of economic inequality in Hong Kong, for it relies solely upon the income figures collected in censuses and by-censuses, whereas the grossly unequal distribution of wealth is not taken into account.

Accompanying rising economic inequality is a reduction in mobility opportunities for the increasingly better educated young people. Although the middle-class sector in Hong Kong has expanded with the enlargement of the service sector, the increase in high status and high paying jobs in that sector failed to match the rapid rise in the number of education-ally qualified people, despite the sudden increase in such jobs arising from the emigration of their occupants because of 1997. Although Hong Kong has become a 'credential' society, the utilitarian value of education has depreciated.

At a time when opportunities for upward mobility for the educated fail to match their expectation, thus causing a sense of plight in the swelling middle class, Hong Kong is witnessing the consolidation of an upper class of the super-rich who increasingly tend to impede the chances of the ambi-tious and the capable to move to the top. The conspicuous consumption of the rich and their luxurious lifestyle are given wide, occasionally scan-dalous, coverage by the mass media. The crass materialism of the upper class coexists with their lack of cultural hegemony, thus depriving the rich of social respect. Instead, their snobbery and social aloofness have alien-ated them from the common folk, who look upon the rich with a mixture of admiration, jealousy and scorn.

The simultaneous appearance of an enduring underclass, an increasingly closed upper class, and a 'crisis-ridden' middle class has produced social disharmony and friction among socio-economic groups, perhaps even a visible sense of class antagonism.[10]

The 1997 issue must be the most powerful factor in loosening the fabric of the Hong Kong community. Anxieties about an uncertain political future have already produced an exodus of people, prominent among whom are the better educated and those with professional and managerial skills.[11] Although to a certain extent the inflow of expatriates has replenished the talent pool, the newcomers' commitment to Hong Kong is fragile. Sensing their collective powerlessness, Hong Kong people are using every means to safeguard the future of themselves and their families, including illegal, illicit or shady methods. Self-seeking overrides mundane moral concerns. Self-interest drives out concern for others. Altruistic feelings, which have never been bountiful in a society of immigrants who have really settled down only since the 1960s, have substantially depleted. The rise of interpersonal and social conflicts on the one hand, and the erosion of respect for authorities of various kinds (who are seen as ineffective in safeguarding Hong Kong's interests in an unpredictable environment and who are seen as eschewing their social responsibilities by their alacrity to abandon Hong Kong) have together produced a social milieu suffused with greed, querulousness, disorientation, sullenness, cynicism, small-mindedness, intolerance and nastiness. These are all signs of a *fin de siècle* mentality. The bashing of the rich, the powerful and the well-known has become a popular pastime. And these sentiments are increasingly exploited by the profit-minded mass media, which in turn exacerbate the divisive tendencies in society.

In short, changes since the 1980s have weakened the authorities, the institutions and the community of Hong Kong. And their impact on people's social and economic attitudes are palpable.

Interpersonal trust

The most conspicuous victim of social change since the early 1980s is the trust people have in one another. Admittedly, interpersonal trust in Hong Kong has never been in abundant supply. In 1985, while 42 per cent of respondents in the survey disagreed or strongly disagreed with the statement that most people could not be trusted, at the same time 57.1 per cent agreed and 2.6 per cent agreed strongly that 'in these days one really doesn't know whom to rely on or trust'.

The situation has apparently worsened since then. In 1991, on the one hand 59.4 per cent of respondents said that they trusted or very much trusted the Hong Kong people, and 54.6 per cent reported that they liked or very much liked them; on the other hand, as many as 57.1 per cent thought that more and more Hong Kong people were using improper methods to pursue their self-interests, with only 9.7 per cent disagreeing. Moreover, 45.1 per cent agreed that Hong Kong people were like a sheet of sand, while 40.6 per cent disagreed; among those who agreed, only 19.3 per cent were optimistic about the possibility of Hong Kong people getting united, but a much larger proportion (53 per cent) were pessimistic.

Over the last few years, interpersonal trust has apparently tumbled. Slightly less than a third (30.1 per cent) of respondents in 1994 agreed that most Hong Kong people could be trusted. About the same proportion (33.5 per cent) disagreed, with 31 per cent taking a neutral position. Interestingly, the older respondents were more trustful of others, which bespeaks a higher degree of social alienation among the younger people. Furthermore, Hong Kong people have an unflattering image of others as members of society. In 1994, very few respondents (14.9 per cent) saw a strong sense of social responsibility among other people. More of them (35.2 per cent) were of the opinion that the sense of social responsibility was weak, while the plurality of them (40.8 per cent) considered it as average. Nonetheless, at the interpersonal level, the situation is much better. When queried as to whether Hong Kong people were willing to help others, the impression of the respondents was obviously better. About half of them (51.6 per cent) gave a positive answer. Only 12.3 per cent denied that Hong Kong people were willing to help, with 34 per cent giving 'average' as the reply.

Hong Kong as a fair society

In the postwar period, economic growth, a fluid class structure, minimal government intervention in society and the expansion of education have afforded a lot of opportunities for social mobility to the people of Hong Kong. These objective factors find their subjective counterparts in the ethos of the Hong Kong Chinese which sets great store on individual competition in a context of free enterprise as the mechanism for success and failure. A fair society is commonly defined as a society with equality of opportunity. In Hong Kong, public demand for equality of opportunity is extremely strong. In 1992a, 96.9 per cent of respondents agreed that every citizen should have equal educational opportunity regardless of his/her income or wealth. 82 per cent were of the opinion that equality meant that all people would have an equal chance to own their business.

Equality of outcome is not seen as a criterion of fairness. For example, in 1992a, most of the respondents (73.8 per cent) did not agree with the statement that 'equality means that all people earn about the same income'. In the same vein, 81.4 per cent of respondents in 1994 disagreed with the view that a fair society was one where everybody had the same income. Even so, the Hong Kong people are not necessarily against a society with equality of outcome. In 1985, it was found that 55.3 per cent of respondents agreed or strongly agreed that a *good* society was one where there was not much difference in income (39.9 per cent disagreed or strongly disagreed).

Under this definition of fairness, it follows that as long as people see plenty of opportunities in Hong Kong, they will consider society to be fair, otherwise a sense of social injustice will arise.

Generally speaking, up to now Hong Kong is still seen as a fair or equal society by the people. In 1985, 49.8 per cent of the respondents considered Hong Kong society to be fair or very fair, while only 38.2 per cent thought otherwise. In 1991, just over half of the respondents (51.9 per cent) rated Hong Kong as a fair society (32.7 per cent did not think so). In 1992a, 61.6 per cent agreed that Hong Kong was an equal society. The idea that Hong Kong is a land of opportunity is still in vogue. In 1985, a large majority of respondents (87.6 per cent) agreed that Hong Kong was a place full of opportunities. 84.2 per cent of respondents in 1986 agreed that in Hong Kong, provided a person had the ability and worked hard, he or she should have the opportunity to improve his or her social and economic status. In 1994, as many as 82.5 per cent of respondents still saw Hong Kong as a society with plenty of opportunities. Particularly important is the fact that people tend to think that they can experience upward social mobility in their lifetime. In 1986, 85.4 per cent of respondents were confident that the chance that they could improve their social status was great or very great. By the same token, 65.6 per cent of respondents in 1994 thought that they had more opportunities than their parents for achieving social success.

As a result of the popular belief in individual initiative under a capitalist system, the rich people in Hong Kong are accorded a legitimate status. 48.1 per cent of respondents in 1986 thought that most rich people obtained their wealth by individual efforts. Only 27.5 per cent explained the success of the rich as resulting from economic exploitation or illicit means. Similarly, the 1985 survey showed that 55.4 per cent of respondents did not agree with the statement that the money of the rich was acquired through exploiting the toiling masses. By the same token, poverty is also explained in individualistic terms. In 1986, 59.7 per cent of respondents used individual factors to account for the poverty of people, and 40.3 per cent employed social causes. In 1993, 52.5 per cent of respondents imputed personal problems as the causes of poverty, whilst 28.8 per cent cited unfair treatment by society.

The widening of the gap between the rich and the poor since the early 1980s has only slightly eroded the legitimacy of the rich. As can be seen in Table 1, in 1994 the public still attributed the success of the rich to individual initiative.

Nevertheless, ominous signs are looming on the horizon. There is lingering and perhaps even growing public unease at the growing economic inequalities in Hong Kong and the worsening public image of the wealthy. In 1986, 67 per cent of respondents saw the gap between the rich and the poor as serious, 16 per cent as average and 17 per cent as not serious. In 1990, 56.4 per cent of respondents castigated wealth distribution in

Table 1 Publicly perceived factors of success for the rich (1994) (in percentages)

	Not important	Average	Important
1 Initiative and risk-taking	6.4	11.6	74.7
2 Family background	3.8	17.8	62.7
3 Diligence	6.0	15.9	74.9
4 Help from 'patron'	18.0	21.0	49.9
5 Capability	4.4	16.6	75.2
6 Exploiting others	29.8	28.2	28.1
7 Illicit means	28.6	29.9	23.3
8 Good luck	20.0	25.4	45.3
9 Good education	18.4	23.9	52.8

Source: Author

Hong Kong as unfair, with only 26.4 per cent praising it as fair. Moreover, in the same survey, 69.7 per cent affirmed that there was conflict between social classes, with 18.2 per cent denying it. In 1992a, just over half of the respondents (50.6 per cent) agreed with the comment that 'Hong Kong's failure is in allowing the rich to get richer, and the poor poorer,' with 42.4 per cent disagreeing. In 1994, 57.3 per cent of respondents were of the opinion that wealth distribution in Hong Kong was unfair, only 12.7 per cent rating it as fair.

The 1992a survey also detects pockets of discontent with regard to incomes earned by particular social groups. 30.1 per cent of respondents considered most business people to be getting more income than they deserved. The corresponding figure with respect to professionals is 22.4 per cent. Most importantly, as many as 33.5 per cent of respondents were of the view that the working people in Hong Kong did not get a fair share of the profits on what they produced.

Moreover, people are not pleased with the sense of social responsibility displayed by their social superiors. In 1994, more than half of the respondents (58.6 per cent) did not think that the rich people had adequately fulfilled their responsibility to society, with only a minuscule 4.9 per cent taking the opposite view. Hence it is not surprising that only 13.9 per cent of respondents thought that the rich should be respected by the people of Hong Kong. Likewise, 51.8 per cent thought that people with social status had not assumed enough responsibility in society, while 7.4 per cent thought otherwise. More tellingly, only 20.7 per cent of respondents said they respected people with social status. And business leaders are the least preferred people as political leaders.

Acceptance of the capitalist system

In the mind of the Hong Kong people, the capitalist system is equated with economic prosperity and social stability. Public aversion to the socialist system in China further bolsters the legitimacy of capitalism in Hong Kong. Compared to other capitalist societies, Hong Kong's capitalist system is characterized by an untrammelled market and *laissez-faire* governance. The adoption of the *laissez-faire* approach to economic development in Hong Kong is the result of a configuration of historical and political factors,[12] but its tremendous success has turned it into a dominant economic ideology in the territory.[13]

There is tremendous support for the capitalist system in Hong Kong. In 1992a, 65.2 per cent of respondents saw the private enterprise system as generally a fair and efficient system. But there is evidence that support for capitalism is dropping slowly. In 1991, 58.4 per cent of respondents agreed that the capitalist system was a good system, with only 5.2 per cent disagreeing. In 1994, those who agreed had dropped slightly to 57 per cent, whilst those who disagreed rose to 10.2 per cent.

By the same token, even though the *laissez-faire* policy of the Hong Kong government is still endorsed by the people, their enthusiasm for it has apparently cooled down. In 1988, 57.5 per cent of respondents were for *laissez-faire*, and 23.8 per cent against it. In 1994, respondents who supported *laissez-faire* had dropped to 42.5 per cent, with 27.7 per cent opposing it and 16.2 per cent replying 'average'.

It is noteworthy that public endorsement for capitalism and *laissez-faire* in general does not imply that governmental intervention in particular areas will be regarded with revulsion. Indeed a strong traditionalist element is still alive and strong in the economic ethos of the Hong Kong people, who encourage or even demand governmental actions to promote economic well-being, redress social ills and rectify egregious inequalities.[14] Despite their acceptance of *laissez-faire* as a principle, the Hong Kong people's demand for public welfare is fairly strong and extensive. Underlying the view that the government is responsible for the provision of welfare is a fairly popular view of welfare as entitlement.[15] In fact, the demand for governmental intervention in the economic sphere has become even stronger since the mid-1980s, reflecting the existence of many problems created by Hong Kong's social and economic changes. Table 2 testifies to the growing public pressure for government action, which is particularly notable in the area of unemployment benefits, which are now unavailable.

It can be foreseen that public pressure on the government will become even stronger in the years ahead, and there is an increasing probability that it will and has to respond to such public demand. In the meantime, *laissez-faire* capitalism as an economic system is less respected and less taken for granted by the Hong Kong people than previously.

Table 2   Support for governmental economic functions in 1988 and 1994

Support for:	*1988*	*1994*
1 Set minimum wage of salary	57.6	60.7
2 Unemployment benefits	74.2	78.5
3 Regulate speculation	66.2	66.4
4 Central provident fund	80.3	80.2
5 Tax the rich more	74.7	76.7

Source: Author

Note: Four-point scale questions were used in the 1988 survey, whereas five-point scale questions were used in the 1994 survey. Therefore, if four-point scales were used in 1994, the figures on the table should be larger.
Figures on the table show the percentages of respondents who supported or strongly supported specific governmental economic functions.

Socio-economic discontent and social integration

The fading of the image of Hong Kong as a fair society and the ending of the sacrosanctity of the capitalist system in the mind of the Hong Kong people has the effect of loosening the socio-economic fabric of the community. It also engenders a moderate level of social and economic discontent, which is registered in survey data. In 1988, 1990, 1991, 1992, 1992b and 1994, the percentages of respondents who were satisfied with the economic conditions of Hong Kong are 57, 42.1, 31.6, 32.2, 35.3 and 39.3 respectively. With respect to social conditions, the figures in 1988, 1991, 1992 and 1994 are 40.9, 17.5, 22.1 and 26.7 respectively. The fraying of the socio-economic fabric has also produced increasing mistrust of social, political and economic authorities and pervasive cynicism.[16]

A substantial majority of people are worried about the rising level of social conflict in Hong Kong. In 1994, more than half of the respondents (65.6 per cent) agreed that there were too many conflicts of various sorts in society, with only 11 per cent disagreeing. There appears to be a search for social order among the people, and the consensus is that law is the linchpin of a viable social order. In 1985 law was perceived by a large number of respondents (41 per cent) as the most important integrative factor in society. The figure rose to 54.4 per cent in 1990. When asked in 1994 what was most crucial in terms of maintaining Hong Kong's future prosperity and stability, 21.1 per cent selected political system, 7.1 per cent rested their hope on political leaders, but the majority (58.7 per cent) picked law.[17]

The importance of law as the keystone of Hong Kong's social order is further manifested in the findings shown in Table 3, which records the degrees of importance accorded by the respondents in the 1994 survey to various factors with respect to their role in enhancing Hong Kong's social order.

The importance placed on morality and law is unmistakable. In fact, in Chinese culture, education and law are two sides of the same coin.

Table 3 Perceived importance of factors for Hong Kong's social order (1994)

Factors	Not important	Average	Important
1 Civic education	1.1	4.0	92.0
2 Harsh laws	8.0	17.1	69.9
3 Political leaders	10.7	21.5	53.9
4 Government	2.3	7.3	85.8
5 Legal system	0.4	3.8	91.4
6 Moral education	0.6	8.0	88.1
7 Democratic system	5.1	22.9	62.1
8 Economic growth	0.8	8.8	86.9
9 Higher living standard	0.5	7.1	90.0

Source: Author

Although an overwhelming majority (90.6 per cent) of respondents in 1994 agreed that people should continue to respect traditional Chinese moral values such as loyalty, filial piety, benevolence and righteousness, they despaired at the decrease in morality in Hong Kong. 51 per cent of them agreed that Hong Kong people's moral standard had declined, with 27.1 per cent disagreeing. Moreover, with the declining performance of the economy and increasing distrust of the government and political leaders, the role of law as the foundation of Hong Kong's social order becomes even more indispensable. The maintenance of the rule of law in Hong Kong is thus of central significance in countering the adverse effects of the fraying of the socio-economic fabric of the community. Although there is still strong respect for law in Hong Kong, unfortunately over the years trust in law has been slowly declining. The percentage of respondents who considered the legal system in Hong Kong as fair is 75.4 per cent in 1985, 69.2 per cent in 1988, 53 per cent in 1990, 51.4 per cent in 1991 and 48.3 per cent in 1994.[18] Moreover, the legal system is also seen as privileging the higher strata. Thus, in 1985, 73.4 per cent of respondents did not think that there would be fair trial in court as the rich would be favoured. The corresponding figures in 1988 and 1994 are 69.2 per cent and 74.9 per cent respectively. Public trust in lawyers and judges has also dropped. The percentages of respondents who trusted lawyers in 1988, 1990, 1991 and 1994 were 63.4 per cent, 60.8 per cent, 52.8 per cent and 33.9 per cent respectively. The corresponding figures for judges in 1988, 1991 and 1994 are 64 per cent, 61.4 per cent and 50 per cent respectively.

They are also increasingly concerned about the erosion of public respect for law. In 1994, although more than half (56.7 per cent) of respondents agreed that Hong Kong people were generally law-abiding (14.5 per cent disagreed), still as many as 30 per cent were of the view that, as compared to three years before, fewer Hong Kong people were law-abiding (15.7 per cent said more people had become law-abiding and 42.3 per cent saw no difference).

In addition, people are not satisfied with the law because they think that it is too lenient with the criminals. Such views were supported by 68.1 per cent of the respondents in 1994, with only 10.2 per cent not supporting.

Even more ominously, people do not have confidence in the legal system after 1997. In 1994 only 8.2 per cent of respondents expressed confidence in the post-1997 legal system, whilst 43.8 per cent had little or no confidence.

Thus, despite the increasing importance of law as the chief integrative factor in Hong Kong's social order and notwithstanding the public trust it still manages to enjoy, its effectiveness as an integrative mechanism has also depreciated.

Paradoxically, in spite of the fraying of the socio-economic fabric of Hong Kong, people's sense of community identification has not suffered. Compared to the past, Hong Kong people today are even more prone to

The fraying of the socio-economic fabric of Hong Kong   437

identify themselves as 'Hongkongese' instead of as 'Chinese'. In 1985, 59.5 per cent of respondents identified themselves as Hongkongese, 36.2 per cent as Chinese. The corresponding figures are 57.2 per cent and 26.4 per cent in 1990, 56.6 per cent and 25.4 per cent in 1991, 53.3 per cent and 32.7 per cent in 1993, and 56.5 per cent and 24.3 per cent in 1994. As the 'Hongkongese' identifiers are the better educated people, it thus appears that better educational opportunities for all have not significantly enhanced the sense of community identification. Be that as it may, Hong Kong's economic achievements have still generated a sense of pride among its people. In 1994, almost half of the respondents (46.5 per cent) were proud to be a part of Hong Kong. Only 18.8 per cent did not feel any pride at all.

By the same token, the fraying of the socio-economic fabric does not seem to affect people's sense of social belonging, which has remained, despite short-term fluctuations, more or less the same. In 1988, 1990, 1991, 1992b and 1994, the percentages of respondents who said that they had a strong sense of social belonging were 67.1, 55.1, 55.1, 66 and 77 respectively. Not surprisingly, people who identify themselves as 'Hongkongese' are more likely to evince a sense of social belonging. It is also noteworthy that in 1994 a substantial majority of them (67 per cent) recognized that they had a responsibility to do something for Hong Kong, even though in reality very few of them took part in social service and charity work. In 1994, 62.2 per cent of respondents could be characterized as apathetic to those activities.

The coexistence of the fraying socio-economic fabric and a stable sense of community identification is also reflected in the lack of statistical correlation between the trust in others and the indicators of community identification. It might be the case that, in the absence of all the social changes which are unfavourable to community solidarity, the sense of community identification should have markedly increased. Or, put differently, social changes since the 1980s might have retarded the growth of community identification in the territory.

Although the sources of community identification are not clear from the survey data, the existence of it at a stable level is instrumental in reducing the adverse impact of the fraying of the socio-economic fabric on Hong Kong.

Conclusion

A decade after the Basic Law Drafting Committee was set up, and five years since the promulgation of the Basic Law, it appears that the socio-economic system as enshrined in the mini-constitution of post-1997 Hong Kong has already lost some gloss in the mind of the people. Survey data collected in the past decade have shown that the socio-economic system of Hong Kong is treated with a certain degree of skepticism by the public. In view of the trend of socio-economic changes in the next two years, at the time when the Hong Kong Special Administrative Region (HKSAR) is set up on 1 July 1997, public enthusiasm for the system will be even lower than today. Obviously it is not auspicious for the HKSAR, the legitimacy of the Basic Law and the post-colonial government of Hong Kong.

For the sake of reassuring the Hong Kong people that China's policy toward Hong Kong will remain unchanged for a long time, the amendment of the Basic Law, which embodies China's policy of 'one country, two systems', is made extremely complicated and difficult. Unlike other constitutions, the Basic Law spells out in meticulous detail the social and economic policies for the SAR government, thus in fact binding the future government to a socio-economic system which flowered in the early 1980s, before the momentous changes that have impinged on Hong Kong since the mid-1980s. Increasingly the Basic Law will become a straitjacket for the SAR government, inhibiting its ability to adapt its policies to the changing environment. As a result of lingering public distrust of China, in all likelihood the SAR government will enjoy a lower level of political support than the colonial government it replaces. There is thus the probability that the Basic Law will prevent the SAR government from mobilizing public support through policy changes in the social and economic sphere. Consequently the legitimacy of the SAR government will be adversely affected.

Acknowledgements

This paper is based on a research project entitled 'Decline of authority, social conflict, and social re-integration in Hong Kong'. The project was generously funded by the Research Grants Council of the Universities Grants Committee. The project is under the auspices of the Social Indicators and Social Development Research Program of the Hong Kong Institute of Asia-Pacific Studies at the Chinese University of Hong Kong. I am grateful to Ms Wan Po-san, Research Officer of the Institute, for rendering assistance to the project in many respects. Special thanks are due to my research assistant, Mr Shum Kwok-cheung, for his help in administering the questionnaire survey and in data preparation.

Notes

1 *The Basic Law of the Hong Kong Special Administrative Region of the People's Republic of China* (Hong Kong: One Country Two Systems Economic Research Institute, 1992) p. 6.
2 Lau Siu-kai, 'Social change, bureaucratic rule, and emergent political issues in Hong Kong', *World Politics* 35(4) (July 1983), pp. 544–62.
3 In all these surveys, only persons over 18 years old were interviewed. The sampling frame is composed of households. The respondents selected for interview are derived from the sample of households through a random selection process. The following is a brief description of these surveys:

The fraying of the socio-economic fabric of Hong Kong   439

a   The 1985 survey was conducted in the summer and autumn of 1985 in Kwun Tong, a heterogeneous industrial-cum-residential community in Hong Kong. The sampling frame used was based on a 2 per cent sample of the complete household list prepared by the Census and Statistics Department for the 1981 Census. The size of the systematic sample was 1,687. In all, 792 interviews had been successfully completed, yielding a response rate of 46.9 per cent.

b   The 1986 survey is also a survey of the Kwun Tong residents. It was executed in the summer and autumn of 1986. The sampling frame was furnished by the Census and Statistics Department with 175,138 households. The final systematic sample was composed of 800 households. At the end of the exercise, 539 completed interviews were obtained. The response rate is 67.4 per cent.

c   The 1988 survey was undertaken in the summer of 1988. The sample used in the survey was prepared by means of a multi-stage design, starting with a sample of 3,488 residential addresses from the computerized Sub-Frame of Living Quarters maintained by the Census and Statistics Department. The sample consists of 649 households. In total, 396 successful interviews were obtained, yielding a response rate of 61 per cent.

d   The 1990 survey was conducted in the summer and winter of 1990. The sample was derived in the same manner as in the 1988 survey. The size of the sample was 613. In all, 390 interviews had been successfully completed, resulting in a response rate of 63.6 per cent.

e   The 1991 survey was implemented in the summer of 1991. The sampling method used was the same as that in 1990. The sample size was 718. In total, 401 successful interviews were obtained, yielding a response rate of 55.8 per cent.

f   The sample for the 1992 survey was drawn by the same method as used in 1991. The sample size for the 1992 survey was 1,568. The 1992 survey was conducted mostly from May to November 1992. At the end of the survey, 868 interviews were successfully completed, yielding a response rate of 55.4 per cent.

g   The sample for the 1992a survey was drawn simultaneously with that of the 1992 survey. The sample size was 1,125. Interviews were conducted mostly from October 1992 to July 1993. 615 interviews were completed, and the response rate is 54.5 per cent.

h   The sample for the 1992b survey was drawn in a similar manner as the sample for the 1992 survey. The sample size for the survey was 3,361. The survey was conducted mostly from December 1992 to February 1993. At the end of the survey, 1,993 interviews were successfully completed, yielding a response rate of 54.9 per cent.

i   The 1993 survey was conducted in the summer of 1993. The sampling method was the same as that used in 1992. The sample size was 1,633. At the end of the survey, a total of 892 successful interviews were completed, yielding a response rate of 54.6 per cent.

4   The sample used in the questionnaire survey was drawn by means of a multi-stage design. The target population of the survey are the Chinese inhabitants in Hong Kong aged 18 years old or over. Since the full list of such adults was impossible to obtain, I used the list of permanent and residential areas prepared and kept by the Census and Statistics Department's computerized Sub-Frame of Living Quarters as my sampling frame. With the assistance of the Department, a replicated systematic random sample 2,072 addresses was selected from the sampling frame. After the exclusion of vacant, demolished,

unidentifiable addresses, addresses without Chinese residents, and addresses eventually unused, the actual sample size was reduced to 1,748.

The next stage of sampling involved the selection of households and eligible respondents by the interviewers. Interviewers were required to call at each address and list all the households residing there. If there were two or more households, only one would be selected according to the random selection table pre-attached to each address assignment sheet. For each selected household, the interviewer was required to list all persons aged 18 years old or over and arrange them in an order according to sex and age. The respondent was then selected from the list by means of a random selection grid (a modified Kish grid) pre-attached to each address assignment sheet.

Face-to-face interviews with structured questionnaires were carried out by interviewers who were recruited from local tertiary institutions and required to attend a half-day briefing session. Fieldwork was conducted mostly from the end of May to July 1994, wherein approximately 95 per cent of the successful interviews were completed, while the rest of the interviews were carried out from August to September 1994. All completed questionnaires were subsequently checked by follow-up phone calls to the respondents concerned as a means of data quality control. Additional data control checks were also made to improve data quality.

At the end of the survey, 997 interviews were successfully completed, yielding a response rate of 57 per cent.

The socio-demographic profile of the respondents in the survey is by and large similar to that of the 1991 Hong Kong census.

5   See particularly Lau Siu-kai, 'Hong Kong's "ungovernability" in the twilight of colonial rule', in Zhiling Lin and Thomas W. Robinson (eds) *The Chinese and Their Future: Beijing, Taipei, and Hong Kong*, Washington, DC: The AEI Press, 1994, pp. 287–314.

6   Census and Statistics Department, *Annual Digest of Statistics, 1994 Edition*, Hong Kong: Government Printer, 1994, p. 111; and Government Information Services, *Hong Kong 1995*, Hong Kong: Government Printer, 1995, p. 63.

7   See Census and Statistics Department, *Hong Kong Social and Economic Trends 1982–1992*, Hong Kong: Government Printer, 1993; and Government Information Services, *Hong Kong 1995*, p. 498.

8   Tsang Shu-ki, 'The economy,' in Donald H. McMillen and Man Si-wai (eds) *The Other Hong Kong Report 1994*, Hong Kong: Chinese University Press, 1994, p. 132.

9   Fung Ho-lup (1995), 'Attitudes toward protest action,' in Lau Siu-kai, Lee Ming-kwan, Wan Po-san and Wong Siu-lun (eds) *Indicators of Social Development: Hong Kong 1993*, Hong Kong: Hong Kong Institute of Asia-Pacific Studies, The Chinese University of Hong Kong, pp. 291–320.

10 See Tsang Wing Kwong, *Educational and Early Socioeconomic Status Attainment in Hong Kong*, Hong Kong: Hong Kong Institute of Asia-Pacific Studies, The Chinese University of Hong Kong, April 1993; *idem*, 'Consolidation of a class structure: changes in the class structure of Hong Kong,' in Lau Siu-kai, Lee Ming-kwan, Wan Po-san and Wong Siu-lun (eds) *Inequalities and Development: Social Stratification in Chinese Societies*, Hong Kong: Hong Kong Institute of Asia-Pacific Studies, The Chinese University of Hong Kong, 1994, pp. 73–121; Benjamin K.P. Leung, '"Class" and "Class Formation" in Hong Kong Studies', *ibid*, pp. 47–71; Thomas W.P. Wong and Lui Tai-lok, *Reinstating Class: A Structural and Developmental Study of Hong Kong Society*, Hong Kong: Department of Sociology, The University of Hong Kong, 1992; and *idem, Morality, Class and the Hong Kong Way of Life*, Hong Kong: Hong Kong Institute of Asia-Pacific Studies, The Chinese University of Hong Kong, November 1993.

11 Ronald Skeldon, 'Immigration and emigration: current trends, dilemmas and policies', in McMillen and Man (eds) The Other Hong Kong Report, pp. 165–86.

12 Stephen Chiu, *The Politics of Laissez-faire: Hong Kong's Strategy of Industrialization in Historical Perspective*, Hong Kong: Hong Kong Institute of Asia-Pacific Studies, The Chinese University of Hong Kong, November 1994.

13 Lau Siu-kai and Kuan Hsin-chi, 'Public attitude toward laissez-faire in Hong Kong', *Asian Survey* 30(8) (August 1990), pp. 766–81.

14 See Lau Siu-kai and Kuan Hsin-chi, *The Ethos of the Hong Kong Chinese*, Hong Kong: Chinese University Press, 1988, pp. 56–65; and *idem*, 'Public attitudes toward political authorities and colonial legitimacy in Hong Kong', *The Journal of Commonwealth and Comparative Politics* 33(1) (March 1995), pp. 79–102.

15 Lau and Kuan, 'Public attitude toward laissez faire in Hong Kong', and Wong Chack-kie, 'Welfare attitudes: one way of explaining the laggard welfare state phenomenon', in Lau, Lee, Wan and Wong (eds) *Indicators of Social Development: Hong Kong 1993*, pp. 205–22.

16 Lau Siu-kai, 'Democratization and declining trust in public institutions', (unpublished paper); and *idem*, 'Decline of governmental authority, political cynicism and political inefficacy in Hong Kong', *Journal of Northeast Asian Studies* 11(2) (Summer 1992), pp. 3–20.

17 See also Berry Hsu, *The Common Law in Chinese Context*, Hong Kong: Hong Kong University Press, 1992.

18 The four-point scale question was used in the 1988 survey, whereas in the other surveys the five-point scale question was used.

Part V
Social Issues and Social Policy

[19]

School knowledge, the state and the market: an analysis of the Hong Kong secondary school curriculum

PAUL MORRIS

This paper explores the differential status and validity accorded to subjects in Hong Kong secondary schools and analyses the structures and processes which maintain the nature of school knowledge. It initially focuses on the central role played by the state and subsequently by the market in which schools compete for pupils. It is argued that the curriculum continues to manifest those characteristics which emerged in the early postwar period, which was characterized by direct state control. The outcome is a curriculum which contains those features associated with: a collection code; closed systems; disciplinary modes of conceptualizing knowledge; and a focus on public knowledge, despite the government's attempts, over the last two decades, to promote a curriculum which displays the opposite features.

A central tenet of sociological analyses of education is that the unequal distribution of power and resources in societies is reflected in the school curriculum with the result that different forms of knowledge are provided to the empowered and to the less privileged. Bourdieu and Passeron (1977: 8) have described this stratification of knowledge in terms of the cultural capital of dominant and subordinate classes. They use the term 'arbitrary' to describe the criteria used to select that capital, which partly echoes Williams's (1961: 153) depiction of the operation of a 'selective tradition' for defining knowledge. Similar portrayals have been provided by Young (1971) and Bernstein (1971: 51–4). The latter focuses on the pedagogic and organizational characteristics of school curricula, and distinguishes between 'collection codes' and 'integrated codes'. The former are characterized by a strong boundary between subjects and a low degree of control of the curriculum by teachers and pupils. An integrated code possesses the opposite features. This parallels the designations of Schrag (1988) and Berlak and Berlak (1981) who identify respectively 'disciplinary' and 'progressive' modes of conceptualizing knowledge and 'public' and 'personal' knowledge. Essentially, collected or disciplinary codes are perceived to have higher social prestige and are variously described as 'more worthwhile' (Young 1971: 34), of 'high status' (Bourdieu and Passeron 1977), as 'highbrow' (Levine 1988: 1) and as 'hegemonic

Paul Morris is a Reader in the Department of Curriculum Studies, Faculty of Education, University of Hong Kong, Pokfulam Road, Hong Kong. He is currently involved in evaluating a Target Oriented Curriculum that was recently introduced into Hong Kong schools. His recent publications include: *The Hong Kong School Curriculum: Development, Issues and Policy*, 2nd edn (Hong Kong University Press 1996), and (co-edited with Anthony Sweeting) *Education and Development in East Asia* (New York: Garland, 1995).

knowledge' (Connell *et al.* 1982: 199). More recent postmodernist analyses have attempted to eschew the economic determinism of much of this work, especially with regard to the link between knowledge, power and social stratification, but they have conceptualized curricula in similar terms. Thus Doll (1994: 14) distinguishes between 'closed systems' that transmit and transfer what is known and knowledge which derives from 'open systems' which attempt to transform society through a focus on social action and visions of a better and more moral society. Usher and Edwards (1994: 25) even argue with regard to the curriculum that: 'trends of interdisciplinarity and experiential approaches to teaching and learning can be seen as changes taking place under the impact of the postmodern and therefore very much part of it'.

An alternative macro-level interpretation of school curricula focuses on the impact of 'global' and 'world system' influences (Ginsburg *et al.* 1990: 482), and argues that educational innovations initiated in the USA and Europe have set the standard for those of other nations. While worldwide trends have provided both the rhetoric and models for policy initiatives this has sometimes had limited impact on the nature and implementation of curricula in specific national settings (Morris *et al.* 1997).

These perspectives have provided insights into both the correspondence between school curricula and social structures and the complex way in which power is manifested and maintained. However, the context of and empirical basis for such analyses of school curricula has primarily been derived from societies where the state is portrayed as operating in a pluralist or corporatist political context and primarily serving the role of reflecting broader socioeconomic interests. In many societies, not only colonial aberrations such as Hong Kong, the state itself plays a central role (Dale 1989), not merely as an agency which reflects and reinforces the prevailing power relationships that exist in society, but also acting to protect and promote its own interests in ways which have an important impact on the nature of education policy generally and school curricula particularly.

This paper explores the differential status and validity accorded to subjects in Hong Kong secondary schools and analyses the structures and processes which maintain the nature of school knowledge. The interpretation initially focuses on the central role played by the state in that process during the postwar period. Subsequently it examines the systems of assessing and allocating pupils, the range of reform initiatives introduced by the government, the characteristics of the curriculum constructed by schools, and the evidence with regard to schools' impact on social, personal and moral education. It is argued that the school curriculum continues to manifest those characteristics which are associated with a collection code: closed systems; disciplinary modes of conceptualizing knowledge; and a focus on public knowledge – despite the government's attempts in the last two decades to promote a curriculum which displays the opposite features. The explanation for this focuses initially on the critical role played by the state in the postwar period in defining the parameters of what knowledge was acceptable, or more precisely not acceptable, for inclusion in the school curriculum. Subsequently this view of knowledge has been incorporated, reproduced and internalized through a highly competitive educational market to the extent that it has achieved a 'taken for granted' status. This has been paralleled by both a reduction in the state's direct control of the curriculum and the promotion of a range of essentially symbolic policies which exhort schools to provide pupils with a broad, integrated and open curriculum. The portrayal provided by this paper is discussed with reference to the changing sources of control of the curriculum and the impact of market forces on the nature of the curriculum.

Antecedents

Three distinct periods are evident. In the first, from about 1945 to 1965, the role of the state in defining the nature of valid knowledge relied on coercion and was primarily designed to counter any direct threats to the legitimacy of the colonial government. In the second period, from about 1965 to 1984, the nature of valid knowledge was primarily defined by the market, but was also influenced by a desire to avoid offending the sensitivities of the People's Republic of China (PRC). In the final period, from 1984 onwards, the definition of valid school knowledge has continued to be defined by the market but has also been substantially influenced by the impending transfer of sovereignty.

In the period since 1945 Hong Kong has experienced a tenfold increase in population and the emergence, from small beginnings, of one of the world's largest trading and financial centres. These changes have been overseen by a colonial political system in which the civil service has performed both the legislative and executive functions of Government and in which the approach generally adopted to public policy has been described as 'laissez-faire' or 'positive non-interventionism' (Sweeting 1995: 107).

Given a stable sociopolitical environment, this approach should have ensured that all, or most, of the requirements for services such as housing, schooling and health were satisfied through private provision. As a concomitant of the policy, the curriculum of privately run schools should have been their own concern and in the postwar period different curriculum models co-existed, leading to a variety of qualifications and certificates, as has remained the case in neighbouring Macau (Bray and Hui 1991). However, the sociopolitical environment did not remain stable, and the response of the Government to threats to public order and to its own tenuous legitimacy has been to expand public provision, particularly of schools and housing and to strengthen its control of the curriculum (Sweeting and Morris 1993).

Within the area of schooling which was mainly provided by the private sector, early provision included schools sponsored by one or other of the warring factions in China, the Chinese Communist Party and the Kuomintang, neither of which viewed Hong Kong's colonial status in a favourable light. To counter the influence of such schools, which had developed their own curricula, the Government built its own rival schools in their vicinity and encouraged the many missionary societies which had fled the civil war in China to establish schools in Hong Kong. It also enacted a complex range of regulations which allowed it to suppress anything perceived as subversive in both the government and private school curriculum. Regulations were introduced in the late 1940s which gave the Director of the Department of Education control over school subjects, textbooks and all other teaching materials, and over any activities in schools which might be thought to be political in nature (Morris and Sweeting 1991). These regulations were used to deregister teachers and to close a communist school in 1949. This was further reinforced by the production of 'model' syllabuses and the establishment of a common system of public examinations and certification.

The outcome of this process was the emergence of a highly centralized and bureaucratic system of control of the curriculum in which schools are provided with syllabuses for permitted subjects, textbooks which have been vetted by officials, recommended teaching guides, and official examination syllabuses. The key purpose of this system of control was to ensure that the content of syllabuses and textbooks was depoliticized. This was achieved by avoiding any content which was concerned with contemporary China, the local context or any 'sensitive' topics. The effect of this policy is that pupils studied the history, geography and literature of other cultures or of distant time periods: in other words, a decontextualized and remote curriculum. Luk (1991: 668) elegantly describes the consequences of this for Chinese culture subjects (language, literature and history):

> Thus generations of Hong Kong Chinese pupils grew up learning from the Chinese culture subjects to identify themselves as Chinese but relating that Chineseness to neither contemporary China nor the local Hong Kong landscape. It was a Chinese identity in the abstract, a patriotism of the 'émigré', probably held all the more absolute because it was not connected to a tangible reality.

Consequently the curriculum, as experienced by most students, consisted of abstract academic content, taught by transmission and examined in English by means which emphasized memory over understanding and reproduction over application to real problems. In essence the curriculum which emerged had all the characteristics of a strong collection code and a 'closed system' of knowledge. The 1960s heralded a significant shift in the nature of the political and economic conditions prevailing in Hong Kong. On the political front it became increasingly evident that Hong Kong existed because it was tolerated by the PRC and the riots of the mid 1960s served to underline Hong Kong sensitivity to events in the PRC. In terms of curriculum policy making this served to place a premium on maintaining a curriculum which avoided sensitive content and issues. From about 1965 rapid industrialization placed the territory in the forefront of the dynamic Asian economies. This was subsequently followed by the emergence of a middle class and growing levels of affluence, and an increase in the numbers of refugees from the PRC. As the economy grew so access to well paid employment, in a transient society where mobility was based on achieved rather than ascribed criteria, was provided by educational qualifications. This placed a premium on the selective and allocative role of schooling and created public pressure for the expansion of educational provision. In 1969 nine years of compulsory education became government policy which was fully implemented in 1979. This served to highlight the inadequacy of the prevailing curriculum and saw the government promote a series of reforms designed to achieve a broadly based and balanced school curriculum which is portrayed in terms akin to an integrated code, open systems and personal knowledge. It is the difference between this policy and practice in schools which throws into sharp relief the different perceptions of the validity of different types of knowledge held by participants in the system of schooling.

Only in the last decade have the culture and contemporary politics of Hong Kong become valid items for inclusion in the school curriculum. Given the impending transfer of sovereignty in 1997, the government has changed the criteria for the selection of valid curriculum knowledge. The study of previously sensitive issues, politics and of aspects of the PRC has become acceptable (though still within certain limits). The most recent manifestation of this involved attempts by the Director of Education in 1994 to discourage references in textbooks to the suppression in 1989 of the student movement in Tiananmen Square. In defence of his policy he argued that twenty years should elapse before events were included in school texts to ensure the necessary objectivity. In effect the criterion for defining sensitivity has shifted from focusing on the concerns of the colonial government to those of the PRC.

The characteristics of curriculum knowledge, especially the bias towards abstract and remote material, is reinforced by the fact that most pupils officially study in English language schools. It is at best difficult to make a lesson concrete and relevant when the medium of instruction is not the students' nor the teacher's mother tongue. Teachers routinely provide verbal instruction in Cantonese, the local dialect, but texts, examinations and written work are mainly in English, creating a major barrier to interest, relevance and motivation. It also encourages teachers to adhere strictly to the material provided by the textbook and by previous examination papers. The significant proportion of time devoted to the study of English is itself problematic, for its use is remote from the immediate needs and interests of most students and it is unnecessary for most adults. (In Hong Kong 97% of the population speak Cantonese.) Unlike, for example, Singapore, India, or many African countries, English does not serve as a compromise *lingua franca*. For parents, the value of English is based on the access it affords in the long term to tertiary education and to the professions.

In spite of this high degree of central control over parts of the curriculum, a significant set of key decisions has been left to be taken at school level, including the specific combinations of subjects to be studied; the time devoted to specific subjects; the mechanisms for the streaming and selection of pupils up to secondary year 5; the language of instruction; and all aspects of pedagogy. These issues have become increasingly critical with the advent of universal junior secondary education in 1979 and the growth of the range of available school subjects (especially computer studies and Putonghua/Mandarin).

However, the most critical influence on the curriculum has been the emergence of an allocative mechanism which has encouraged the operation of a market in which schools compete for pupils. The impact of this market has effectively superseded direct control by the state. This has allowed the Government to revert to its traditional *laissez faire* role and allowed it to minimize its statutory control of missionary and benevolent bodies whose willingness to provide education has partly depended on their freedom to promote their specific beliefs. The effect of the emergence of a market for pupils has been to ensure that issues of valid knowledge, valid transmission of knowledge and valid evaluation affect schools at the individual level rather than reaching them more diffusely through decisions taken at national level.

Selection and allocation of students to schools

Access to each level of schooling and to different quality schools is determined by pupils' performance in public examinations, which are organized around traditional academic subjects. Consequently a school's status, and that of its members, is largely determined by its results. Further, this status will affect its future intake of pupils. In an uneasy compromise between allowing parental choice to operate and attaining egalitarianism, the Government has set up geographical nets. On leaving primary school pupils are placed in one of five bands according to a test of their academic aptitude, which is derived from a combination of internal school assessments and academic aptitude tests in mathematics and Chinese.

Band 1 comprises those who have performed best whilst the Band 5 pupils have performed worst. Parents must opt for schools in the network defined by where they live. A proportion of pupils in Band 1 are allocated first according to their parents' choice of school; then a proportion of pupils in Band 2 are allocated to their first choice, and so on. By the time the Band 5 pupils are allocated most schools in a net are full and these students are allocated to the least popular schools. When the schools in a geographical net are unable to cater for all the pupils in that district it is the Band 5 pupils who will be allocated to schools in another district. In effect the role and impact of parental choice is strong for Band 1 pupils and very weak for Band 5 pupils. Consequently, schools that are perceived in the community as more successful will be able to attract primarily Band 1 and Band 2 pupils and so maintain their future examination results and the banding of their intake. Schools cannot, however, only fill their quota by selecting high band students. Preference can be given to siblings and to those with the appropriate religious affiliation. There is also a tendency for some parents to opt for schools which are perceived as good but not the most prestigious, as this will maximize the chance of getting selected within the first round of the exercise. The effect is to widen the range of achievement of students entering popular schools, but to leave the unpopular schools with a narrow range of low achievers.

The more prestigious schools also have a further opportunity to reinforce their reputation at the end of secondary year five where a market operates for pupils who wish to study in forms six and seven. Only pupils who achieve good results in the Hong Kong Certificate of Education Examination (HKCEE) will be allowed to continue to form six, with the result that many pupils will have to change schools to pursue their studies; so the more prestigious schools can recruit pupils who have done well from other schools.

Two factors are perceived by parents as the key determinants of a school's status: its academic record, which is primarily a function of pupils' performance in public examinations, and the school's policy on the language of instruction. Schools which claim to use English have generally been accorded a higher status than those which use Chinese. The pressure to obtain a place at a 'good' school and for pupils to proceed as far as possible through formal education is reinforced by the very strong link between qualifications and lifetime earnings, large income disparities, and by the weak influence of ascribed characteristics on social mobility in a refugee society. More recently the desire to emigrate has also served to heighten the status of English proficiency.

Any school which decides to opt out of this cycle of competing to attract 'better' students and provide a more diversified and balanced curriculum suffers from a compounded problem. First, its status will drop as it places less emphasis on preparing students for examinations. But, more importantly, parental choice will ensure that it recruits students of ever lower levels of academic achievement and a greater proportion of students with behavioural problems.

Precisely the same problem affects schools which publicly state that they will use Chinese as the language of instruction. This was well illustrated by one school in the late 1980s which changed to using Chinese on the grounds that this would improve the quality of students' learning and of teaching. Three years later the school reversed its policy as the 'quality' of the student intake was perceived to have dropped. Parents of academically more able pupils were not willing to send them to a Chinese-language school. Consequently, the impact on any one school of exercising its power to implement a broader curriculum and to provide mother tongue instruction is far-reaching. In this situation most schools adopt a 'wait and see' policy analogous to the pricing behaviour of a firm in an oligopolistic market. As a local principal (Ha 1993: 15) expressed it:

> As a principal, I regard the provision of two PE [physical education] Lessons per week as insufficient. A better arrangement would be three to four lessons per week. Nevertheless, the realization of this 'ideal' is constrained by a number of factors, including the insufficiency of space and teaching staff, not to mention the demands of other academic subjects (especially for those classes which are facing public examinations). The problem cannot be solved by any one school independently.... This is indeed a problem for the whole system.

A similar problem faces individual teachers. If a teacher presents the official syllabus at the recommended speed, and covers the syllabus, then any failures can be attributed to the students. However, if a teacher attempts to adjust the syllabus to the perceived needs of the students, or introduces other models of teaching, the locus of attribution shifts.

Official policy and its rhetoric

As indicated above, curriculum policy makers have not been blind to the problems associated with an academically oriented curriculum, particularly with regard to the needs of the majority of students. A wide range of reforms have been introduced by the government since 1969 in an attempt to create a broader curriculum, to promote less didactic styles of teaching, to encourage schools to use Chinese as the language of instruction, and to improve assessment practices. Overall these reforms were designed to move the curriculum away from its collection orientation and focus on public knowledge towards a more integrated orientation and towards progressive modes of conceptualizing knowledge. In what follows the nature of the recommended and observed curriculum is analysed with reference to its aims; the distribution of time allocated to subjects; the assessment of pupils; attempts to promote personal, social and moral education; and the introduction of integrated subjects.

The aims of schooling

Although the aims and nature of the curriculum have not been set out formally in policy documents until relatively recently (Education and Manpower Branch 1993, CDC 1993a, b, c, d) the same ideas have appeared piecemeal in earlier documents. Broadly derived from UK sources (Department of Education and Science 1985) they identify five aims: *intellectual, communicative, social and moral, personal and physical, aesthetic development*; seven areas of learning: *mathematical, scientific, technological and practical, social and moral, personal, physical, aesthetic*; and three elements of learning: *knowledge, skills, attitudes*. The official guidelines (CDC 1993c: 14) state that:

> In the planning of the curriculum, schools carry the responsibility to ensure that the specific curriculum aims [listed above] are achieved. In order to provide schools with a conceptual framework and guidelines for consideration in their curriculum planning, it is suggested that an all-round curriculum should consist of all the major areas of learning and experience as well as their major elements of learning ... to provide a well balanced general education suitable for all students at this level, whether or not they continue their formal education beyond secondary level three.

The guidelines refer to a 'common core curriculum' consisting of a range of subjects together with cross-curricular activities. Unfortunately the concept of a common core is not well defined and is used inconsistently. For example it is confused with a 'common course' (CDC 1993c: 3):

In the development of the junior secondary curriculum, one milestone was the introduction of the concept of a common-core curriculum. The 1974 White Paper 'Secondary Education in Hong Kong over the Next Decade' recommended, *inter alia*, that all children should follow a common course of general education throughout the nine years of free and compulsory education.

Moreover, the freedom of schools to choose combinations of subjects ensures that courses are not common, as is noted in the same document (CDC 1993c: 19):

> About half of the schools offer Social Studies as an integrated subject while the rest offer history, geography and economic and public affairs as alternatives.

The concept of a core implies the existence of a non-core space which schools are free to use, for example, to match parts of the curriculum to the perceived needs of different groups of students. However, it is acknowledged that the recommended core leaves little room for manoeuvre (CDC 1993c: 20):

> Since the suggested time allocation for the constituent subjects amounts to 42 periods a week, it leaves virtually no room for any new subject which may have been developed to meet the changing needs of society. As a result of the overcrowding of the curriculum, some schools do not place due emphasis on the cross-curricular studies such as Civic, Moral and Sex Education.

The distribution of time to school subjects

Data for the actual allocation of time to different areas of the curriculum in 1990 and 1994, derived from a survey, undertaken by the author, of 44 schools (a 10% random sample of the population of schools) are contrasted with the recommended allocations in table 1. Pre-vocational schools, for which a higher proportion of time for 'cultural, practical and technical' subjects is recommended, have not been included. The recommended allocations are already strongly oriented to languages, mathematics and sciences, which may well reflect the impact of prevailing practices on official policy. In practice, they are exceeded at the expense of the lower status subjects. The drift towards even greater allocations to the high status subjects in 1994 reflects the growing popularity of Putonghua/Mandarin and computer studies, which has exacerbated the problem. 'Other learning activities' (those of a cross-curricular nature, topical issues or religious/ethical education) are particularly at risk.

    The overall pattern indicates that schools place the greatest weight on those subjects which are derived from the traditional academic disciplines, that is, high-status knowledge. Cultural subjects, such as music and physical education, come low in the hierarchy of esteem and so are starved of resources. Ng (1994) found that over 40 secondary schools (about 10%) had ceased offering music as a subject, and that many schools now offered it for only one period per week (as opposed to the two periods recommended). Similarly, the Education Department recommends that physical education be allocated two periods per week but this is often not achieved (Hong Kong Sports Development Board 1993). Further, many teachers of

Table 1. Comparison of subject selection and time allocation (S1–3).

Subject group	Suggested time allocation (%)	Actual time allocation (%)		Subject selection
		(1990)	(1994)	
Languages	35–40	34·4	36·4	Chinese, English, Putonghua/Mandarin
Maths and science	20–25	27·1	28·5	Mathematics, science, computer literacy
Humanities	15–20	17·5	16·4	Geography, history, Chinese history, economic and public affairs
Cultural, practical and technical	15–20	17·5	15·6	Physical education, art and design, music, home economics/design and technology
Other learning	5	3·5	3·1	Cross-curricular activities

subjects such as music and physical education are expected to 'give up' lessons to more academic subjects when public examinations loom. In response, teachers of music and physical education attempt to raise the status of their subjects by preparing students for various public competitions (such as the Speech and Drama Festival and the Music Festival) and by calling for more academic and publicly assessed syllabuses.

Assessment weighting

The hierarchy of esteem in which subjects are held is also reflected and reinforced by the way schools allocate marks within their internal examinations. Table 2 shows how the marks from specific subjects contributed to the final internal school assessment of students at the end of secondary year 3 in the same 44 schools. These examinations, which are critical for students' overall marks, are used to decide which combination of subjects they can study in secondary years 4/5 and the stream in which they will be placed. A pupil who performs well will usually be channelled into the science stream whilst poor performers are allocated to the humanities or commercial streams. A similar phenomenon occurs at the end of secondary year 5 when schools compete with each other to recruit the best secondary year 5 pupils to their secondary years 6/7 courses. Consequently many weaker pupils have to change schools to continue their education beyond secondary year 5.

Table 2 confirms the importance attached to languages, mathematics and sciences and the very low status of cultural, practical and technical

Table 2. Contribution of subject groups to the overall internal examination mark at secondary level 3.

Subject group	Subject	%
Languages	Chinese English Putonghua	36·8
Mathematics and science	Mathematics Science Computer literacy	28·6
Humanities	History Geography Chinese history Economic and public affairs	26·6
Cultural, practical and technical	Physical education Art and design Music Home economics/design technology	8·0

subjects. The overall data obscure some individual differences between schools. For example, many schools awarded grades for music and art but did not incorporate them in the overall mark or grade.

The very clear message that students receive in most schools from the distribution of time and marks to different school subjects is that what really counts is the study of languages, mathematics and the sciences. A student who is more talented or interested in the humanities and cultural subjects is bound to do fairly poorly in the internal examination of most schools given the way the marks are allocated. In contrast one whose interest and capabilities are in languages or sciences will receive a relatively high internal examination mark. Given its divergence from the rhetoric of public policy statements, this might be viewed as a 'hidden curriculum' (Giroux 1981) but it is only hidden to the most superficial observation. Teachers, parents and especially students are very aware of what counts as valid knowledge and of the consequences of acquiring or not acquiring it.

Taken by themselves, the proportions of time allocated to subjects and their assessment weightings do not characterise a curriculum fully. The nature of what is taught within subjects is also important. Insight into this can be gained from the reactions by the schools to the variety of curricular innovations designed to broaden the curriculum. These include the promotion of cross-curricular studies, integrated subjects and more formative systems of assessment.

Cross-curricular studies

Cross-curricular studies are promoted through the Education Department's guidelines for civic, moral, sex and environmental education. These attempt to describe how social, moral and personal education can be achieved within the prevailing organization of the secondary school curriculum and through the content of existing school subjects. Essentially, the intention is that subject teachers will 'permeate' their curricula with themes and issues which would develop social, personal and moral education.

Although official documents tend to paint a rosy picture, it is hard to find evidence that the various guidelines on civic, sex and environmental education have made a significant impact on practice. A survey of history teachers (Tang and Morris 1989: 48) led to the conclusion that, despite official documents claiming high levels of implementation:

> The picture that emerges is that the majority of History teachers were not aware of or were unclear about the guidelines and their school policy towards civic education. The overall situation evidenced in this study is that 65% of the sample of History teachers had not attempted to implement the guidelines. Of those who had attempted implementation the level of use was low.

With regard to the impact of sex education in schools, Ng (1994: 422) comments:

> The cruel fact is that, despite a lot of statistics [and] 'a lot of work' cited by the educational bodies, they are not meeting the educational needs. ... The obvious problem is that the 'sex education' efforts in Hong Kong so far have been superficial, consisting of empty words much more than effective action or support. Memoranda, guidelines, resources, lectures and theories were produced with no concern paid to their practicality or feasibility.

Fung and Lee (1993: 4) conclude from their analysis of the implementation of the Environmental Education Guidelines (CDC 1992) that:

> ... a fairly high percentage of the schools surveyed have neither appointed a coordinator nor set up a committee for the promotion of environmental education. Moreover, only a small number of schools have circulated the guidelines and have discussed them among staff. With such a low level of institutional commitment it is doubtful that the good intentions of the guidelines could be fully realized.

Clearly, cross-curricular guides which are not seen to link to public examinations and high-status knowledge do not exert a powerful influence on the implemented curriculum. This is especially true when teachers feel that they have inadequate time to cover the existing examination syllabuses. Where teachers have not even read the guidelines, implementation can only be fortuitous.

Integration

Both to reduce the time allocated to discrete school subjects and to promote a more cross-disciplinary and inquiry-oriented pedagogy, syllabuses for integrated subjects have been developed and promoted in the areas of science and social science. In each case the initial conception was that they should be compulsory subjects. Integrated science would replace biology, chemistry and physics. Social studies would replace geography, history, and economic and public affairs. Derived from the Scottish Integrated Science Project, integrated science was designed to offer a broad course in science for all students, incorporating an emphasis on learning through guided discovery and a strong orientation towards practical applications.

Social studies was intended to incorporate elements of history including Chinese history, geography, economics and so on. The initial conception was very much that of overseas models, particularly from the USA, with their 'attention to process [inquiry and discovery], critical thinking, ... contemporary issues and relevance' (Seixas 1993: 235). It was intended to be utilitarian, oriented towards less academic students and taught in Chinese (Morris *et al.* 1997). Although the fates of these subjects have been very different they both help to illuminate the priorities operating in schools. In terms of adoption by schools, the two syllabuses have been unequally received. Integrated science is taught in virtually all schools, whereas social studies is taught in no more than 20%. If the focus shifts to implementation of the original aims, the story is rather different. Neither shows significant implementation of the pedagogic aims.

In the case of integrated science, the major aim was to bring about learning through 'guided discovery'. As the Report (Education Department 1974) on the pilot study for the first year of the programme shows, most teachers indicated that they could not manage guided discovery with the resources available and given the class sizes which were (and still are) customary. The official response was to stop asking such questions (McClelland 1991). The second year was evaluated solely on the basis of written tests, while the third year was not evaluated in any way before being recommended for adoption. Later small-scale evaluations indicate that the content of the first two years is taught predominantly through teacher talk. In the third year many schools abandon integrated science and embark on the single-subject syllabuses intended for senior secondary level. Rather than three years of an integrated course followed by two years of separate subjects, they reverse the pattern. This is a direct consequence of the hierarchy of esteem in which subjects are held. Given a wide range of aptitudes and abilities among their students, teachers would rather extend the time devoted to high-status knowledge than reduce the academic content.

The high level of adoption of integrated science owes less to any commitment to its stated aims than to widespread dissatisfaction with general science or single-subject sciences which preceded its introduction and to a perception that the content can provide a reasonable foundation for science at senior secondary level. Although the syllabus recommended radical changes in pedagogy, it could be taught without their implementation. Also, it was introduced at a time when integrated science, at this level of schooling, was enjoying a world-wide 'band-wagon' effect.

The low level of adoption of social studies provides an instructive contrast. A strongly perceived need to provide a platform for later studies led to rejection, or watering down, of many of the original intentions and to a syllabus in which:

> Chinese history had managed to extricate itself, maintaining itself as a discrete subject, while the remaining contributing subject areas, geography, economics, history and public affairs, had roughly equal shares of the 'territory'. The needs of non-academic pupils, as originally perceived by the curriculum policy-makers, had been subordinated to those of more academic pupils as the designers attempted to satisfy the demand from each subject for adequate linkage to the critical examination syllabus in secondary years 4 and 5. This, in turn, led to the inclusion of an excessive amount of material. Social Studies did not emerge as an integrated subject, but rather as an excessively large collection of topics from different subjects. (Morris *et al.* 1997)

Disincentives to adoption included many of the features which Fullan (1989) associates with well-intentioned but unsuccessful innovations: the operational difficulties of coordinating and timetabling teachers; the low status accorded to the integrated subject by teachers, pupils and parents; the sensitive or controversial nature of some of the topics included in the syllabus; the insecurity of teachers who saw their role as derived from teaching a single subject; and the perception that the syllabus had failed adequately to take into account the views of teachers and the practicalities of classrooms. The reaction to social studies has been paralleled by the reaction to a more recent proposal to introduce liberal studies to broaden the curriculum of the matriculation years (Wan 1991).

Assessment

Although the junior secondary stage of schooling no longer ends with a public examination, the 'backwash' effects of public examinations two years later are powerful and ever-present. Almost all of those who complete secondary year 3 go on to senior secondary schooling and take these examinations, which act as the first gate-keepers to well-paid careers.

Government has attempted to reduce assessment pressure by abolishing a public Junior Secondary Education Assessment examination. However, the results of school-based examinations are used to allocate students to senior secondary schools so the pressure to succeed remains strong, and the pressure on schools to be seen to produce success is equally strong.

A recent proposal has been to introduce a 'Target Oriented Curriculum' (TOC) intended to address problems recognized as being associated with current assessment methods and practices. Clark *et al.* (1994) argue that these include: fragmentation and specialization; a failure to operationalize the stated aims of the curriculum; over-emphasis on rote memorization and on the linear mastery of decontextualized skills; the view that students are born with a fixed amount of intelligence; and a lack of differentiation in teaching, learning and assessment. School assessment is described as relying too much on gap-filling and multiple-choice questions designed to put a class into rank order.

The proposed means to address the problem consists of a progressive framework of criterion-based targets, initially confined to Chinese, English

SCHOOL KNOWLEDGE, THE STATE AND THE MARKET   343

and mathematics. The goal, as outlined by Clark *et al.* (1994) is to encourage more individualized, active and self-initiated learning; develop a less compartmentalized and more integrated curriculum; shift away from norm-based assessment; avoid over-assessment; and promote inquiring, conceptualizing, reasoning, problem solving and communicating. The TOC initiative can be seen as an attempt to shift both assessment and pedagogy towards more responsive, differentiated and formative modes. It is clear from the rhetoric of official documents and the nature of the proposed reforms that the official curriculum promotes a form of assessment consistent with a curriculum which embodies an integrated code.

The essentially worthwhile nature of its goals ensured that the scheme initially received general support from the local educational community. However, as the scheme moved closer to implementation, support in principle shifted to concern and subsequently to hostility. It is unclear at present what the outcome will be with regard to TOC as it is being modified in the light of responses from schools and other interested parties. Further attempts are being made to encourage its adoption. However, it is evident that it is going to be an uphill struggle to get it implemented in a form which allows its intentions to be realized.

Initial responses have tended to focus on characteristics of the documentation and of the strategy of curriculum development such as: excessive complexity, lack of clarity, perceived excess of costs over benefits, and, the planners' failure to include a significant number of teachers with a variety of experiences in the planning process or to provide concrete examples of the scheme's operationalization. Even if these criticisms can be met, a deeper level of opposition can be predicted, for implementation of the scheme as intended would require radical changes of attitudes to the nature of valid knowledge, pedagogy and assessment.

Although the scheme maintains subject boundaries, its focus on targets which fit with the existing level of achievement of each individual student would be difficult to reconcile with the widespread current practice of whole class instruction. It would be extremely difficult to reconcile with five years of secondary schooling, preparing pupils for the public examinations which currently are the culmination of this stage of schooling, and its only means of certification.

Given the prevailing separation of powers between Government and schools, and in spite of the considerable resources used to initiate, develop and promote them, proposals for reform are no more than advisory and hortatory, so that a massive disjuncture can develop between what is proposed and what is practised. This is, of course, recognized by officials but has been rationalized by increasingly drawing upon the rhetoric of school-based curriculum development, with its focus on teacher empowerment and professionalism. Teachers are regularly reminded that (CDC 1993c: 34):

> In view of the varied abilities of pupils resulting from compulsory education, school heads and teachers are strongly advised to modify and adapt the common-core curriculum to make due allowance for the individual needs and capabilities of every pupil.

In addition (CDC 1993c: 42,36), 'Teachers, as professionals, are expected to choose the most appropriate approaches and methods for their pupils'. However, as the same document explains, there are limits to the degree of adaptation: 'Teachers are encouraged to develop curriculum projects that help to complement or supplement the centrally designed curriculum'.

Naturally, there are areas in which policy and practice are in reasonable harmony, perhaps because policy reflects practice rather than attempting to lead it. However, non-mandatory curriculum policies essentially serve a symbolic function, demonstrating the Government's good will in important areas of public policy.

Homework

The nature and prevalence of homework offers further insight into attitudes to the curriculum and to learning. If teachers and parents take the view that knowledge exists independently of the learner, and that learning is a direct result of effort, then anything which increases effort and time on task is justifiable. In Hong Kong, homework is not the preserve of secondary schooling, but is characteristic even of pre-school life. Principals of pre-schools argue that, if the teachers do not set homework, parents will send their children to places which do. At every level, its demands are heavy as the following quotation (Speak 1992: 73) from the suicide note of a 10-year-old boy poignantly illustrates:

> Everyday, there are many homeworks. They are not only in large quantity, but also difficult to do. Each recess is only engaging for 10 minutes. If getting one day more holiday, (I) will be given 10 odd homeworks. Especially in long vacation, the homeworks will be more. (I) can get no rest in any day. Dictations, quizs and examinations will be more. Though after 12 O'clock in every night I still have to revise my homeworks. I can't go to bed until 1 O'clock odd. At 6.50 hours in the morning, I have got to get up. (I) am so hard. I do wish no studying.

The separation of decision-making powers between the state and schools is often mirrored within schools with regard to the control of homework. Various policy documents describe the appropriate quantity of homework and indicate how it should be scheduled. In practice individual teachers usually have the autonomy to decide its quantity and frequency. The status of a subject, students' exam results and a teacher's dedication are all associated with large amounts of homework. Consequently, in the absence of an effective school policy, students can be inundated with routine exercises and tasks requiring memorization. While this contributes to maintaining a culture of diligence and perseverance, it carries a cost, for it helps to restrict the range of desirable outcomes of schooling to those associated with examination success.

Outcomes of schooling

The bulk of research studies designed to assess and compare pupils' learning in Hong Kong has focused on high-status knowledge. They have demonstrated levels of achievement in science and mathematics similar to or better than those in other countries. This partly reflects the degree of specialization and the proportion of time devoted to mathematics and sciences. Thus, Wolf (1982: 647) comments with reference to the interpretation of international test performance scores:

> A great deal of specialization can occur at the upper secondary level. In Hong Kong, for example, there is heavy specialization and students in their last three years of secondary school take only three school subjects. One common combination is mathematics, physics and chemistry. With such few courses there is opportunity for much more instruction and classes generally meet for about 400 minutes a week. In contrast, a student in a US high school is fortunate to have 250 minutes of instruction a week in a science course.

There is comparatively little analysis of pupils' achievement in areas associated with aesthetics, personal, social and moral education. The few data available do indicate a cause for concern and support the contention that the curriculum has failed to address the development of knowledge, attitudes and skills which are not central concerns of a collectivist curriculum. Wong (1990) reported poor development of aesthetic sensitivity. Low levels of physical activity and fitness have been reported by To (1985). Students' upper body strength was markedly weaker than their lower body strength (possibly a consequence of their having to climb stairs frequently during the course of a school day in a typical multi-storey Hong Kong school).

With regard to the social and political domain various commentators have noted low levels of political awareness. Lau (1990) has described Hong Kong people as tending to identify with narrow familial and other parochial interests, with an absence of a strong sense of communal solidarity. Bond (1993) argues that Hong Kong students display a complex identity which is characterized by a low level of identification with Hong Kong and a tendency to ascribe to themselves the more desirable features of both traditional Chinese and Western identities. This lack of a coherent sense of civic identity or community values amongst Hong Kong's youth is especially pertinent given the return of sovereignty to China in 1997. Essentially this area is characterized by a vacuum which could be readily filled by a government with a clear sense of purpose.

A survey of health-related behaviour (Day *et al.* 1994) indicated low levels of trust of adults by students and confirmed the weak level of identification with Hong Kong. A study of secondary school students' attitudes to sex and sex education (CDI 1995) suggested that their main sources of information about sex were the mass media and their peers and that their knowledge of 'sensitive' topics was poor. The survey also indicated that many boys held very traditional views towards sex roles and that many girls had a low 'self-image'.

A curriculum biased strongly towards achievement in high-status subjects is not necessarily appropriate for lower achievers, nor for those whose interests and capabilities point them in other directions. As pointed out earlier, such students are likely to be classified as lower achievers because of the bias in subject weightings. The timetable bias ensures that those skills they have are less than fully developed. In a recent survey (CDC 1995) unsatisfactory academic achievement was cited as their main concern by 40% of all students. Yung (1994) examined the levels of academic achievement of a sample of 391 such students in secondary year 3. He found that, while 70% of the sample were attaining at or below the primary year 4 level in Chinese, English and mathematics, as many as 50% had non-verbal reasoning scores near the average for their age group. This suggests that their poor performance cannot be wholly explained by intellectual capability and that the nature of the curriculum must be viewed as problematic. Paradoxically, the reforms (CDI 1993) designed to cater for these pupils have attempted to increase the time they spend on academic subjects.

Conclusions

The curriculum which prevails displays those features which were actively promoted in the postwar period when the state exercised direct control of the school curriculum. The outcome was the emergence of an abstract and academic curriculum which displays all the features of a collectivist code, a focus on public knowledge and disciplinary modes of conceptualizing knowledge. The advent of mass schooling and the rapid growth of the economy from the 1960s created an environment in which those features were appropriated and reinforced as schooling provided the key source of upward socioeconomic mobility in a rapidly growing economy. Parents, pupils and schools competed to obtain and provide access to high status knowledge. In parallel with the emergence of the resulting educational market the state has promoted a wide range of essentially exhortatory reforms designed to move the curriculum to a broader, more integrated and more relevant orientation. The result is a massive disjuncture between 'policy' and its implementation which allows the interests of the state to be protected.

In terms of the role of the state as an influence on the nature and status of school knowledge the situation saw a shift from an emphasis on state control to a state-centred model in which the locus of overt influence shifted but any threat to the stability of the state continued to be averted. As was most evident in the postwar period the state played a key role in defining the nature of curricular knowledge and its motives derived primarily from the desire to avoid threats to its stability. In the period since 1965, with the advent of mass schooling and of a highly competitive school system, the role of the state has been less direct as schools have adopted and reinforced the collectivist code established in the postwar period. In parallel the state has busied itself with promoting a variety of mainly pedagogic reforms, albeit with little impact on schools, and these have mainly served a symbolic role of demonstrating their concern to improve schooling. The situation has thus seen a shift from the use of coercion and regulation by the state to one where schools have developed a system of self-regulation. Foucault (1979) asserted that the exercise of coercion and repression are signs of the failure of power. In this respect the Hong Kong Government must be viewed as highly successful.

Finally, the autonomy given to schools to decide the nature of the school curriculum, and the emergence of a market in which schools compete for pupils, have not, as Gerstner (1994), and Caldwell and Spinks (1992) suggest, resulted in the emergence of a diverse range of innovative programmes designed to meet the needs of pupils and of discerning parents. The experience in Hong Kong would suggest that the operation of market forces has compelled schools to be responsive to the expectations of parents. However, this has resulted in a relatively uniform and narrowly constructed school curriculum as individual schools compete to attract the academically most able pupils. While the parents of academically able pupils might be viewed as empowered this does not extend to the parents of pupils whose talents do not match those promoted in the curriculum. The gap between the intended and implemented curriculum exists because the latter is a more accurate reflection of social and community expectations and the signals of the market, and the former reflects the state's desire to be seen to be promoting worthwhile visions of schooling for all.

References

BERLAK, H. and BERLAK, A. (1981) *Dilemmas of Schooling: Teaching and Social Change* (London: Methuen).

BERNSTEIN, B. (1971) On the classification and framing of educational knowledge. In M. F. D. Young (ed.), *Knowledge and Control: New Directions for the Sociology of Education* (London: Macmillan), 47 – 69.

BOND, M. H. (1993) *Between the yin and yang: the identity of the Hong Kong Chinese*. Professorial Inaugural Lecture Series, No. 19, Chinese University Bulletin, Supplement 31.

BOURDIEU, P. and PASSERON, J. C. (1977) *Reproduction in Education, Society and Culture*, trans. by Richard Nice. (London: Sage).

BRAY, M. and HUI, P. (1991) Macau. In C. Marsh and P. Morris (eds), *Curriculum Development in East Asia* (London: Falmer), 181 – 201.

CALDWELL, B. and SPINKS, J. (1992) *Leading the Self Managing School* (London: Falmer).

CLARK, J., SCARINO, A. and BROWNELL, J. (1994) *Improving the Quality of Learning* (Hong Kong: Hong Kong Bank Language Development Fund/Institute of Language in Education).

CONNELL, R. W., ASHENDEN, D. J., KESSLER, S. and DOWSETT, G. W. (1982) *Making the Difference* (Sydney: Allen & Unwin).

CURRICULUM DEVELOPMENT COUNCIL (CDC) (1992) *Guidelines on Environmental Education in Schools* (Hong Kong: Hong Kong Government Printer).

CURRICULUM DEVELOPMENT COUNCIL (CDC) (1993a) *Guide to the Kindergarten Curriculum* (Hong Kong: Hong Kong Government Printer).

CURRICULUM DEVELOPMENT COUNCIL (CDC) (1993b) *Guide to the Primary Curriculum* (Hong Kong: Hong Kong Government Printer).

CURRICULUM DEVELOPMENT COUNCIL (CDC) (1993c) *Guide to the Secondary Curriculum 1 – 5* (Hong Kong: Hong Kong Government Printer).

CURRICULUM DEVELOPMENT COUNCIL (CDC) (1993d) *Guide to the Sixth Form Curriculum* (Hong Kong: Hong Kong Government Printer).

CURRICULUM DEVELOPMENT COUNCIL (CDC) (1995) *A Study on the Development of Civic Awareness and Attitudes of Pupils of Secondary Schools in Hong Kong* (Hong Kong: CDC).

CURRICULUM DEVELOPMENT INSTITUTE (CDI) (1993) *Draft Report of the Working Group on Support Series for Schools with Band 5 Students* (Hong Kong: CDI).

CURRICULUM DEVELOPMENT INSTITUTE (CDI) (1995) *A Study on Knowledge and Attitudes of Secondary School Pupils on Sex and Sex Education* (Hong Kong: CDI).

DALE, R. (1989) *The State and Education Policy* (Milton Keynes: Open University Press).

DAY, J., BACON SHONE, J. and LAW, S. W. (1994) *The Health Related Behaviour Survey in Hong Kong* (Hong Kong: Social Science Research Centre, University of Hong Kong).

DEPARTMENT OF EDUCATION AND SCIENCE (DES) (1985) *The Curriculum from 5 – 16* (London: HMSO).

DOLL, W. E. (1993) *A Post-Modern Perspective on Curriculum* (New York: Teachers College Press).

EDUCATION DEPARTMENT (1974) *Report on First Year Trial of Integrated Science Project* (Hong Kong: Education Department).

EDUCATION AND MANPOWER BRANCH (1993) *School Education in Hong Kong: A Statement of Aims* (Hong Kong: EMB).

FOUCAULT, M. (1979) *Discipline and Punish: The Birth of the Prison* (Harmondsworth: Penguin).

FULLAN, M. (1989) Planning, doing and coping with change. In B. Moon, P. Murphy and J. Raynor (eds), *Policies for the Curriculum* (London: Hodder & Stoughton), 183 – 211.

FUNG, Y. W. and LEE, C. K. (1993) Environmental education in Hong Kong secondary schools. Paper tabled at the UNESCO Conference: Overcoming the Barriers to Environmental Education through Teacher Education (Brisbane, Australia: Griffith University).

GERSTNER, L. V. (1994) *Reinventing Education* (New York: Dutton).

GINSBURG, M. B., COOPER, S., RAGHU, R. and ZEGARRA, H. (1990) National and world-system explanations of educational reforms. *Comparative Education Review*, 34 (4): 474 – 496.

GIROUX, H. (1981) *Ideology, Culture and the Process of Schooling* (Philadelphia: Temple University Press).

HA, T. (1993) Sport in schools - how important? In Hong Kong, Sports Development Board, *Sport in Schools: The Future Challenge: A Seminar for School Principals* (Hong Kong: Hong Kong Sports Development Board), 11 – 15.

HONG KONG, SPORTS DEVELOPMENT BOARD (1993) *Sport in Schools: The Future Challenge: A Seminar for School Principals* (Hong Kong: Hong Kong Sports Development Board).

LAU, S. K. (1990) Decolonization without independence and the poverty of political leaders in Hong Kong. Occasional Paper No. 1 (Hong Kong: Hong Kong Institute of Asian Pacific Studies, Chinese University of Hong Kong).

LEVINE, L. (1988) *Highbrow/Lowbrow: The Emergence of Cultural Hierarchy in America* (Cambridge: Harvard University Press).

LUK, H. K. (1991) Chinese culture in the Hong Kong curriculum: heritage and colonialism. *Comparative Education Review*, 34 (4), 650 – 68.

MCCLELLAND, J. A. G. (1991) Curriculum development in Hong Kong. In C. Marsh and P. Morris (eds), *Curriculum Development in East Asia* (London: Falmer).

MORRIS, P. and SWEETING, A. (1991) Education and politics: the case of Hong Kong from an historical perspective. *Oxford Review of Education*, 17(3), 249 – 67.

MORRIS, P., MCCLELLAND, J. A. G. and WONG, P. M. (1997) Explaining curriculum change: the case of social studies. *Comparative Education Review*, 41 (1).

NG, M. L. (1994) Sexuality in Hong Kong. In D. McMillen and S. W. Man (eds), *The Other Hong Kong Report* (Hong Kong: Chinese University Press), 415 – 28.

SCHRAG, F. (1988) *Thinking in School and Society* (London: Routledge & Kegan Paul).

SEIXAS, P. (1993) Parallel crises: history and the social studies curriculum in the USA. *Journal of Curriculum Studies*, 25 (3), 235 – 50.

SPEAK, M. (1992) The children's view. In M. Speak (ed.), *Children in Sport*, proceedings of a conference organized by the Department of Physical Education, University of Hong Kong and the Hong Kong Sports Development Board (Hong Kong: Sports Development Board), 72 – 3.

SWEETING, A. (1995) Educational policy in a time of transition: the case of Hong Kong. In T. Wragg (ed.), *Research Papers in Education: Policy and Practice*, 10 (1), 101 – 29.

SWEETING, A. and MORRIS, P. (1993) Educational reform in post-war Hong Kong: planning and crisis intervention. *International Journal of Education Development*, 13(3), 201 – 16.

TANG, C. K. and MORRIS, P. (1989) The abuse of educational evaluation: a study of the evaluation of the implementation of the civic education 'Guidelines'. *Educational Research Journal* (Hong Kong), 4, 41 – 9.

TO, C. Y. (1985) *Physical Fitness of Children in Hong Kong* (Hong Kong: Chinese University of Hong Kong, School of Education).

USHER, R. and EDWARDS, R. (1994) *Postmodernism and Education* (London: Routledge).

WAN, K. K. (1991) A study of curriculum innovation in Hong Kong: the case of liberal studies. Master's thesis, University of Hong Kong.

WILLIAMS, R. (1961) *The Long Revolution* (New York: Harper Torchbooks).

WOLF, R. M. (1982) International test performance. In M. C. Alkin (ed.), *Encyclopedia of Educational Research*, 6th edn, Vol. 2 (New York: Macmillan), 642 – 52.

WONG, M. A. (1990) The music curriculum in the primary and secondary schools of Hong Kong. Doctoral thesis, Teachers College, Columbia University.

YOUNG, M. F. D. (1971) An approach to the study of curricula as socially organized knowledge. In M. F. D. Young (ed.), *Knowledge and Control: New Directions for the Sociology of Education* (London: Collier-Macmillan), 19 – 46.

YUNG, K. K. (1994) Nonverbal reasoning ability. Paper presented at the 1994 Hong Kong Educational Research Association Conference.

[20]

The Chinese Society and Family Policy for Hong Kong

Nelson W. S. Chow

SUMMARY. As in other industrial societies, family in Hong Kong has undergone rapid and drastic change since the 1970s. This paper uses the development of governmental family policy as a vehicle for examining the various topics and methods that have been of interest to family scholars in Hong Kong. Considerable attention is given to the suggestions of a White Paper (policy paper) published in March 1991 by the Hong Kong Government on the future development of social welfare. The paper also suggests objectives regarding family that may be used to guide further development of family policy in Hong Kong. *[Article copies available from The Haworth Document Delivery Service: 1-800-342-9678. E-mail address: getinfo@haworth.com]*

THE CHINESE SOCIETY AND THE NEED FOR A FAMILY POLICY

As 98% of the population in Hong Kong is made up of ethnic Chinese, culturally it is not unlike other societies with a predominant Chinese population (Latourette, 1964). A common feature shared by all Chinese societies is the important place given to the family, which is also regarded as the most fundamental institution in

Nelson W. S. Chow is Professor at the Department of Social Work and Social Administration of The University of Hong Kong.

[Haworth co-indexing entry note]: "The Chinese Society and Family Policy for Hong Kong." Chow, Nelson W. S. Co-published simultaneously in *Marriage & Family Review* (The Haworth Press, Inc.) Vol. 22, No. 1/2, 1996, pp. 55-72; and: *Intercultural Variation in Family Research and Theory: Implications for Cross-National Studies* (ed: Marvin B. Sussman, and Roma Stovall Hanks) The Haworth Press, Inc., 1996, pp. 55-72. Single or multiple copies of this article are available from The Haworth Document Delivery Service [1-800-342-9678, 9:00 a.m. - 5:00 p.m. (EST). E-mail address: getinfo@haworth.com].

© 1996 by The Haworth Press, Inc. All rights reserved.

society (Baker, 1979). Recent research shows that although the family system in present-day Chinese societies still possesses the same characteristic of being founded upon blood relations with members descending from the same ancestor, it has changed greatly from its traditional forms and purposes (Fei, 1984). In traditional China, the family and kinship system conferred upon its members distinct rights and responsibilities but these entitlements and obligations have so much been weakened nowadays, as a result of industrialization and urbanization, that the belief that the family is the best provider of care and support for its members has been cast into doubt (Chow, 1990).

The necessity of formulating a family policy in Hong Kong was not raised until the late 1970s when evidence increasingly showed that, similar to other industrial societies, the family system in Hong Kong was undergoing rapid and drastic changes (Wong, 1975). The number of divorces, which had remained extremely low or almost non-existent, began to exceed 1,000 cases in 1980 and increased to 5-6,000 cases a year at the end of the decade. The Population Census conducted in 1981 found for the first time that "unextended nuclear families" accounted for over half of the total number of households. The traditional belief that children in Chinese families normally preferred to live with their parents, even after marriage, was therefore not necessarily true. Since the early 1980s, profit-making child care centers and nursing homes have also mushroomed, indicating a rising demand for such services and a weakening of the caring functions of the family system for the young and the old.

In view of the changes in both the structure and functions of the family system, a Working Party was set up in 1980 by the Hong Kong Council of Social Service, a co-ordinating body of non-government organizations in Hong Kong, to examine the need for a family policy (Hong Kong Council of Social Service, 1986). The Working Party concluded a year later that welfare services in support of the family, especially child care services, should be increased in quantity and in variety. It did not, however, recommend the formulation of a family policy. It was perceived at that time that most Chinese in Hong Kong would still regard matters related to the family as their own business and might not be receptive to the idea of a family policy. Furthermore, the Working Party admitted that even if such a policy were to be formulated, no consensus could be reached on what the policy should contain or how it should best be implemented.

What then is a family policy? Kamerman and Kahn (1978), in their edited book on family policy, documented 14 countries as having either explicit, like France, or implicit, like United Kingdom, family policies. They accepted that there was no consensus "on exactly what is happening to the family in the industrialized world, how family change should be regarded, or what is meant by family policy" (Kamerman & Kahn, 1978, p. 1). And they defined family policy as "referring to deliberate actions taken toward the family, such as day care, child welfare, family counseling, family planning, income maintenance, some tax benefits, and some housing policies" (p. 3). The definition concentrates on actions taken to support the family. It "means everything that government does to and for the family" (p. 3). The explicit and implicit family policies of the 14 countries discussed in the book of Kamerman and Kahn were illustrations of what governments could do to and for the family.

The conclusion drawn is that in a modern industrial society, not only is the family system having difficulty in functioning as a self-sufficient unit, but it also requires the intervention of the state. The purpose of state intervention is either to replace some of the functions formerly performed by the family or to assist it in performing its roles. There is no fixed form of the kind of state intervention required, as societies differ in their cultural and political heritage, but what is certain is that no country, in the process of industrialization, can leave its family system unassisted (Wilensky & Lebeaux, 1965). So far as Hong Kong is concerned, the sufficiency of the Chinese family system to meet all the needs of its members has been questioned as early as in 1966 by Gertrude Williams, a consultant invited by the Hong Kong Government to come to Hong Kong to determine the feasibility of establishing social welfare legislation. While recognizing the traditional Chinese family system as possessing "a built-in social security system" (Williams, 1966, p. 12), Williams observed that "whilst the family provides help during an emergency, the exigencies of the industrial urban life now lived by most people does not allow of prolonged and continuous help" (p. 14). More recent research by Law finds that not only are most families in present-day Hong Kong unable to provide prolonged help, but also they are in need of assistance from outside (Law, 1991b). Social and economic changes which Hong Kong has undergone in recent years have therefore called for a re-examination of the need for a government policy for the family.

CHINESE CULTURAL VALUES AND THE FAMILY

In a White Paper (policy paper) published in March 1991 on the future development of social welfare, the Hong Kong Government stated that "The family unit is a vital component of society . . . In Hong Kong, high values continue to be attached to the family unit to an extent which cannot be matched by any other institution" (Hong Kong Government, 1991, p. 19). Such importance attached to the family is based on the view, popularly accepted by the Chinese in Hong Kong, that it is primarily through the family system that people learn their proper behavior towards each other (Chow, 1991). A change or even a weakening of this important role of the family in socializing its members would therefore not only affect the relationships within the family but also the stability of the larger community. The insistence on the family as a vital institution implies that a loosening of this position would bring havoc to the society.

The functions of a traditional Chinese family system are often described as including the following (Yang, 1959): the proper expression of sexual desire, the education of the children, the care of the young and the old, the organization of productive activities, the inheritance of properties, the discipline of deviant behavior, the cultivation of proper relationships and the worship of ancestors. Obviously not all of the above functions are still performed by Hong Kong families. For example, the organization of productive activities is now the domain of commercial institutions while others, like the care of the young and the old, have also taken very different forms (Tao, 1981).

Although industrialization and urbanization have often been cited as the two major phenomena directly contributing to the change in functions of the traditional Chinese family system, it is more likely that a change of historic values has had a more powerful influence on how family members should interact with one another. These values include such notions as filial piety, chastity and faithfulness. In brief, according to traditional teaching, children are expected to be pious towards their elders, wives to be chaste towards their husbands and all to be faithful towards one another. These cardinal values of the family system have not all disappeared in present-day Chinese societies. However, it is erroneous to assume that they still form the bases for the organization and activities of families in Hong Kong.

The question now facing Hong Kong, so far as families are concerned, is to what extent are the above functions and values regarding the family still in effect (Lau & Kuan, 1988)? If the myths about family system values are still held dear by the people in Hong Kong and the families are largely functioning according to traditional expectations, there would not be any need for a family policy. On the other hand, if the traditional values no longer dominate and the families in Hong Kong are taking on new values and beliefs and are changing rapidly their functions, then there would be a case for the formulation of a family policy. The question is where does Hong Kong now stand?

THE HONG KONG FAMILY SYSTEM UNDERGOING CHANGES

The family system in Hong Kong is obviously undergoing rapid changes. The White Paper published in 1991, though emphasizing the family as a vital component of society, also accepted that "in this rapidly developing society [of Hong Kong] the traditional roles of the different constituents which make up the family unit and in particular the role of women, are changing" (Hong Kong Government, 1991, p. 19). The White Paper further elaborated the changes occurring in the family system in Hong Kong as follows:

> The number of single parent families is increasing as marital separations and divorces become more common. The continuing movement of families to the New Towns affects their

ability to provide care and support for other family members who are not part of the relocation process. Relocation in itself can be stressful for those involved and the adoption of foreign values and concepts is adding to the challenges facing the traditional cohesiveness of the family. Child abuse and neglect are increasing and the problem of battered spouses continues. The increased participation of women in the local work force presents problems relating to child care. Increased emigration is also a feature which has implications for the demand for social welfare services. (Hong Kong Government, 1991, p. 19)

The changes described by the White Paper are also supported by the findings of the Population Census conducted in 1991. It has already been mentioned that nuclear families have long become the norm rather than the exception, with the average size of the household decreasing from 3.9 persons in 1981 to 3.4 persons in 1991. Figures on the number of single parent families are still not available from the 1991 Population Census but a "social indicator survey" carried out in 1988 by a group of social researchers in Hong Kong found that single-parent families accounted for 6.6% of all households (Lau, Lee, Wan, & Wong, 1991, p. 43), or 104,426 single-parent families out of the 1,582,215 households in 1991. The needs of single-parent families for social support may not necessarily be greater than others but a recent study on single-parent families in Hong Kong found that when compared with those having both parents present, they were usually worse off financially; and a higher percentage of their children also showed learning problems (Law, 1991a). The rising divorce rate is probably the major factor contributing to the increase in single-parent families. However, it is also known that some of the single-parent families are brought about by the absence of the women whom Hong Kong residents married in China but are still waiting to come to Hong Kong to join their husbands. Anyway, it appears quite definite that the number of single-parent families in Hong Kong will increase further.

The geographical relocation of families brought about by the development of new towns in Hong Kong is another factor which has contributed to the loosening of family ties. In 1991, 41.9% of the households in Hong Kong were located in the new towns of the

New Territories, as compared to 26% in 1981 (Census & Statistics Department, 1991, p. 68). As most of the families which moved to the new towns are made up of young couples and their children, this has resulted in a reduction in the number of "vertically extended nuclear families" or more commonly known as "three-generation families." In 1991, the number of "vertically extended nuclear families" stood at 10.7% of all households, a decrease from 13.6% in 1981 (Census & Statistics Department, 1991, p. 59). There are of course other reasons accounting for the decreasing rate of co-residence between married children and their parents but a study conducted in 1987 in Tuen Mun, one of the new towns in Hong Kong, found that the new families moved there have often left their elderly parents in the old urban districts (Chow, 1988). Another study on the housing needs of the elderly in Hong Kong found that some of the old housing estates have as high as 20% or even more of their residents aged 60 and above, nearly double that in the general population (Hong Kong Housing Authority, 1989). Hence, with the development of new towns in Hong Kong, families are being split up and fewer members are now available in the household to provide care and support for each other.

The family system is still perceived as a vital component of society because the Hong Kong Government believes, as stated in the 1991 White paper, that "It [the family] provides an intimate environment in which physical care, mutual support and emotional security are normally available . . . " (Hong Kong Government, 1991, p. 19). Studies on family functioning have found that women in general are most important within the family to provide the needed physical care and support (Bulmer, 1987). But more and more women in Hong Kong are now joining the labor force, numbering more than 1 million out of the total of 2.8 million economically active persons in 1991, their caring roles in the family are expected to diminish. The fact that more than 80,000 imported domestic maidservants are now employed in Hong Kong indicates that many families have found it necessary to spare the women their household responsibilities. The increasing percentage of women employed in the labor market has also produced other far-reaching effects on the family system, such as the change in the decision-making process and the differentiation of sex roles within the family, but relevant data are still to be gathered.

So far, discussions have concentrated on the negative rather than the positive effects brought about by demographic changes to the family system in Hong Kong. But the "social indicator survey" mentioned previously found that, despite the decreasing size of the family, the prevailing preference of married children to set up their own families and the increasing percentage of women staying in their jobs after marriage, the overwhelming majority of the respondents still "found home a comfortable and pleasant place" (Lau, Lee, Wan, & Wong, 1991, p. 51). However, when enquired about some of the traditional values held dear regarding the family, the respondents were less certain in their answers. For example, while as many as 62.1% of respondents felt that they would not object to newly married couples living away from their parents, 23.2% were not sure if this was right (Lau, Lee, Wan, & Wong, 1991, p. 43). There was also a split in response to whether or not children were obliged to support their parents. As for the sex roles of husbands and wives, respondents were again unsure of the traditional norm that women should stay at home while men should go out to work.

What can be concluded from the above discussion is that while most values traditionally held dear by the people in Hong Kong are not rejected outright, they certainly do not have the same binding power on behavior as in the past. Census data also show that most people in Hong Kong are increasingly modifying their practices to suit the requirements of a modern industrial society. For example, while ideologically newly married couples would still feel obliged to live with their parents, housing conditions in Hong Kong are such that young couples often find it more preferable to set up their own families and put aside the traditional values.

In view of the changes which have occurred in the family system in Hong Kong, the question which needs to be asked is: To what extent could one totally disregard the traditional values in formulating policies for the family? There is certainly no easy answer as the people are also split in their opinions on the values that they should hold regarding the family. This uncertainty is also reflected in the 1991 White Paper on social welfare when it emphasized, on the one hand, the importance of the family as the most fundamental unit in society and accepted, on the other, that the family had changed so much in its functions that it needed strengthening. These two views about the family might not necessarily contradict each other but it certainly makes the task of formulating a family policy in Hong Kong a difficult one.

EXISTING POLICIES ON THE FAMILY IN HONG KONG

In 1965, when the Hong Kong Government issued its first White Paper on social welfare, services in support of the family were almost non-existent (Hong Kong Government, 1965). This did not imply that the family system was unimportant; the belief guiding the Government at that time was that the family system was such a significant source of help and support to its members that what was needed was to preserve this tradition. Hence the policy as stated in the 1965 White Paper was to "help families to remain intact as strong natural units and to care for (and not to abandon) their children and handicapped or aged members" (Hong Kong Government, 1965, p. 10). Furthermore, the White Paper stated that:

> It is of the greatest importance that social welfare services should not be organized in such a way as to make it easier for socially disruptive influences to gain a hold over the community, or to accelerate the breakdown of the natural or traditional sense of responsibility—for example by encouraging the natural family unit to shed on to social welfare agencies, public or private, its moral responsibility to care for the aged or infirm. (Hong Kong Government, 1965, p. 5)

Although the 1965 White Paper was not in favor of a specific policy for the family, the Social Welfare Department of the Hong Kong Government began in the mid-1960s to group under one umbrella the various welfare services provided for the family, such as public assistance, adoption service, care and protection service for women and juveniles, services for the infirm and the aged, and located them in "family services centers" of the different districts. The purpose of setting up these "family services centers," as stated in another White Paper on social welfare published in 1973, was to provide "comprehensive services to families in need" (Hong Kong Government, 1973, p. 9) and there was no intention of turning this arrangement into a specific policy for the family.

Welfare services provided for the family have hardly experienced any change since 1973 except with the addition of family life education programs in 1979. As stated in another White Paper on social welfare published in 1979, the general aim of the family life education programs was "to preserve and strengthen the family as a unit" (Hong Kong Government, 1979, p. 19) which probably means maintaining as far as possible a continuation of the traditional roles of the family in providing care and support for its members. The aims of the family welfare services, and thus the needs and functions of the families, were not looked at again until the Government was preparing the 1991 White Paper. Some of the changes and problems encountered by the family system in Hong Kong which this White Paper has identified have already been mentioned. Instead of assuming that the family system is performing among of its traditional functions, the Government has accepted for the first time the conclusion of the research findings that such phenomena as child abuse, domestic violence and marital conflicts, so common in other industrial societies, were also threatening the family system in Hong Kong (Young, 1985). It was thus stated in the 1991 White Paper that:

> It is apparent that many of the problems traditionally associated with social needs will continue to warrant attention but it is equally clear that new needs and problems are emerging which will require greater efforts in the context of the future provision of family and child care services. (Hong Kong Government, 1991, p. 20)

What then should be done to address the emerging needs and problems of the family? Taking the 1991 White Paper as a whole, those that have been suggested differed very little from those that were already in existence. However, the 1991 White paper was obviously more positive in its attitude towards the purposes of the welfare services provided for the family. It stated that the philosophy underlying the provision of social welfare in Hong Kong was to "encourage men and women as individuals to develop their capacities to the full and to be active and productive members of society while meeting their commitments to family and society" (Hong Kong Government, 1991, pp. 15-16). It nevertheless maintained that "In the course of the future development of welfare services, emphasis will continue to be placed on the importance of the family unit as the primary provider of care and welfare, and thus on the need to preserve and support it" (Hong Kong Government, 1991, p. 16). It is therefore not the intention of the Government to change the structure of the family in Hong Kong, nor to alter any of the functions it still manages to perform. The 1991 White Paper implies, however, that there may be a need to strengthen the family system as its functions are diminishing or are being shared increasingly with societal institutions and organizations. As for the specific policies regarding the provision of welfare services for the family, the White Paper stated:

> The overall objectives of family welfare services are to preserve and strengthen the family as a unit and to develop caring interpersonal relationships, to enable individuals and family members to prevent personal and family problems and to deal with them when they arise and to provide for needs which cannot be met from within the family. With these objectives in mind, support services have been developed to assist families when they are unable to discharge their caring and protective functions satisfactorily. (Hong Kong Government, 1991, p. 19)

The questions one would raise are: Could the emerging needs and problems of the family system in Hong Kong be adequately dealt with by the above policy on family welfare services (Craven, Rimmer, & Wicks, 1982)? To what extent could the existing family welfare services achieve the purpose of preserving the family as a unit and developing caring relationships? How could individuals and family members be helped to prevent personal and family problems and to deal with them when they arise (Rossiter & Wicks, 1982)? In short, can one be content with a policy which aims at not more than preserving, and at best strengthening, a social institution and its functions which are admittedly undergoing drastic changes (Zimmerman, 1988)?

The provision of welfare services for the family now consists of:

family life education, family casework and counseling, medical social service, home help service, clinical psychological service, various child care services, and shelter and assistance to such marginal groups as street sleepers and drug abusers. Except for the family life education programs, which aim at educating "the public on the importance of family life and how it can be sustained" (Hong Kong Government, 1991, p. 20), the welfare services are mainly remedial with the purpose to help families and their members to overcome problems and difficulties. These services are undoubtedly very much needed, and they are often severely scarce in supply, but could they meet the emerging needs and problems of the family system in Hong Kong? Could they indeed achieve the purpose of strengthening the family? The need for a comprehensive family policy will be discussed later but it suffices to point out here that the contribution of the Government is now inevitable in any attempt to strengthen the family system in Hong Kong.

A FAMILY POLICY FOR HONG KONG

Once the Hong Kong Government has accepted its responsibility to meet the needs of the families, some sort of family policy, in accordance with the definition given by Kamerman and Kahn (1978), can be said to be already in existence. The Hong Kong Government is, however, not ready for an "explicit" family policy as it has turned down in 1990 a similar request for the formulation of a youth policy, for fear that an "explicit" policy would imply a greater government commitment (Central Committee on Youth, 1988). Can one then take the policies embodied in the 1991 White Paper on the development of welfare services for the family as representing an "implicit" family policy?

According to Kamerman and Kahn (1978), the idea of a family policy may be used as a field "in which certain objectives regarding the family are established," or as an instrument "for goal attainment" (Kamerman & Kahn, 1978, pp. 5-6), or as perspective or criterion for social policy choice. In addition to being unready for an "explicit" family policy, it appears that the Hong Kong Government is also not prepared to formulate any objective for the family. It should be mentioned that the present Hong Kong Government will cease its rule over the Territory in 1997, and it obviously has the worry that any objective which it now lays down for the family might not be acceptable to the future Hong Kong Special Administrative Region Government. Likewise, the present Government would be hesitant to use the family as an instrument to achieve goals stated in various public policies for fear that it would be accused of using the family as an agent of social control, contrary to the wishes of the future rulers.

Hence, apart from providing support for its members and to perform certain traditional functions, such as transmitting societal values and exercising control over the behavior of its members, the Hong Kong Government has on the whole refrained from giving the impression that it is trying to manipulate the functions of the family or adding to it any responsibility which it is not already performing. Such fears of intruding into the affairs of the family arise partly from the traditional belief that family matters should be resolved within the family and partly from the general philosophy of the Hong Kong Government to do only the minimum, especially in this transitional period when the sovereignty over the Territory will soon change hands (Lau, 1982).

Finally, would a family policy be necessary in Hong Kong that would be used as "perspective or criterion" for public policy choices? When the 1991 White Paper was in its drafting stage, a suggestion had been made for all public policies, especially those closely affecting the quality of life of the people, to carry a "statement" on their impact on the family. The suggestion was, however, not accepted by the Government on the ground that such "statements" would probably be too vague to be useful. As a result, there exist in Hong Kong only certain expectations of what functions the family should perform and there is hardly any objective or goal for it to achieve. If a family policy is defined as "what the government is doing for the family," the best that one can say about the situation in Hong Kong is that a wide range of welfare services are provided either to prevent or to remedy problems encountered by the family. The functions to be performed by the family are, however, taken for granted; there is no deliberate action to either change or modify them. The attitude of the Hong Kong Government is that the family should by and large be left on its own and it is even believed that any interference would probably do more harm than good. So, in a way, a family policy is non-existent in Hong Kong; even if one counts the provision of welfare services as "what the state does for the family," the policy can only be described as "implicit," or "reactive" as the intention is to patch up what the family fails to perform.

A reactive "family policy," or no policy at all, may not be unacceptable, especially if the family system can be assumed to be functioning well; but there is evidence that the inaction of the Government has already produced the following ill effects (Rhind, 1989). First, as welfare services are provided to remedy what the family fails to perform, they have hardly any positive objectives of their own. For example, the increased number of married women going into gainful employment has recently created a demand for after-school service, but there is little debate on the purpose of the service or who should be responsible for its provision. Second, since welfare services have been developed at different times to meet the divergent needs of the family, they could hardly come together to form a coherent whole and sometimes may even contradict one another in their effects. For example, the assessment of eligibility for public assistance takes the income of the entire household into consideration and effectively excludes many needy elderly from receiving the allowance who are residing with their married children. Hence, in order to be eligible for public assistance, some of these needy elderly have taken the option of living away from their married children and consequently depriving themselves of the companionship of their children. Without the benefit of a family policy to act as "perspective" for policy choices, it is not surprising to find that existing welfare measures often produce contradictory effects (Chow, 1990).

The third ill effect is the inability of the Government to respond to the changing needs of the family. It has been mentioned that although the 1991 White Paper succeeded in identifying the emerging needs and problems of the family, it failed to push forward any concrete suggestion to either prevent or remedy the situation. The best it could recommend was "to develop family welfare services in line with the increasing diversity of family-related problems" (Hong Kong Government, 1991, p. 21). The absence of viable suggestions may be the result of a number of factors but as long as positive measures to strengthen the family system are yet to be devised, few new services to meet the changing needs and problems of the family will be forthcoming.

The conclusion that one can draw is that the family system in Hong Kong, though remaining the most fundamental unit in society, is no longer capable of providing its members with all the care and support they need. A survey conducted in the early 1980s on the life-style of the elderly in Hong Kong found that the care and support of the elderly had already become a responsibility of the family as much as that of the Government (Chow & Kwan, 1983). It is thus imperative for plans of action to be drawn to support the family. The devise of such plans recognizes, on the one hand, that the family system in Hong Kong is no longer a self-sufficient unit and accepts, on the other, that public intervention into the family system is both necessary and desirable. Research shows that people in Hong Kong are ideologically receptive to such changing roles of the Government on the family and they would not regard this as a negation of the traditional family values (Chow, 1992). Hence, though the time may not be appropriate for the formulation of an "explicit" family policy, it is definitely essential for the Hong Kong Government to put under a coherent plan of action all the efforts that are now made to support the family. In formulating such a plan, some common objectives regarding the family have first to be agreed upon. The following is a suggested list of such objectives:

1. The family should be preserved as a coherent unit for the provision of care and welfare to its members.
2. Family members, while having a responsibility to care for one other, should also have the right to societal support when necessary.
3. The family unit is a vital component of society and can only function effectively when it is closely linked with other institutions.
4. The family is a place where members find comfort and security and not fear and oppression.
5. Welfare measures should be introduced to enhance the functioning of the family and not to inhibit or substitute it.

6. The interests of the individual and of the family, though not identical, often complement each other in bringing greater fulfillment to all.

The above list is not meant to be exhaustive. As they stand, they will not only bring together under one coherent policy the various welfare services but also harmonize what seem to be conflicting old and new ideas about the family.

FAMILY POLICY FOR THE FUTURE

The devising of a family policy is never an easy task. As Hong Kong develops into a mature industrial economy, its family system cannot avoid changes. Studies on the family system in Hong Kong as well as Population Census findings all indicate that the family system in Hong Kong is undergoing changes similar to those detected in other industrial societies. The family system in Hong Kong is not just getting smaller in size but is also affected by increasing incidence of separation, divorces, domestic violence and desertion. Evidence shows that the Chinese families in Hong Kong are losing much of their traditional functions. At the same time, the Government is increasingly being pressurized into providing more and more welfare services to support the family.

While the Hong Kong Government must necessarily do more for the family, it is, however, reluctant to come up with any specific policy for the family. It emphasizes, on the one hand, the importance of preserving the family as a vital unit of society and admits, on the other, the necessity of supporting it with various welfare services. While these may not necessarily contradict each other, the Government is so obscure in its actions that overall objectives are often lacking in what it is now doing for the family.

The formulation of a family policy in Hong Kong should not be seen as a negation of the traditional values regarding the family; it can be a positive step to ensure that ideas held dear by the people are enshrined in some broad objectives to support the family. The discussion in this paper concludes that, in the face of rapid social and economic changes, a "reactive" approach which has so far been adopted by the Hong Kong Government towards the family is no longer sufficient. It is thus argued that only the formulation of a family policy, even an "implicit" one, as represented by a plan of action, would be more effective in promoting a better family life for the people in Hong Kong.

REFERENCES

Baker, H. D. R. (1979). *Chinese family and kinship.* London: Macmillan.

Bulmer, M. (1987). *The social basis of community care.* London: Allen & Unwin.

Census and Statistics Department, Hong Kong Government (1991). *Hong Kong 1991 population census-summary results.* Hong Kong: Hong Kong Government Printer.

Central Committee on Youth. (1988). *Report on youth policy.* Hong Kong: Hong Kong Government Printer.

Chow, N. W. S. (1983). The Chinese family and support of the elderly in Hong Kong. *The Gerontologist,* 23(6), 353-366.

Chow, N. W. S. (1988). *Social adaptation in new towns: A report of a survey on the quality of life of Tuen Mun inhabitants.* Resource paper series no.12. Hong Kong: Department of Social Work & Social Administration, University of Hong Kong.

Chow, N. W. S. (1990). Social welfare in China. In Elliott, D., Mayadas, N. S., & Watts, T. D. (Eds.). *The world of social welfare* (pp. 219-231). Springfield, Illinois: Charles C. Thomas.

Chow, N. W. S. (1990). Social welfare. In R.Y.C. Wong & J. Y. S. Cheng (Eds.). *The other Hong Kong report 1990* (pp. 429-443). Hong Kong: The Chinese University Press.

Chow, N. W. S. (1991). Does filial piety exist under Chinese communism? *Journal of Aging & Social Policy,* 3(1/2), 209-225.

Chow, N. W. S. (1992). Family care of the elderly in Hong Kong. In Kosberg, J. I. (Ed.). *Family care of the elderly-Social and cultural changes* (pp. 123-138). Newbury Park, California: SAGE Publications.

Chow, N. W. S. & Kwan, A. Y. H. (1983). *The life-style of the elderly in low income families in Hong Kong.* Hong Kong: Department of Social Work, The Chinese University of Hong Kong.

Craven, E., Rimmer, L., & Wicks, M. (1982). *Family issues and public policy.* London: Study Commission on the Family.

Fei, X-T. (1984). On changes in the Chinese family structure. In Chu, D. S. K. (Ed.). *Sociology and society in contemporary China* (pp. 32-45). New York: M. E. Sharpe.

Hong Kong Council of Social Service. (1986). *Family and child care services-An overall review and future development.* Mimeograph. Hong Kong: Hong Kong Council of Social Service.

Hong Kong Government. (1965). *Aims and policy for social welfare in Hong Kong.* Hong Kong: Hong Kong Government Printer.

72   *Intercultural Variation in Family Research and Theory*

Hong Kong Government. (1973). *Social welfare in Hong Kong: The way ahead.* Hong Kong: Hong Kong Government Printer.

Hong Kong Government. (1979). *Social welfare into the 1980s.* Hong Kong: Hong Kong Government Printer.

Hong Kong Government. (1991). *Social welfare into the 1990s and beyond.* Hong Kong: Hong Kong Government Printer.

Hong Kong Housing Authority. (1989). *Executive summary of the report of the working party on housing for the elderly.* Hong Kong: Hong Kong Housing Authority.

Kamerman, S. B., & Kahn, A. J. (Eds.) (1978). *Family policy.* New York: Columbia University Press.

Latourette, K. S. (1964). *The Chinese, their history and culture.* New York: Macmillan.

Lau, S. K. (1982). *Society and politics in Hong Kong.* Hong Kong: The Chinese University Press.

Lau, S. K., & Kuan, H. C. (1988). *The ethos of the Hong Kong Chinese.* Hong Kong: The Chinese University Press.

Lau, S. K., Lee, M. K., Wan, P. S., & Wong, S. L. (Eds.) (1991). *Indicators of social development: Hong Kong 1988.* Hong Kong: The Chinese University Press.

Law, C. K. (1991a). *Needs of single parent families: A comparative study.* Hong Kong: Hong Kong Family Welfare Society.

Law, C. K. (1991b). *An evaluation research report on family resource center.* Hong Kong: Hong Kong Family Welfare Society.

Rhind, N. (Ed.). (1989). *Strengthening families.* Hong Kong: Hong Kong Family Welfare Society.

Rossiter, C., & Wicks, M. (1982). *Crisis or challenges?-Family care, elderly people and social policy.* London: Study Commission on the Family.

Tao, J. (1981). Growing old in Hong Kong: problems and programs. In J. F. Jones (Ed.). *The common welfare: Hong Kong's social services* (pp. 107-116). Hong Kong: The Chinese University Press.

Wilensky, H., & Lebeaux, C. N. (1965). *Industrial society and social welfare.* New York: Free Press.

Williams, G. (1966). *Report on the feasibility of a survey into social welfare provision and allied topics in Hong Kong.* Hong Kong: Hong Kong Government Printer.

Wong. F. M. (1975). Industrialization and family structure in Hong Kong. *Journal of Marriage and the Family, 37*(4), 985-1000.

Yang, C. K. (1959). *The Chinese family in the communist revolution.* Cambridge, MA: Harvard University Press.

Young, K. P. H. (1985). *A study on single parent families in Hong Kong.* Hong Kong: Department of Social Work, The University of Hong Kong.

Zimmerman, S. L. (1988). *Understanding family policy: Theoretical approaches.* Newbury Park, California: SAGE Publications.

[21]

THE STRUCTURING OF SOCIAL MOVEMENTS IN CONTEMPORARY HONG KONG

TAI-LOK LUI and STEPHEN W.K. CHIU*

Many researchers in the field of Hong Kong politics, regardless of the differences in their approach to the study of political development, have made the observation that

> The persistence of the colonial constitutional order has been accompanied by remarkable political stability. Hong Kong has never experienced any large-scale revolt or revolution. On the contrary, it is reputed for its lack of serious disputes. (Kuan 1979)

In his seminal work on society and politics in Hong Kong, Lau calls "the existence of political stability under highly destabilizing conditions" in Hong Kong a "miracle" of the 20th century (Lau 1982, p. 1). In a recent review of the study of social conflict and collective actions in Hong Kong, Leung notes that "[a]lthough a rapidly modernizing society under colonial rule, Hong Kong has been exceptional in having been spared the frequent turmoil and instability that have plagued other countries of a similar socio-economic and political status. Since they have not been a particularly salient feature of the society, social conflict and social movements have rarely been the subject of inquiry in studies of Hong Kong" (Leung 1996, p. 159).

Of course, few observers of Hong Kong politics would deny the existence of social conflict and social movements in contemporary Hong Kong. Rather, they argue that "conflicts will be confined in scale because, under normal conditions, it is extremely difficult to mobilize the Chinese people in Hong Kong to embark upon a sustained, high-cost political movement" (Lau 1982, p. 20). Given that most local collective actions have not been able to present a forceful challenge to the colonial state and thus do not constitute a serious threat to the stability of the existing political order, social conflict and social movements are relegated to a position of secondary importance, if not of total insignificance, in the analysis of Hong Kong politics.

However, this so-called politically quiescent society has, since the 1970s, witnessed wave after wave of collective actions — from student activism, urban protests, to organized actions of civil service unions — indicating a change in the

*   The authors are Associate Professors in the Sociology Department of The Chinese University of Hong Kong. An earlier version of this paper was presented at the International Workshop on "Hong Kong: Polity, Society, and Economy Under Colonial Rule", Sinological Institute, Leiden University, The Netherlands, 22-24 August 1996, and the Workshop on "The Social History of Hong Kong", Osaka University of Foreign Studies, Japan, 1-2 March 1997. The authors would like to thank the conference participants and the anonymous reviewer for their comments.

98

parameters of the political arena under colonial rule. While these collective actions have not shaken the social basis of political stability in Hong Kong, their significance, as pointed out by a number of authors[1], has gone far beyond the issues and domains of social life which gave rise to them, and they have had repercussions for Hong Kong politics as a whole.

This article explores the development of social movements in contemporary Hong Kong in the context of historical changes in Hong Kong society. In particular, we highlight the effects of the changing political opportunity structure; state-society relations; how changes in the framing of collective action have shaped social movements in Hong Kong; and how the latter, in turn, have restructured the institutional environment of Hong Kong politics.[2] We argue that popular mobilization and collective action constitute important components of social life in Hong Kong. They are constituted and constitutive of the changing political parameters, state-society relations, and public discourse on politics.

The 1966 and 1967 Riots

The 1966 Kowloon disturbances and the 1967 riots were a watershed in the configuration of Hong Kong politics.[3] They best sum up the changing contours of Hong Kong politics before and after the mid-1960s. The 1966 Kowloon disturbances were a series of demonstrations, marches, riots, and street violence triggered by a hunger strike in opposition to a fare increase by the Star Ferry[4], lasting from 4 to 9 April 1966. As pointed out by the Commission of Inquiry (1967) of the 1966 Kowloon disturbances,

> the direct causes seem to lie in the escalation of events from the much publicized opposition and petitions concerning the Star Ferry fare increase (Sept. 1965 — April 4th 1966), to the 'hunger strike' by one man (April 4th), to his defiance of authority (April 5th), to the organized march ending in more serious defiance and clashes with authority (April 5/6th), to the further demonstrations merging into riots (April 6/7th). (p. 118)

The significance of the 1966 Kowloon disturbances lies in the fact that they symbolized the arrival of a new generation ready to express their hopes and frustrations. As noted in the *Report of Commission of Inquiry: Kowloon Disturbances 1966*, "[t]here is evidence of a growing interest in Hong Kong on the part of youth and a tendency to protest at a situation which their parents might tacitly accept" (p. 129).

---

[1] See, for example, Cheung 1987; Chiu and Lui (forthcoming); Lui 1994.

[2] On relevant theoretical literature, see, for example, McAdam *et al.* 1996.

[3] For a summary of the two incidents, see Scott 1989, Chapter 3.

[4] For the details of the 1966 Kowloon disturbances, see Commission of Inquiry 1967.

99

The actions were "spontaneous and unco-ordinated and [...] there appeared to be no central organization or control" (p. 112). More importantly, "there was no indication of any political or triad control or exploitation of the situation" (p. 112). It was the suggestion of the Commission of Inquiry that the underlying cause of the disturbances was "failure of communication" between the public and the government. Whether the analysis presented by the Commission of Inquiry is convincing or not is not our concern here. The important point is that the sudden outbursts of anger by the young people had caught the colonial administration totally unprepared. The Commission of Inquiry had to recognize that "with a new generation growing up who have never had experience outside Hong Kong it is important to develop avenues for participation in the life of the community and to give expression to young peoples' zeal for service" (p. 126). The disturbances symbolize the first major, spontaneous attempt by the post-war baby-boomers to express their discontents openly. Although their demands were diffuse and not well articulated, their sense of uneasiness was clear.

In May 1967, while the colonial administration was still working on new programs to address issues brought up in the 1966 Kowloon riots, 21 men were arrested at a plastic flower factory in San Po Kong. This incident was soon followed by further clashes between Communist supporters and the police, and riots broke out. Confrontations soon gave way to other forms of collective action, from work stoppages, strikes, and boycotts to terrorist attacks (for example, bomb attacks). The 1967 riots had clear and specific political objectives: their origins "lay in the Cultural Revolution in China" (Scott 1989, p. 96), and local Communist supporters used them to challenge the colonial rule. However, the participants in the confrontational actions organized in the early stage of the riots were by no means confined to local supporters of Communist China.[5] The subsequent development of these anti-colonial actions into terrorism in fact brought about a split of opinion on this political matter among the local population. Indeed, as remarked by Scott:

---

[5]   There are no publicly available documentary records of the participants of the 1967 riots. In a recent interview, Mr. Yuk Sing Tsang, a leader of the pro-China political group, Democratic Alliance for the Betterment of Hong Kong, gave an account of the 1967 riots and the imprisonment of his siblings: "[My brother] was a Form 6 student at St. Paul's, and was a timid boy. ...I was obviously leftist by then ...I had no idea what my brother was doing ...He printed some leaflets at home calling for the reform of the school curriculum, denouncing the British for the Opium War, and so on. He got a pile of those leaflets and handed them out during lunchtime at school. ...[W]hen classes resumed, the riot police were there. ...My brother was arrested straight away, tried, and sentenced to two years. ...A couple of months later, my fifteen-year-old sister was also arrested. She was a Form 3 girl at Belilios Public School. She was in the playground with thirteen other girls, and when the school bell rang they refused to go back to the classroom. ...[The headmistress] called the police, and all fourteen were tried, and found guilty of breaching the emergency legislation in force during 1967, and sent to the women's prison in Lai Chi Kok for one month". See Blyth and Wotherspoon 1996, pp. 96-97. St. Paul's and Belilios Public School are prestigious secondary schools in Hong Kong. Both this fact and Mr. Tsang's account of the two incidents indicate that non-Communist students were also mobilized in 1967.

100

[t]here can be little doubt that by December 1967 the communists had lost whatever public sympathy the labour disputes had initially generated. ...Ironically, in the light of communist objectives, the end-result of the disturbances was to increase the support for, and the legitimacy of, the existing order. Faced with a choice between communism of the Cultural Revolution variety and the, as yet, unreformed colonial capitalist state, most people chose to side with the devil they knew. (Scott 1989, p. 104)

Of course, one has to be cautious in interpreting the change of the popular mood in the mid-1960s. There had not been any sudden swing of support from one political and ideological camp to another. As noted by many observers, at that time, Hong Kong was considered by many local Chinese as a "life boat" in a sea of political turmoil. On the one hand, their emigration to the Colony was essentially an attempt to stay away from the flux and change of China's politics. On the other, there is little evidence to show that the Hong Kong Chinese accepted colonial rule as a legitimate political order.[6] Until the start of the negotiations over Hong Kong's future in 1982, the 1967 riots can be seen as a temporary end to the discourse on "Chinese politics" in Hong Kong which had prevailed in the post-war decades. This discourse viewed Hong Kong politics as an extension of the Communist-Nationalist struggle, and was based on a fear that China would intervene in Hong Kong.[7] The 1966 disturbances and the 1967 riots marked the end of an era and the beginning of new one — a (temporary) farewell to politics phrased within the framework of "Chinese politics" and the start of a phase where political demands were perceived as spontaneous, issue-driven, and non-ideological. Politics had now been localized.[8]

---

[6]   It could be argued that, at least in that period of time, legitimacy was not a real political issue for the Hong Kong Chinese making a living under colonial rule. In his seminal work on Hong Kong politics, Lau noted that "[t]he colonized Chinese people came to Hong Kong to subject themselves voluntarily under the rule of an alien colonial administration" (Lau 1982, p. 7). He then moved on to look at the characteristics of Hong Kong society as a minimally integrated socio-political system, bypassing the question of legitimacy. When most of the Hong Kong Chinese saw themselves as migrants or refugees, legitimacy of political rule was simply irrelevant. Hughes also made a note on this Hong Kong mood: it is "one of masterly expedience and crisis-to-crisis adjustment and recovery. It is partly a gambler's mentality, partly fatalism. As in Shanghai, no foreigner came to Hong Kong to make a home there; he came to make a living and get out. Nor does any Chinese live in Hong Kong against his will" (Hughes 1976, p. 129). This gradually changed in the 1970s when more and more demands were put to the colonial administration and a new government-people relationship took shape.

[7]   One way to look at this question is to compare and contrast the interpretation of the relevance of Chinese politics to the riots in 1956 and 1966. For details of the riots in 1956, see Hong Kong Government 1956. About the approach of "firmness without provocation" in handling Chinese politics in Hong Kong by the colonial administration, see Tsang 1995, pp. 290-294.

[8]   The emergence of a form of localized politics was recognized, though not explicitly, in the official report on the 1966 Kowloon disturbances. See Commission of Inquiry 1967.

101

The disturbances and riots of the mid-1960s led to the drawing up of a new political agenda, especially for the new generation. Wai Luen Lo, a local expert on the development of literature in Hong Kong, recalls that most of her friends in the 1950s and 1960s had a rather negative evaluation of life in the Colony and few identified with either Hong Kong or China (Lo 1996).[9] But the shocks arising from the bank runs in 1965, the 1966 disturbances, and the 1967 riots had given rise to a series of new questions. Young people began to rethink their relations with Hong Kong — what was this place? What should be done? How did they see themselves as Chinese growing up in a colony? (Lo 1996, p. 62) The effect of the events in the mid-1960s on the ideological and political minds of the post-war baby-boomers is a question awaiting more serious research and documentation. Existing data related to this issue is so scarce that it would be premature to draw any definitive conclusion.

Our conjecture is that the 1966 and 1967 riots did have a significant impact on Hong Kong society in terms of a) loosening the existing institutional structure of the colonial administrative state and b) re-framing issues for public debate. The former can best be observed in the government's moves toward strengthening the communication channels between the bureaucracy and the grassroots, notably the establishment of the City District Officer Scheme. This type of administrative reform did not, however, open new channels for political participation, as its main purpose was to strengthen communication between the bureaucracy and the non-elite public.

Nevertheless, such moves, together with other reforms in response to social problems exposed during the riots, for example labor conditions, contributed to the creation of a new political and social climate, which promoted public discussion about improving the current state of affairs and "of the problems which were still to be overcome"[10], and helped bring popular frustrations and discontents into the public domain. An atmosphere was created which facilitated a confluence of the younger generation's quest for a Hong Kong identity and the redressing of grassroots' grievances. For example, the status of the Chinese language and the suggestion to institute an ombudsman system were among the issues brought into the public domain at that time. This turn towards more emphasis on social concern and criticism was also evident in various activities organized by young intellectuals and students.[11]

---

9   Also see Wong 1992a and 1992b for a review of related research on the changing perception of life in Hong Kong.

10   *Hong Kong Annual Report 1968*. The title of the leading article in this report is "Progress".

11   On the gradual change of orientation of *Zhongguo xuesheng zhoubao* (The Chinese Student Weekly), see Lo 1996, pp. 54-73.

102

The Rise of Protest Actions

The late 1960s was a period of reaction to the issues brought up by the disturbances and riots in 1966 and 1967. While the colonial government tried hard to find official means to deal with issues highlighted by the disturbances and riots, ranging from government-sponsored dances for young people and the establishment of the City District Officer Scheme to the attempt of launching the Hong Kong Festival, the new generation had already taken the initiative of identifying social problems which were previously hidden behind the façade of growing affluence. College students' involvement in public affairs began with conflicts over campus issues (the university reform campaign at the University of Hong Kong and the protest actions against the dismissal of Chu Hai College students), and later developed into the call for "getting out of the ivory tower" and more active intervention in community issues. But the real event which kicked off the student movement was the territorial dispute concerning the status of the Diaoyutai Islands.[12]

Although the "Defend Diaoyutai Movement" was an "imported" movement, inspired by student actions abroad, it very quickly became one of the most large-scale campaigns in the history of Hong Kong's student movement. While inspired by Chinese nationalism, it was not a replica of earlier Communist-Nationalist tensions. While the two governments did try to intervene and articulate (or de-articulate) the issues to suit their own political agendas, the movement was more about the expression of student nationalistic sentiments than about the political projects of individual governments.

The movement had different layers of meaning. Firstly, it showed that colonialism failed to provide the new generation with a framework for their quest for identity after the disturbances and riots in the 1960s. Nationalism, mainly in cultural terms, could give them a source of identity. The fads of Chinese nationalism on college campuses generated by the movement were a good indication of the appeal of nationalism to the new generation. Secondly, at that historical conjuncture, nationalism (being well suited to the political vision of Communist China after the "Defend Diaoyutai Movement") served both as a source of identity formation and as an ideology for the critique of colonial administration. Young people of different political persuasions, ranging from cultural nationalism to radicalism criticizing colonialism and capitalism, could work together under the same nationalist umbrella.

Only at a later stage did this multi-layered nationalism reveal its internal contradictions, leading to an internal political and ideological struggle within the student movement. The struggle between the two ideological camps of nationalism and social activism could be seen as an outcome of the tension within the project of identity formation in a colonial setting. While nationalism promised to transcend colonialism and lead its supporters to a grander national project, it was also very remote from daily life under colonialism. This was especially true in the context of

---

[12]   On the development of the student movement, see Observers of Far Eastern Affairs 1982; and the Hong Kong Federation of Students 1983.

103

Hong Kong in the 1970s: Communist China had no plans to intervene into local affairs after the 1967 riots, and was more concerned with maintaining the *status quo* while promoting nationalism within the Colony. The social activist camp, very much inspired by New Left radical ideology, attempted to address local issues with the objective of exposing and publicly voicing the hidden pains and discontents of people living under the colonial regime. The ideological struggle between the two student camps became evident during the "Anti-corruption, Arrest Godber Movement" and the gap between them subsequently continued to widen until the fall of the "Gang of Four" in Communist China.

The change in Party line after the death of Mao was a blow to the pro-China nationalist camp within the student movement and ideological struggle was replaced by a more locally oriented approach to social intervention. While the ideological critique of colonialism and capitalism was still important as a guideline, by the late 1970s, the student movement was one among many factions participating in the emerging pressure group politics. To be sure, student bodies were active members of the leadership of two important movements — the "Yaumati Boat People Action" (a series of protest actions against the government's resettlement policy and the arrest of protesters), and the "Golden Jubilee Secondary School Incident" (a series of actions triggered by alleged corruption in school management and against the subsequent action taken by the government to close down the secondary school). But the ideological and organizational leadership was primarily in the hands of the emerging pressure group leaders.

We can look at the twists and turns in the development of the student movement from different angles. From one angle, it was a question of the rise and fall of a social movement in terms of mobilization. But it can also be seen as a failure to make use of the new social and political climate created by the events in the mid-1960s and establish a framework for nationalism. In view of space limitations we shall not dwell upon these questions. What interests us here is the framing of political demands by various social movements in the 1970s. In this regard, the student movement can be seen as the vanguard of the social movements of the 1970s in the sense that of all the social movement organizations and political groups (with the possible exception of the Trotskyite Revolutionary Marxist Alliance), it was the student movement that articulated a radical ideological framework to challenge the colonial social order. Implicit in most of the students' organized actions, ranging from social actions to protest government policy to the cultural critique of everyday life, there was an underlying agenda of criticizing the colonial government and the capitalist economy.[13] While the criticisms of colonialism and capitalism had never taken the form of a political program, they did constitute a kind of tacit understanding among social movement organizers. In the context of colonial Hong Kong in the 1970s, this tacit understanding much facilitated joint action.

Before elaborating on this, we shall first look briefly at the growth of contentious actions in the 1970s. The early part of the decade witnessed several waves

---

[13]   *Idem.*

of collective action. While the student movement addressed the broader ideological and political issues of that period, urban protests and industrial actions in the public sector were driven by community-based and work-related interests. When we look at the social movements in the 1970s, two characteristics can be noted. Firstly, most of the collective actions were protest actions.[14] This partly reflected the limited resources of the movement — the main strategy was to rally the support of third parties for the purpose of exerting pressure on the government, which showed that their resources for mass mobilization were limited and that they had a relatively weak bargaining position *vis-à-vis* the colonial state. Secondly, it was an outcome of the so-called "consultative democracy" arrangement. Prior to the reform of local administration (i.e. the establishment of district boards and the related elections) in the early 1980s, the channels for open political participation were confined (through election) to the Urban Council. More importantly, within this "consultative democracy" framework, the administrative state was politically insulated from society, and depoliticization was the ruling strategy of the colonial administration (Lau 1982; Harris 1978). While the elitist interest groups could gain access to the government through the appointment of representatives or related persons to consultative bodies and exert political influence on the bureaucrats, political demands made by the general public were channeled to the non-institutional arena. Simply put, the structure of the colonial state and the system of representing political interests forced those who wanted to voice popular claims and demands to organize protest actions (Lui 1984; Lui and Kung 1985; Jenkins and Klandermans 1995).

By the end of the 1970s, something like a "social movement industry" was taking shape. The proliferation of different types of collective action had greatly broadened the scope of contentious politics. A variety of interests and latent groups had been mobilized and their claims and demands were recognized. Protest groups and pressure groups were formed to sustain mobilization.[15] In a way, the early activism of the student movement in organizing collective action, and the subsequent decline in importance of students in leading popular mobilization, revealed the growth of social movement organizations and the formation of a "social movement industry". The increased importance of pressure groups such as the Hong Kong Professional Teachers' Union and the Society for Community Organization, and the formation of *ad hoc* alliances for joint action under the leadership of these pressure groups, illustrated a change towards a consolidation of social protest through pressure group politics.

Our discussion of the institutional configuration of social protest is also useful to understand the rise of pressure group politics in the late 1970s and early 1980s. In essence, pressure group politics was more a continuation of than a break with

---

14　　See Lui and Kung 1985; and Lipsky 1968.

15　　On the formation of protest groups and pressure groups concerning community politics, see Hong Kong Council of Social Service, *Community Development Resource Book*, various years.

105

protest actions in the early 1970s (Lui 1989). Despite the fact that some of them were coopted into the colonial administrative system through appointment to advisory committees (mainly on an individual and not on a group basis), most pressure groups were active outside formal institutional politics.[16] Indeed, the fact that most activist groups were "outsiders" helped create some kind of tacit understanding among them. In the joint actions organized in the late 1970s and early 1980s, they could easily get together to form *ad hoc* organizations for a common cause. Though ideological differences among the different groups still played a role, on the whole they had little difficulty in making common claims and staging jointly organized protest actions.

The affinity among these groups was largely a consequence of the restricted opportunity for political participation in that period. The closed political system created a common understanding among the activists because they shared the experience of being rejected and sometimes repressed by the colonial state. Restricted entry into the formal channels of the polity gave rise to an oppositional force active in the non-institutional political arena. Some of these groups (for example, the student activists), were critical of colonialism and/or capitalism. Others, such as residents' organizations, did not have an elaborate ideological program but were equally critical of the colonial administration which was not responsive to their demands. By the early 1980s and on the eve of the Sino-British negotiations over Hong Kong's future, there was a loosely knit network of pressure groups, social movement organizations, and grassroots protest groups constituting an oppositional force resisting the colonial state.

Paradoxically, while various social movements brought up different types of claims and demands, the capability of the colonial state in meeting these challenges facilitated the restructuring of colonial hegemony in terms of governmental responsiveness and administrative efficiency. The accommodation of popular demands within the political parameters of the growing administrative state was of paramount importance in re-structuring Hong Kong's state-society relations of that period. As noted earlier, the colonial administration in the 1970s maintained its insulation from society by rejecting the idea of political reform and of re-structuring the channels of political participation.[17] The colonial administration recruited the emerging young professionals and executives into the major decision-making bodies to replace some members of the old elite (Tang 1973). Furthermore, it strengthened its position by increasing the state's responsiveness to local demands, combatting corruption,

---

[16]   The best example of the attitude of the colonial administration towards local pressure groups is the comment made in the report (1979) of the Standing Committee on Pressure Groups (SCOPG). See Lee 1987, p. 134.

[17]   A rather revealing paper on the government's position on re-structuring the channels of political participation in the post-riot years is its *White Paper* 1971.

106

broadening the scope of social services, and improving administrative efficiency.[18]

In the 1970s, then, a paradoxical development took place where the attempts to challenge the colonial authority and the capitalist order were gradually replaced by a demand for rights and entitlements (an attitude which was very different from the earlier conception of "Hong Kong as a lifeboat"), giving the colonial state an opportunity to develop its hegemony through its response to these demands. The colonial state was subsequently perceived as an efficient administration which could meet the needs of the population and provide them with an institutional framework enabling the Hong Kong Chinese to prove their livelihood through entrepreneurship and/or credentialism.[19] The basis of this hegemony was the belief held by many people that as long as the administrative state was able to uphold law and order and the legal framework, and to attend to the basic needs of the local population (for instance, mass housing), they were best off being left alone and free to pursue their own goals and careers.

Thus, the social movements in the 1970s had made an interesting turn. In the beginning, the quest for identity was paramount, in response to the issues brought up during the events of 1966 and 1967, and subsequently re-framed by the student movement into a quest for cultural nationalism. Thereafter, identity politics gradually faded into the background and was replaced by a series of collective actions centering on the allocation of resources.[20] The attempts to embarrass the colonial authority with demands and actions and the anti-colonial rhetoric paradoxically ended up confirming the role of the colonial state as the major agency for resource-allocation to cope with rapid economic and social development. The problems were now framed in issue-specific and functional terms. While the social movements certainly contributed to the liberalization of Hong Kong's civil society and the bringing of popular demands to the public arena, their scope was much confined. This was in accordance with the broader context of de-politicization in the 1970s, when contentious politics was replaced with a concern for specific policies and concrete administrative issues. The colonial nature of the administrative state was left unscathed.

A New Political Order

The Sino-British negotiations over Hong Kong's future and the subsequent agreement between the two governments on returning the colony to China on 1 July 1997

---

[18] For an interesting review of the strengthening of the administrative state in the 1970s, see Scott 1989.

[19] This is the hegemonic ruling strategy of the colonial administrative state. For an interesting (and symbolic) account of this strategy, see the last address by the governor at the opening of the Legislative Council session (Patten 1996).

[20] We thank Tak-Wing Ngo for alerting us to this change in the direction taken by social movements in the 1970s.

107

brought drastic changes in both the political agenda and the parameters of Hong Kong politics. The agreement signaled the beginning of the decolonization process.[21] Although initially, the pressure groups, social movement organizations, and grassroots protest groups had their doubts about participating in formal institutional politics[22], they were quickly drawn into electoral politics, first at the level of elections to the district boards and Urban and Regional Councils, and later in direct and indirect elections to the Legislature. The new question was: how can a new political order be instituted within the parameters of "decolonization without independence" and the diplomatic politics between Britain and China? People prepared to formulate politics for the transitional period. In the realm of *Realpolitik*, the 1980s was a period of political contention through electoral politics.

Studies of social conflicts in 1975-1991 have shown that from 1984 onward, there was a drastic increase of conflicts related to political issues (i.e. those concerning constitutional matters and issues about political and civil rights).[23] Prior to 1984, constitutional matters rarely appeared on the agendas of local social movements. This was not due to political indifference but to the fact that prior to the political reforms of the 1980s, the question of democratization was regarded by most activists as being remote from political reality, since it was unlikely to have any practical meaning in the face of a closed colonial administration. The growing importance of political issues after 1984 was a result of the rise of new political opportunities brought about by decolonization, and of the increased attention for political participation in formal institutional politics on the part of pressure groups, social movement organizations, and grassroots protest groups. The struggle for democracy became the major concern of the activists in the 1980s and 1990s.[24] The aims were to deepen political reform before 1997 and democratize the political structure of the future Special Administrative Region (SAR) government.

The new political opportunities had a double-edged effect on the development of social movements in Hong Kong. On the one hand, former pressure groups, social movement organizations, and newly established political groups formed a loosely defined group of democrats on the basis of previous collaborative experience and some tacit understanding of the need to fight for the democratic cause. This was the result of the new opportunities for political intervention in the sphere of electoral politics and in the designing of Hong Kong's future political structure. Overcoming their initial reservations, activists quickly formed new political groups for the purpose of preparing for elections at different levels, and voicing their opinions to the

---

[21]   Whether or not the initiation of political reforms (from the establishment of the district boards to the introduction of popularly elected members to the Legislative Council) was part of the British Government's preparation for decolonization is beyond the scope of our discussion here.

[22]   Lui 1989 and 1994; Lui and Kung 1995.

[23]   Cheung and Louie 1991; Chui and Lai 1994.

[24]   For a discussion of the democratic movement, see Sing (forthcoming).

108

Chinese government in regard to the blueprints for the transition and the post-1997 administration.[25] Many of these groups actively participated in the democracy movement for the purpose of securing the establishment of a more democratic political structure before 1997. In the early 1980s, many activists saw the 1997 question as an opportunity for societal mobilization — to place topics which had previously been suppressed on the political agenda for public discussion.[26] The move towards the establishment of a representative government in 1985 and 1988 meant that, for the first time in colonial history, Hong Kong elected members to the Legislature, albeit only indirectly, through functional constituency and an electoral college. This brought about the further politicization of pressure groups and social movement organizations. Political parties were subsequently formed for consolidating the existing networks of activists and concerned groups.

On the other hand, participation in formal political institutional politics also gave rise to divisions among the groups involved. The twists and turns during the Sino-British talks about political reform, the political structure of the future SAR government, and the emphasis on convergence towards a social and political system which China would find acceptable, posed new questions to the political groups and social movement organizations. Division ensued in regard to the choice between pragmatism (i.e. accepting the parameters prescribed by China) and continuing to play the role of an oppositional force (especially after the June 4th Incident in 1989). The loosely formulated consensus found among activist groups in the 1970s lost its relevance and the solidarity among the so-called democrats was weakened. Previous informal political networking was replaced by formal party participation and inter-organizational linkages.

A gradual separation took place between grassroots' mobilization and community action on the one hand, and party politics on the other.[27] After a short period of active participation in local elections, some pressure groups and grassroots protest groups changed their strategy, assuming a low profile in the 1991 and 1995 elections to the Legislative Council.[28] The position of the democrats' leaders became more problematic as a result of the separation. Most of them had started their political careers in organizing protest actions and social movements in the 1970s and 1980s. Their close connections with social movement organizations created expectations from the grassroots that they would continue to play the role of leading popular mobilization against government policies. While they assumed a double role in Hong Kong politics — they were both the leaders of protest actions and politicians as-

---

[25]   On the formation of political groups in the early 1980s, see Cheng 1984.

[26]   For an impression of the activists' mood in the early 1980s, see Tsang 1982.

[27]   Ho (forthcoming); Leung 1986; Lui 1990.

[28]   See, for example, Lui 1993. The discussion about the incorporation of popular mobilization and protest action into party and electoral politics reflects the peculiarities of social movements and political groups in Hong Kong. See Lui 1996.

109

suming an oppositional position in the elected bodies at different levels — it also became clear that grassroots' mobilization was different from election politics. The rapid development of electoral politics and the increased concern for parliamentary struggle had led to a "hollowing out" of political organizations at the grassroots' level. This problem was also found among local unions.[29]

The mass actions which took place before and after the June 4th Incident did not really change the picture. While more than one million people joined the street rallies and marches in protest against the suppression of the student movement in Beijing, the pro-Chinese democracy movement quickly "fell from the peak" after the crackdown.[30] Nor did the criticism over the political reform program put forward by the Governor, Mr. Chris Patten, trigger another round of pro-democracy popular mobilization. As Hong Kong approached 1997, it was increasingly difficult to mobilize the public and stage open confrontational action against China.

All in all, the changes since the Sino-British negotiations had not really brought social movements into institutional politics. To be sure, the 1980s and 1990s witnessed the participation of pressure groups, social movement organizations, and protest groups' leaders in electoral politics, but this did not necessarily imply the political transformation of the social movements. While it is true that popular demands were brought to public discussion in the electoral bodies through collective action, they were mediated by party and electoral politics.

The odd situation in Hong Kong was that, on the one hand, there was a kind of party politics operating in a political institutional setting which did not allow parties to assume decision-making power. This setting made party and electoral politics a kind of oppositional politics, in close connection with grassroots social movements. On the other hand, the agenda and room for maneuver of this kind of oppositional politics were significantly restricted by the decolonization process. The very fact that the future of Hong Kong politics had to be accommodated within the broader framework of Chinese politics and that the crafting of the future SAR blueprints was restricted to diplomatic talks between the Chinese and British governments made it very difficult for the opposition to convince the masses of the viability of a form of alternative politics which could go beyond the restrictions imposed by the existing framework of decolonization.

At the same time, the approaching issues brought about by the 1997 question drove almost all active political participants to concentrate on political matters, especially those concerning China-Hong Kong relations. Issues which were most relevant to grassroots' mobilization were not successfully articulated in the 1997 political agenda, reinforcing the separation of party and electoral politics from social movements and popular mobilization.

Moreover, the social movements in Hong Kong were unable to consolidate popular solidarity by focusing on major contradictions such as ethnic conflicts or

---

[29]   Chiu and Levin (forthcoming).

[30]   Wong (forthcoming).

110

institutions, for example religion. Despite growing tensions among political groups, pressure groups and social movement organizations, the lastnamed still maintained loose and often informal connections with various political groups, especially those which were broadly categorized as the democratic camp (Lui, 1993 and 1996).

The overall picture was that social movements played a rather limited role in the transition to 1997. Under the shadow of "decolonization without independence" and the dominant position of China on Hong Kong matters, the collective identity of being "Hong Kong locals" and the belief in a liberal and open socio-political order, which were partly an outcome of the rise of collective actions in the 1970s, did not become the basis for political action. In a sense, the agenda once articulated by social movements and collective actions in the 1960s and 1970s was abandoned precisely in a political context where such issues were extremely pertinent.

Concluding Remarks

In contrast to the claim that Hong Kong has been politically stable thanks to the colonial constitutional order, we have seen that Hong Kong has seen wave after wave of social movements. These movements were both constituted and constitutive of the political environment: on the one hand, they were socially constructed, but on the other, they also contribute to the constitution of Hong Kong politics by restructuring the public political discourse and creating political opportunities. They were by no means marginal in the constitution of political life in Hong Kong, but an important factor in the structuring of state-society relations and the discourse on politics.

We are well aware of the existence of gaps in our analysis, which are mainly due to the lack of documentation and previous secondary studies. Our discussion has, as a result, been conjectural. We do hold, however, that there is a need for more thorough examination of the interaction among the changing political opportunity structure, state-society relations, and the public discourse on politics, in order to increase our understanding of the development of social movements in Hong Kong. In this article, it has at least been shown that the claim that Hong Kong has been politically stable in the sense of "lacking serious disputes" (Kuan 1979) thanks to benevolent British rule, is incorrect.

REFERENCES

Blyth, Sally, and Ian Wotherspoon, 1996. *Hong Kong Remembers*. Hong Kong: Oxford University Press.

Cheng, Joseph Y.S. (Ed.), 1984. *Hong Kong: In Search of a Future*. Hong Kong: Oxford University Press.

Cheung, Anthony B.L., 1987. "Xin zhongchan jieji de maoqi yu zhengzhi yingxiang" (The New Middle Class: Its Emergence and Political Influence), *Ming Pao Yue Kan*, No. 253, pp. 10-15.

111

Cheung, Anthony B.L., and K.S. Louie, 1991. "Social Conflicts in Hong Kong, 1975-1986". Occasional Paper No. 3, Hong Kong Institute of Asia-Pacific Studies.

Chiu, Stephen W.K., and Tai-lok Lui (Eds.), (forthcoming). *Social Movements in Hong Kong*. Armonk: M.E. Sharpe.

Chiu, Stephen W.K., and David Levin, (forthcoming). "Private Sector Unionism". In Chiu and Lui (forthcoming).

Chui, Ernest, and O.K. Lai, 1994. "Patterns of Social Conflicts in Hong Kong in the Period 1981 to 1991". Mimeograph.

Commission of Inquiry, 1967. *Kowloon Disturbances 1966: Report of Commission of Inquiry*. Hong Kong: Government Printer.

Harris, Peter, 1978. *Hong Kong: A Study in Bureaucratic Politics*. Hong Kong: Heinemann Asia.

Ho, K.L. (forthcoming), "Housing Movements". In Chiu and Lui (forthcoming).

Hong Kong Annual Report 1968. Hong Kong: Government Printer.

Hong Kong Council of Social Service (Xianggang shehui fuwu lianhui), various years. *Community Development Resource Book* (Shequ fazhan ziliao huibian).

Hong Kong Federation of Students (Xianggang zhuanshangxuesheng lianhui) (Ed.), 1983. *Xianggang xuesheng yundong huigu* (A Review of the Hong Kong Student Movement). Hong Kong: Wide Angle Publications.

Hong Kong Government, 1956. *Report on the Riots in Kowloon and Tsuen, October 10th to 12th, 1956*. Hong Kong: Government Printer.

Hughes, Richard, 1976. *Borrowed Place Borrowed Time: Hong Kong and Its Many Faces*, Second Revised Edition. London: André Deutsch.

Scott, Ian, 1989. *Political Change and The Crisis of Legitimacy in Hong Kong*. Hong Kong: Oxford University Press.

Jenkins, Craig, and Bert Klandermans (Eds.), 1995. *The Politics of Social Protest*. London: UCL Press.

Kuan, Hsin-chi, 1979. "Political Stability and Change in Hong Kong". In T.B. Lin, R.P. Lee, and U. Simonis (Eds.), *Hong Kong: Economic, Social and Political Studies in Development*. New York: M.E. Sharpe.

112

Lau, Siu-kai, 1982. *Society and Politics in Hong Kong*. Hong Kong: The Chinese University Press.

Lee, Ming Kwan, 1987. "Yali tuanti yu zhengdang zhengzhi" (Pressure Groups and Party Politics). In Ming Kwan Lee, *Bianqian zhong de Xianggang zhengzhi he shehui* (Hong Kong Politics and Society in Transition). Hong Kong: Commercial Press.

Leung, Benjamin K.P., 1996. *Perspectives on Hong Kong Society*. Hong Kong: Oxford University Press.

Leung, C.B., 1986. "Community Participation: The Decline of Residents' Organizations". In J. Cheng (Ed.), *Hong Kong in Transition*. Hong Kong: Oxford University Press, pp. 354-371.

Lipsky, M., 1968. "Protest as a Political Resource", *American Political Sciences Review*, Vol. 62, No. 4, pp. 1144-1158.

Lo, Wai Luen, 1996. *Xianggang gushi* (Hong Kong Story). Hong Kong: Oxford University Press.

Lui, Tai-lok, 1984. "Urban Protests in Hong Kong". Unpublished M.Phil thesis, University of Hong Kong.

————— 1989. "Yali tuanti zhengzhi yu zhengzhi canyu" (Pressure Group Politics and Political Participation). In Joseph Cheng (Ed.), *Guoduqi de Xianggang* (Hong Kong in the Transitional Period). Hong Kong: Sanlian shudian, pp. 1-18.

————— 1990. "Fanpu guizhen" (Back to Basics: Rethinking the Roles of Residents' Organizations). *Community Development Resource Book*, pp. 12-14.

————— 1993. "Two Logics of Community Politics". In S.K. Lau and K.S. Louie (Eds.), *Hong Kong Tried Democracy*. Hong Kong: Hong Kong Institute of Asia-Pacific Studies, pp. 331-344.

————— 1994. "Mishi yu jiju zhuanbian zhengzhi huanjing de Xianggang minzhòng yundong" (The Path of Development of Hong Kong's Popular Movements), *Xianggang shehui kexue xuebao* (Hong Kong Journal of Social Sciences), No. 4, pp. 67-78.

————— 1996. "Zuo shemma?" (What is to be Done?), *Gan yan* (Outspoken) (a bulletin published by the Democratic Party), Hong Kong, June 1996.

113

Lui, Tai-lok, and James K.S. Kung, 1985. *Chengshi zongheng* (City Unlimited: Community Movement and Urban Politics in Hong Kong). Hong Kong: Wide Angle Publications.

McAdam, Doug, McCarthy, John, and Zald, Mayer (Eds.), 1996. *Comparative Perspectives on Social Movements*. Cambridge: Cambridge University Press.

Observers of Far Eastern Affairs (Liandong shiwu pinglun she) (Ed.), 1982. *Yundong chunqiu* (The Student Movement). Hong Kong: Observers of Far Eastern Affairs.

Patten, Christopher, 1996. *Hong Kong: Transition — The 1996 Policy Address*. Hong Kong: Government Printer.

Sing, Ming, forthcoming. "The Democracy Movement in Hong Kong Under the Shadow of 1997". In Stephen Chiu and Tai-lok Lui (Eds.), *Social Movements in Hong Kong*. Armonk: M.E. Sharpe.

Tang, Stephen, 1973. "The Power Structure in a Colonial Society: A Sociological Study of the Legislative Council in Hong Kong (1948-1971)". Unpublished senior B.Soc.Sci. thesis, Sociology Department, The Chinese University of Hong Kong.

Tsang, Shu-ki, *et al.*, 1982. *Wuxingqi xia de Xianggang* (Hong Kong Under the Red Flag). Hong Kong: Twilight Books.

Tsang, Steve (Ed.), 1995. *A Documentary History of Hong Kong: Government and Politics*. Hong Kong: Hong Kong University Press.

White Paper 1971. *White Paper: The Urban Council*. Hong Kong: Government Printer.

Wong, P.W. (forthcoming). "The Pro-Chinese Democracy Movement in Hong Kong". In Chiu and Lui (forthcoming).

Wong, Thomas W.P., 1992a. "Discourses and Dilemmas: 25 Years of Subjective Indicators Studies in Hong Kong". In S.K. Lau *et al.* (Eds.), *Indicators of Social Development: Hong Kong 1990*. Hong Kong: Institute of Asia-Pacific Studies, pp. 239-268.

────── 1992b. "Personal Experience and Social Ideology". In S.K. Lau *et al.* (Eds.), *Indicators of Social Development: Hong Kong 1990*. Hong Kong: Institute of Asia-Pacific Studies, pp. 205-237.

[22]

CHOI PO-KING

The Politics of Identity

The Women's Movement in Hong Kong

This paper examines the women's movement in Hong Kong in the context of the political transformation of the 1980s and 1990s. This period is significant in that it witnesses the unfolding of two related political processes, namely, decolonization and the transition of sovereignty to the People's Republic of China. These two processes have a great impact on the women's movement, as they do on other aspects of social development.

The sources on which this paper is based include publications and other printed materials of major local women's groups in Hong Kong. More important, this paper relies on in-depth interviews of twelve core members of local women's (and mostly feminist) organizations that have appeared since the mid-1980s. These include the Hong Kong Federation of the Women's Center (established 1985), the Association for the Advancement of Feminism (AAF, established 1984), Hong Kong Women Christian Council (HKWCC, established 1987), Hong Kong Women Workers' Association (HKWWA, established 1989), and Queer Sisters (established 1995). I also interviewed women who worked with the Women Affairs Committee of the Hong Kong Confederation of Trade Unions (1990). These interviews, which took place in the form of relatively free and relaxed conversations, were conducted between July and September 1995, and they are important in that they help reveal more deeply set beliefs and thoughts of the people concerned, as well as the development of these beliefs over time. Pseudonyms are used whenever I directly quote the women I interviewed.

Translation © 1998 M. E. Sharpe, Inc., from the Chinese text. This paper is abridged and translated from "The Women's Movement in Hong Kong: The Construction of Identities and Its Contradictions," a chapter in *Praxis à la Hong Kong* (in Chinese), ed. Man Si-wai and Leung Mei-yee (Hong Kong: Youth Literary Press, 1997), pp. 319–59.

"Return to the Grass Roots": Decolonization in the Women's Movement

Hong Kong witnessed the emergence of a distinct local identity and culture in the 1980s, following the maturation of the first postwar generation of Chinese who have grown up in an unprecedented period of social and political isolation from mainland China.[1] This process of decolonization and localization had a great impact on the women's movement, shaping its nature and direction.

The earliest feminist organization, the Hong Kong Council of Women, was set up as early as 1947. Its membership had a high percentage of expatriates, and the few Chinese who participated were English-speaking. Christine, a Chinese woman whom I interviewed, had been active in the council in the late 1970s and 1980s. She recalled that the council had made serious attempts at localization in the 1970s, and these resulted in the establishment in 1985 of two service centers, namely, Harmony House, a shelter for battered women and their children, and the Hong Kong Federation of the Women's Center.[2] Christine observed, however, that until the 1980s, feminism had not aroused much interest among upper-middle-class Chinese women, who might have been attracted to the English-speaking Council of Women, and since the 1980s, a younger generation of Chinese women feminists chose instead to set up their own women's groups and to start their own local women's movement.

To this new generation of women, the designation of a movement being "local" carries several meanings. First, it means an identification with one's own language and culture. This emerges in my interview with Ah Kuen, one of the founding members of the AAF. In trying to explain why she and her colleagues did not accept the invitation to join the Hong Kong Women's Council, she said the following:

> People from the Hong Kong Women's Council would ask why local women like us have to start their own organization? . . . There are some practical issues here. We feel that the council uses a lot of English in their meetings, minutes, and so on. That's probably because it has a lot of expatriate members, so English is the working language there. We feel, however, that we should return to our own culture. There is a belief behind the establishment of the AAF, namely, that we have to start an indigenous organization.

This idea of "returning" to one's own culture is not peculiar to AAF but is rather common among other women's organizations. Margaret, of the HKWCC, for example, stressed that theirs was "a local and not a foreign Christian women's organization. Locals and foreigners maintain a relationship of exchange. Our sharing is international in scope, but our membership is predominantly local. This is a healthy relationship." Similarly, Ah Dik and Ah Yan of Queer Sisters, a group that challenges conventional modes of sexuality, also pointed out that the reason they had left a predominantly lesbian group, Women Like Us, was that they found this latter group too English-centered and felt it lacked interest in related local developments.

A second meaning to "local" for these women activists is an orientation toward the grassroots. Such an orientation, which has its roots in the students' and related social movements of the 1970s, permeates the ideologies and work goals of the feminists. This is hardly surprising, since many of them had themselves participated in these movements. Thus Ah Wai, who had joined the AAF at an stage stage, described the organization's orientation in this way:

> Most of the founders of AAF have participated in social movements before. The basic orientation of these social and students' movements was to start working from the grass-roots level: support those in the grass roots and apply class analysis to social situations. We, the AAF, had this same orientation from the very start. In other words, in dealing with women's issues we had to confront the issues of women at the grass-roots level. If the problems of these women are not resolved, women's issues in general will not be resolved. . . . Most of our members feel that the AAF should concern itself with grass-roots women . . . I do not mean that each of us has to participate in grassroots organization, but we all feel that this is what the AAF should do, that this is our priority.

Similarly Margaret, who had participated in community concern and development projects in the 1970s, under the aegis of a minority arm of the Christian Church influenced by the Third-World ecumenical Christian Movement and liberation theology of Latin America, summarized her experience: "These involvements furnished me with a belief and basis for my religion: we should stand by the needy, the poor, and the oppressed. This belief has had a profound influence on me."

Naturally Margaret brought this into the HKWCC, which she helped found in 1987. Space does not permit a detailed description of such a grass-roots orientation, nor its concomitant work style, which I found common to the local women's organizations founded in the 1980s and 1990s. The interesting point to note here is that this orientation was used repeatedly by my respondents to distinguish between their own (local) movement from that of the expatriates. AAF's Ah Yee and Ah Kuen said, in two separate interviews, that they were different from the Hong Kong Council of Women in that the latter did not have specific concern for grassroots women, while the AAF aimed at fighting for their rights. Even Ah Dik and Ah Yan of the Queer Sisters, established not long before the interview took place, emphasized that the reason they left the group Women Like Us was that "there are too many expatriates, and they are either too high class or are middle class."

Obviously, in the eyes of local Chinese, race is related to social class. This probably is an association naturally embedded in colonial history, and one constantly invoked in the process of decolonization: Colonizers (Westerners) are the privileged class, whereas the colonized people (the Chinese) form the lower stratum. This perception is, of course, not totally accurate today, as a minority of Hong Kong Chinese have already made their way into the world-class rich, and a Chinese professional class has clearly emerged. The point is, however, not the accuracy of the perception but how this shapes the goal of decolonization and localization. In Hong Kong's women's movement, "decolonization" is constructed as a process of returning to the grass roots.

In order to explain this social construction, one probably has to look beyond the women's movement into the wider cultural context of Hong Kong. Rey Chow, in discussing the works of Luo Dayou, a songwriter who originates from Taiwan but has made Hong Kong his home, talks about the difficulty of "postcolonial self-writing" for Hong Kong, sandwiched as it is between the two colonizers, Britain and China.[3] She observes that Hong Kong experiences an "existentialist angst" arising from "its permanent colonial taint," which makes her "quest for China" a fundamentally futile one, because "the more Hong Kong tries, the more it reveals its lack of 'Chinese-ness,' and the more it is a deviation from the norm. The past would follow Hong Kong like an unshakable curse of inferiority."[4]

In other words, Hong Kong could not pretend to be able to recover its "cultural authenticity" ("Chinese-ness") in the process of decolonization. And being a thoroughly modern city, it does not have the option of nativism either. Thus restricted historically, the students' movement of the 1970s chose to identify with and "return to" the working classes as the basis of its anticolonial stance, and, of course, the mainstream nationalistic faction within it also chose specifically to "return to the socialist motherland."[5] It is perfectly natural, therefore, that the women's movement of the 1980s also took up a "return to the grass roots" as its basic standpoint with regard to localization. But the predominantly middle-class and intellectual slant of the movement was increasingly being highlighted by the nascent women workers' movement that appeared in the 1990s.[6] This posed a challenge to the "grass roots" imaginary constructed in the process of decolonization, thus calling for perhaps a new "postcolonial, self-writing," to use Rey Chow's expression.

Closely connected to the grass roots imagery of decolonization is the strategy taken up by the women's movement. Well schooled in students' movements and related social movements of the 1970s, the first generation of feminists were well acquainted with the strategies of so-called peripheral politics favored by pressure groups outside the power center. These strategies include organizing protest campaigns, creating liaisons with grass-roots groups, gaining media exposure, and so on. With the emergence of parliamentary politics in the 1980s, lobbying of legislators and electoral candidates also became popular among women's groups, as a way of furthering their goals.

These strategies of "peripheral politics" and public lobbying were in great contrast to those traditionally used by the Hong Kong Council of Women. According to Christine, many members of the council before the mid-1980s were either related to or were themselves influential personages in the colonial government. With what Christine called "political clout," these women were accustomed to lobbying in private, by "pulling strings," as it were, in order to effect desired changes in government policy or to secure resources for new projects. By the mid to late 1980s, the membership changed to include more professionals than those related to the higher echelons of the government, and what Christine called "mainstream politics" subsided. The council disbanded in 1995, and its splinter group, AWARE, now placed much greater emphasis on self-empowerment of women in lower-middle-class communities. As the curtain falls on the British colonial government, "mainstream politics" in the women's movement increasingly shifted to the pro-Beijing camp, as we shall see later.

The Challenge of 1997

To many Hong Kong women's groups, the Fourth U.N. Conference on Women in Beijing in the fall of 1995 was a sobering occasion that brought various identity and political anxieties into the open.

Contrary to the norm, Hong Kong was represented by two separate and mutually non-communicating delegations in the Non-Government Organizations Forum (NGO Forum) of the Conference. These two were: the Hong Kong NGO Working Group on the Fourth World Conference (Hong Kong NGO), formed with the blessing of the Hong Kong–based New China News Agency and the Beijing government, and the Hong Kong Women's Coalition for Beijing '95 (Coalition '95), made up of local women's groups. The wide gap and tense relations between the two brought home all too clearly the political tensions and threats associated with the return of sovereignty to China in 1997.

Simply put, the Hong Kong NGO was Beijing-oriented, whereas the Coalition '95 had a strong local base. In terms of their respective agendas, the Hong Kong NGO steered clear of any challenge to the existing political and class structure, attaching itself, as it did, to a "consensus" framework. By contrast, Coalition '95 did not shy away from direct questioning of the existing patriarchal and class structures, and its feminist ideals were more in the forefront.[7] The gap and tensions between these two delegations were highly revealing of the contradictions between Hong Kong and the mainland, not only with regard to political relations but also in terms of cultural identity and sexual politics, that is, the direction of the women's movement itself.

In fact the mainland–Hong Kong contradictions within the women's movement had become evident as early as the fall of 1993, when the Hong Kong Federation of Women was inaugurated. The founding of the federation and its leaders were closely linked to the Beijing government, featuring among them members of the high-powered Chinese People's Political Consultative Conference, the National People's Congress, the Beijing-appointed Preliminary Working Committee for the Special Administrative Region, the Beijing-appointed Hong Kong Affairs Advisers, the pro–Beijing Democratic Alliance for the Betterment of Hong Kong, as well as the now defunct Basic Law Consultative Committee. With its close relations with prominent pro-Beijing capitalists, too, the federation was well financed, a situation that was constantly denied to the locally founded women's groups.[8] The appearance of the federation brought a sense of urgency to the misgivings local women's groups had of the so-called 1997 issue. Such misgivings may, I think, be summarized in four major aspects, as outlined below:

First, the emergence of the women's movement was closely related to the broadening of space for political discussion and participation in society. Most of the women activists I interviewed believed that the movement had benefited from the introduction of some sort of parliamentary politics, in the sense that greater room for maneuver had been created for lobbying for policy changes. The unprecedented "civic forum," which came into being in the 1980s and 1990s, allowed various interest groups to make their political appearance, and one recurrent theme in this forum was human rights. The movement for women's rights took its place in the forum, anchoring, in particular, on this theme of human rights. It is for this reason that local women's groups came out in strong support of democratization of the political structure. The return of sovereignty to Beijing signified to them an unwelcome imposition of authoritarianism and therefore retrenchment of the democratization process. As put by Ah Kuen, one of my respondents, the "June 4 Complex" also stood in the way of the resolution of contradictions between the mainland and Hong Kong, thereby contributing to the gap and tension between the pro-Beijing and the local camps in the movement.

Second, the imposition of authoritarian rule meant, for the women's movement, a danger that the feminist agenda would be swept away entirely, a theme that cropped up repeatedly in my interviews. Susan, member of the AAF, for example, felt that until the time of her interview, there was still ample space for various women's groups, because these were targets for Beijing's "United Front" tactics before the 1997 takeover. She thought, however, that, in due course, some groups or individuals deemed "radical" would be singled out for attack, and this might lead to fragmentation of the movement. After 1997, women's groups would, according to Susan, have to take up a more conformist stance in order to survive authoritarian rule. For Susan, this would entail self-restraint to the point of "castration."

To Christine, who had had richer experience in dealing with groups and individuals on the mainland, the threat of state authoritarianism to the women's movement was just as real. She observed the tragic subordination of the feminist agenda to the so-called foremost issue (*tai chin tai*) of building up national strength in post-1949 China, which had left no room for any autonomous women's movement. Her worry was that the women's movement in Hong Kong would suffer the same fate. The only hope she could harbor now hinged on a more open and accepting Beijing government, and a greater degree of autonomy allowed for the SAR (Special Administrative Region) government.

For the women's groups in Hong Kong, unfortunately, the steely power of state authoritarianism, as well as the conservatism of the pro-Beijing women's delegation, was already being felt in the Fourth U.N. Conference in Beijing.[9] For example, during the opening session of the Hong Kong Tent, base of the pro-Beijing Hong Kong NGO, there was a discussion about "second wives" kept by Hong Kong men working on the mainland.[10] Wan Shaofen, officiating guest and deputy minister of the Central United Front Bureau of the Chinese Communist Party, suggested "strengthening the education" of men and women on both sides of the border, so as to increase their self-respect and thus enable them to avoid such matters. Such a stance of individualizing problems and dodging structural issues was in great contrast to that taken up by the Coalition '95. Lacking a fixed base, member groups of this coalition undertook itinerant activities, such as exhibitions, talks, and street theater, in order to address problems arising from class and sex inequalities.[11] Participants from this coalition also took up highly sensitive issues, like alternative sex orientation and the plight of "comfort women."[12]

Third, the 1997 takeover threatens to substitute "mainstream politics" (effecting changes via networking with influential personages) for the "peripheral politics" commonly adopted by the local women's movement. "Mainstream politics" was involved in the establishment of the Hong Kong Federation of Women in 1993. The federation was a top-down effort, put together by women who worked closely with the New China News Agency and the Beijing government. The contrast between such networking strategies and those of "peripheral politics," characteristic of local women's groups, was brought out clearly in the NGO Forum of the 1995 Conference on Women in Beijing. For example, the Hong Kong NGO had as its base a sturdy tent on the premise, sponsored by Hong Kong Telecom and the Hong Kong Tourist Association. This, and other hotel venues, housed the various occasions it organized.[13] A preliminary conference had been organized in Hong Kong the preceding spring, and Huang

Qizao, deputy director of the All-China Federation of Women, was invited to address it.[14] During the opening session of the Hong Kong Tent at the NGO Forum itself, high-ranking officials who were invited included Huang, as well as Wang Fengchao, deputy director of the Hong Kong and Macao Office of the State Council, and Wan Shaofen.[15] In contrast, Coalition '95 had no assigned base nor scheduled times and places of meeting. What its member groups did was to conduct a sort of "guerrilla warfare" in giving talks, distributing pamphlets, and staging street theater whenever and wherever they could.[16]

The fourth threat of 1997 concerns the unique identity of Hong Kong. Among my respondents, Ah Wai put it most succinctly:

> After 1997 our concern for Chinese development and the state of women will become much greater, and so the demand on the Hong Kong women's movement will increase. By that time the women's movement in Hong Kong cannot be strictly a local movement but will have to confront Chinese women. . . . We might know about issues concerning women in China, but if we want to speak out, on what identity (*san fun*) would we speak? We can't speak out on our own, without linking up with other women on the mainland. *I think our future identity* (san fun) *will be very confused, and we might not be able to grapple with our own role.* . . . Some of us feel that, at present, we should keep quiet about issues of mainland Chinese women, and let Chinese women speak out for themselves. . . . But after 1997, can we still do this? Can we still keep silent and refrain from criticizing the state? *On the other hand, Hong Kong has its own identity, and the problems we face might not be the same as those faced by women on mainland China. How can we bridge this gap, and what will our identity be?* (my emphasis)

Ah Wai's anxiety regarding the identity of Hong Kong can, I think, be better understood in the context of the students' movement of the 1970s and 1980s. With the waning of the high tide of "returning to the socialist motherland," student activists of the late 1970s and early 1980s began to focus more on Hong Kong society itself. Founders of the women's movement, themselves having been nurtured in the students' movement of this latter era, naturally found the 1997 takeover a severe threat to their local cultural identity, which they dearly held. Of course such a threat was not only pitted against the women's movement but against society as a whole.

Conclusion

We have seen how the process of decolonization and the return of sovereignty to China affect the women's movement of Hong Kong, with respect to cultural identity, sociopolitical orientation, and strategies. At this point we might conclude, by raising yet another problem, namely, that the threat of the women's movement itself is being submerged by "macro politics." In short, demands for greater liberalization and democratization of the polity, in response to the threat of state authoritarianism signified by 1997, have resulted in deep political cleavages in society. On the one hand, the women's movement has emerged in the wake of a democracy movement. On the other hand, the general feeling among women activists was that the women's voice and feminist agenda should not be drowned out by the mainstream political struggle for democracy. Ah Yee, of the AAF, put it this way:

> It is clear that the AAF supports the democratic movement . . . we support it in principle, though very few of us actually participate in such organizations. But, in the end, how involved should we be in politics? We know we cannot separate the women's movement from politics, but the two differ. We should not be sucked into the political current, be submerged by it. If we allowed this to happen, we would not be able to uphold the uniqueness of the women's movement; we would only fall back into the previous trap expressed in the slogan, "women's liberation follows the workers' liberation."

Ah Yee's worry is understandable, given the history of male domination in all sorts of movements, whether those of the students, the community, the workers, or the gays. It is certainly not easy to persevere in the women's moment or to uphold the women's agenda amid wider social contradictions and sweeping political changes. On a more abstract note, given the close link between the women's movement and its political context, it is also difficult to separate out the ideological bases and strategies of the women's movement, on the one hand, and those of the wider political movement, on the other. How the women's movement resolves this dilemma remains to be seen.

Notes

1. See Po-King Choi, "A Search for Cultural Identity: The Students' Movement of the Early 1970s," in Anthony Sweeting, ed., *Differences and Identities: The Educational Argument in Late-Twentieth-Century Hong Kong* (Hong Kong: Faculty of Education, University of Hong Kong, Education Papers No. 9, 1990), pp. 81–107. See also Choi Po King, "Popular Culture," in Richard Y. C. Wong and Joseph Y. S. Cheng, eds., *The Other Hong Kong Report, 1990* (Hong Kong: Chinese University Press, 1990), pp. 537–63.

2. The Hong Kong Council of Women closed down in 1995. Harmony House and the Hong Kong Federation of the Women's Center became independent shortly after their establishment in 1985, as the council originally intended, and these are still going strong at the time of this writing. During the Gulf War in the early 1990s, political differences resulted in the formation of a splinter group, and this became the Association of Women for Action and Research (AWARE). AWARE is still in existence, and its members are both expatriates and English-speaking Chinese. It directs its resources toward supporting the local women's movement, and it now plays a consultative and supportive role for the Hong Kong Federation of the Women's Center, which opened a second base in Tai Po, apart from its long-standing one in Lai Kok Estate, Cheung Sha Wan.

3. Rey Chow, "Between Colonizers: Hong Kong's Postcolonial Self-Writing in the 1990s," *Diaspora: A Journal of Transnational Studies* 2, no. 2 (Fall 1992): 151–70.

4. Ibid., 163.

5. See Po-King Choi, "A Search for Cultural Identity: The Students' Movement of the Early 1970s," ibid.

6. See, for example, the vehement attack launched by To Kit-lai, chief executive of the Hong Kong Women Workers' Association, of the AAF, in a conference entitled "Strategies of the Hong Kong Women's Movement," organized by the AAF itself. While commenting on the day chosen for the conference (Saturday morning and afternoon), a time highly inconvenient for women workers, To said: "Are you reserving privileges and the right of growth only for a small circle of women? Is consciousness-raising for women only restricted to intellectuals? Is this because grassroots women can't understand, can't be enlightened? Where is the class struggle we have been talking about?" (To Kit-lai, "Women Workers' Movement: Cul-de-sac and the Way Out," in *Proceedings of The Conference on the Strategies of the Hong Kong Women's Movement* [in Chinese] [Hong Kong: Association for the Advancement of Feminism, 1995], pp. 6–9.)

7. See Lisa Fischler, "Women and Representations: The Construction of Gender Identities in Hong Kong," presented at the conference entitled "Managing Culture: Chinese Organizations in Action," October 20–22, 1995, Hong Kong University of Science and Technology. I am indebted to the author for allowing me to use her as yet unpublished paper. See also Ho Chui-fan, "The Challenge of the Hong Kong Women's Movement: Reflections on the Beijing Women's Conference," in *Our Thoughts and Feelings on Returning from Huairou, Beijing: Experience of the Fourth U.N. World Conference and NGO Forum* (Hong Kong: Hong Kong Womens' Christian Council, 1996), pp. 24–25.

8. For a more detailed description of the federation at its founding, see Choi Po-king, "Women," in Choi Po-king and Ho Lok-sang, eds., *The Other Hong Kong Report, 1993*, pp. 360–400, especially p. 397.

9. For related comments made by members of the Coalition '95 concerning the display of state power in the conference, see Wong Mei-yuk, "After Huairou," and Lung Ngan-ling "Four Obstructing Forces to the Women's Movement, as Witnessed in the NGO Forum," both in *Our Thoughts and Feelings on Returning from Huairou, Beijing*, pp. 14 and 9–10, respectively.

10. "In the Discussion of Hong Kong Men's Second Wives, at the World Conference on Women, Chinese Official Said Education of Women Should Be Strengthened," in *Tin Tin Daily News*, August 31, 1995.

11. Ho Chui-fan, ibid.

12. Wu Lou-sai, "I Said What Chinese People Could Not Say," in *Our Thoughts and Feelings on Returning from Huairou, Beijing*, p. 8; and Cheung Choi-wan, "Lesbians' Rights Are Women's Rights, Too," in *Seeing the World from Women's Eyes: NGO Forum of the Fourth World Conference on Women* (Hong Kong: Association for the Advancement of Feminism, 1996), pp. 26–27.

13. Fischler, "Women and Representations."

14. Program of "Forum on the Vision of Hong Kong Women for the Twenty-first Century," organized by the Hong Kong NGO Working Group of the Fourth World Conference on Women, 1995, Hong Kong, February 18–19, 1995.

15. Fischler, "Women and Representations."

16. Rey Chow, "Between Colonizers."

[23]

Siu-lun Wong and Janet W. Salaff

Network capital: emigration from Hong Kong[1]

ABSTRACT

In this paper, we argue that it would be fruitful to regard personal networks as a form of capital capable of generating economic returns by drawing on our research findings on the recent wave of emigration from Hong Kong. By putting network capital on a par with economic and cultural capital, we seek to identify its distinctive features in terms of institutionalization, capacity, moral economy, and processes of conversion and reproduction. In substantiating our argument, we present some quantitative evidence from our survey data on the uneven distribution of kinship ties which can be mobilized for emigration among different occupational classes. We then make use of our in-depth interview data to show that there is a qualitative variation too in the type of networks used by different occupational classes for emigration purposes. We conclude by reflecting on the implications of the concept of network capital for the study of migration, class formation, and the global economy.

KEYWORDS: Personal networks; emigration decisions; class situations

Personal networks are useful but elusive assets.[2] Individuals rely on them for contacts in seeking jobs and other opportunities, for quick assistance when they are under pressure, for daily needs such as childcare and other practical help, and for emotional support (see e.g. Granovetter 1974; Grieco 1987; Wellman and Wortley 1990; Poel 1993). Like people in other societies, the Chinese value networks and devote much attention to the cultivation of personal bonds which they call *guanxi [kuan-hsi]*. The importance attached to *guanxi* by the Chinese is well recognized by scholars. The systematic study by Ambrose King on this topic can be taken as representative. He points out that 'network building is used (consciously or unconsciously) by Chinese adults as a cultural strategy in mobilizing social resources for goal attainment in various spheres of social life' (King 1991: 79. See also Smart 1993; Yang 1994). However, his main concern is with the cultural logic of network building, and he has not attended to the uneven distribution of *guanxi* as a resource within Chinese society.

Networks affect decisions about major changes in life events, such as the decision to emigrate. How Chinese families draw upon networks when they relocate themselves is an important area of research. From the vantage point of emigration, we can develop a differentiated model of emigrant networks. A study of the emigration decisions of Hong Kong Chinese is an excellent place to begin, because Hong Kong emigrants come from diverse social backgrounds. In this paper, we shall compare how working and middle classes, who structure their personal networks differently, use their contacts and connections when they decide to move across the seas. Our findings suggest that it would be fruitful to regard networks as a form of capital, that is, an asset 'capable of conferring strength, power and consequently profit on their holder' (Bourdieu 1987: 4; see also Bottomore 1983: 60–4).

This paper is divided into three parts. In the first part, we outline the major findings of our research on emigration from Hong Kong and develop the notion of network capital. In the second part, we present in more detail the quantitative evidence on how networks affect migration decisions of different occupational groupings by drawing on our survey data. Then in the third part, we make use of our qualitative interview data to discuss and to illustrate how potential migrants from a range of socioeconomic backgrounds perceive the role of networks in their emigration decisions.

NETWORK CAPITAL

In our research, we discover three significant features in the use of personal networks for emigration purposes in Hong Kong. First, there is a quantitative variation in terms of occupational class.[3] On the whole, the higher the class position of the family, the larger the number of social ties which can be mobilized for emigration. Options increase as one moves up the social ladder. Second, there is a qualitative variation too in the type of networks used by members of different occupational classes. Working-class emigrants tend to depend heavily on family and kinship ties, while affluent emigrants are more inclined to activate diverse bonds of friendship. Third, members of the lower middle class, whose livelihood hinges on bureaucratic careers and wages, have the lowest emigration propensity. It seems that the assets they possess are the least mobile and transferable.

These findings suggest that it may be fruitful to regard networks as a form of capital. This idea is akin to the concept of social capital as proposed by scholars such as Pierre Bourdieu and James Coleman. But Bourdieu (1987: 4) refers to social capital as 'resources based on connections and group membership' without further elaboration. Coleman (1988: S95), pursuing a different theoretical objective, uses the concept in a wide sense to include obligations and expectations, information channels, and social norms. Here we are confining ourselves more specifically to the mobilization of personal networks to generate economic returns. By putting forth the notion of network capital, we are trying to elaborate on Bourdieu's idea and relate *quanxi* and connections directly to the question of social inequality and class formation (for another elaboration on Bourdieu's idea, see Erickson 1996). When network capital is put on a par with other forms of capital, such as economic and cultural capital, we may come to a better appreciation of the diversity and fluidity in the class structure of a Chinese community such as Hong Kong. It would also lead us to identify at least three analytically distinct class situations which correspond to the three forms of capital, namely the entrepreneurs with networks as asset, the capitalists with economic means of production as property, and the professionals with knowledge and skill as resource.[4] These class situations are of course only theoretical constructions or ideal types. In reality, they tend to overlap and seldom exist in the pure form.

Institutionalization

In comparison with economic and cultural capital, network capital is the least institutionalized form of asset. On the whole, economic capital is institutionalized in the form of property rights, and cultural capital in the form of educational qualifications (Bourdieu 1986: 243). Both forms depend heavily on the reliability of social institutions or system trust. Network capital, on the other hand, is basically a diffused asset. It may sometimes take the institutional form of associations of various kinds, but it is generally lodged in reciprocal relations that may or may not be upheld by the parties concerned. In order to reduce uncertainty and to reinforce mutual obligations, personal trust plays a more prominent role as a cementing force in the accumulation of network capital. Therefore, relatively speaking, network capital is less dependent on system trust though it can never be completely free of this form of trust as resources such as classmate networks are derived from reliable educational institutions. (On the distinction between personal and system trust, see Luhmann 1979; Wong 1991).

Because of the different degrees of institutionalization, the three forms of capital tend to have distinctive patterns of geographical mobility and would be drawn to different destinations. The movement of economic capital would typically follow the logic of comparative advantage. In the case of Hong Kong, for example, industrialists in the cotton spinning sector had a tendency to diversify their investments into South-east Asia, Latin America and parts of Africa where labour costs were relatively low and textile quotas were available (Wong 1988: 39).

The movement of cultural capital is affected by the recognition of credentials and the compatibility in educational systems in host countries. Consequently, as revealed in our study, the most popular destinations for the present wave of educated migrants from Hong Kong are English speaking countries such as Canada, the USA, and Australia.

Network capital, being less dependent on system trust, has a greater scope for diffusion and a better ability to transcend boundaries. It tends to spread with the Chinese Diaspora through personal connections. It can venture into territories with shaky institutional frameworks for business operations, such as the People's Republic of China and Vietnam, and still manage to flourish (see Smart and Smart 1991; Sung 1992; Wong 1995).

Network Capacity

After contrasting network capital with other forms of capital, it is necessary to examine the heterogeneous nature of personal networks and its implications more closely. Different types of personal bonds exist, with various capacities in facilitating mobility and economic competition. There are kin and non-kin ties, and there are strong and weak linkages (Granovetter 1982). In our study, we have found that reliance on kin ties and strong linkages are more characteristic of the working class. For members of the affluent class, they are actually disinclined to make use of such ties and often refuse help from family members and relatives.

It is our hypothesis that there exist a restricted and elaborated style of network construction among our respondents. Dependence on kin relations and strong ties is the hallmark of the restricted network style, and flexible use of non-kin relations and weak ties is the key feature of the elaborated network style. An elaborated network style is useful in economic competition because it generates greater access to sources of information and provides more autonomy for action. It can create networks rich in 'structural holes', that is, networks with relationships of low redundancy. Ronald Burt (1992: 21) asserts that such 'optimized' networks have two design principles. The first is efficiency, achieved by concentrating on the primary contact and allowing relationship with others in the cluster to weaken into indirect relations. Second is effectiveness, attained through differentiating primary from secondary contacts in order to focus resources on preserving the former.

Moral Economy

The 'design' principles as set out by Burt alert us to what may be called the moral economy of network capital. In the attempt to optimize benefits, individuals would have to be calculative in manipulating relations to their favour. Granovetter seems to be conscious of this moral ambivalence inherent in network construction when he states wryly

> Lest readers of SWT [Strength of Weak Ties] and this chapter ditch all their close friends and set out to construct large networks of acquaintances, I had better say that strong ties can also have some value. (1982: 113; see also Burt 1992: 262)

This defensive statement reveals the basic reason why those who are skilled at networking, such as entrepreneurs, tend to incur popular hostility and resentment in a society (see Yang 1994: 51–64; Chu and Ju 1993: 133–4; 150–3). They would appear to be too cunning and pragmatic. They spurn the sacredness of personal relations, turning ends into means. Thus they are open to charges of undermining social solidarity and eroding group allegiance. These are the dark sides of network capital.

Conversion and Reproduction

Another source of hostility towards network construction can be traced to the sites of tension with other forms of capital, especially with cultural capital. In Hong Kong society, studies have shown that people tend to seek advancement through two major channels of mobility: the entrepreneurial route of starting one's own business, and the credential route of acquiring professional qualifications (T.W.P. Wong 1991: 164–5). Both network capital and cultural capital are apparently valued and sought after. Yet other studies have revealed a strong anti-capitalist sentiment and deep distrust of entrepreneurs among the educated professionals (Wong 1994: 230–2). Evidently friction and rivalry exist between carriers of network and cultural capital.

Such a tension draws our attention to the problem of conversion and reproduction of various forms of capital. The conversion of cultural capital into network capital in the process of migration is relatively well documented by now. Research on small factory owners in Hong Kong has found that many of these entrepreneurs were immigrants from China with high educational attainment. But their credentials were not recognized in Hong Kong, thus forcing them to seek advancement through industrial endeavours instead (Sit and Wong 1989: 97–100). In the present wave of emigration from Hong Kong, the educated professionals are facing a similar barrier overseas where their qualifications and experience are not fully recognized. A substantial number of them are thus turning themselves into entrepreneurs by setting up small businesses in destination countries such as Australia (see Lever-Tracy et al. 1991).

However, the direction of conversion is by no means one way only. We have found that there is nearly a universal concern expressed by our respondents for their children's education. Hong Kong Chinese entrepreneurs, whether potential or actual, share with others the same preoccupation with the cultivation of cultural capital for themselves and among their offspring (see Ong 1992). Thus there appears to exist a cyclical process of intergenerational conversion of network into cultural capital and *vice versa*. But how precisely is network capital reproduced in the family and passed through the generations? What role does gender play in particular in the accumulation and transmission of this type of capital? We know very little about these issues and clearly more research is needed. The study of emigration from Hong Kong, as we attempt to show in the following sections, should provide a good vantage point in scrutinizing the multifarious features of network capital.

EMIGRATION DECISIONS AND KIN NETWORKS

The Setting

Situated at the mouth of the Pearl River delta in southern China, Hong Kong had long been a city of migrants with massive outflow and inflow of people. Before the Second World War, Hong Kong served as the major port for the Chinese living in Guangdong and other provinces to venture abroad (Sinn 1995). Following the Chinese Communist victory in 1949, over one million refugees left the mainland and sought shelter in the territory. Some brought machinery and know-how from Shanghai and other Chinese coastal cities, and formed a local entrepreneurial class that mainly produced textiles to start the industrialization there (Wong 1988). Others became low-cost labour. With the influx of these immigrants of diverse social backgrounds, Hong Kong transformed itself from an entrepot into a manufacturing centre.

As sojourners in their own city, Hong Kong people move readily to places where they can find work. The first wave of emigrants were mainly poor. The Western economies drew on unskilled immigrant workers, and Great Britain had not yet set up barriers to Commonwealth members. Poorly educated New Territories village men left in large numbers to open and work in restaurants in Great Britain (Watson 1975). Gradually, the skills of those who went abroad changed. As Hong Kong society developed a strong middle class tied intellectually to Western nations, many sent their children abroad to study. The students often remained, and these links prompted more people to go back and forth. In the 1980s, Western immigration policies favoured young, well educated, English-speaking professionals, technicians and managers with financial means (Skeldon 1990–1; Kwong 1991).

More recently, in response to Hong Kong's anticipated reversion to China in 1997, many have applied to emigrate abroad (Wong 1992; Salaff and Wong 1994). Since 1989, about 60,000 Hong Kong Chinese emigrated annually, with Canada, the USA, and Australia as the main destinations. But we find that while political concerns are widespread, connections not attitudes determine who plan to exit and where they go. Emigrants activate their personal networks to leave Hong Kong.

Source of Data

We begin with survey data on kin networks and their association with the emigration decisions of people from different occupational classes. The statistical data come from a survey of 1552 Hong Kong respondents conducted in Hong Kong in 1991. The year after the survey, we chose 30 respondents from the sample for an in-depth analysis of how social ties affect the emigration decisions of their families, and talked with them at length several times from 1992 through 1997. The qualitative interviews went further than the survey questionnaire and located the wider kin circles, friends and colleagues of both husband and wife that applied to emigrate. We obtained these respondents' views of the importance of networks in emigrating, and how these networks entered in their plans. (For details about the survey and the interviews, see Skeldon 1995)

In the table based on the survey findings, we divided the respondents into four major groups according to occupation. For the convenience of presentation, we referred to them as different classes in common sense terms. The affluent class included managers and professionals. Then there were the small businessmen making up the petty bourgeois class. The lower middle class consisted mainly of white-collar workers. Finally, the working class was composed of manual and seasonal workers. We classified our married respondents' families by the highest occupation of either spouse.

In choosing the families for our qualitative study, we sought to interview equal numbers of emigrant and non-emigrant families. An emigrant family is one in which any member has actually submitted an immigration application form to a foreign country. But this is used only as a convenient operational definition, because in reality the situation is very complex and the distinction between emigrant and non-emigrant families is far from clear cut. Many families in Hong Kong are in fact what Myron Cohen (1970) has called 'dispersed' families with members and relatives already living abroad yet still maintaining a shared economy.

We tried to match the emigrant and non-emigrant families by occupational background, ranging from the administrators to the manual workers. In our discussion of these families, because of the small number of cases, we reduced the occupational class classification from four to three. We did not classify the petty bourgeoisie as a separate category. Rather, since we had a lot of information about the qualitative sample, it enabled us to place the families of the small businessmen into either the lower middle class or the affluent class as appropriate.

Survey Findings

Many Hong Kong families have kin abroad. In Table I, we can see that the 1552 respondents in the survey have an average of about 0.67 to 0.38 family members abroad, depending on their class background. The working class has the fewest family members abroad and the affluent class has the most. On the whole, those that plan to emigrate have twice as many kin overseas as those that do not plan to leave.

When we look at the differences among the class groups, we find a trend. Emigrants in three of the four class groups have more kinship ties overseas than do non-emigrant households. We find the most dramatic difference in the number of kinship ties between emigrant and non-emigrant households in the working class. The petty bourgeois businessmen, mainly with family businesses, follow. The lower middle class shows a small difference, while the affluent class actually reverses the trend.

However, the survey questions did not ask how kinship ties actually affect

TABLE I: *Average kinship ties of survey respondents by emigration status and class*

Class	Average number of kin living abroad		
	Emigrant	Non-emigrant	Average
Affluent class	0.61	0.87	0.67
Petty bourgeois	0.69	0.54	0.56
Lower middle class	0.53	0.50	0.50
Working class	0.72	0.36	0.38

Source: 1991 Emigration Survey

the emigration decisions. Ties of friendship and acquaintance were not covered. Further, the survey collected only limited data about the family as a unit. For instance, the survey asked questions about the number of the respondent's family members abroad, but not specifically about those of their spouse's kin. To address such questions, we have to turn to the in-depth interviews.

SOCIAL NETWORKS AND FAMILY STRATEGIES

Among the families that we interviewed over time, those who had applied to emigrate had more emigrant siblings than the non-emigrants (see Table II). In addition, we asked emigrant applicants if they had other friends and colleagues that had already gone abroad. We expected that having emigrant friends would also propel our respondents to emigrate. Those with friends and colleagues who had already emigrated could count on them for help or tips about life abroad. Those with friends that were applying could form a reference group of like minded people, who would influence each other that it was a right decision to exit. Other factors such as family size and family life cycle may also affect the propensity to emigrate, but we shall not delve into them in this paper.

We found that friends were important to future emigrants: 13 of the 16 applicants had friends, neighbours, or co-workers that had emigrated and with whom they kept in touch. In contrast, non-emigrants did not keep in touch with acquaintances abroad. Indeed, few could think of any close friends who had emigrated. These suggestive findings, taken together with our survey material on a larger sample of Hong Kong respondents, highlights the importance of looking further into the relations abroad of emigrant and non-emigrant families. In the following analysis of our case materials, we assume the existence of a coherent emigration strategy for each family though we are aware that individual strategies may collide with the family strategy and that individual family members do not necessary share the same views and uphold the same decisions. In our interviews with the families, we tried to obtain the views of both spouses. But since

TABLE II: *Emigrant kin of emigrants and non-emigrants*

Respondents	Emigrant siblings as % of total no. of siblings
Emigrants	
Working class	36
Lower middle class	46
Affluent class	33
Non-emigrants	
Working class	12
Lower middle class	15
Affluent class	27
Total	
Working class	24
Lower middle class	24
Affluent class	46

Source: Interviews 1992–1997

husbands are expected to play the public role as heads of household in Hong Kong, the adult male views are often more vocal.

Working Class Families

In Table I, we find that emigrant working-class families have almost double the number of kinship ties of non-emigrant working-class families. Our in-depth interviews confirm the survey findings. Working-class families emigrate with an eye to those kin who have already left. Their ties to kin run deep even while they live in Hong Kong. Kin assistance is widespread and multi-stranded among these working-class Chinese.

Given the many ties of assistance among the working class in Hong Kong, we can understand why emigrants follow kin abroad. We interviewed 10 working-class families, of whom four had applied to emigrate. Three sought to emigrate through kin support. Some started with complete dependence. One route for folks without wealth is emigration through joint family action. To even think of emigrating, their close kin have to help them to find jobs.

Having kin abroad not only helps people to find a job, it also provides practical help to emigrate. Siblings and in-laws may buy homes together in the new country. Or, they may put together an investment package to qualify for a visa. For instance, we spoke with a part-time restaurant helper, married to a truck driver, and the mother of two teenage daughters and an autistic son. Her older brother sponsored her for an American visa, and offered her a job. Family reunification was the legal basis for their application. Further, she could only consider emigration because she expected to find work in her brother's Boston restaurant. As she recounted, so had others in her family before her

> By working in the kitchen of Brother's restaurant, Father earned more than US $1,000 dollars a month. He said that he could earn as much in one month as he could working in China for several years! But now he is old. He does not work any more . . . Altogether, we are six brothers and sisters, and I'm the youngest. All have applied to go to the United States. My third sister is most enthusiastic about emigrating. She is a housewife and her husband works in a Chinese restaurant here in Hong Kong. They have money. My elder brother said he would help her buy a house to prepare for her coming. He'll rent it out until she arrives.

Few working-class families have ever lived outside of Hong Kong or the Chinese mainland. They have little first hand experience of other places. They closely depend on those they know in foreign lands to give them information about life abroad. In most cases, these are kin. Those working-class folks we met who applied to emigrate applied to just one country. This was largely because they could count on kin support in only one place.

Echoing the survey results, we found in our interviews that working-class non-emigrants tended to have fewer kin overseas than the emigrants. Yet there were still many non-emigrants who did have siblings abroad. When we asked those with siblings abroad why they decided not to follow them, their answers were mainly that they did not have enough money and their siblings could not offer them work.

We expected that having emigrant friends would also propel our respondents to emigrate. But we found that working-class folks did not get the chance to emigrate from friends. While they might find comfort in learning about the plans of their workmates and acquaintances, they could not lean too heavily on these contacts. Their friends, who had jobs like theirs, were not able to give them a job that was good enough to warrant emigrating, or one that would get them a visa. Further, these working-class friends had their own kin to support.

A young clerk in a dry cleaning store, married to a bank messenger, told us of her novel way to contact a former neighbour and a former co-worker in Toronto and San Francisco. In the Christmas of 1992, she used the fax machine in the store to send greetings to these friends. 'It's unusual, and it only costs 10 [Hong Kong] dollars!' But she felt,

> We can't depend on our friends to emigrate. Actually, I don't like San Francisco very much. So I think I'd travel there, but not settle down. You know, it is difficult to find a job over there and we have few relatives and friends who could take care of us.

In sum, dense ties provide a link to the job, which is a necessity to working-class emigrants. Most have dense ties only to kin. Migrant kin that already have a foothold abroad may have opened small businesses serving the ethnic community (Zhou 1992). If they feel obliged to offer their close kin jobs, these relatives can join them. We believe that this is one of the few ways the working-class families can succeed in emigrating. For without jobs in the working-class community, they have little chance to survive in the distant world, that is currently in recession, and whose language they speak only poorly.

Lower Middle-Class Families

Our study contains 10 lower-middle-class couples. Most worked in bureaucracies in the public and private sectors. Four families had applied to emigrate. The way they applied was distinctive, for while lower-middle-class emigrants had kin abroad, they were less likely to draw on these kin to emigrate. None applied for family reunification. Their main strategy was to take advantage of their civil service status to apply as special emigrants to Singapore or England. A civil servant, a former policeman, and a teacher pursued this option. Kinship connections did not help them.

The reason that few used kin ties to emigrate was not because they had none. In Tables I and II, we find that they did have emigrant kin. Rather, the nature of their family economies were separate from kin. The lower middle class cannot depend on kin to emigrate because the kinds of jobs they hold are not obtained through kin. They secure their livelihood on the basis of the cultural capital at their disposal, in the form of achieved qualifications obtained through advanced study and legitimated by diplomas. These nurses, technicians, civil servants, assistant engineers and school teachers earn good salaries and can look forward to promotions through fixed steps on career ladders.

These lower-middle-class workers are not particularly keen to emigrate for a number of reasons. First, they cannot count on getting similar jobs abroad. Some work for the Hong Kong government and other public bodies, and they cannot qualify for this kind of work elsewhere. None have the necessary licenses or certification to start again in their lines of work overseas. Second, they earn too little for most investment categories of immigrants. They also lack the higher education or specialized training to qualify as 'other independent' emigrants to Canada and Australia, the countries to which most Hong Kong people seek to enter. Third, they have enjoyed a spurt of income improvement in recent years, which they fear they cannot match abroad. Partly this is because both spouses work to maintain a dual-income household economy. With few grand economic hopes for themselves, they are not eager to reject the solid living standard they enjoy for an uncertain life abroad.

Only one of these lower-middle-class couples used its kinship ties in an application to emigrate, but failed. Unable to get jobs in their line of work from kin, the lower middle class use other channels. A civil servant status helps the most. Singapore and England remain options for those that have quasi-political civil service jobs. We interviewed a woman who was a clerk in the correctional services, married to a man now in England, who had been a constable. As a young man, her husband had studied karate in his uncle's martial arts studio. He then studied with the instructor of the world famous martial artist Bruce Lee, became proficient, and began to teach in a community centre. He joined the police force. While at a training course, he met a British constable, also an admirer of Bruce Lee, who had come to Hong Kong to meet Lee's coach and train with him. In early 1989, the British constable invited him to Manchester to teach karate, and arranged for him to have working papers.

This couple combined channels of help from friends, kin, and the government bureaucracy. The husband had been close to his kin in Hong Kong. He received concrete help in finding work abroad from a colleague. But they took the crucial emigration step through the wife's civil service connection. In the wake of the June 4th incident in 1989, the constable became afraid of remaining in Hong Kong. As a member of the correctional services, his wife qualified for the British Nationality Scheme for herself, with eligibility extended to her immediate family. She and their son had emigrated to Manchester to join her husband.

Those who have no kin abroad are reluctant to exit. But even with kin abroad, those holding modest salaried posts may not join them. They worry that they cannot easily fit into an enterprise that their foreign kin might run. If the work relationship falters, they will be left without a base of their own and will have no economic recourse. Indeed, these lower-middle-class workers pride themselves on being independent.

A land inspector in the Hong Kong civil service, married to a part-time sewing operator, decided not to emigrate even though his sister's daughter asked them to join her family in moving to Australia and offered him a job in the firm her family hoped to open

> I tell my niece I won't go. Her husband is a professor of pharmacy in the United States. He is doing some research and patented a new medicine. They are very rich. They want to set up a pharmacy shop in Australia, and ask me to go there to do some clerical job. But I don't want to rely on my relatives. I am independent. I never seek help from others.

Affluent Families

In Table I above, we find that while the affluent families have more kin abroad than any other group, those that emigrate do not have more kin than non-emigrants. Indeed, the relationship is reversed. This points to a phenomenon that is worth exploring: that the affluent class does not depend on kin to emigrate as much as do the working class.

The affluent class includes businessmen and professionals. Two of the ten we interviewed worked with kin in flourishing family firms. The rest had little occupational contact with kin. Regardless of whether they worked closely with kin or not, they sought emigration in similar ways.

Eight of the ten affluent families we interviewed were emigrants in a technical sense, because they already had visas to exit Hong Kong. However, they told us that they did not intend to use them unless 1997 precipitated a crisis. Their visas were for 'insurance purposes'. Several had the option to emigrate to more than one country. Only one, an unmarried member of a family that had members living in Alberta, was granted a visa in the family reunification category. The rest applied to emigrate under the business or 'other independent' categories, or as civil servants under the British Nationality Scheme. While several of these affluent respondents had visas to places where they had kin, they were not economically dependent on these relatives. Instead, they were more likely to stress their networks of schoolmates and former colleagues.

It is true that well-placed kin living abroad can offer positions to those they sponsor. But our affluent respondents did not depend on these jobs. To be sure, some used the offer of a convenient 'paper' job to satisfy the requirements of the emigration authorities. They did not actually expect to take up the jobs so offered.

The kinship contacts of these affluent couples are neither spatially nor socially concentrated, but are spread out. Most have emigrant kin in many different countries. Their relatives are more widely placed around the globe than those of the working class. They are not tied tightly to their kin contacts, as they have other network resources. Here, friends, colleagues, and classmates are crucial.

A site engineer, who received his diploma in the Hong Kong Polytechnic after six years of part-time study, was employed by a large Hong Kong construction firm. On the side, he had opened a small interior design firm with his classmates as shareholders. Yet another Polytechnic classmate emigrated to Vancouver

> I bought my Richmond house there with his help. I never even went there; he helped me care for it. My classmate told me not to emigrate because the economy for construction was so bad. But if I do emigrate, I'll probably work with him at the start.

Affluent emigrants feel these emigrant friends are important, and they sometimes are more important than emigrant kin. Depending on classmates for help is common for our affluent respondents. Friends are already established and can give the newcomer a first job. The job can be permanent, or if it does not work out, it provides a temporary shelter while the new immigrant settles in and learns the ropes. Or, in the same trade, the classmate can give the newcomer information that leads to a job.

The managing director of a printing company had got visas from several countries. His sister was living in Canada, but he excluded that country as an emigration possibility. 'It's too far for our children to travel there, and we don't like the climate,' his wife said. 'If we haven't even bothered to visit her in Toronto, how could we want to live there?' They chose Singapore as their emigration destination. There, the managing director had many classmates and business associates. Singapore is also closer than Toronto to their China-based business. They had already bought a house there, as part of their emigration plan. Another of the managing director's brothers would emigrate to Singapore as a teacher, but he claimed he was not influenced by this relation. He chose to apply for Singapore papers because it had more flexible emigration conditions, and because his classmates who were there could help him work out his business arrangements.

When affluent individuals describe working with former classmates, it is clear they are thinking of an egalitarian relationship. They do not feel that by agreeing to work together, they are depending on friends. It is understood that each side is ready to put up capital. Their classmates know one another's skills and talents. Such a relationship with friends does not make them feel like they are getting a handout. For this reason, many potential emigrants from the affluent class prefer to rely on friends than kin.

CONCLUSION

Emigration is a selective and transformative process. People mobilize diverse forms of resources to move from place to place. As their assets and endowments vary, they make up different types of migrants. Our findings suggest that future analyses of international migration should distinguish between not only rich and poor, literate and illiterate migrants, but also migrants with disparate network capacities. The outflow of professionals and entrepreneurs from the Pacific Rim region has altered the scene of international migration which used to be dominated by labour migrants. This calls for more sophisticated theories and models to enable us to have a better understanding of the new situation.

The present wave of emigration from Hong Kong has often been characterized as a middle-class phenomenon. By developing the notion of network capital, we try to demonstrate the need to revise the conventional class analysis in which property and education are over-emphasized and personal connections are neglected. The fuzzy concept of the middle class should perhaps be replaced with the more precise categories of capitalists, professionals, and entrepreneurs, each pursuing its own distinctive strategies of competition and resource accumulation.

Network capital has its microscopic and macroscopic dimensions. In this paper, we focus on the significance of personal ties and linkages at the individual and family level. When aggregated, such ties and linkages would constitute the collective asset of a society. Seen in this light, the popular alarm about capital flight and brain drain from Hong Kong as a result of the recent outflow of people is probably misplaced. With the multitude of social ties being activated and extended by numerous families in the process of migration, the international linkages of Hong Kong as a node in the global economy are enhanced rather than diminished. Yet, as we have indicated in the theoretical discussion, network capital is a mixed blessing fraught with moral ambivalence. In the shadow of the intricate web of overseas linkages lurk popular resentment and hostility among the host communities. That is where the real danger lies for Hong Kong and its emigrants.

(Date accepted: June 1997)

Siu-lun Wong
Centre of Asian Studies
The University of Hong Kong
and
Janet W. Salaff
Department of Sociology
The University of Toronto

NOTES

1. This paper is a product of the research projects on 'Emigration from Hong Kong: Tendencies and Impacts' and 'Emigration from Hong Kong: Families, Networks, and Returnees' funded by the Hong Kong Research Grants Council. Besides the authors, other co-investigators include Daniel Wai-wah Cheung, Yiu-kwan Fan, Kit-chun Lam, Elizabeth Yuk-yee Sinn and Ronald Skeldon. We are grateful to Eric Fong for his contributions to the analysis of the survey data, and to Tak-wing Chan, Katharyne Mitchell, Don Nonini, Aihwa Ong, Mayfair Yang and the two anonymous reviewers for their comments. We thank the following bodies for support: The Hong Kong Research Grants Council; The University Grants Committee of Hong Kong; The Canada-Hong Kong Project, Joint Centre on Asian Pacific Studies, University of Toronto/York University; Initiatives Fund, Institute for International Programs, University of Toronto; and Centre of Urban and Community Studies, University of Toronto. The Department of Sociology and Centre of Asian Studies, University of Hong Kong, gave us on the ground assistance.

2. We are mainly dealing with informal networks based on kinship or friendship ties in this paper. These personal networks should be distinguished from other types of social networks, such as the more formal intercorporate relations and interlocking directorships, which are known for their capacity in generating economic returns (see e.g. Mizruchi and Schwartz 1987; for a concise overview of social network analysis,

see Scott 1991) but with which we are not concerned here.

3. For our survey and interviews, we adopt an operational definition of class in terms of occupation. We use only occupation as the indicator of social class partly to avoid cumbersome technical complexities of index construction, and partly because other proxies such as education and income are found to be less reliable for classification purposes in Hong Kong. For a fuller justification, see Salaff and Wong (1994) and Tsang (1994).

In our theoretical discussions, we mainly adopt Max Weber's approach to class analysis by emphasizing the possession of assets which can generate economic returns on the market and affect life chances. As John Scott (1996) points out, the distinction between 'class situation' and 'social class' is fundamental to the Weberian approach. In this paper, we attempt to differentiate among various 'class situations' that consist of specific causal components of life chances. We leave the question of 'social classes', that is, the formation of actual social groupings for collective action, for future analysis (see Lee and Turner 1996 for the recent debate on class analysis).

Since the operational definition is used as a classificatory device to guide us in the initial interpretation of the data and we arrive at the analytical definition only at a later stage in the research, the class categories derived from the two definitions do not match completely.

4. We follow Joseph Schumpeter's

definition of entrepreneurs as those who carry out innovations and new combinations of factors of production. In Schumpeter's view, entrepreneurs should be distinguished from capitalists who are owners of money, credit or other material goods. He observes that '[risk] obviously always falls on the owner of the means of production or of the money-capital which was paid for them, hence never on the entrepreneur as such ... A shareholder may be an entrepreneur. He may even owe to his holding a controlling interest the power to act as an entrepreneur. Shareholders per se, however, are never entrepreneurs, but merely capitalists, who in consideration of their submitting to certain risks participate in profits.' (Schumpeter 1961: 75, original emphases)

BIBLIOGRAPHY

Bottomore, T. (ed.) 1983 *A Dictionary of Marxist Thought*, Cambridge: Harvard University Press.

Bourdieu, P. 1986 'The Forms of Capital' in J. Richarson (ed.) *Handbook of Theory and Research for the Sociology of Education*, New York: Greenwood Press.

—— 1987 'What Makes a Social Class? On the Theoretical and Practical Existence of Groups', *Berkeley Journal of Sociology* 32: 1–17.

Burt, R. 1992 *Structural Holes: the Social Structure of Competition*, Cambridge: Harvard University Press.

Chu, G. C. and Ju, Y. 1993 *The Great Wall in Ruins: Communication and Cultural Change in China*, Albany: State University of New York Press.

Cohen, M. L. 1970 'Developmental Process in the Chinese Domestic Group', in M. Freedman (ed.) *Family and Kinship in Chinese Society*, Stanford: Stanford University Press.

Coleman, J. S. 1988 'Social Capital in the Creation of Human Capital', *American Journal of Sociology* 94 (Supplement): S95–S120.

Erickson, B. H. 1996 'Culture, Class, and Connections', *American Journal of Sociology* 102(1): 217–51.

Granovetter, M. 1974 *Getting a Job: A Study of Contacts and Careers*, Cambridge: Harvard University Press.

—— 1982 'The Strength of Weak Ties: A Network Theory Revisited', in P. V. Marsden and N. Lin (eds) *Social Structure and Network Analysis*, Beverly Hills: Sage.

Grieco, M. 1987 *Keeping It In The Family: Social Networks and Employment Chance*, London and New York: Tavistock Publications.

King, A. Y. C. 1991 'Kuan-hsi and Network Building: A Sociological Interpretation', *Daedalus* 120(2): 63–84.

Kwong, P. C. K. 1991 'Emigration and Manpower Shortage', in R. Y. C. Wong and J. Y. S. Cheng (eds) *The Other Hong Kong Report, 1990*, Hong Kong: The Chinese University Press.

Lee, D. J. and Turner, B. S. (eds) 1996 *Conflicts About Class: Debating Inequality in Late Industrialism*, London and New York: Longman.

Lever-Tracy, C. et al. 1991 *Asian Entrepreneurs in Australia: Ethnic Small Business in the Indian and Chinese Communities of Brisbane and Sydney*, Canberra: Australian Government Publishing Service.

Luhmann, N. 1979 *Trust and Power*, Chichester: John Wiley and Sons.

Mizruchi, M. S. and Schwartz, M. 1987 *Intercorporate Relations: The Structural Analysis of Business*, Cambridge: Cambridge University Press.

Ong, A. 1992 'Limits to Cultural Accumulation: Chinese Capitalists on the American Pacific Rim', *Annals of the New York Academy of Sciences* 645: 125–43.

Poel, M. 1993 *Personal Networks: A Rational-Choice Explanation of Their Size and Composition*, Lisse: Swets and Zeitlinger.

Salaff, J. W. and Wong, S. L. 1994 'Exiting Hong Kong: Social Class Experiences and the Adjustment to 1997', in S. K. Lau et al. (eds) *Inequalities and Development: Social Stratification in Chinese Societies*, Hong Kong: Hong Kong Institute of Asia–Pacific Studies, The Chinese University of Hong Kong.

Schumpeter, J. A. 1961 *The Theory of Economic Development: An Inquiry into Profits, Capital, Credit, Interest, and the Business Cycle*, Cambridge: Harvard University Press.

Scott, J. 1991 *Social Network Analysis: A Handbook*, London: Sage Publications.

—— 1996 'Class Analysis: Back to the

374 *Siu-lun Wong and Janet W. Salaff*

Future?', in D. J. Lee and B. S. Turner (eds) *Conflicts About Class: Debating Inequality in Late Industrialism*, London and New York: Longman.

Sinn, E. 1995 'Emigration from Hong Kong before 1941: General Trends', in R. Skeldon (ed.) *Emigration from Hong Kong*, Hong Kong: The Chinese University Press.

Sit, V. F. S. and S. L. Wong 1989 *Small and Medium Industries in an Export-Oriented Economy: The Case of Hong Kong*, Hong Kong: Centre of Asian Studies, University of Hong Kong.

Skeldon, R. 1990–1 'Emigration and the Future of Hong Kong', *Pacific Affairs* 63(4): 500–23.

Skeldon, R. (ed) 1995 *Emigration From Hong Kong*, Hong Kong: Chinese University Press.

Smart, A. 1993 'Gifts, Bribes, and *Guanxi*: A Reconsideration of Bourdieu's Social Capital', *Cultural Anthropology* 8(3): 388–408.

Smart, J. and Smart A. 1991 'Personal Relations and Divergent Economies: A Case Study of Hong Kong Investment in South China', *International Journal of Urban and Regional Research* 15: 216–33.

Sung, Y. W. 1992 'Non-institutional Economic Integration Via Cultural Affinity: The Case of Mainland China, Taiwan and Hong Kong', Occasional Paper No.13, Hong Kong Institute of Asia–Pacific Studies, Hong Kong: The Chinese University of Hong Kong.

Tsang, W. K. 1994 'Behind the Land of Abundant Opportunities: A Study of Class Structuration in Hong Kong', in B. K. P. Leung and T. Y. C. Wong (eds) *25 Years of Social and Economic Development in Hong Kong*, Hong Kong: Centre of Asian Studies, University of Hong Kong.

Watson, J. L. 1975 *Emigration and the Chinese Lineage: The Mans in Hong Kong and London*, Berkeley: University of California Press.

Wellman, B. and Wortley, S. 1990 'Different Strokes from Different Folks: Community Ties and Social Support', *American Journal of Sociology* 96: 558–88.

Wong, S. L. 1988 *Emigrant Entrepreneurs: Shanghai Industrialists in Hong Kong*, Hong Kong: Oxford University Press.

—— 1991 'Chinese Entrepreneurs and Business Trust', in G. Hamilton (ed.) *Business Networks and Economic Development in East and South-east Asia*, Hong Kong: Centre of Asian Studies, University of Hong Kong.

—— 1992 'Emigration and Stability in Hong Kong', *Asian Survey* 32(10): 918–33.

—— 1994 'Business and Politics in Hong Kong During The Transition', in B. K. P. Leung and T. Y. C. Wong (eds) *25 Years of Social and Economic Development in Hong Kong*, Hong Kong: Centre of Asian Studies, University of Hong Kong.

—— 1995 'Business Networks, Cultural Values and The State in Hong Kong and Singapore', in R. A. Brown (ed.) *Chinese Business Enterprise in Asia*, London and New York: Routledge.

Wong, T. W. P. 1991 'Inequality, Stratification and Mobility', in S. K. Lau et al. (eds) *Indicators of Social Development: Hong Kong 1988*, Hong Kong: Institute of Asia–Pacific Studies, The Chinese University of Hong Kong.

Yang, M. M. H. 1994 *Gifts, Favors and Banquets: The Art of Social Relationships in China*, Ithaca and London: Cornell University Press.

Zhou, M. 1992 *Chinatown: The Socioeconomic Potential of an Urban Enclave*, Philadelphia: Temple University Press.

Name Index

For Product Safety Concerns and Information please contact our EU
representative GPSR@taylorandfrancis.com Taylor & Francis Verlag GmbH,
Kaufingerstraße 24, 80331 München, Germany

Printed and bound by CPI Group (UK) Ltd, Croydon, CR0 4YY
01/05/2025
01858467-0001